ISBN 978-1-5279-4923-2
PIBN 10003346

METEOROLOGY

A TEXT-BOOK ON THE WEATHER, THE CAUSES OF ITS CHANGES, AND WEATHER FORECASTING

FOR THE STUDENT AND GENERAL READER

BY

WILLIS ISBISTER MILHAM, Ph.D.

FIELD MEMORIAL PROFESSOR OF ASTRONOMY
IN WILLIAMS COLLEGE

New York

THE MACMILLAN COMPANY

1912

PREFACE

THIS book owes its existence to a course on meteorology which has been given by the author in Williams College for the last eight years. This course is a Junior and Senior elective course with three exercises a week during a half year. A syllabus, covering both the text-book used and the added material, was prepared for the course. This was at first mimeographed, then revised and printed. Later it was again revised and reprinted. This book follows the order of topics in this last syllabus very closely, and is thus essentially a resumé of the material which has been gathered for the course.

This book is essentially a text-book. For this reason, the marginal comments at the sides of the pages, the questions, topics for investigation, and practical exercises have been added. A syllabus of each chapter has been placed at its beginning, and the book has been divided into numbered sections, each treating a definite topic. The book is also intended for the general reader of scientific tastes, and it is hoped that these earmarks of a text-book will not be found objectionable by him. It can hardly be called an elementary treatise, but it starts at the beginning and no previous knowledge of meteorology itself is anywhere assumed. It is assumed, however, that the reader is familiar with the great general facts of science. References have been added at the end of each chapter. These include pamphlets and articles in the periodical literature as well as books. These are the first things which a student would naturally look up in order to gain further information. In appendix IX an attempt has been made to summarize the literature of meteorology. Here the books are arranged in alphabetical order without regard to age or value. Both the metric and English system of units and the Fahrenheit and centigrade thermometer scales have been used in the book. It seems unnecessary in quoting facts and data from many sources to change everything to conform to one set of units. In appendices I, II, and III, the English and metric systems of units and conversion tables have been added. The facts of meteorology have now become so general and accepted that there can be but little that is new in such a book. The originality must lie in the arrangement, use, and perhaps interpretation of these facts. Whenever a distinctive idea has been introduced by some investigator, credit is always given in the text when this idea is quoted. Such credit has never been intentionally omitted by the author.

Although this book has assumed a considerable size, it can lay no claim to completeness. No single volume can be a treatise containing all known facts to date and all the explanations which have been offered. Four aspects or applications of meteorology have been entirely omitted. These are:

(1) Mathematical Meteorology.

(2) Meteorology applied to living things; including phenology and the influence of climate on man.

(3) Meteorology and medicine; including climate and disease.

(4) A History of Meteorology; including a biography of the men who have contributed much to its development.

To each of these subjects a long chapter could be devoted, and they could easily be expanded so as to become large books. Such topics as "Meteorological Apparatus," "The Daily, Annual, and Irregular Variation in the Various Meteorological Elements," "The Isothermal Layer," etc., could be easily treated at such length as to become a large book. In fact, each chapter in this book could be expanded into a sizable volume. This book, then, makes no attempt at completeness, but it does attempt to give a fairly full presentation of the present state of the science, and also to point the way for further acquisition of information on the part of him who desires it.

In the preparation of this book, the author is particularly indebted to the United States Weather Bureau and to Professor Willis L. Moore, its chief. Every opportunity was given to make use of what is probably the largest meteorological library in the world; tables of data to illustrate various points were supplied; and free permission to reproduce and quote much that has appeared in government publications was given. The author is under great obligations to many persons who have helped him in various ways: to Professor William J. Humphreys, Professor of Meteorological Physics, United States Weather Bureau, who has read the entire manuscript and made many helpful and valuable suggestions; to Edward H. Bowie of the forecast division, Preston C. Day and Maitland C. Bennett of the climatological division, C. Fitzhugh Talman in charge of the library, and Cleveland Abbe, Jr., of the library division, for assistance and suggestions during the final revision of this book while in Washington; to Professor Cleveland Abbe, who has read a portion of the manuscript and whose kindly interest is an inspiration to any one who is teaching meteorology or doing research in that subject; to George

T. Todd, local forecaster in charge of the Albany station, and Herbert E. Vail, the first assistant. These last gentlemen have read the entire manuscript, made many suggestions, and guarded it against minor mistakes in connection with the routine work of the United States Weather Bureau. They have also responded with unfailing cheerfulness and promptness to the many calls for data in connection with the Albany station.

W. I. M.

WILLIAMS COLLEGE,
WILLIAMSTOWN, MASS., July, 1911.

ERRATA IN MILHAM'S METEOROLOGY

Page 70, line 3 from bottom, "Bornstein's" should be "Börnstein's."

Page 84, line 13 from top, "cosmische" should be "kosmische."

Page 111, line 20 from bottom, "méteorologie" should be "météorologie."

Page 165, line 1 from top, "Dové" should be "Dove."

(The same change should be made on p. 165, l. 18 from top; p. 168, l. 8 from top; p. 185, l. 4 from bottom; p. 266, l. 2 from bottom; p. 293, l. 3 from bottom.)

Page 188, line 20 from top, "Barometre" should be "Baromètre."

Page 188, line 20 from top, "Met." should be "Mét."

Page 188, line 21 from bottom, "Temperatur verhältnisse" should be "Temperaturverhältnisse."

Page 188, line 19 from bottom, "Repetorium" should be "Repertorium."

Page 188, line 5 from bottom, "phere" should be "phère."

Page 257, line 2 from bottom, "21" should be "12."

Page 262, line 24 from top, "Hebertson" should be "Herbertson."

Page 262, line 18 from bottom, "Sanderra, Maso' Miguel" should be "Saderra Maso', Miguel."

Page 262, line 17 from bottom, "Niederschlägsverhältnisse" should be "Niederschlagsverhältnisse."

Page 263, line 18 from top, "Niederschlägsverhältnisse" should be "Niederschlagsverhältnisse."

Page 263, line 26 from top, "servico" should be "servicio."

Page 266, line 12 from bottom, "baguois" should be "baguios."

Page 320, lines 17 and 18 from top, the "1" is misplaced.

Page 338, line 10 from bottom, "J. F." should be "J. P."

Page 339, line 2 from top, "Greeley" should be "Greely."

Page 351, line 1 from top, "Fisher" should be "Fischer."

Page 351, line 5 from top, "sailors" should be "Sailors'."

Page 354, line 6 from top, "storm" should be "storms."

Page 364, line 16 from bottom, "came" should be "come."

Page 375, line 11 from bottom, "Pottsdam and Lindenburg" should be "Potsdam and Lindenberg."

Page 375, line 10 from bottom, "Mont Souris" should be "Montsouris."

Page 375, line 7 from bottom, "Santis" should be "Säntis."

Page 377, line 5 from bottom, "Van" should be "van."

Page 381, line 10 from bottom, "prediction" should be "predictions."

Page 402, line 9 from bottom, "Garriot" should be "Garriott."

Page 422, line 2 from bottom, "preponderence" should be "preponderance."

Page 423, lines 12, 14, and 16 from top, "Van" should be "van."

Page 423, line 20 from top, "prevision" should be "prévision."

ERRATA IN MILHAM'S METEOROLOGY

Page 423, line 20 from bottom, "Osterreich" should be "Österreich."
Page 440, line 6 from top, "Erkunde" should be "Erdkunde."
Page 440, lines 23 and 25 from bottom, "M'adie" should be "McAdie."
Page 441, line 8 from top, "attineze" should be "attinenze."
Page 441, line 11 from top, "Buckner, Edward" should be "Brückner, Eduard."
Page 482, line 13 from top, "Luftelectrizität" should be "Luftelektrizität."
Page 482, line 14 from top, "atmospherische Electrizität" should be "atmosphärische Elektrizität."
Page 482, line 20 from bottom, "cosmischen" should be "kosmischen."
Page 482, line 17 from bottom, "Lemstron" should be "Lemström."
Page 482, line 17 from bottom, "boreale" should be "boréale."
Page 519, line 8 from bottom, "Andre" should be "André."
Page 520, line 8 from top, "meteorologists" should be "meteorologist's."
Page 521, line 6 from bottom, "Koppen" should be "Köppen."
Page 522, line 11 from top, "Muller" should be "Müller."
Page 523, line 16 from bottom, "Meteorologiques" should be "Météorologiques."
Page 524, line 10 from top, "Petersbourg" should be "Pétersbourg."
Page 524, line 2 from bottom, "practique" should be "pratique."
Page 524, line 2 from bottom, "Precision" should be "Précision."
Page 525, line 16 from top, "Atlantschen" should be "Atlantischen."
Page 526, line 4 from top, "Verdunstrung" should be "Verdunstung."
Page 526, line 11 from top, "canans" should be "canons."
Page 527, line 9 from top, "Erganzungscheft" should be "Ergänzungsheft."
Page 527, line 17 from bottom, "staaten" should be "Staaten."
Page 527, line 16 from bottom, "America" should be "Amerika."
Page 527, line 9 from bottom, "varaus" should be "voraus."
Page 528, line 13 from top, "Osterreich" should be "Österreich."
Page 528, line 11 from bottom, "Edward" should be "Eduard."
Page 529, line 16 from top, "Mühy" should be "Mühry."
Page 529, line 21 from top, "dell Italia" should be "dell' Italia."
Page 529, line 25 from top, "Veranderungen" should be "Veränderungen."
Page 529, line 25 from top, "Eisgeit" should be "Eiszeit."
Page 530, line 1 from top, "Lemstrom" should be "Lemström."
Page 530, line 4 from top, "Atmospherische" should be "Atmosphärische."
Page 537, line 12 from bottom, "1886" should be "1866."
Page 538, line 7 from top, "Schnee Krystalle" should be "Schneekrystalle."
Page 538, line 8 from top, "Psychromter-Tofeln" should be "Psychrometer-Tafe'n."
Page 538, line 8 from top, "bibliograph" should be "bibliography."
Page 538, line 2 from bottom, "physichen" should be "physischen."
Page 539, line 5 from top, "Atmospherique" should be "Atmosphérique."
Page 539, line 7 from top, "Schwidler" should be "Schweidler."
Page 539, line 14 from bottom, "Beiblätter" should be "Beiblatter."
Page 540, line 13 from bottom, "Luftschiffahrten" should be "Luftschiffahrt."
Page 541, line 4 from bottom, "Comtes" should be "Comptes."
Page 543, "Baguois" should be "Baguios."
Chart xxvi, "Greeley's" should be "Greely's."

CONTENTS

PART I

ILLUSTRATIONS

CHARTS [1]

[1] The fifty charts are placed together at the end of the book.

METEOROLOGY

METEOROLOGY

CHAPTER I

INTRODUCTION — THE ATMOSPHERE

INTRODUCTION

1. The science of meteorology. — The natural sciences deal with the phenomena of the world of nature about us. As examples of these varied phenomena one might mention the fall of a stone, the rusting of iron, the growth of a plant, the change in the phases of the moon, the formation of a cloud, or the erosion of a valley by a stream. These are all occurrences in or phenomena of this world of nature, and it is thus the province of some one of the natural sciences to treat fully each one of these phenomena.

Definition of a natural science.

B   1

Since these phenomena are so numerous, complex, and varied, they are divided among several natural sciences. Physics and chemistry are the two fundamental natural sciences because they treat matter and energy, the two components of the material world, in the abstract. They are also fundamental because so many of their facts, laws, principles, and methods are used in the other sciences. Biology with its many subdivisions includes all those phenomena where life is involved. The remaining phenomena, those of inanimate matter, are divided up between astronomy, meteorology, and geology; astronomy treating the heavenly bodies beyond the earth, meteorology the earth's atmosphere or envelope of gas, and geology the earth itself.

Enumeration of the natural sciences.

Meteorology is thus one of the natural sciences, since it has a group of phenomena to investigate. It treats of the condition of the atmosphere, its changes of condition, and the causes of these changes. Its duty is to arrange the facts in an orderly way so that the relation of cause and effect can be traced and generalizations can be formed. Meteorology is often defined briefly as the study of atmospheric phenomena. Since so many of the laws and principles of physics find their application in the atmosphere on a stupendous scale, meteorology is also often defined as the physics of the atmosphere.

Definition of meteorology.[1]

2. **Outline history of meteorology.** — The annual change from summer to winter and back again from winter to summer, with all the attendant changes in vegetation, the daily change from the heat of the day to the cool of the night, the falling of rain and snow, the coming of a thunder shower, all such things must have profoundly occupied the mind of man from remote ages. The antiquity of many weather proverbs and of much of our weather lore also shows that meteorological observations and generalizations were among the first acts of an intelligent race.[2]

Meteorology an old science.

The word meteorology comes from the Greek. Socrates is spoken of in one of Plato's dialogues as " a wise man, both a thinker on supra-terrestrial things and an investigator of all things beneath the earth." [3] The word for supra-terrestrial fur-

Origin of the word meteorology.

[1] The abbreviation used by the U.S. Weather Bureau for meteorological is met'l, and this form will be used in this book whenever the word is abbreviated.

[2] In the British Museum in London there are Babylonian clay tablets dating from about 4000 B.C. which contain weather proverbs. One, for example, reads : "When a ring surrounds the sun, then will rain fall." See HELLMANN, " The Dawn of Meteorology," *Quarterly Journal of the Royal Met. Society,* October, 1908, vol. XXXIV, No. 148, p. 227.

[3] Σωκράτης, σοφὸς ἀνήρ, τάτε μετέωρα φροντιστής, καὶ τὰ ὑπὸ γῆς ἅπαντα ἀνεζητηκώς. — *Apologia Socratis,* cap. II.

nishes the root for the word meteorology.[1] About fifty years later (350 B.C.) Aristotle wrote the first treatise [2] on meteorology, consisting of four books or parts, containing a large amount of information of very mixed value. Not only were the things considered to-day under the term meteorology treated in this book, but the appearance of the stars, comets, meteors, earthquakes, northern lights, the composition of matter, and other things as well.

3. Four periods may be recognized in the history of meteorology. The first lasts from antiquity until about 1600 A.D. The observations were crude, nearly all made without the help of instruments, and they were often inaccurate and much influenced by superstition and imagination. The explanations were often fantastic and supernatural. *The four periods in the history of meteorology and their characteristics.*

The coming of the second period was brought about by the invention of instruments for making observations. The two most important instruments in meteorology were invented at about this time, the thermometer in 1590 by Galileo and Sanctorius of Padua, and the barometer in 1643 by Torricelli; and in 1653 Ferdinand II, Grand Duke of Tuscany, established stations throughout northern Italy for making careful observations of meteorological phenomena. The chief characteristics of this period are a larger number of observations and a great increase in accuracy.

The third period begins a little before 1800 and runs until about 1850. Its great characteristic is the attempt to give logical explanations for the various phenomena which were now being observed with ever increasing accuracy.

The fourth period is the modern period, and it began about 1850. Shortly after this the various governments began establishing weather bureaus, interest in meteorology increased markedly, and great advances began to be made. The three great characteristics of this period are: ceaseless activity in gaining information, the utmost accuracy in observing all possible phenomena, and the rigid testing of all explanations and hypotheses.

4. Utility. — The utility of meteorology may be noted in two entirely different directions. First, there is the financial saving caused by the timely forecasting of those weather conditions which do damage to commerce, agriculture, products in transit, *Two lines of usefulness.*

---

[1] μετέωρα = supra-terrestrial; λόγος = description or treatise. Thus meteorology was originally a treatise on supra-terrestrial things.

[2] The Greek title was τὰ μετεωρολογικά.

and businesses of many sorts; and, secondly, there is the educational advantage in the study of meteorology. The storms on the coast and also those on the lakes and rivers do immense damage to shipping and commerce and often cause an appalling loss of life. The late frosts of spring and the early frosts of autumn work havoc with the farm produce which may be exposed. Sudden changes in temperature from hot to cold or from cold to hot may cause the complete loss of produce or merchandise which is being transported by boat or train from one point to another. Traffic on railroad and street car lines may be delayed or even stopped entirely by a heavy, unexpected fall of snow. The timely warning of the approach of these damage-causing weather conditions allows every precaution to be taken and has resulted in a tremendous financial saving. The total cost in maintaining the Weather Bureau now amounts to about a million and a half dollars, while the most conservative estimates place the saving to this country brought about by timely forecasts at many times the cost of maintaining the bureau.

Financial saving by timely forecasts.

Mark W. Harrington, a former chief of the United States Weather Bureau, says in the preface of his book, *About the Weather :* [1] " The more than twelve hundred thousand dollars expended every year by the Government may perhaps be considered an exorbitant price to pay for learning what weather we are likely to have for the coming twenty-four hours, but the truth is that no public investment is so immediately and so immensely profitable as that applied to the maintenance of the Weather Bureau.

Harrington's opinion.

" Not only are cyclones in the West, but the floods of streams and rivers, especially those of the Mississippi Valley, are foretold, ' and incalculable saving of life and property,' as Mr. J. E. Prindle says in a report on the subject, 'results from their warnings. Before the days of the bureau,' he proceeds, ' the West India hurricanes came unannounced, and sometimes two thousand lives were lost in a single storm. Under the warnings of the Weather Bureau three such storms have passed in succession without the loss of a single life, and the property destroyed in one storm would support the service for two years. At Buffalo, in the winter of 1895–1896, by forecasting six very severe storms, one hundred and fifty vessels, valued at seventeen million dollars and carrying eighteen hundred persons, were held safely in port by the warnings.' "

[1] Reprinted from HARRINGTON's *About the Weather*, copyright, 1899, by D. Appleton and Company.

5. There is probably no subject which so fully occupies our attention and is of so much importance to us as the weather. And yet there is probably no subject about which the ordinary person knows less and where ignorance and superstition are more universal than in connection with the weather and the causes of its changes. Meteorology has long since dispelled the mystery connected with weather and weather changes. It is thus the duty of our schools and colleges to produce an educated public which can distinguish between truth and error, fact and superstition. It is also a real pleasure to know the causes of the changes and to appreciate the mechanism back of the ever shifting panorama which constitutes our weather.

The desirability of an educated public.

There are many facts to be arranged in an orderly way so that the relation of cause and effect can be traced. Many observations have been made so that generalizations and laws can be derived. Meteorology thus has a well-developed descriptive side which offers training in exactness of statement and description and in meteorological reasoning. It also has a mathematical side which is being developed more and more at the present time. This develops extreme exactness in thought and expression along abstract lines. Apparatus must be tested and improved, and there are many observations to be made. It thus has a laboratory side which gives increased familiarity with the details of a subject, sense training, and skill in manipulation. Meteorology thus offers to the student training along three different lines, and these are the three kinds of training offered by any one of the natural sciences.

The educational value of meteorology.

6. In discussing the utility of meteorology attention should be called to how intimately the weather enters into every aspect of human life. Our plans are daily made and unmade on account of the weather. Our very moods depend upon the weather. As mentioned above, untold damage to property and even loss of life may be caused by storms, frosts, hail, lightning, and sudden changes in temperature. The student of criminology tells us that certain crimes are more prevalent at one season of the year than at another or with one particular type of weather. Theft, for example, is more common during the winter, and the cause is not far to seek. During the winter there is less employment than during the summer, and incomes are smaller. In addition the cold makes food, clothing, and shelter more imperative. The result is a decided seasonal variation in this form of crime. The physician tells us that certain diseases show a well-marked seasonal variation and are more prevalent with one kind of weather than

The varied influences of the weather.

another. Meteorological statistics are being made use of more and more by lawyers not only in damage cases but also in criminal cases. A single illustration will suffice. A certain burglary case turned largely on the certain identification of a person seen to come from a building during the early hours of the morning. The observer was in a near-by building across the street. It was put in evidence by the defense that on the morning in question, according to the observations of the Weather Bureau, a dense fog hung over the city so that positive identification at the distance in question would have been impossible.

7. Relation of meteorology to the other natural sciences. — The relation of meteorology to physics is a close one. Many of its terms and facts are used, and so many of its laws and principles find their application on a large scale in the atmosphere that meteorology is sometimes defined as the physics of the atmosphere. A course on elementary physics at least should precede a study of meteorology. Such terms as the following must be used from time to time, and but little space can be given to defining or illustrating them : mass, volume, density, velocity, acceleration, rotation, revolution, force, inertia, centrifugal force, gravitation, gravity, weight, pressure; atom, molecule, ether, solid, liquid, gas ; sound; heat, temperature, expansion, specific heat, latent heat, conduction; light, reflection, transmission, absorption, radiant energy; magnetism, electricity.

Relation of meteorology to physics.

But two groups of facts are borrowed from astronomy. They are, first, the facts concerning the position of the earth with reference to the sun at different times of year, and, secondly, the facts concerning the rotation of the earth on its axis. These will be presented at the appropriate place. (See section 36.)

Relation of meteorology to the natural sciences.

Meteorology touches chemistry in but one place, namely, in the discussion of the composition of the atmosphere.

Almost no facts or principles are taken from biology or geology.

THE ATMOSPHERE

8. The atmosphere and its properties. — The atmosphere [1] may be defined as the envelope of gas surrounding the earth. It is an odorless, colorless, tasteless gas and when at rest one might almost doubt its substantiality. When it is in motion, however, in the form of wind, and hinders walking and even overturns trees and houses, there can be no doubt of its existence.

Definition of the atmosphere.

[1] From the Greek : ἀτμός = vapor or gas ; and σφαῖρα = sphere.

Since air is a gas, it must have all the physical properties of gases. Some of these properties which will be met with later are the following: (1) A given quantity of air will occupy all the space open to it. (2) If allowed to expand, it will become cooled; and conversely, if compressed, it will be heated. (3) If the temperature is kept constant, the volume of a given quantity of air will vary inversely as the pressure, a larger pressure thus producing a smaller volume. (4) If the temperature changes, either the volume or the pressure will change according as the other is kept constant.[1]

Properties of a gas.

One cubic centimeter[2] of pure dry air under standard conditions[3] has a mass of 0.0012927 gram. In the English system of units it requires about 13 cubic feet for 1 pound.

9. **Composition of the atmosphere.** — The four major constituents of pure dry air in order of amount are nitrogen (N), oxygen (O), argon (Ar), and carbon dioxide or carbonic acid gas (CO_2). The following table shows the percentage composition by volume and by weight. The atomic weights are also added.

The four major components of air.

	N	O	Ar	CO_2
Vol. %	78.04	20.99	0.94	0.03
Weight %	75.46	23.19	1.30	0.05
Atomic weight	14.04	16.00	39.9	

The various determinations made by different investigators with different apparatus and in different parts of the world will, of course, differ slightly from each other and from the above figures. This is due to errors of observation and perhaps to a very small real difference in the composition of the air. The first decimal place in the case of both N and O is certainly correct.

Hydrogen is also one of the permanent constituents of the atmosphere, but the quantity is extremely small at the earth's surface. It is now usually assumed to be about 0.01 per cent by volume, but it was formerly considered to be less than this and thus not considered among the major components of the air.

The rarer components.

[1] The relation between volume, pressure, and temperature is expressed by the formula $PV = RT$, where P denotes the pressure, V the volume, and T the absolute temperature reckoned from 273 below zero Centigrade or 491 below the freezing point Fahrenheit. R is a constant whose value depends upon the units chosen, and the starting points.

[2] The use of the metric system cannot be avoided. For the tables and equivalents see Appendix I.

[3] Standard conditions are pressure 760 mm., temp. 0° C., standard gravity.

Argon was first separated from nitrogen by Rayleigh and Ramsay in 1894, and it is now known that what was formerly considered argon was not a single substance, but had several other gases mixed with it. Helium, neon, krypton, and xenon have been separated from it. The amount of these rare gases in the atmosphere is about 4.1, 12, .05, and .006 parts in a million respectively.

Air is a mechanical mixture of its component gases and not a chemical compound. This is proved by the following considerations: (1) The relative proportions of the components are not those of their combining or atomic weights. (2) When liquid air is allowed to boil off, some components leave faster than others so that the percentage composition changes. (3) When the components are mixed together, no change in temperature or volume occurs. (4) The index of refraction has the average value of its components, and not a unique value. (5) The composition of air dissolved in a liquid is not the same as that of the free air.

Air a mechanical mixture.

10. The composition of the air is remarkably constant. In the case of oxygen, for example, all the reliable determinations which have ever been made fall between 20.81 and 21.00 vol. per cent. In the case of carbon dioxide the amount varies from 0.036 to 0.0304 according to different determinations. In closed rooms the amount of CO_2 is of course larger; 0.07 per cent is usually considered the limit of good ventilation. In closed rooms occupied by many people, particularly in sleeping rooms, values from 0.24 to even 0.95 have been found. It is a popular mistake to think that the air exhaled from human lungs contains a very large amount of CO_2. The following table gives the percent composition of inhaled and exhaled air:

Constancy of the components of the air.

	INHALED	EXHALED
N	78.04	78.04
O	20.99	16.03
Ar	0.94	0.94
CO_2	0.03	4.40

It will be seen that exhaled air contains less than five per cent of CO_2.

There are two reasons why the composition of the air remains so nearly constant. (1) The air is very mobile. It is being constantly mixed and transported great distances by the wind. (2) Gases diffuse easily, so that any irregularity would quickly eliminate itself even if there were no wind.

Reasons for the constancy.

11. Although the composition of the air is so uniform all over the earth's surface and to any height at which man can live, this is far from the case when great heights above the earth's surface are considered. **Composition at great heights.** Since air is only a mechanical mixture of its gaseous components, each behaves as if the others were not present. This means that the heavier gases will be held closer to the earth's surface, and the lighter components will predominate at great heights. This is, of course, based on the assumption that the atmosphere is not mixed by the wind, but that each gas is free to distribute itself in accordance with its density. Accordingly, from a knowledge of the composition of air at the earth's surface and the temperature at different elevations, the percentage composition at any height can be computed. The following table computed by Humphreys[1] gives the percentage composition at various heights. In Fig. 1 these results are shown graphically.

HEIGHT IN KILOMETERS	N	O	Ar	CO_2	H	HELIUM
5	77.89	20.95	0.94	0.03	0.01	0.00
15	79.56	19.66	0.74	0.02	0.02	0.00
30	84.48	15.10	0.22	0.00	0.20	0.00
50	86.16	10.01	0.08	0.00	3.72	0.03
80	22.70	1.38	0.00	0.00	75.47	0.45
100	1.63	0.07	0.00	0.00	97.84	0.46
150	0.00	0.00	0.00	0.00	99.73	0.27

It will be seen that at the height of 150 kilometers, or less than 100 miles, the atmosphere should be composed almost entirely of hydrogen with a little helium. This is in good agreement with the observed fact that the spectrum of shooting stars shows prominently the hydrogen and helium lines (section 24). The outside of the atmosphere of the sun is also composed largely of hydrogen and helium.

12. The minor constituents of the atmosphere, often spoken of as impurities, are water vapor, nitric acid, sulfuric acid, ozone, organic and inorganic particles, and minute traces of several other things. Of these the water vapor and the particles and perhaps ozone are the only ones which deserve attention. **The minor constituents of the atmosphere.**

The amount of water vapor in the atmosphere is always small, as it never exceeds 4 per cent and the amount is constantly changing with every change in the weather. Yet it is one of the most important components, for without it both plant and **Water vapor.**

[1] Mount Weather Bulletin, vol. II, p. 66, 1909.

animal life would be an impossibility.   The discussion of its various
forms, such as dew, frost, fog, cloud, rain, hail, and snow must form a
large portion of every book on meteorology.

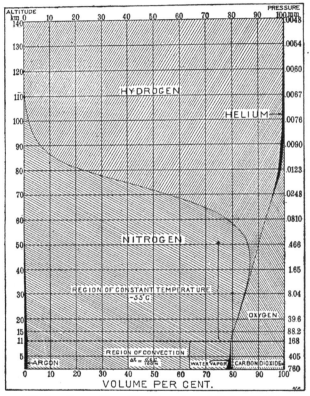

FIG. 1. — Composition of the Atmosphere at Various Heights.
(After HUMPHREYS in Mount Weather Bulletin, Vol. II.)

13.   The organic particles include bacteria and the spores of plants, and
these minute organisms are scattered throughout the atmosphere.   It is
Organic   estimated that even on high mountains and over the oceans
particles.   there is at least one in every cubic meter of air.   In the streets
of our cities the number probably runs up to 3000 per cubic meter and in
crowded houses and hospital wards it probably reaches at least 80,000.

· The inorganic particles are usually spoken of as dust. These dust particles are much more numerous than the organic particles and are for the most part entirely invisible to the naked eye. A few of the giant members of the family may be seen when carried up **Inorganic particles.** by the wind from the earth's surface or when a beam of sunlight falls through a small opening into a darkened room. At the present time the number of these particles in a given volume can be determined by means of Aitken's dust counter. This in- **Aitken's** genious instrument, as illustrated in Fig. 2, consists of a **dust** pump P by means of which the air in the box B may be **counter.** suddenly rarefied. This is provided with a graduated part G at the lower end so that a known quantity of air may be withdrawn from B if desired. The box B is about a centimeter thick with other dimensions in proportion. At the bottom is a glass plate divided by lines usually into square millimeters and illuminated by the mirror M. At the top is a lens L for viewing and magnifying the spaces on the glass plate. At the sides are pieces of filter paper saturated with water. There are two stop cocks at C and C'. The working principle of the apparatus is this: Whenever the air in B is suddenly rarefied, it becomes colder and can no longer hold all the water vapor in it. (See section 183.) This water vapor collects on the dust particles and forms a fog which slowly settles and collects on the glass plate. The number of these small water drops can then be counted, and this is the number of dust particles which were present. The method of conducting this experiment is at once ap-parent. By repeated rarefactions all dust must be removed from the box B. A known quantity of the air to be investigated is then introduced. The instrument is shaken so as to mix the air

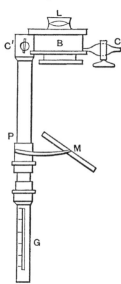

FIG. 2. — Aitken's Dust Counter.

(From ARRHENIUS, *Lehrbuch der kosmischen Physik*.)

with the dust-free air already in the box and saturate it with moisture. The next rarefaction will coat the introduced dust particles with water and cause them to collect on the glass plate, where they may be counted.

Many determinations have been made with this or similar instruments

in all parts of the world and at various heights on mountains and under
The amount many conditions. The following table gives the results ob-
of dust un- tained by Fridlander for the various oceans and at various
der various
conditions. heights on the Bieshorn:

	Dust Particles Per Cubic Centimeter	Elevation	Dust Particles Per Cubic Centimeter
Atlantic Ocean	2,053	6,700	950
Pacific Ocean	613	8,200	480
Indian Ocean	512	8,400	513
		10,665	406
		11,000	257
		13,200	219
		13,600	157

In a dusty city 100,000 dust particles per cubic centimeter is by no means
large, and it has been found that a single puff of cigarette smoke contains
about 4,000,000,000 particles.

There are four chief sources of the dust of the atmosphere. (1) It is
blown up from the earth by the wind. (2) It is injected into the atmos-
Sources of phere by volcanoes while in eruption. During the volcanic
dust. explosion in Krakatoa between Sumatra and Java in 1883 it is
estimated that dust and steam were thrown into the air to a height of
nearly twenty miles, and the presence of this dust could be detected in
sunset colors all over the world for more than three years. (3) Shooting stars put an immense quantity of dust into the upper atmosphere as
a result of their combustion and disintegration. (4) Ocean spray when
evaporated adds dust to the atmosphere, especially fine particles of salt.

Atmospheric dust plays an important part in at least four ways.
(1) It is one of the chief causes of haze. (2) It probably serves as cen-
Effects of ters of condensation for all fog particles and raindrops. It
dust. was once thought that condensation was impossible without
it. (3) It is the cause of the sunrise and sunset colors and perhaps
of the blue color of the sky. (4) It is the cause of twilight. These
various effects of the dust will be fully considered later.

14. Ozone is a peculiar form of oxygen, in that its molecule is composed
of three atoms of oxygen linked together, while the ordinary oxygen mole-
Properties of cule consists of two. This third atom has a tendency to
ozone. leave the molecule, and to this is due the oxidizing power
of ozone and its value as a sanitary agent. Ozone is usually con-

sidered a powerful disinfectant and is supposed to be particularly useful where organic matter is decomposing. It is even stated by some that the bracing effect of certain climates and of certain types of weather is due to a slightly larger amount of ozone present in the air. In this last instance, however, a lower temperature, the absence of moisture, and an increase in the amount of electricity in the air probably play a far more important part in determining one's feelings than any change in the amount of ozone.

The discovery of ozone is usually attributed to Schönbein in 1848, although a few investigators probably detected its peculiar odor earlier. It is this peculiar odor which gives it its name.[1] The amount present in the atmosphere is extremely small, usually about one part in a million. The amount shows a decided *The quantity of ozone.* daily and annual variation and irregular fluctuations which are closely correlated with the type of weather. There is much more during the winter than during the summer.

The quantity of ozone present in the air is determined by means of its oxidizing power on certain chemical compounds. Potassium iodide is ordinarily used, and a known quantity of this substance is added to a paste formed by dissolving a known quantity of starch in water. This is then spread on pieces of paper and exposed to the air a known length of time. From the depth of the blue color which results from the decomposition of the potassium iodide, the amount of ozone is estimated. The determinations are none too accurate, even when a standard paste and a standard color scale are used, as other things affect the decomposition of the potassium iodide slightly.

Ozone is formed in the laboratory by allowing electricity to discharge through oxygen or air, and here the peculiar odor can be readily detected. In nature it may be formed by electrical *Formation of ozone.* discharges, or by the action of ultra-violet light on oxygen, or possibly in connection with the evaporation of water.

15. Offices and activities of the atmosphere and its components. — The atmosphere as a whole has the following offices and activities: (1) It disseminates bacteria and the spores of plants. It also carries the seeds of some plants, particularly those provided with down like the thistle, long distances. (2) It makes flight *Activities of the atmosphere as a whole.* possible in the case of birds and insects and even some animals. (3) It furnishes power to sailing vessels and windmills. (4) It transports moisture, thus making possible animal and plant life

[1] From the Greek ὄζω = I smell.

on large portions of the earth's surface. (5) It produces sand mounds or sand dunes and is the cause of "weathering." (6) It produces waves on bodies of water. (7) It makes sound possible.

16. The various components of the atmosphere all have their own individual offices and functions.

Nitrogen is the inert component. When taken into the lungs of animals, it appears to have no physiological effects. Plants also are unable to use it. It serves simply to dilute the oxygen, which is the active element in the atmosphere. Nitrogen does not readily form compounds, and to this may be due the fact that it makes up such a large part of the atmosphere and is found to such a slight extent in the compounds in the earth's crust.

Oxygen is the active, energizing component. All animals derive their energy and power to do work from the oxygen. It is taken into the lungs,

The functions of the various components of the atmosphere. makes its way into the blood, and joins with the tissues of the body, liberating energy. As a result of this process, large amounts of CO_2 are added to the atmosphere by animals. Oxygen is also consumed whenever combustion takes place.

A small amount of oxygen is also lost to the earth's atmosphere by forming compounds with certain substances in the earth's crust. Oxygen is supplied to the atmosphere chiefly by green plants. A small amount comes from volcanoes and other vents in the earth.

Carbon dioxide is also an important component of the earth's atmosphere, for without it plant life would be impossible. The sap which comes up from the roots consists largely of water with certain organic and inorganic substances in solution. The green cells in the leaves in the presence of sunshine have the power of taking the CO_2 from the air and combining it with the sap, thus building the complex molecules which make up its own tissues, at the same time liberating oxygen. CO_2 is supplied to the atmosphere from many sources. It comes from volcanoes and vents in the earth's crust. The water of the ocean contains a large amount of it. Meteors when consumed in the atmosphere add a small quantity. A large quantity is added as the result of combustion, but the largest quantity is put into the atmosphere by animal life and the slow decay of vegetation. It is estimated that nearly a billion tons of coal are now burned annually. This alone would put into the atmosphere nearly four billion tons of CO_2 each year. Slightly more CO_2 is found over the oceans than over the land, more in the southern hemisphere than in the northern, more over cities than in the country, and more at night than during the day. The reasons for this are apparent.

Oxygen and carbon dioxide are thus held in a state of equilibrium by means of plant and animal life. The animal consumes O and liberates CO_2, while the plant consumes CO_2 and liberates O. Should the quantity of CO_2 increase, plant life would become more luxuriant and animal life would be hindered. Should the quantity of oxygen increase. animal life would become more active and exhilarated and plant life would be stunted. In each case the tendency would be towards a restoration of equilibrium.

Argon, hydrogen, and the rare gases of the atmosphere like nitrogen have no individual functions. The sources and effects of the secondary components of the atmosphere have already been considered.

17. **Atmosphere of other heavenly bodies.** — There are many indirect methods of determining the amount of atmosphere possessed by the various heavenly bodies which make up our solar system. The giant planet Jupiter, with an equatorial diameter of 90,190 miles and a temperature certainly above that of boiling water, has an immense atmosphere which is constantly filled with clouds. Saturn, Uranus, and Neptune seem to be in about the same condition. Venus has an atmosphere somewhat less bulky than that of the earth. Mars. an older, colder planet with a diameter of 4352 miles, has only one twelfth as dense an atmosphere as the earth. while the moon with a diameter of 2163 miles has practically no atmosphere at all. Thus great diversity is found in the amount of atmosphere. The determining factors seem to be temperature, size, and perhaps age. For a full discussion of this topic the reader must be referred to books on astronomy.

The atmospheres of the various heavenly bodies.

18. **Evolution of the atmosphere.** — If during the early history of the earth it was molten to any extent, or even if the surface temperatures were high, the water of the oceans and the volatile mineral substances of the earth itself must have been in the atmosphere as vapor or gas. The earth must then have possessed an immense atmosphere laden with vapor and clouds. As the earth cooled, the water and other substances would be precipitated upon the earth, thus reducing the atmosphere to its present bulk.

Early history of the atmosphere.

If, however. as some prefer to believe, the earth grew very slowly by accretion, each small mass of matter as it was joined to the earth bringing its quantum of gas, it may be that the surface temperatures were never high and that the atmosphere grew as gradually in bulk as the earth itself.

19. Whatever may have been the initial bulk or condition of the atmosphere, there are several reasons for believing that great changes in composition have taken place during geological times. The **Changes in composition during geological ages.** luxuriant plant growth during the carboniferous age when the great coal beds were laid down is usually explained by assuming a much larger amount of CO_2 in the atmosphere. In fact some go so far as to state that there was probably extremely little oxygen in the atmosphere at that time, as it had all been used in forming compounds such as water and the oxides of the earth's crust. They attribute our present supply of oxygen entirely to the very luxuriant plant life of that time. Great changes have also taken place in the temperatures of the earth. Glaciation has extended down the Mississippi Valley as far as Kansas, and traces of luxuriant vegetation have been found in northern countries where it is at present impossible. One of the explanations sometimes given is to attribute it to changes in the composition of the atmosphere. So great is the influence of composition on temperature that the statement is sometimes made that if the amount of CO_2 in the atmosphere were multiplied by four, the vegetation of Florida would be found in Greenland.

20. Future of the atmosphere. — At present a small amount of gas is lost to the atmosphere by escape into space beyond the gravitational control of the earth. A small amount, particularly oxygen, **Atmosphere at present constant in bulk and composition.** is also lost in the formation of compounds which become part of the ocean or the solid earth. The atmosphere gains a small amount of gas from meteors and from volcanoes and other vents in the earth. The amount of O and CO_2 in the atmosphere is kept constant by the balance between plant and animal life. It thus seems that the future will see no such changes in bulk or composition as have been witnessed in the past. A very gradual diminution in the amount of the atmosphere is, however, to be expected.

The sun, however, may eventually grow cold. The only two explanations of the maintenance of the sun's heat which have withstood modern investigation are that the heat is maintained by slow **The future disappearance of the atmosphere.** contraction or by the presence of radio-active material. However the outpour of heat may be maintained, it must eventually come to an end. As the sun grows cold, so must the earth. A glance at the accompanying table of the boiling points of the components of the atmosphere shows the inevitable results.

Water, 100° C.
CO², — 78°
O, — 183°
Ar, — 187°
N, — 194°
H, — 253°

When the temperature reaches — 78° C. the CO_2 will come out of the atmosphere, and plant and animal life must cease. As the temperature falls lower, the various components will come out in order as their boiling points are reached, until finally a cold, dark, airless world will be revolving about a dying sun. And a strange world it will indeed be, for the sky will be black instead of blue, the stars will be visible in the daytime, shadows will be entirely black, sound will be impossible, and the constant bombardment by meteors will make life in the open more dangerous than on a modern battlefield.

The Pressure and Height of the Atmosphere

21. **Gravity and its effects.** — All objects near the earth's surface are pulled downward by what is called the force of gravity. Gravity in magnitude and direction is in reality the resultant of three forces: the attraction of the earth as a whole, the attraction of surrounding objects, and the centrifugal force due to the earth's rotation. It is ordinarily considered that the direction of the force of gravity is towards the center of the earth and that *Gravity, its definition, magnitude, and direction.* its magnitude is constant all over the earth. These statements are, however, only approximately true. It is only at the equator and poles of the earth that the direction of gravity is directly towards the earth's center. At other places it is nearly but not exactly towards it. The magnitude also varies slightly with latitude and elevation. Two effects of this ever present force deserve attention. (1) All masses, since they are acted on by gravity, will have weight and exert a downward pressure on whatever supports them. (2) Since a fluid does not resist a *Effects of* change of form, it will set itself with its lower surface con- *gravity.* forming to the shape of the containing vessel and with its upper surface at right angles to the direction of gravity.

22. **Geosphere, hydrosphere, atmosphere.** The great mass of the earth is solid at least in the outer crust and has a very uneven surface. It is covered in part by the oceans, which under the influence *Geosphere,* of gravity conform with their under surfaces to the irregular- *hydro-* ities of the solid earth and have an upper free surface at right *sphere,* angles to gravity. Both solid earth and fluid oceans are sur- *atmosphere.* rounded by an envelope of gas. These three are often spoken of as the geosphere, the hydrosphere, and the atmosphere.[1]

[1] From the Greek : γῆ = earth ; ὕδωρ = water ; ἀτμός = gas ; σφαῖρα = sphere.

c

The surface of the hydrosphere is considered a level surface, and all elevations are reckoned from it. When the dimensions of the earth are stated, it. is always the dimensions of the hydrosphere that are given. Its form is not that of a perfect sphere, but that of a sphere flattened at the poles and known as an oblate spheroid. The equatorial diameter of the hydrosphere exceeds the polar by 27 miles.

The following are a few numerical facts concerning the geosphere, hydrosphere, and atmosphere.

Polar diameter of hydrosphere	7899.580 miles
Equatorial diameter of hydrosphere	7926.592 miles
Area of oceans	$\frac{3}{4}$ of earth's surface
Mass of oceans	$\frac{1}{4340}$ of earth
Mass of atmosphere	$\frac{1}{200000}$ of earth

23. Since the atmosphere is substantial, that is, possesses mass, and is acted upon by gravity, it must have weight and exert a downward pres-

Pressure of the atmosphere. sure. The pressure of the atmosphere is simply the weight of the column of air above the point in question. With elevation above the earth's surface the pressure must thus grow less. The average pressure of the atmosphere at the surface of the hydrosphere is 14.7 pounds per square inch. It is usually measured, however, not in pounds per square inch, but by the length of the balancing or equivalent mercury column, as will be fully discussed in Chapter IV.

24. Height of the atmosphere. — Since a gas tends to expand and occupy all the space open to it, no theoretical limit can be set to the

No theoretical limit. height of the atmosphere. We cannot picture the free surface of a gas, and must think of the earth's atmosphere as growing gradually thinner and thinner with elevation until it merges with that mere trace of gas which fills interplanetary space. The question of the height of the atmosphere may be put, however, in a more practical form. To what height above the earth's surface does a sufficient quantity of air extend to give us any indication whatever of its presence? This is sometimes called the sensible height of the atmosphere, and there are several methods of detecting the presence of air at considerable heights.

The five ways of determining the sensible height of the atmosphere. (1) Twilight is caused by the reflection of sunlight from the dust and perhaps moisture particles of the upper air after the sun has gone below the horizon of the place in question. Diffraction may also play a part. The cause of twilight is illustrated in Fig. 3. By noting the duration of twilight and knowing the dimensions of the earth, the height of the air producing the twilight may be determined. It has been found that a sufficient quantity of air for deflecting an appreciable amount of sunlight extends to a height of 63 kilometers. (2) The ordinary height. of clouds is from a few meters to perhaps 15 kilometers, but on certain rare occasions, at night, particularly in high latitudes in midsummer, faint luminous clouds have been observed as high as 83 kilometers [1] above the earth's surface. (3) The Aurora Borealis, or northern lights, is supposed to be due to the electrical discharges in the rarefied gases of the upper atmosphere.[2] The height of the aurora has

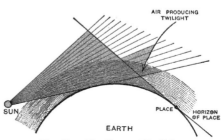

FIG. 3. — The Cause of Twilight.

been determined and is found in some cases to vary from 60 to 200 kilometers.[1] (4) Meteors are masses of matter from pinhead size up which are flying haphazard through space and often enter the earth's atmosphere with velocities from 12 to 50 miles per second. The heat caused by the resistance of the air raises them to incandescence and they become visible as shooting stars. The height at which these become visible has often been determined, and the larger values vary from 240 to 300 kilometers.[1] Thus there is sufficient quantity of air above this height, even, to make a meteor incandescent. (5) From observations of eclipses of the moon it is also possible to determine the extent of the atmosphere. This method also gives elevations as great as 300 kilometers.

As a general conclusion, then, it may be stated that a sufficient quantity of air extends to a height of 300 kilometers to give us an indication of its presence. It should be noted in this connection that the highest mountain does not rise above 10 kilometers. The greatest height attained by a manned balloon is not over 11 kilometers (10.3 kilometers or 6½ miles by Dr. Berson and Professor Süring in 1901), and unmanned balloons and kites have not gone above 29 and 8 kilometers respectively.

[1] These heights are determined trigonometrically, by means of surveying instruments. It is the regular problem of determining the height of an inaccessible object. Simultaneous observations of direction and altitude from the two ends of a base line several miles long are necessary.

[2] The exact nature of the aurora is so little understood and the various values for height which have been determined are so discordant, that it is questionable whether observations of the aurora are of any value in determining the extent of the atmosphere.

The earth's atmosphere cannot extend more than 21,000 miles and turn with the earth as it rotates on its axis. At this distance centrifugal force due to the rotation and gravitational attraction balance so that the air would be abandoned.

The middle layer of the atmosphere is at a height of 3.6 miles. That is to say, there is as much air above as below this level. If the air were of the same density throughout as at the earth's surface, it would have a height of about 5 miles. This is sometimes called the height of a homogeneous atmosphere.

The Meteorological Elements

25. The meteorological elements. — The condition of the atmosphere at any particular time and place is completely determined by six things.

The six meteorological elements. These are called the meteorological or weather elements. They are temperature, pressure, wind, humidity, clouds, and precipitation. Dust and atmospheric electricity are sometimes included. For example, at 8 A.M., July 4, 1908, at Williamstown, Mass., the temperature was 72°, the lowest temperature during the night had been 65° and it occurred a little before sunrise; the pressure was 29.47 inches; the wind velocity was two miles per hour from the east; the air contained 7.10 grains of moisture per cubic foot and held 86 per cent of what it could; the clouds were stratus and were moving slowly from the east; the sky was totally covered with clouds; no rain was falling or had fallen during the night.

The exact condition of the atmosphere at any time and place can thus be stated by giving numerical values to different phases of the meteorological elements.

26. Weather and climate. — Weather is defined as the condition of the atmosphere at any time and place and is thus best described by giving the numerical values for the meteorological elements. Climate is generalized weather. It is concerned more with the average rather than the particular values of the meteorological elements. **Definition of weather and climate.** The term is only used in connection with larger areas and longer periods of time. Thus one should speak of the weather on December 25, 1910, in New York City, but of the winter climate of New England.

27. Periodic and irregular variation and normal values. — The numerical values of the meteorological elements are by no means constant, but are always undergoing change or variation. There **Periodic and irregular variation.** are two kinds of change or variation, periodic and irregular. Whenever in the course of a variation the initial value repeats itself after the lapse of approximately equal intervals of time, the variation is said to be periodic. If the changes are irregular or haphazard as regards amount or time of occurrence, the variation is said to be irregular. Now the meteorological elements are undergoing both kinds of variation simultaneously, and sometimes two or more periodic variations are present at the same time. For example, it usually grows warmer during the morning and early afternoon and then cooler during the rest of the afternoon and night. This is a periodic variation in temperature. A sudden thunder shower may, however, lower the temperature twenty degrees or more during a few minutes. This is an irregular variation. Sometimes the irregular variations are of such magnitude as to cloak or render almost imperceptible the periodic variations. (See Figs. 4 and 5.) There are no examples in the realm of meteorology of a pure undisturbed periodic variation except for a short time. Examples of this must be taken from mechanics or physics. The amount of snow which falls during the winter, the highest or lowest temperature which occurs on a definite date for successive years, the number of thunder showers during a year, are all examples of an irregular variation only. In all the changes of temperature and pressure, and in fact of all the meteorological elements from moment to moment, we have examples of periodic and irregular variation combined.

28. Whenever observations of any phase of any one of the meteorological elements have been made for a considerable time, it often is desirable to summarize the observations. Such a summary ordinarily contains four things, the average value, usually called the normal value, the greatest value, the least value, and the average departure from normal. The meaning of these and the method of determining their numerical values can be best illustrated by an example. The total number of thunder showers observed at Albany, N.Y., for successive years is shown on p. 22.

The four things given in summarizing observations.

The average number of thunder showers is 22. This is usually spoken of as the normal number of thunder showers per year. The largest number is 35 in 1910, and the smallest number is 7 in 1890. The year 1910, with 35, shows a departure from normal of 13. The departure for 1909 is 1, for 1908 is 9, etc. When these departures are averaged, the result is called the average departure from normal. Its value here is 6. Thus the normal value 22, the greatest value, 35, the least value, 7, and the average departure from normal, 6, form a convenient summary of the observations and give a complete picture of what may be expected at Albany as regard the number of thunder showers during a year. This, by the way, is an

Year	Number of Thunder Showers	Year	Number of Thunder Showers
1884	24	1898	26
1885	22	1899	22
1886	14	1900	28
1887	14	1901	30
1888	9	1902	28
1889	12	1903	22
1890	7	1904	28
1891	9	1905	21
1892	32	1906	32
1893	23	1907	25
1894	23	1908	31
1895	18	1909	21
1896	17	1910	35
1897	22		

illustration of irregular variation, and most of the observations which are summarized in this way are examples of irregular variation only.

29. **Graphical representation.** — The variation of a quantity, whether periodic, irregular, or both, can be readily pictured by plotting its various

Graphical representation.

values to scale. This is called graphical representation, and the curve which represents the variation is called the graph. The values are plotted by choosing equal distances on one of the axes, usually the horizontal or X-axis, to represent time and by laying off equal distances on the other axis to represent the values of the quantity. The points are then located in accordance with the observations and are connected by a broken line or a continuous curve.

The following examples will illustrate this method:

Place : Williamstown, Mass.

Time	June 12, 1907 Temperature	June 9, 1906 Temperature	Time	June 12, 1907 Temperature	June 9, 1906 Temperature
2 A.M.	48	64	6 P.M.	70	68
4 A.M.	44	63	8 P.M.	62	66
6 A.M.	43	63	10 P.M.	55	65
8 A.M.	51	68	Midnight	50	64
10 A.M.	62	73	2 A.M.	47	63
Noon	68	80	4 A.M.	44	62
2 P.M.	70	83	6 A.M.	43	60
4 P.M.	71	72			

Figure 4 represents the typical undisturbed daily variation in temperature. Figure 5 represents the typical variation disturbed by a thunder shower between 3 and 5 o'clock in the afternoon.

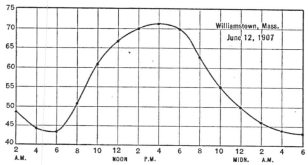

FIG. 4. — Typical Undisturbed Daily Variation in Temperature.

Many other examples of graphical representation will be found scattered through the book. Usually it is the variation of some quantity

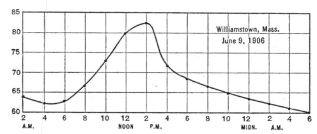

FIG. 5. — Daily Variation Disturbed by a Thunder Shower.

with the time that is represented. If any two quantities are so related that, as one changes, the other changes also, the relation can be represented graphically in this way.

THE THREE METHODS OF INVESTIGATION

30. There are three methods of reaching conclusions and deriving general laws. These are known as the inductive, the deduc- The inductive, and the experimental method. In the inductive method tive method. many facts and observations are generalized to get the underlying law.

It proceeds from particulars to the general. All general statements based upon meteorological observations are thus examples of the inductive method. Meteorology is usually presented largely from the inductive standpoint.

In the deductive method one starts with general principles or laws and determines what ought to take place in a particular instance. This must **The deduc-** then be compared with the observed facts and an agreement **tive method.** stamps the whole work as correct. In Chapter IV, C, will be found a good example of deductive reasoning. We will start there with two fundamental facts, determine what ought to be the pressure and wind direction at different points on the earth's surface and then compare this with the results of observation.

The experimental method is essentially the laboratory method. Here the conditions of the experiment are changed and the results noted. In **The experi-** this way new facts, relations, and laws may be found, and all **mental** assumptions and hypotheses may be rigidly tested. The **method.** experimental method is of only limited application in meteorology because all of the phenomena furnished by nature are on such a stupendous scale that man is powerless to change the conditions.

THE PLAN OF THE BOOK

31. The first chapter contains introductory material and a discussion of the composition, pressure, and height of the atmosphere. The second **The con-** chapter is devoted to the study of the heating and cooling of **tents of the** the atmosphere, as this is of such fundamental importance. **book.** The meteorological elements are then considered in order in Chapters III to V inclusive. Chapter VI is given up to the study of the different kinds of storms. Two chapters are then devoted to the practical side of meteorology, namely, weather bureaus and their work and weather prediction. Part II, consisting of five chapters, is devoted to special subjects not always included in meteorology proper. They could be omitted without destroying the unity or completeness of the book. These subjects are climate, floods and river stages, atmospheric electricity, atmospheric optics, atmospheric acoustics.

Since the book is intended primarily as a text-book, the syllabus at the beginning of each chapter, the marginal topics, and the questions, topics **The aim of** for investigation, exercises, and references at the end of each **the book.** chapter have been added. It is also hoped that the intelligent reader who has not had special training in this science will be able to find here a clear and concise picture of the modern aspects of the subject. This volume does not pretend to be a compendium or compilation of existing knowledge. It is hoped, however, that in the references to the literature the way has been pointed out to a more complete knowledge of the subject on the part of him who desires it.

QUESTIONS

(1) How is a natural science defined? (2) Mention several natural phenomena and state the natural science to which each belongs. (3) Enumerate the natural sciences. (4) What is the province of each? (5) Why are physics and chemistry fundamental? (6) Define met'y. (7) Of what does it treat? (8) Why is met'y one of the oldest sciences? (9) What is the origin of the word met'y? (10) When, where, and by whom was the first treatise written? (11) What did it contain? (12) Give the dates of the four periods in the history of met'y. (13) What brought about each new period? (14) State the characteristics of each period. (15) Name the two lines of usefulness of met'y. (16) Discuss the question of the financial saving. (17) What is the duty of schools and colleges as regards the public? (18) Why should the public be educated in met'y? (19) Along what three lines does a natural science offer training? (20) State the character of each line of training. (21) State some of the varied influences and effects of the weather. (22) State the relation of met'y to physics. (23) State its relation to the other natural sciences. (24) Define the atmosphere. (25) What is the origin of the word? (26) What are the characteristics of air? (27) State some of the physical properties of gases. (28) What are the four major constituents of pure dry air? (29) State approximately the percentage composition. (30) How much hydrogen is found in the atmosphere? (31) Name the four rare gases in the atmosphere. (32) Give the proofs that air is a mechanical mixture of its components. (33) How constant is the composition of the air? (34) How much CO_2 does exhaled air contain? (35) Why is the composition so constant? (36) What change takes place in the composition at great heights? (37) Why? (38) What facts tend to justify this theoretical conclusion? (39) Name the minor constituents of the atmosphere. (40) How much water vapor is present? (41) Why is water vapor important? (42) How many organic particles are in the atmosphere? (43) What is included under this head? (44) Are the inorganic particles visible? (45) Describe Aitken's dust counter. (46) State the working principle. (47) How is a determination made? (48) How numerous are dust particles? (49) Does the number vary? (50) What are the sources of atmospheric dust? (51) What are some of the effects of this dust? (52) What is the nature and use of ozone? (53) What are the offices and activities of the atmosphere as a whole? (54) What are the functions of the nitrogen? (55) What are the functions of the oxygen? (56) How is it lost to and gained by the atmosphere? (57) What are the functions of CO_2? (58) How is it lost to and gained by the atmosphere? (59) Is the amount constant everywhere? (60) How are CO_2 and O held in equilibrium? (61) Describe briefly the atmosphere of the other heavenly bodies in our solar system. (62) What determines the amount? (63) What was the early history of the atmosphere? (64) What changes have taken place in geological times? (65) Is the atmosphere changing now as regards bulk and composition? (66) What is the probable future of the atmosphere? (67) Define gravity. (68) What is its direction? (69) State some of its effects

on matter. (70) Define geosphere, hydrosphere, and atmosphere. (71) State the origin of the words. (72) What is the shape of the hydrosphere? (73) Why does the atmosphere exert a pressure? (74) How much is the pressure? (75) Why does the pressure change with elevation? (76) Why is there no theoretical limit to the atmosphere? (77) What is meant by the " sensible " height of the atmosphere? (78) Describe the five methods of determining it. (79) What heights have been found? (80) How high have balloons and kites ascended? (81) At what height is the middle layer located? (82) Define the met'l or weather elements. (83) Name the six met'l elements. (84) How may the condition of the atmosphere at any time and place be described? (85) Define weather and climate. (86) Have the met'l elements constant values? (87) Name and define the two kinds of variation or change. (88) Illustrate by means of examples the various kinds of change or variation. (89) What four things are computed when observations are summarized? (90) Define what is meant by normal values. (91) What is graphical representation? (92) How is a graph constructed? (93) Name the three methods of deriving general laws. (94) Describe and illustrate the inductive method. (95) Describe and illustrate the deductive method. (96) Describe and illustrate the experimental method. (97) What is the relative importance of these three methods in met'y? (98) What is the order of the subjects treated in this book? (99) What is the aim of the book?

TOPICS FOR INVESTIGATION

(1) The financial saving caused by the Weather Bureau.
(2) The relation of weather to disease.
(3) The relation of weather to crime.
(4) The extent to which meteorology is taught in the schools and colleges.
(5) The agreement of the various determinations of the composition of the air.
(6) The history of the discovery of argon.
(7) The history of the discovery of the rarer gases in the air.
(8) The dust of the atmosphere.
(9) Changes in the earth's atmosphere during geological times.
(10) Meteors. (See any text-book on astronomy.)
(11) Determination of the maximum and minimum of graphs. (The usual method is to draw a line connecting the middle points of the chords parallel to the X-axis and terminated by the curve.)

PRACTICAL EXERCISES

(1) Copy from some source or compile a few tables showing the relation between the prevalence of certain kinds of disease and the weather. Plot the graphs representing these relations.
(2) If physical and chemical apparatus is available, experiments may be performed to show the composition and pressure of the atmosphere, and the properties of gases.
(3) Write out a definite, complete description of the weather at several different times and places.
(4) Copy from some source or compile observations of some quantity subject to irregular variation only and summarize them in accordance with section 28.
(5) Plot several graphs from series of observations of some kind and state the kinds of variation depicted.

REFERENCES [1]

AITKEN, " Observations of Atmospheric Dust," W. B. Bulletin 11, p. 734.

ARRHENIUS, *Lehrbuch der kosmischen Physik*, Composition of the atmosphere, pp. 473–490.

KASSNER, *Das Wetter und seine Bedeutung für das praktische Leben*, Relation to practical life, pp. 114–144.

MOORE, JOHN W., *Meteorology*, 2d ed., The Influence of Weather on Disease, pp. 407–457.

RAMSAY, *The Gases of the Atmosphere; the History of their Discovery.*

[1] For further information concerning the books and for further references see Appendix IX.

CHAPTER II

THE HEATING AND COOLING OF THE ATMOSPHERE

THE NATURE OF MATTER, HEAT, TEMPERATURE, AND RADIANT ENERGY

32. In order to have a clear conception of those fundamental processes which are operative in the heating and cooling of the atmosphere, one must understand something about the nature of matter, heat, tempera-

28

ture, and radiant energy. This subject will be briefly sketched here, but the reader must be referred to text-books on physics for a more detailed treatment.

A piece of wood, for example, may be divided into several portions. These portions may in turn be subdivided, and the process may be continued until the resulting portions are scarcely visible in a **Molecule** powerful microscope. The natural question at once arises **and atom.** as to whether this process of subdivision could be carried on indefinitely provided the mechanical means were at hand for accomplishing it. The answer is that there is a smallest particle which can exist and still retain the properties of the substance in question. This smallest particle is called the molecule. Molecules in turn are composed of atoms which may be like or unlike, few or many. There are known to us nearly eighty different kinds of atoms corresponding to the number of so-called elements. It is now thought that the atom in turn is composed of perhaps thousands of particles which may even, under certain conditions, escape from the atoms themselves. These particles are spoken of as corpuscles or electrons. Atoms are held together by the force of chemical affinity to form molecules. Molecules are held together by cohesion or adhesion to form masses, and all masses exert a gravitational attraction upon each other.

The whole of space is supposed to be filled with a substance called luminiferous ether, having certain properties that account for the facts which are observed. The planets have moved through it for **The ether.** ages, without showing appreciable retardation; it must, then, be practically frictionless. Waves are transmitted with tremendous velocities; it must, therefore, be highly elastic. It has the same properties at all points and in all directions; it must, then, be homogeneous. Ether waves come to us from the remotest depths of space; it must, therefore, be all-pervading. We picture, then, the whole universe as filled with this frictionless, highly elastic, homogeneous, all-pervading substance.

The molecules are not in contact with each other, but are embedded in this ether as dust particles are suspended in dusty air or as sand particles are suspended in muddy water. These molecules are **Heat and** not at rest but are in constant motion, colliding with their **temperature.** neighbors, rebounding, and quivering. Since the molecules are in motion, they possess energy, and this energy of the molecules is called heat. Heat, therefore, is not a substance, but a form of energy. The heat which a body possesses is thus simply the sum total of the kinetic energy of its molecules. A careful distinction must be made between heat and tem-, perature. Temperature is determined by the velocity of motion on the part of the individual molecules. If a body is heated, the molecules move more rapidly and hit harder. There is now a more active motion on the part of the molecules and the temperature is thus higher. The sum total of the energy of the molecules is also greater and the body thus possesses a greater amount of heat. The following numerical data for hydrogen gas under normal conditions will give a more concrete and definite picture of the construction of matter: The number of molecules per cubic centimeter is 21×10^{18}, each molecule with a diameter of 5.8×10^{-8} centimeter, and a mass of 4.6×10^{-24} gram. These molecules are moving with an average velocity of 1.9×10^5 centimeters per second over a path whose average length is 9.7×10^{-6} centimeter, before colliding with another molecule. The average number of collisions per second is 9,480,000,000.

Whenever a molecule collides with another and rebounds and quivers, it becomes the center of an ether wave which spreads out spherically in **Radiant** every direction. These waves have various lengths, that is, **energy.** distances from crest to crest, and various intensities. They all proceed with a velocity of 186,330 miles per second in ether in which no material molecules are embedded. If a body is heated, a larger number of waves is sent out and shorter waves are emitted as well as the longer ones. Terrestrial bodies at ordinary temperatures are emitting waves having lengths from, say, 0.001500 to 0.000270 centimeter. The waves sent out by the sun are shorter, having lengths varying from 0.000270 to 0.000019. Those between 0.000270 and 0.000075 are spoken of as the infra-red or heat rays; those between 0.000075 and 0.000036 are light waves running from red to violet; those between 0.000036 and 0.000019 are called the ultra-violet rays. The limits are by no means definite. The figures have been given simply to indicate approximately the limits between which the great majority of the wave lengths lie. The energy conveyed by ether waves is spoken of as radiant energy. All bodies, then, which are emitting ether waves are sending out radiant energy.

The statements above have been made as if they were all undoubted facts. Many, however, are merely highly probable hypotheses. The **Hypotheses.** probability of the correctness of an hypothesis increases with the number and complexity of the facts of observation which are explained and correlated by it, and as it becomes more and more evident that no other hypothesis can explain as much. ·

The Sources of Atmospheric Heat

33. There are three sources from which heat might be supplied to the atmosphere; namely, the sun, the earth's interior, and the stars and other heavenly bodies.   There are two ways of showing that *The three* the amount of heat supplied to the atmosphere by any other *possible* source than the sun is relatively negligible.   The heat sup- *sources.* plied by the earth's interior would be the same at the poles as at the equator, the same by night and during the winter as by day and during the summer.   Thus the changes, at least, in the temperature *All but the* of the atmosphere cannot be accounted for by heat from the *sun negli-* earth's interior.   Again, the stars shine by day as well as by *gible.* night and on all portions of the earth's surface.   Thus again, the changes in temperature cannot be explained on the basis of heat received from the stars.   Direct measurements have also been made of the amount of heat received from these sources, and the general conclusion is that the total amount of heat supplied by all other sources than the sun is not sufficient to change atmospheric temperatures by 0.25° Fahrenheit.   It is the sun, then, which controls absolutely the heating of the atmosphere and we must look to it for an explanation of all the varied changes in atmospheric temperatures.

Insolation

34. Amount. — The radiant energy, that is the energy in the form of ether waves, received from the sun is given the special name of insolation. Some idea of the amount of insolation can be gained by con- *Definition of* sidering the size and condition of the sun itself.   The sun is an *insolation.* immense globe, having a diameter of 866,000 miles and an average surface temperature of 10,000 degrees Fahrenheit.   More than 325 worlds like ours could be strung like beads on a string around the equator of the sun.   The best pictures of the intense incan- *Illustrations.* descence of the sun's surface can be gained by realizing that each square yard of the sun's surface is constantly emitting 140,000 horse power of energy.   The earth receives but a minute portion (one two-billionth part) of this, and yet the energy received in the course of a year would be sufficient to melt a layer of ice 241 feet thick covering the whole earth.

35. Variation with latitude and time of year. — The amount of insolation received from the sun at any particular time and place depends upon three things. (1) *Nearness to the sun.*   The amount of radiant energy

received from any hot body varies inversely as the square of the distance
of the point in question from the body.   Thus the amount of insolation
varies inversely as the square of the distance from the sun.

The three factors which determine the amount of insolation received.
(2) *Directness of the rays.*   When the sun's rays fall upon a
surface obliquely they are spread out over a larger area than
when they fall perpendicularly.   The result is that the
amount of energy received by each unit of surface is far less
for oblique than for perpendicular rays.   The accompanying
figure will illustrate the principle.   Let AB be the width of a beam falling
perpendicularly upon the surface MN.   The total amount of energy will

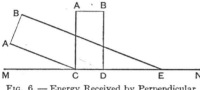

FIG. 6. — Energy Received by Perpendicular
and Oblique Incidence.

be concentrated upon the space
whose width is CD.   Suppose a
beam of the same width AB falls
obliquely upon the surface.   Its
energy will be spread out over a
surface having a width of CE.
The amount of energy received
by each unit of area of CD will
be seen at once to be far greater
than that received by each unit of area of the surface CE.   It can be
readily demonstrated by trigonometry that the amount of energy received by a unit area in the case of oblique incidence equals the amount
received by perpendicular incidence multiplied by the sine of the angle
of elevation of the sun.   By angle of elevation is meant the distance in
degrees of the sun above the horizon.   (3) *Duration.*   The amount of
energy received from a radiating body is, of course, directly proportional
to the duration of the radiation.

36.   In order to understand the variations · in these three factors
which determine the amount of insolation received, one must study the

The characteristics of the earth's revolution about the sun.
revolution of the earth about the sun.   The earth's path
around the sun is not an exact circle, but an ellipse with the
sun in one focus.   This path or orbit of the earth lies in a
plane called the plane of the ecliptic, and the earth completes
the circuit in $365\frac{1}{4}$ days.   The average distance of the earth
from the sun is 92,900,000 miles, but the actual distance varies about
1,500,000 miles in either direction in the course of a year.   The
earth is nearest to the sun on January 1 and at its greatest distance on
July 1.   The axis of the earth is not perpendicular to the plane of the
ecliptic, but makes an angle of $66\frac{1}{2}°$ with it, and this axis remains parallel to
itself as the earth revolves about the sun.   Figure 7 illustrates this revo-

lution. The north pole and the northern hemisphere of the earth are turned most directly towards the sun on June 21 and away from the sun on Dec. 21. Figure 8, which indicates the position of the earth on June 21 and Dec. 21, shows this change in presentation to the sun's rays.

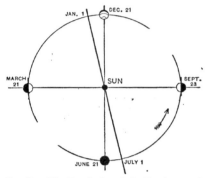

Since the earth is nearly 3,000,000 miles nearer the sun on January 1 than on July 1, more insolation must be received in January than in July. The difference amounts to about 7 per cent.

Changes in distance.

The change in the presentation of the earth to the sun's rays causes an apparent migration of the sun northward from Dec. 21 until

Changes in directness.

Fig. 7. — The Revolution of the Earth around the Sun.

June 21 and southward from June 21 until Dec. 21. The sun is directly overhead at noon on March 21 at the equator, on June 21 at the tropic of Cancer, on September 23 at the equator again, and on December 21 at the tropic of Capricorn. This migration of

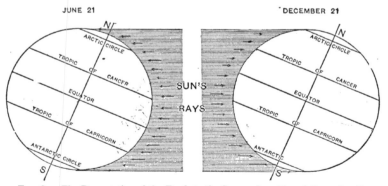

JUNE 21   DECEMBER 21

Fig. 8. — The Presentation of the Earth to the Sun on June 21 and December 21.

the sun through $47°$ $(2 \times 23\frac{1}{2})$ causes decided changes in the directness of the sun's rays and in the length of the day. The elevation of the sun at noon for any place in the northern hemisphere whose latitude

D

is ϕ can be shown to be $90° - \phi$ on March 21 and September 23, $90° - \phi + 23\frac{1}{2}°$ on June 21, and $90° - \phi - 23\frac{1}{2}°$ on December 21. Thus the noon elevation of the sun changes by $47°$ during the year, and this causes a great change in the insolation received at any one point. The length of the day also changes markedly throughout the year. In the latitude of New York State the length of the day varies from a little

Changes in   more than fifteen hours during the summer to a little less than

Duration.   nine hours during the winter. The accompanying table gives the greatest possible duration of insolation for various latitudes.

Latitude	0°	17°	41°	49°	63°	66° 30'	67° 21'	69° 51'	78° 11'	90°
Duration	12$^{hr.}$	13$^{hr.}$	15$^{hr.}$	16$^{hr.}$	20$^{hr.}$	24$^{hr.}$	1 mo.	2 mo.	4 mo.	6 mo.

37. Since the three factors that determine the amount of insolation received have different values for different latitudes and times of year, it follows that the insolation received varies with both time and latitude. For the same place it is

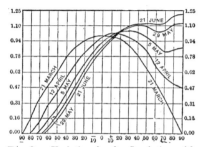

FIG. 9. — Variation in the Insolation with Latitude at five Different Dates. (after WIENER.)

different at different times of year ; for the same date it is different at different places, that is, different latitudes, on the earth's surface.

Figure 9 shows the amount of insolation received for the various

Variation   latitudes at five differ-

with lati-   ent times between the

tude.   20th of March and the

21st of June. The unit of insolation is the amount received in a day at the equator on March 21.

On the 21st of June, for example, the amount of insolation received at the North Pole is greater than that received at the equator. As far as nearness to the sun is concerned, the amount received at both the equator and the pole would be the same; as far as directness is concerned, at the equator the sun at noon has an elevation of $66\frac{1}{2}°$ while at the pole its elevation is but $23\frac{1}{2}°$. Due to this cause, much more energy would be received at the equator. The duration of insolation, however, is twelve hours at the equator and twenty-four hours at the pole. When the combined effect of the three factors is determined it is found that the amount of energy received at the pole is greater than that received at the equator.

Figure 10 shows the amount of energy received at three different latitudes on the earth's surface during the year. At the equator there are two maxima, one on the 21st of March and the other on the 23d of September. These are the dates when the sun is directly over head at noon and the insolation thus falls most directly. The March maximum is greater than the September one because the earth is nearer to the sun at that time. The duration and directness are the same in both cases. The amount of insolation received at the North Pole is zero until the 21st of March, when it rises rapidly to a maximum on the 21st of June, then drops back to zero again on the 23d of September.

Variation with time.

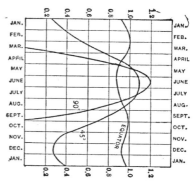

FIG. 10. — Annual Variation in the Insolation Received at three Different Latitudes.

38. **Distribution over the earth.** — The distribution of insolation over the earth is shown by the accompanying table, which gives the values of insolation for four different dates and six different latitudes. The unit is the amount of insolation received in a day at the equator on the 21st of March.

Distribution given as a table.

LATITUDE	0°	20°	40°	60°	90°	− 90°
March 21	1.000	0.934	0.763	0.499	0.000	0.000
June 21	0.881	1.040	1.103	1.090	1.202	0.000
Sept. 23	0.984	0.938	0.760	0.299	0.000	0.000
Dec. 21	0.942	0.679	0.352	0.000	0.000	1.284

Since the amount of insolation varies with both latitude and time, this relation cannot be expressed in the ordinary way by means of a graph. Figure 11, however, from DAVIS's *Elementary Meteorology*, represents this relation. The time is plotted along the horizontal axis and latitude along the other axis. The amount of insolation for any definite time and latitude is indicated by the length of the perpendicular to the plane of the axes. The surface which passes through the ends of these perpendiculars thus represents the variation in insolation. The curves given in Fig. 9

Graphical representation of insolation received.

represent the intersection with this surface, of planes perpendicular to the time axis at the dates in question. The curves given in Fig. 10 are the intersection with this surface of planes passed perpendicularly to the latitude axis at the latitudes in question.

In all of the foregoing sections, 35 to 38 inclusive, it is the distribution of the insolation at the outer limit of the atmosphere which has been

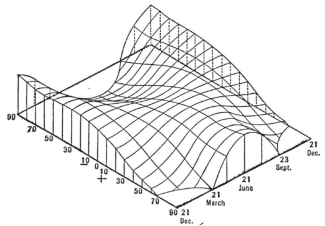

Fig. 11. — Variation in the Insolation with Latitude and Time.
(From Davis's *Elementary Meteorology.*)

considered. The distribution over the earth's surface would be the same provided the earth had no atmosphere or provided the earth's atmosphere allowed all of the insolation to pass through it. The question as to what portion is absorbed in the atmosphere and what portion actually reaches the earth's surface will be discussed in sections 43 and 44.

THE INTERRELATION OF MATTER AND RADIANT ENERGY

39. Reflection. — Whenever ether waves strike a material medium and are turned back, they are said to be reflected. This is entirely analogous to the reflection of a sound or water wave from a rigid barrier. There are two kinds of reflection; regular or mirror reflection, and irregular or diffuse reflection. In the case of regular reflection the angle made by the incident ray with the perpendicular to the reflecting surface is equal to the angle

Definition.

The two kinds of reflection.

made by the reflected ray.   In the case of diffuse reflection the reflected rays pass off in every direction and are scattered.   Burnished metal and ground glass are examples of these two kinds of reflectors.   The best reflector known is burnished silver, and this reflects about 98 per cent of the incident rays when these are perpendicular to the surface.   The insolation from the sun on reaching the earth falls upon a variety of things, such as pure air, dusty air, clouds, water, etc.   These Reflection things arranged in order of reflective power are as follows : of insola- water, snow, cloud, dusty air, earth, pure air.   Of these pure tion. air reflects practically nothing, while snow and water reflect from 30 to 50 per cent.

40. **Transmission.** — Whenever ether waves are allowed by a body to pass through it they are said to be transmitted.   A water analogy would be the passage of an ocean wave through the meshes of a fish Definition. net suspended in it.   The best transmitter known is rock salt. All transmission is selective in character ; that is, each body allows waves of certain lengths to pass through it more readily than others.   In the case of the atmosphere, the longer waves are the ones which Selective are most readily transmitted.   Glass transmits well the light transmis- waves, but does not transmit as readily the longer and shorter sion. ether waves.   Rock salt is the best transmitter because it transmits well ether waves of practically all lengths.   The various things upon which insolation may fall arranged in order of excellence as regards The trans- transmission are as follows : pure air, dusty air, water, snow, mission of cloud, earth.   Of these the earth transmits practically noth- insolation. ing, while pure air transmits more than 90 per cent of the insolation which falls upon it.

41. **Absorption.** — Absorption takes place whenever an ether wave enters a body and is destroyed by it.   A water analogy for this process can also be given.   Suppose a pond to be covered with logs which Definition. are in contact with each other.   If a water wave strikes these logs, the first ones are set in motion, and the wave loses intensity and soon ceases to exist.   The logs grind against each other and by friction are soon brought to rest.   The energy of the water wave has Absorption been transformed into energy of motion on the part of the logs. of insola- The best absorber known is carbon.   The various things upon tion. which insolation may fall, arranged in order of excellence as regards absorption, are as follows : earth, snow, cloud, water, dusty air, pure air.

42. There are four effects of absorption.   (1) *It may heat the body.* In this case the energy of the ether waves has been used up in exciting

a more vigorous motion on the part of the molecules.   (2) *It may cause vision.*   Whenever ether waves of certain wave lengths enter the eye **The four effects of absorption.** and fall upon the rods and cones of the retina, they impart a stimulus which results in vision.   In the human eye waves of greater length than 0.000075 centimeter or shorter than 0.000036 centimeter produce no effect.   The color perceived ranges from red to violet, depending upon the length of the wave. (3) *It may cause chemical reaction.*   When ether waves are absorbed by certain compounds, the molecular activity which is excited is so great that the molecules break apart and new compounds are formed.   This is the basis of photography.   Certain salts, usually silver salts, on the photographic plate are decomposed by the light rays, and thus a permanent impression on the photographic plate is made.   But a small amount of ether energy is used up in this way, however.   A very large amount of insolation is consumed by plants, as described in section 15, when the green cells using the energy of insolation combine the sap and CO_2 of the atmosphere, making the complex molecules from which its own tissues are built up.   The longer ether waves are usually more efficacious in causing heat, the intermediate ones in producing vision, and the shorter ones in causing chemical decomposition.   (4) *It may cause change of state.*   If insolation falls upon a body of water, a large amount of energy is used in evaporating the water, that is, in changing it from the liquid to the gaseous state without changing its temperature.   Energy used in this way is called latent heat.

43. Actinometry.[1] — An actinometer is an instrument for measuring the insolation received from the sun.   Actinometry treats of the use

The fundamental formula. of this instrument in making the measurements.   The fundamental formula in actinometry is $H = Ca^l$. H here represents

FIG. 12. — Showing the Contrast in the Thickness of Air passed through by Vertical and Oblique Rays.

the amount of energy received on a unit of surface at right angles to the sun's rays. a is the percentage transmitted by the atmosphere when the rays fall vertically. Its value would be unity for perfect transmission. l is the thickness of the atmosphere, considering the vertical thickness as unity. C is a

[1] For a bibliography of the articles on solar radiation see *Bulletin of the Mount Weather Obs.*, Vol. 3, part 2, pp. 118–126.

constant, ordinarily known as the " solar constant." The thickness of
the layer of air through which the sun's rays pass increases rapidly as
the sun nears the horizon.   Figure 12 shows well the contrast in the
length of path for vertical and oblique rays.   The following table gives
the various values of l for the different altitudes of the sun: —

Altitude of sun	0°	5°	10°	20°	30°	50°	70°	90°
Thickness of the atmosphere in units	35.5	10.2	5.56	2.90	1.99	1.31	1.06	1.00

It will be seen that the thickness of air traversed by the sun's rays
when rising or setting is more than 35 times the thickness when the
sun is directly overhead.

The term a^l is equal to one if l is zero or if a is equal to one. a
equaling one means that the atmosphere is transmitting all of the insolation. l equaling zero corresponds to a zero thickness of
the atmosphere, that is, to no atmosphere at all.   In either
case H would equal C. C is then the amount of energy received by a given unit area at right angles to the sun's rays,
provided there were no absorption in the atmosphere, or provided the
atmosphere were not present.

The meaning of the solar constant.

If the elevation of the sun above the horizon be known, l can at once be
computed from the table given above.   Absolute or relative values of
H can be determined in several different ways.   Seven ways
will be simply mentioned here, and the reader must be referred to special articles on the subject for a more detailed
treatment.   The various methods make use of calorimetrical
apparatus, a pyrheliometer, a thermoelectric couple, a bolom-
eter arrangement, chemical decomposition, a black bulb thermometer
in vacuo, or photographic paper.   The black bulb thermometer will be
considered later in section 70.   The method by means of photographic
paper, although not especially sensitive, is so ingenious as to be worthy of
a few words of explanation.   The blackening of an ordinary piece of
photographic paper when exposed to the rays of the sun depends upon
the intensity of the sun's rays and the time of exposure.   The ratio of
the times of exposure required to produce the same amount of blackening for two different values of intensity gives the ratio of those intensities.
Thus relative values of H can be readily determined.

The methods of determining the energy received.

If H has been determined for two different values of l, by means of
the equation given above, C and a can be computed.   There
are two ways of getting different values of l.   One is by
taking simultaneous observations from a mountain top and a·

The method of determining C and a.

near-by valley, another is to take observations at different times of day, thus getting different thicknesses of atmosphere.

Numerous measurements have been made under the most varied conditions and at many places. The different values for C are not in good *The values* agreement, but the values ordinarily used at present are *of C and a.* two or three calories per square centimeter per minute. A calorie is one one-hundredth of the amount of heat required to raise the temperature of one cubic centimeter of water from the freezing to the

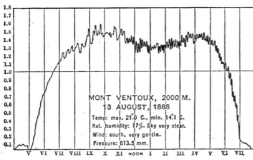

FIG. 13. — Insolation on Mont Ventoux.
(From HANN's *Lehrbuch der Meteorologie.*)

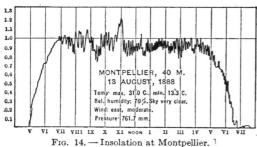

FIG. 14. — Insolation at Montpellier.]
(From HANN's *Lehrbuch der Meteorologie.*)

boiling point. When the sky is clear, the value of a, the transmission coefficient, is about 75 per cent. Under various conditions, however, values as low as even 10 or 20 per cent have been observed. The best modern values of the solar constant are without doubt those which have been obtained since 1904 by Mr. C. G. Abbot, Director of the Astrophysical Observatory of the Smithsonian Institution. His results are in much better agreement and would seem to indicate a value a little less than 2 (between 1.9 and 2.0).

44. Figures 13 and 14 show well the effects of atmospheric absorption. *The effect* The amount of insolation received on a surface having an area *of atmos-* of a square centimeter and kept constantly at right angles to *pheric ab-* *sorption on* the sun's rays is here given for the same day, August 13, 1888, *the insola-* on Mont Ventoux with an elevation of 2000 meters, and at *tion re-* *ceived dur-* Montpellier with an elevation of 40 meters. If the atmos- *ing the day.* phere had transmitted all the incident insolation, the value

would have been equal to the solar constant at both places from the moment of sunrise until sunset. The graphs show well the gradual rise in value of the insolation during the morning and the gradual falling off during the afternoon, due to the varying thickness of the atmosphere through which the rays were coming. The largest value on Mont Ventoux is 1.6, while the largest value at Montpellier is 1.2. This difference is due to the absorption of the 1960 meters of air which is their difference in elevation. It will also be noticed that the maximum values were received several hours before noon. The reason for this is that the air is always less transparent in the afternoon than during the morning.

The accompanying table from Angot gives the amount of insolation received at the earth's surface in the course of a year at various latitudes, first on the assumption that the atmosphere transmits all The effect of the insolation, and secondly on the assumption that 60 per of atmos- cent is transmitted. The unit of insolation here used is the pheric ab- sorption on amount of insolation received at the equator during one day, the insola- on the 21st of March. It will be seen at once that the dim- tion re- ceived at inution in the amount of insolation at the pole is much more different marked than at the equator. It will also be seen from latitudes. Figs. 13 and 14 and the table that not more than one third of the insolation received from the sun actually reaches the earth's surface on any particular day. The transmission coefficients for various places on the earth's surface during any given day probably vary from 10 per cent to perhaps 90 per cent. Sixty per cent would seem to be a large average for the whole earth, but even with that figure, as seen from the table, barely more than one third of the insolation reaches the earth's surface.

	Equator	10°	20°	30°	40°	50°	60°	70°	80°	Pole
a = 1.0	350.3	345.5	331.2	307.9	276.8	239.8	199.2	166.3	150.2	145.4
a = 0.6	170.2	166.5	155.1	137.6	115.2	90.6	67.4	47.7	33.5	28.4

45. Behavior of the ocean as regards reflection, transmission, and absorption. — The ocean reflects about 40 per cent of the insolation which falls upon it, and transmits the remainder to considerable The ocean depths, but eventually absorbs all the insolation which changes but is transmitted. The ocean changes extremely little in tem- little in tem- perature perature between day and night, and there are five reasons for between day this : (1) So large an amount, namely, 40 per cent, of the in- and night. solation is reflected and thus lost as far as heating the water is concerned. (2) The insolation which is absorbed is transmitted to considerable depths, and thus it is not a thin surface layer, but a layer of considerable depth that is involved. (3) A considerable amount of evaporation takes place from the surface of the ocean, and the insolation is used in causing this change of state and not used in causing a rise of temperature on the part of the water. (4) It requires a larger amount of heat to raise the temperature of a given quantity of water than any other substance. The technical expression for this is that the specific heat of water is larger than that of any other substance. For this reason a very small rise in temperature takes place as the result of absorbing a considerable amount of insolation. (5) The water of the ocean is in continual motion. Thus again it is not a small surface layer which is involved, but a considerable amount of water. On account of these five reasons, the temperature of the ocean rises very little during the day. Because it is in continual motion, transmits readily, and has a high specific heat, it cools but little during the night. The behavior of the ocean may then be briefly summarized by stating that its temperature change between day and night is extremely small, never amounting to more than one or two degrees.

46. Behavior of the land as regards reflection, transmission, and absorption. — Dry ground reflects but a few per cent of the insolation which falls upon it, transmits practically none, and thus absorbs nearly

The temperature change of ground between day and night is large.

all. The rise in temperature of dry ground under insolation is very great. In the first place, it absorbs nearly all of the insolation, and, being opaque, this takes place in a thin surface layer. Its specific heat is far less than that of water, and, being a solid, there can be no mixture as in the case of the ocean. Since the ground is a good absorber, it is also a good radiator and thus for similar reasons it cools rapidly at night. If the ground is wet or covered with vegetation, the change in temperature between day and night is much less than for dry ground, but always greater than for the ocean.

47. Behavior of the atmosphere as regards reflection, transmission, and absorption. — Pure air reflects but a minute quantity of the

The change in temperature of the atmosphere between day and night.

insolation which falls upon it, absorbs as little, and transmits practically all. As a result it changes temperature but very little between day and night. Dusty air reflects more and absorbs more, but is still one of the best transmitters. The atmosphere contains more dust, water vapor, and carbon dioxide near the earth's surface, and thus the change in temperature between day and night increases with nearness to the earth's surface. The reflection and absorption on the part of the atmosphere are both selective. In the case of reflection it is the shorter waves which are most effectively scattered, while in the case of absorption it is the longer ether waves which are chiefly concernèd, and this is particularly the case when the amount of water vapor and carbon dioxide is large. The temperature change in the atmosphere between day and night due to its behavior as regards insolation would probably be larger than the temperature change of the ocean, but never as large as that of the ground.

The contrast between the temperature of the ground and the temperature of the air above it is well brought out by the following typical example. The observations were made in 1895 at Tiflis with a latitude of $41° 43'$ and an elevation of 410 meters. The averages of the temperature observations (centigrade) made during three months at twelve different times of day are here given. The air temperatures were taken 3 meters above the ground. *The contrast between air and ground.*

Time	1 A.M.	3	5	7	9	11	1 P.M.	3	5	7	9	11
			December		January		February					
Ground	0.2	−0.2	−0.5	−0.8	3.0	10.3	13.2	10.9	4.3	1.9	1.2	0.6
Air	1.5	1.1	0.8	0.5	2.0	4.6	6.6	7.3	5.6	3.8	2.7	2.1
Difference	−1.3	−1.3	−1.3	−1.3	1.0	5.7	6.6	3.6	−1.3	−1.9	−1.5	−1.5
			June		July		August					
Ground	19.2	18.1	17.6	23.1	34.7	45.1	49.0	45.4	35.8	26.1	22.3	20.5
Air	18.9	18.0	17.5	19.4	22.4	24.8	26.3	26.9	26.3	23.8	21.5	20.1
Difference	0.3	0.1	0.1	3.7	12.3	20.3	22.7	18.5	9.5	2.3	0.8	0.4

It will be seen that during the day, both in summer and in winter, the temperature of the ground is much higher than the temperature of the air. The summer difference is two or three times the winter difference. During the night the ground is slightly colder than the air in winter and slightly warmer in summer.

The contrast between the temperature (centigrade) of the ocean and the temperature of the air above it is well shown in the following table, computed by Hann for the Atlantic Ocean from the observations of the *Challenger*. It corresponds to about $30°$ north latitude. The air is thus slightly warmer than the ocean by day and slightly cooler at night. The ocean can thus play *The contrast between air and water.* no part in heating the atmosphere by day or in cooling it at night.

TIME	1 A.M.	3	5	7	9	11	1 P.M.	3	5	7	9	11
Ocean	19.8	19.7	19.8	19.8	20.0	20.1	20.1	20.2	20.1	20.0	19.9	19.8
Air	18.9	18.9	19.0	19.2	19.6	20.2	20.6	20.6	20.3	19.7	19.3	19.0
Difference	0.9	0.8	0.8	0.6	0.4	−0.1	−0.5	−0.4	−0.2	0.3	0.6	0.8

The two typical examples which have just been given hold for the interior of large bodies of land and for the open ocean. Near the seashore a combination of the two would be found.

CONDUCTION AND CONVECTION

48. Conduction. — It is a matter of everyday experience that if one end of a bar of metal is heated, the other end also becomes hot. The

Definition. molecular agitation set up in the heated end of the bar is transmitted from molecule to molecule until the other end is reached. This transmission of heat from one part of a body to another or from one body to another by means of the molecules is known as conduction. All of the metals, particularly silver and copper, are good conductors of heat. All of the substances with which one has to do in meteorology, such as air, ground, water, cloud, etc., are very poor conductors.

The ground during the day may become very warm on account of the insolation received. At night it cools off rapidly, due to radiation. This

Air and ground poor conductors. diurnal oscillation in temperature, however, does not penetrate to a depth of more than two or three feet and requires many hours to penetrate to even this depth. The temperature changes between winter and summer do not penetrate to a depth of more than thirty or forty feet, and nearly six months elapse before the lower layers are reached. Thus at a depth of thirty or forty feet the highest temperatures occur in winter and the lowest in summer.

The air is also a very poor conductor. It might remain in immediate contact with the ground, which is many degrees hotter than itself, for a whole day, and yet, due to conduction alone, only the lower two or three feet would become perceptibly heated.

49. Convection. — Convection takes place only in liquids and gases. It is a transference of heat from one point to another by means of a circu-

Definition. lation of the liquid or gas itself. This process is of such great importance in meteorology that it deserves careful consideration. Let *AB* in Fig. 15 represent a long tank filled with fluid which is heated in the middle and at the bottom. The layer of fluid at

m becomes heated by conduction.   It expands and thus raises the
layers of fluid directly above it.   This causes a slight bulging of the upper surface.   Gravity acting upon this causes   Illustration.
the fluid to flow to the ends of the tank, thus decreasing the pressure
in the middle and in-
creasing it at the ends.
This increase of pres-
sure at the ends drives
in the colder fluid to-
ward the middle, thus
forcing the warmer fluid
to rise.   If the supply
of heat is continued, a

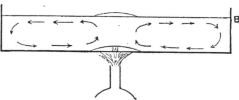

FIG. 15. — Diagram Illustrating Convection.

permanent convectional circulation will be established, which will tend
to equalize the temperatures throughout the tank.

50. **Convection in the Atmosphere.** — Since the atmosphere is fluid,
convection ought to take place in it provided it is heated at the bottom and at one place more than at another.   The equator is   Where con-
constantly heated m. re than the polar regions.   There ought   vection
then to be a permanent convectional circulation between equa-   ought to
take place
tor and pole.   It will be seen later that this is the cause of   in the at-
the general wind system of the globe.   The continents are   mosphere.
warmer than the adjacent oceans during the summer and colder in
winter.   There ought then to be a convectional circulation between the
continents and the near-by oceans.   It will be seen later that this is the
cause of the monsoons.   The land is heated more than the adjacent
ocean during the day, while at night the land is colder than the adjoining
ocean.   There ought thus to be a convectional circulation between the
land and the adjoining ocean.   It will be seen later that this is the cause
of the land and sea breeze.   If, due to irregularities, one place happens to
be heated more than another, there ought to be a minor convectional
circulation between this heated locality and surrounding places.   Evidences of such local circulations will be given in section 52.

51. There is one great difference between convection in a liquid and in
a gas.   If a given quantity of liquid (say water) is by convec-   The differ-
tion forced to rise, no temperature change takes place, pro-   ence be-
vided the rise is sufficiently rapid to make the effect of conduc-   tween convection in a
tion and radiation negligible.   If, however, a given quantity   liquid and
of gas rises, it arrives at regions where the pressure is less.   in a gas.
It therefore expands and grows cooler, even if conduction and radiation

are having no effect upon it. In order to make the illustration more definite, let us consider a cubic foot of water which is raised three thousand feet. If the rise has been sufficiently rapid so that conduction and radiation have added or subtracted only a negligible amount of heat, then there has been no change in temperature whatever. If a cubic foot of air rises three thousand feet, it has expanded and grown markedly cooler without the addition or substraction of heat by radiation or conduction. These temperature changes are called adiabatic, because they take place without the interchange of heat with the surroundings. The

Air cools 1.6° F. for 300 ft. amount of cooling for air is 1.6° Fahrenheit for a rise of 300 feet, or 1° centigrade (more exactly 0.993°) for 100 meters. This assumes that the air remains unsaturated with moisture. Thus, if a cubic foot of air rises three thousand feet, it cools sixteen degrees without gaining heat from or losing heat to its surroundings.

In the case of descending air the converse is true. The air is compressed and a corresponding rise in temperature takes place.

52. Evidences of convection. — There are three natural phenomena of more or less general occurrence which are evidences of convection or that

The three evidences of local convection. the conditions suitable for convection are present. These are mirages, dust whirlwinds, and cumulus clouds. The mirage is most common in hot, dry, desert countries. During the day the surface of the ground becomes excessively heated by the sun's rays. The layer of air in immediate contact with it,

Mirage. by conduction from the hot ground, also becomes considerably heated. The heated air expands and thus becomes lighter than the layers of air immediately above. If an observer is located at a moder-

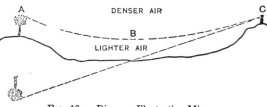

FIG. 16. — Diagram Illustrating Mirage.

ate elevation above this heated layer of air, rays of light coming to him from a distant object will follow the curved path *ABC* in Fig. 16. The reason for the curved path is that the rays of light are bent to or from the normal as they enter layers of different density. They are bent towards the normal if the density is greater, and away from the normal if it is less. The observer considers the object to be in the direction of the ray as it enters his eye, thus sees the inverted image of the object, and imagines it to be the reflection of the object in a body of water between himself and the object.   This phenomenon is called mirage.   The word is of French origin, meaning reflection. It is of common occurrence in hot, dry, dusty regions during the hotter parts of the day.   It does not indicate that convection is taking place, but that conditions suitable for convection are present.

Dust whirlwinds are also common in hot, dry, dusty countries.   A layer of heated air near the earth's surface is forced to rise at some point. The whirl is due to the fact that the indraft of cooler air to- **Dust whirl** ward the point of rise is not aimed directly at it.   This is due **winds.** to unevenness in the surface or local barriers.   The result is, the whirl is formed, which readily builds itself up and remains permanent.   The direction of rotation is accidental, perhaps determined to a slight extent by the direction of the rotation of the earth on its axis.   These dust whirlwinds sometimes rise to a height of a thousand or more feet.   They seldom occur over a very uneven surface for the reason that the mixing of the air is too great to permit a definite whirl.   These whirlwinds are also seldom seen over a vegetation-covered surface.   In the first place, such a surface is not heated as much during the day as is the dry, barren ground.   In the second place, on account of the absence of dust, the whirl is less readily seen should it occur.   Miniature whirlwinds of this kind are often seen on hot summer days along a dusty roadway.

Cumulus clouds are common in winter as well as in summer and especially in the eastern part of the United States.   They are masses of white cloud, looking like exploded cotton bales, usually with rounded, **Cumulus** domelike summits and flat bases.   They occur on what are **clouds.** popularly known as " fair north air days."   They are simply the heads of rising air columns.   A pocket of warm, moist air is forced to rise, due to convection, and as it rises, it expands and is cooled.   If the rise is to a sufficient height, it may become cooled to such a degree that it can no longer hold the moisture.   This condenses in the form of cloud.   The reason for the flat bases is that the pockets of rising air have approximately the same temperature and amount of moisture over considerable areas, and as a result they reach the stage at which they can no longer hold their moisture at about the same height.   By watching carefully the growth and change in form of a cumulus cloud, one can readily see that it is the top of a rising air column.

Vertical Temperature Gradients

53. Vertical temperature gradient. — It is a well-known fact that the temperature is different at different heights above the earth's surface.

Definition. This change in temperature with elevation is called the vertical temperature gradient, and it is usually expressed as a certain number of degrees Fahrenheit for 300 feet, or a certain number of degrees centigrade for 100 meters. The facts in connection with the vertical temperature gradient have been obtained in three ways: by means of mountain observatories, balloon ascensions, and kites.

The vertical temperature gradient may be obtained by taking simultaneous observations in a valley and at different points on the sides and

Mountain observations. top of a mountain. This method is at present no longer used, because the motion of the air and its temperature are influenced by the mountain itself. The temperature on the top of a mountain is not the same as in the open air at the same elevation.

Balloon ascensions for scientific purposes have now been made for more than a century. The older observations, however, are of comparatively little value because the instruments were poor and

Observations obtained from balloons. chiefly because they were not properly exposed and ventilated. A new era commenced with the year 1887, when Assmann's ventilated thermometer (see section 67) was used for the first time, and more attention was paid to the exposure of the instruments in general. Balloon ascensions for the purpose of obtaining meteorological observations have been made, particularly in Europe, and in France and Germany more than in any other countries. In the years 1888 to 1899 inclusive 75 balloon ascensions were made near Berlin and nearly a thousand ascensions have now been made near Paris. The Berlin ascensions are fully described and the observations carefully discussed in a three-volume work by Assmann, Berson, and others, entitled *Wissenschaftliche Luftfahrten*. Figure 19 is taken from page 188 of the second volume of this work, and Fig. 17 shows the equipment of a balloon used for these purposes. Smaller, sounding balloons carrying self-registering instruments have also been used at many stations for obtaining observations at greater heights than those to which a manned balloon can penetrate.

Concerning the use of kites for making meteorological observations, Professor Cleveland Abbe, in his *Aims and Methods of Meteorological Work*, writes:

FIG. 17. — Balloon Equipped for Meteorological Observations.
(From ASSMANN'S *Wissenschaftliche Luftfahrten.*)

FIG. 18. — Kite Equipped for Meteorological Observations.

" In order to obtain records from the air within a mile or two of the earth's surface, the kite was first employed by Alexander Wilson of Glasgow in 1749. In 1885 the writer urged the renewed application The use of of the kite, and since the meeting of the International Confer- kites. ence on Aërial Navigation at Chicago, in 1893, it has become an important meteorological apparatus.   Professor Marvin's construction of the Hargrave cellular kite, or box kite, is fully described in various publications of the United States Weather Bureau; the standard size used by him carries about 68 square feet of supporting surface.   The line is of the best music wire, whose normal tensile strength is 210 pounds.   The Marvin reel, on which the wire is wound, is a modification of the Thomson and Sigsbee deep sea sounding apparatus; it keeps an automatic record of the pull on the wire, both as to its intensity and direction in altitude and azimuth.   The reeling of the wire in and out is done either by hand or by a small gas engine.   The meteorological record at the kite is kept on one sheet of paper by means of the Marvin Meteorograph.   This keeps a continuous record of the time by means of an accurate chronograph, of the atmospheric pressure by means of a Bourdon aneroid, of the temperature of the air by a metallic thermometer, of the relative humidity by a hair hygrometer, and of the velocity of the wind by means of a small Robinson anemometer.   This complete apparatus is inclosed in an aluminum case to protect it from accident, and is lashed securely within the front cell of the kite so that it receives the full force of the wind, and undoubtedly gives a reliable record of the temperature.   The entire meteorograph weighs about two pounds.   Although kites sometimes break away, yet no injury has occurred to any meteorograph in the course of 1500 ascensions."   Figure 18 shows a Marvin Meteorograph and a kite fully equipped for securing observations.

54. The observations made in these three ways all show that the vertical temperature gradient varies markedly with the time of day.   It is also somewhat different at different times of year and varies The character of the with the type of weather and slightly with the location of the acteristics place on the earth's surface.   It must not be thought that the of the gradient. change in temperature with altitude is always a regular decrease by the same number of degrees for each 300 feet.   The temperature sometimes increases with elevation.   Again, it may remain constant for a considerable distance, or a marked change in temperature may take place with a very small change in elevation.   Figure 19 shows the actual vertical temperature gradient obtained at Berlin on the 19th of October, 1893, by means of a balloon ascension.   The ascension took place at 10.18

E

A.M., the balloon reached the highest point at 1.45 P.M., and descended to the ground at 4.20 P.M. The figure thus represents the vertical temperature gradient between 10.18 in the morning and 1.45 in the afternoon.

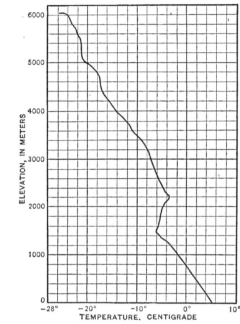

FIG. 19. — Graph Illustrating Vertical Temperature Gradient.

(From ASSMANN'S *Wissenschaftliche Luftfahrten.*)

The average value of the vertical temperature gradient **The average value.** for all times of day, for all seasons of the year, and for all places and weather, is ordinarily considered to be 1° Fahrenheit for 300 feet, or 0.6° Centigrade for 100 meters.

In order to understand the change in the vertical temperature gradient **The three characteristic gradients.** during the day, three characteristic gradients must be first considered. These three gradients are the average vertical temperature gradient which occurs usually about 9.00 in the morning and 8.00 in the evening, the gradient which exists at the time of minimum or lowest temperature, and the gradient which occurs at the time of maximum or highest temperature. These three gradients are represented graphically in Fig. 20 for a typical October day. The maximum temperature during this day is assumed as 60° F., the minimum temperature as 30° F., and the average temperature for the day as 45° F. This average temperature would occur about 9.00 in the morning and 8.00 in the evening. The straight line *B* in the figure represents the average vertical temperature gradient and is plotted with a fall of 1° F. for 300 feet. Thus at a height of 9000 feet above the earth's surface the temperature would be 30° lower than at the earth's surface. The gradient *C*, which represents the vertical temperature gradient at the time of the maximum temperature, follows the average vertical temperature gradient closely in the upper atmosphere, but departs from it more and more as the earth's surface is approached. The gradient A, which represents the vertical temperature gradient at the time of the minimum temperature, also follows the average gradient

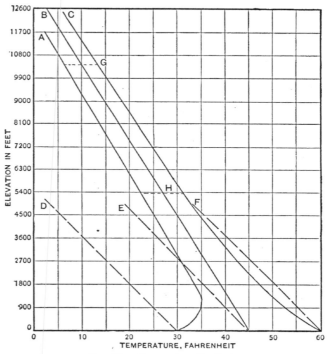

FIG. 20. — Temperature Gradients.

closely in the upper atmosphere but has a sharp bend at low altitudes, usually less than 2000 feet. The normal change in the vertical temperature gradient during the day is as follows: At the time of the minimum temperature it approximates to the form A. It then straightens out and at about 9.00 in the morning is following closely the average vertical temperature gradient, which is represented graphically by the straight line B. It then curves in the other direction, and by the time the maximum temperature is reached it has assumed the form C. It then

The daily variation in the vertical temperature gradient.

straightens out and by about 8.00 in the evening has reached the average vertical temperature gradient again. It then swings over until, when the time of minimum temperature is again reached, it has approximately the form A.

The daily range of temperature is defined as the difference between the highest and lowest temperature which occurs during the day. It will be

Range.

seen in the figure that the daily range decreases rapidly with elevation. At the earth's surface it was taken as 30°; at the height of a mile, as indicated at H, it is only 10°, while at the height of two miles, as indicated at G, its value is about 6°.

It is not necessary to take observations at great elevations in order to ascertain these facts. The observations taken at the top and base of the Eiffel Tower in Paris, at an elevation of 302 and 2 meters respectively, show the same results. The daily range decreases by 62 per cent in winter and by 45 per cent in summer for this difference in elevation of only 300 meters; and for 12 hours during the night in winter and for 8 hours during the night in summer the temperature at the top averages higher than at the base.

The average vertical temperature gradient must not be confused with the adiabatic rate of cooling of unsaturated rising air. The one is 1°

The adiabatic rate of cooling contrasted with the vertical temperature gradient.

Fahrenheit for 300 feet, while the other is 1.6° Fahrenheit for 300 feet. The vertical temperature gradient states that on the average there is a decrease in temperature of 1° F. with each 300 feet of elevation. According to the adiabatic rate of cooling, rising air by expansion cools, when not saturated with moisture, 1.6° F., for every 300 feet. Should air with a temperature of 30° rise, it would cool adiabatically due to expansion 1.6° for 300 feet and the behavior of a quantity of air rising with this initial temperature is shown graphically by D in Fig. 20. The behavior of quantities of air rising with initial temperatures of 45° or 60° is shown by E and F respectively.

55. The isothermal layer.—The facts which have just been stated

Characteristics of the vertical temperature gradient between three and six miles.

concerning the vertical temperature gradient, its average value, and the daily change in it apply only to the lower portion of the atmosphere, say, from the earth's surface to a height of two or three miles. Between the height of two miles and six miles or somewhat more, the characteristics of the vertical temperature gradient are quite different. Here it is much more regular and the average value is somewhat greater than near the earth's surface. The changes in the gradient between day and night, summer and winter, and with the weather, are all extremely small. Of course, at any given height, the temperatures are lower in winter than in summer, but the gradient is almost exactly the same. It is also true that, at the same height, the temperatures are usually higher when the weather is fair than when a storm or area of low pressure is present, but the gradient changes very little with the weather.

After a height of six miles, or in some cases much more is reached, the temperature seems to remain constant with elevation, or may, indeed, increase slightly with altitude. This layer in which the tem- The iso-perature remains so nearly constant is usually spoken of as thermal the isothermal layer or the warm stratum of the atmosphere. layer. This subject has been investigated especially by Teisserenc de Bort, Assmann, and in this country by Humphreys and Rotch, the director of the Blue Hill Observatory near Boston. In an article in the *Monthly Weather Review* of May, 1908 Professor Rotch writes:

" This inversion of temperature was first discovered by M. Teisserenc de Bort with the sounding balloons sent up from his observatory at Trappes, near Paris, France, in 1901, and almost simultaneously by Professor Assmann from similar German observations. Since then almost all the balloons which have risen more than 40,000 feet above central Europe (that is, near latitude 50°) have penetrated this stratum, without, however, determining its upper limit. Teisserenc de Bort early showed that its height above the earth, to the extent of 8000 feet, varied directly with the barometric pressure at the ground. Mr. Dines gives the average height of the isothermal layer above England as 35,000 feet, with extremes of nearly 50 per cent of the mean. Observations conducted last March by our indefatigable French colleague, Teisserenc de Bort, in Sweden, just within the Arctic Circle, show that the minimum temperature occurred at nearly the same height as at Trappes, namely, 36,000 feet, although Professor Hergesell, who made use of sounding balloons over the Arctic Ocean near latitude 75° N., during the summer of 1906, concluded that the isothermal stratum there sank as low as 23,000 feet.

"During the past three years the writer has dispatched 77 sounding balloons from St. Louis, Mo., U.S.A., latitude 38° N., and most of those which rose higher than 43,000 feet entered the inverted stratum of temperature. This was found to be somewhat lower in summer, but the following marked inversions were noted last autumn: October 8, the minimum temperature of '— 90° F., occurred at 47,600 feet, whereas at the maximum altitude of 54,100 feet the temperature had risen to − 72° ; October

10, the lowest temperature of − 80° was found at 39,700 feet, while − 69° was recorded at 42,200 feet, showing a descent of nearly 8000 feet in the temperature inversion within two days. The expedition sent out jointly by M. Teisserenc de Bort and the writer, on the former's steam yacht *Otaria*, to sound the atmosphere over the tropical Atlantic during the summer of 1906, launched sounding balloons both north and south of the equator within the tropics, and although some of these balloons rose to nearly 50,000 feet, they gave no indication of an isothermal stratum. In fact, the paradoxical fact was established that in summer it is colder 10 miles above the thermal equator than it is in winter at the same height in north temperate regions. This results from the more rapid decrease of temperature in the tropics and the absence of the numerous temporary inversions which, as Mr. Dines has pointed out, are common in our regions below 10,000 feet. If, therefore, as seems probable, the isothermal or relatively warm stratum does exist in the tropical and equatorial regions, it must lie at a height exceeding 50,000 feet, from which height, as the data quoted show, it gradually descends toward the pole, at least in the northern hemisphere."

This isothermal layer exists, then, at a great height over the equator, has an average height of about six miles over the middle latitudes, and comes still nearer to the earth's surface in polar regions. Its average temperature during the summer is about − 60° F., while its average winter temperature is about − 71° F. When the weather is fair and is controlled by an area of high pressure, it is at a greater height and about 14° F. colder, as an average for all the quadrants of a high than when the weather is stormy and an area of low pressure is dominant. In this respect its temperate is the opposite of the air temperature between a height of two and six miles. The isothermal layer also begins at a lower altitude in winter than in summer. Its upper bounding surface has never been determined.

56. Many articles dealing with the isothermal layer will be found in the periodical literature since 1900, and the reader must be referred to these for a full treatment of the subject. No single universally accepted explanation of all the facts in connection with the isothermal layer has yet been given. In what follows a possible explanation of some points is presented.

A possible explanation of the isothermal layer.

Since the height of the isothermal layer is always more than five miles above the earth's surface, the amount of water vapor present must be extremely small, if not entirely negligible; furthermore the amount of carbon dioxide is much less than at the earth's surface. Thus of the three ingredients which are the chief causes of the absorption and radiation on the part of the atmosphere the dust alone remains. At a height of five miles or more the dust which exists in the atmosphere can come from but one source. The dust blown up from the earth or that which comes from the evaporation of ocean spray or from volcanoes cannot, except under unusual conditions, penetrate to this height. The dust which exists above five miles must come primarily from the meteors. This is put into the atmosphere at a height of 50 miles or more and as it slowly settles through the atmosphere it must cause the distribution of dust to be uniform below this height. Now the one process which is operative in heating and cooling the atmosphere at a height of more than five miles is the absorption of radiation by the dust particles and the radiation of heat by them to space above and to the earth below. No account is here taken of the adiabatic changes in temperature of the whole layer due to its rising or falling. The temperatures would be different, but the layer would still remain isothermal. It can be shown mathematically that the radiation received by a body at a given distance above an infinite plane (and the earth's surface may be considered as such) is independent of its height above that plane. Thus the amount of insolation from the sun and the amount of radiation from the earth absorbed by the dust particles will be independent of the distance above the earth's surface. One should thus expect that each given volume of air having in it the same number of dust particles would gain by absorption the same amount of heat and would lose at night by radiation the same amount of heat. But the density of the air grows steadily less with elevation. We have thus the same amount of heat applied to a smaller quantity of air, and one would thus expect an increase of temperature with elevation and decreasing density. The dust particles are probably not the only absorbers and radiators. Due to the action of ultra-violet light and the aurora borealis, the ozone content of the upper air may be fairly large. Ozone absorbs certain wave lengths readily, and the other gaseous constituents of the atmosphere may play a small part.

57. **Inversion of temperature — nocturnal stability.** — It was a fact of early observation that on the still, clear nights of winter and during the clear, frosty nights of the late spring or early fall, the tempera- Definition ture was somewhat higher on a mountain top or at a small of inversion. elevation above the earth's surface than on the earth's surface itself. When the temperature increases instead of decreases with elevation, it is spoken of as an inversion of temperature. It was formerly thought that inversions of temperature were of comparatively rare occurrence. It is

now known, however, that they occur on practically all nights. . When an inversion of temperature occurs, the vertical temperature gradient has the form of A in Fig. 20.

During the night the atmosphere is entirely stable. If a quantity of air with a temperature of, say, 30° on a night when an inversion of temperature exists should be pushed up a short distance, it would expand and cool at the rate of 1.6° F. for 300 feet, and find itself cooler than its surroundings. It would then immediately drop back again or, more accurately stated, it would never have spontaneously commenced to rise at all. Thus at night the atmosphere is entirely stable.

Why the atmosphere is stable at night.

58. Diurnal instability — conditions of convection. — During the day the atmosphere may find itself in a state of unstable equilibrium. Suppose when the vertical temperature gradient has the form C in Fig. 20, a quantity of air with a temperature of 60° commences to rise. It will expand and grow cool at the rate of 1.6° F. for 300 feet. As shown by the straight line F in the figure it will continually find itself, in spite of the cooling, warmer than its surroundings, and will continue to rise. It will rise until it has cooled to the temperature of its surroundings ; or, expressed graphically, it will rise until the straight line F intersects the vertical temperature gradient, C.

Why the atmosphere is unstable during the day.

The conditions for convection may now be summarized. (1) The atmosphere must be heated at the bottom. (2) The atmosphere must be heated at one point more than at surrounding places. (3) If a quantity of air which starts to rise is to continue its ascent, the average vertical temperature gradient must be greater than 1.6° for 300 feet.

The three conditions of convection.

How the Atmosphere is Heated and Cooled

59. The present chapter may be summarized and the principles which have been stated may be illustrated by considering the processes which are operative in the diurnal heating and cooling of the atmosphere near the earth.

In the heating of the atmosphere five processes must be considered. These are in order: the absorption of insolation, the absorption of the radiant energy from the earth, conduction from the earth, mixture by means of the wind, and convection. As was seen in the study of actinometry under the most favorable conditions, when the sun's rays fall vertically, only 90 per cent of the insola-

The five heating processes.

tion is transmitted and the remaining 10 per cent is absorbed. If the rays do not fall vertically and under unfavorable atmospheric conditions, a much larger percentage of the insolation is absorbed by the atmosphere. The dust, water vapor, and carbon dioxide are the chief absorbers, and they increase in quantity with nearness to the earth's surface. As a result a rise in temperature of perhaps a degree would occur in the upper atmosphere and a rise in temperature of perhaps 3 or 4° would occur near the earth's surface.

The earth becomes warmed by absorbing the insolation which reaches it and radiates its heat in the form of long ether waves to space. These longer ether waves are absorbed even more readily than the shorter ether waves which constituted the insolation. This constitutes the second source of heating, and here again the amount of absorption and the resulting heating would be greater nearer the earth's surface. In connection with conduction it was seen that only a layer 2 or 3 feet deep of the atmosphere would be heated by being in contact for many hours with the warm ground. This process, then, would be effective only in heating the layer of air close to the ground. The wind, however, so thoroughly mixes the various layers of air which are in contact with each other that the heat imparted by conduction to the 2 or 3 feet nearest the earth's surface would be distributed over perhaps two or three thousand feet near the ground. Since this lower layer is heated by conduction, convection would take place, and this circulation would transfer heat from the earth's surface to altitudes of from one to several miles. As a result of these five heating processes the temperature of the air is raised but little at great altitudes above the earth's surface, and the amount of heating increases rapidly as one approaches the earth's surface. The vertical temperature gradient C in Fig. 20 has thus been explained.

There are three processes operative in the cooling of the atmosphere, namely, radiation to the cool ground and to space, conduction, and mixture by the wind. The air, filled with dust, water vapor, **The three** and carbon dioxide, radiates heat as well as it absorbs it. The **cooling pro-** ground at night cools rapidly and the air loses heat by radi- **cesses.** ation both to the open sky and to the cooling ground. Conduction, again, would cool the two or three feet of air in immediate contact with the earth's surface, and the wind, although on an average less at night than during the day, would mix this layer with the two or three thousand feet of air above it. Since convection is not operative at night, the marked cooling must take place very near the earth's surface. The vertical temperature gradient A at the time of lowest temperature has thus been explained.

QUESTIONS

(1) Describe the molecular structure of matter. (2) How many kinds of atoms are there? (3) Describe the structure of the molecule and atom. (4) Describe the luminiferous ether. (5) What properties must it have? (6) How are molecules related to each other? (7) Define heat and temperature. (8) How are ether waves caused? (9) Define radiant energy. (10) Distinguish between an hypothesis and a fact. (11) Name the three possible sources of atmospheric heat. (12) Prove in two ways that all other sources of heat than the sun are relatively negligible. (13) Describe the size and condition of the sun itself. (14) Give illustrations of the amount of energy received from the sun. (15) Define insolation. (16) Upon what three things does the amount of insolation received depend? (17) Explain how the amount of insolation received depends upon the obliqueness of the rays. (18) Describe the earth's orbit around the sun. (19) What is the effect of the change in distance of the earth from the sun? (20) What are the two effects of the change in presentation of the earth to the sun's rays? (21) How great is the change in directness during the year? (22) State the change in duration at various places during the year. (23) Explain why insolation varies with latitude and time. (24) How may this relation of insolation to latitude and time be expressed graphically? (25) Define reflection. (26) Describe the two kinds of reflection. (27) Arrange the substances upon which insolation may fall in order of excellence as regards reflection. (28) Define transmission. (29) What is meant by selective transmission? (30) Contrast rock salt and glass as transmitters. (31) Arrange the substances upon which insolation may fall in order of excellence as regards transmission and absorption. (32) Give the water analogy for the three processes, reflection, transmission, and absorption. (33) What are the four effects of absorption? (34) Define actinometry. (35) State the fundamental formula in actinometry. (36) What is represented by each letter in the formula? (37) Define the solar constant. (38) In how many ways may absolute and relative values of H be determined? (39) Describe the photograph paper method of determining relative values of H. (40) Describe the method of determining C and a. (41) What values have been found for C and a? (42) State the effect of atmospheric absorption on the insolation received at the earth's surface in different latitudes. (43) What is the behavior of the ocean as regards reflection, transmission, and absorption? (44) Why is the temperature rise during the day extremely small? (45) Why does the ocean cool but little during the night? (46) State the behavior of dry ground as regards reflection, transmission, and absorption. (47) Why is the rise in the temperature of dry ground under insolation so large? (48) What is the effect of dampness or vegetation on the temperature change? (49) State the behavior of the atmosphere as regards reflection, transmission, and absorption. (50) Name the three chief absorbing components of the atmosphere. (51) How do the temperatures of the air and the ground compare during the day and at night? (52) How do the temperatures of the air and the ocean compare during the day and at night? (53) Define conduction. (54) Name several poor and several good conductors. (55) To what height would the air be heated by conduction alone? (56) Describe and explain convection in a liquid. (57) Where should convection be expected to take place in the atmosphere? (58) State the difference between convection in a liquid and in a gas? (59) What is meant by the adiabatic rate of cooling of rising air? (60) Name the three evidences of local convection. (61) Describe and explain the mirage. (62) Describe and explain dust whirlwinds. (63) Describe and explain cumulus clouds. (64) What is meant by vertical

temperature gradient? (65) Name the three ways of obtaining the vertical temperature gradient. (66) Why are mountain observations no longer used? (67) Describe the use of balloons for obtaining observations. (68) Describe the use of kites for making meteorological observations. (69) With what does the vertical temperature gradient vary? (70) What is the numerical value of the average vertical temperature gradient? (71) Describe the three characteristic gradients which occur during the day. (72) Define daily range. (73) How may the daily range be represented graphically? (74) How may the adiabatic rate of cooling be represented graphically? (75) What is meant by the isothermal layer? (76) At what height does it occur? (77) What are its characteristics? (78) Define inversion of temperature. (79) Explain why the atmosphere is unstable during the day. (80) State the three conditions of convection. (81) Name the five processes which are operative in the heating of the atmosphere. (82) Describe in detail the effect of each process. (83) Name the three processes operative in the cooling of the atmosphere. (84) State the effect of each of the three processes.

TOPICS FOR INVESTIGATION

(1) The nature and characteristics of the atom.

(2) The amount reflected, transmitted, and absorbed by the different things upon which insolation may fall.

(3) The methods of obtaining the value of the solar constant.

(4) Changes in the temperature of the ocean between day and night.

(5) Changes in the temperature of the ground between day and night.

(6) The method of computing the adiabatic rate of cooling of air.

(7) Mountain observatories.

(8) Balloon ascensions for scientific purposes.

(9) The use of kites in meteorology.

(10) The variation in the vertical temperature gradient.

(11) The isothermal layer.

PRACTICAL EXERCISES

(1) Contrast the amount of insolation received at several different places and at several different times by working out the exact value of the three factors.

(2) Determine by the photographic paper method the transmission coefficient of the atmosphere on some cloudless day.

(3) Observe carefully, or better photograph, some cumulus clouds.

` (4) Note the time when the cumulus clouds disappear in the late afternoon.

(5) If possible, compare the observations of temperature made on some near-by mountain top and at some valley station.

REFERENCES

ASSMANN, BERSON, and others, *Wissenschaftliche Luftfahrten* (3 vols.).
ROTCH, *Sounding the Ocean of Air.*

For recent articles and references on the isothermal layer see:

GOLD, E., and HARWOOD, W. A., *The present state of our Knowledge of the upper Atmosphere as obtained by the Use of Kites, Balloons, and Pilot Balloons,* 8°, 54 pp., London, 1909.
The Mount Weather Bulletin (particularly articles by Humphreys).
See Appendix IX for other books on upper air investigation.

CHAPTER III

THE OBSERVATION AND DISTRIBUTION OF TEMPERATURE

THE DETERMINATION OF TEMPERATURE

60. Thermometry.—Thermometry, as the word indicates, has to do **Definition.** with the determination of temperature, and the instrument used is called a thermometer.[1] There are three systems of ther-

---

[1] $\theta \acute{\epsilon} \rho \mu \eta$ = heat; $\mu \acute{\epsilon} \tau \rho o \nu$ = measure.

60

mometry or thermometric scales, namely, the Fahrenheit, the Centigrade, and the Réaumur.   The two standard temperatures in each system are the same, namely, the melting point of ice and the boiling point of water.   In each case the ice and water must be pure and the pressure must have the standard value.   These two standard temperatures are numbered differently in the three systems. In the Fahrenheit system the melting point is numbered 32 and the boiling point 212; in the Centigrade system the melting point is numbered 0 and the boiling point 100; in the Réaumur system the melting point is numbered 0 and the boiling point 80.   The intervals between the two standard temperatures are thus 180, 100, and 80 respectively and in the ratio of 9 to 5 to 4.   The following formula thus indicates the relation between the three systems where F, C, and R denote the three corresponding temperatures :

The Fahrenheit, Centigrade, and Réaumur scales.

The interrelation between the three scales.

$$\frac{F-32}{9} = \frac{C}{5} = \frac{R}{4}.$$

$F - 32$ represents the F. interval above the melting point.   Since the C. and R. thermometers have the melting point indicated by zero the intervals are expressed directly by C and R.   A temperature expressed in any one of the three systems can be immediately changed into the corresponding temperature in the other systems by means of this formula.[1] For example, Let $F = 60°$; the equations then become :

$$\frac{60-32}{9} = \frac{C}{5} = \frac{R}{4}; \quad \frac{28}{9} = \frac{C}{5} = \frac{R}{4}; \quad C = 15\tfrac{5}{9} \text{ and } R = 12\tfrac{4}{9}.[2]$$

All temperatures below the zero of the scale are indicated by a minus sign and are read " so many degrees below zero."   Thus 7 degrees below zero Fahrenheit is indicated by $-7°$ F.

---

[1] The formula $\frac{F-32}{9} = \frac{C}{5}$ may be modified so as to make the mental computation of. the Centigrade temperature for the Fahrenheit easier.

$$C = \tfrac{5}{9}(F-32) = (F-32) \cdot 555 \cdots = (F-32)\left(\tfrac{1}{2} + \tfrac{1}{10}\cdot\tfrac{1}{2} + \tfrac{1}{100}\cdot\tfrac{1}{2}\cdots\right).$$

The rule would be: subtract 32 from the Fahrenheit temperature; take one half of it; add to this one tenth of itself, one one-hundredth of itself, etc.   Thus for 60° Fahrenheit:

$$60 - 32 = 28$$
$$\tfrac{1}{2} \text{ of } 28 = 14$$
$$C = 14$$
$$1.4$$
$$.14$$
$$\underline{\quad .01 \text{ etc.}}$$
$$C = 15.55 +$$

[2] In Appendix II is given a table for converting Centigrade into Fahrenheit and *vice versa*.   In Appendix III, a graphic comparison of the two scales is given.

6r. The thermometer was invented by Galileo and Sanctorius of Padua in 1590. Sanctorius had been a pupil of Galileo, and later became professor of medicine at the same university. Figure 21 represents the thermometer used by Sanctorius for determining the feverishness of his patients. It consists simply of a glass globe opening into a narrow tube and partly filled with fluid. This is then inverted and dipped into a vessel of fluid. The bulb is taken in the hand and the feverishness is indicated by the height at which the

The invention of the thermometer.

liquid stands in the tube. If this early form of the thermometer was invented by Galileo and adopted by Sanctorius, or whether it was the combined invention of Galileo and Sanctorius, is uncertain. Within a few years, /however, the instrument was inverted, and it then became a thermometer of essentially the same form as at present. For nearly 100 years after the invention of the thermometer all was confusion and uncertainty, for various fluids were used and various systems of graduation were employed. In order to get a constant temperature various devices were used. Some used the temperature of the water of a certain spring as constant. The temperature of the subcellar of the Observatory of Paris was also used. Out of this chaos the three systems of thermometry emerged, because the founders of the systems made instruments of such high quality and in such large numbers that they came to be recognized as standard instruments.

Confusion and uncertainty for many years.

Fig. 21. — The Original Thermometer of Sanctorius.

The Fahrenheit thermometer originated at Dantzig about 1714. The two great improvements introduced by Fahrenheit were the use of mercury as the fluid, and the use of two known temperatures for the graduation. The reason for choosing 32 as the melting point of ice and 212 as the boiling point of water is uncertain. It is known that he had traveled in Iceland, and it may be that the zero of his thermometer scale was the lowest temperature which he had ever experienced. It may be that he used the temperature of the human body for standardizing his instrument, intending to have this 100°. The average temperature of the human body, however, is only 98.6°. A mixture of salt and snow had been used by the Italian thermometer makers for getting the zero point of their instruments. The temper-

The Fahrenheit thermometer.

ature of this is nearly zero, and it may be that Fahrenheit used this for determining the zero. Fahrenheit was also a maker of astronomical instruments, and it may be that in order to use his machine for dividing circles for graduating the stem of the thermometer, it was necessary to have the interval between the melting and boiling points 180.

The Centigrade thermometer was invented by Celsius and Linnæus at the University of Upsala in Sweden in 1742. The two fixed points are numbered 0 and 100 respectively. In the original thermometers the freezing point may have been numbered 100 and the boiling point 0, so that a distinction should be made between the Celsius and the Centigrade thermometer. Within a few years, however, the Centigrade system of numbering the fixed points came into general use. The Centigrade thermometer.

The Réaumur thermometer was invented in 1731 by the French physicist bearing that name. The Fahrenheit and Centigrade scales are both used in scientific work. The Fahrenheit scale is used popularly in all English-speaking countries. The Centigrade thermometer is used popularly in France and portions of Germany. The Réaumur thermometer is used popularly in Russia and portions of Germany. No thermometer is used popularly in the country in which it was invented. The Réaumur thermometer.

62. Thermometers. — A thermometer consists essentially of a glass capillary tube of uniform diameter or bore called the stem, which is attached to a bulb. The bulb is usually cylindrical in form and of about the same diameter as the stem. This is a matter of convenience, however, and many good thermometers have nearly spherical bulbs. The bulb and part of the stem are filled with some fluid, usually mercury, and above this there is The description and working principle of a thermometer.
a vacuum. The principle of a thermometer is that for a given increase in temperature the fluid in the bulb expands many times (in the case of mercury seven times) as much as the glass itself. Thus if the bore is uniform, to equal increments of temperature will correspond equal changes in the length of the column of fluid in the stem. The thermometer proper is attached to a case in order to protect it from injury and to facilitate its being attached to various objects. The presence of the milk glass at the back of the thermometer is to furnish a better reflector for reading the instrument. The curved front of the stem serves to magnify the liquid within, thus making it more easily visible. Figure 22 represents thermometers of various forms for determining temperature.[1]

[1] The electrotypes for this figure and for many other illustrations of meteorological apparatus appearing in this book were very kindly furnished by Mr. Henry J. Green, 1191 Bedford Avenue, Brooklyn, N. Y.

The first step in the construction of a thermometer is the making of the stem by a glass blower. This must then be tested in order to determine whether the cross section of the bore is uniform. This is done by inserting a small

The various steps in the construction of a thermometer. thread of mercury, and observing its length as it is moved up and down the stem. If it changes in length, it indicates a larger or smaller bore. If the stem has stood this test, the thermometer bulb, and the bulb and tube for filling it, are next added by the glass blower. It then appears as indicated in Fig. 23. In order to fill it, the upper bulb is filled with mercury. By alternately heating and cooling the lower bulb the fluid is drawn down into the thermometer and it is then sealed at the top at the appropriate temperature. Next the strains in the glass incident to the construction of the thermometer must be removed. Formerly the thermometers were laid away for from three to ten years in order to allow the strains to disappear with time. At present, however, the strains are removed by an annealing process which consists in alternately heating and cooling the thermometer. The thermometer is then exposed to the two standard temperatures, and the indications noted. The interval

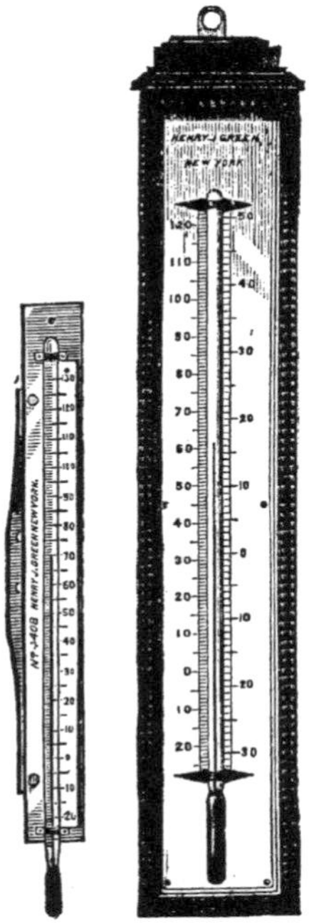

FIG. 22. — Thermometers of Various Forms.

between the two fixed points is then divided into the appropriate number of degrees by means of a dividing machine, and the scale may be extended in either direction from the fixed points.

There are several essentials in a good thermometer. In the first place, a suitable fluid must be used. This fluid

The essentials in a good thermometer. must not freeze at ordinary temperatures, and it must not be decomposed by the action of light or at moderately high temperatures, and it must not vaporize or boil at any ordinary temperature. Mercury is almost universally used for the higher temperatures. For the lower temperatures, since mercury freezes at -39.4 F., alcohol

FIG. 23. — A Thermometer in the Making.

or some light oil is ordinarily used. The larger the bulb the more sensitive the thermometer, but the time required for the thermometer to indicate a new temperature is, however, increased by increasing the size of the bulb. The size of the bulb should be such as to make the thermometer sufficiently sensitive but not unduly sluggish. The bore must be uniform throughout and the fixed points must have been accurately determined. If the graduation is placed on the backing of the thermometer and the thermometer is held against this by a wire fastening, these wires are apt to become loosened with time and the thermometer sags in its fastening, thus causing error. Every good thermometer is graduated on the stem itself.

63. The inaccuracies in determining a temperature (more especially an air temperature) may be divided into two groups: first, avoidable blunders; and secondly, errors in the construction and use of the thermometer which cannot be eliminated without careful *The inaccuracies in* testing and elaborate experimental methods. There are four *determining* avoidable blunders; (1) a good thermometer must be used *a temperature.* in determining the temperature. It is a waste of time to attempt to make an accurate determination of the temperature with an inferior instrument. (2) The error of parallax must be avoided. By parallax in general is meant the displacement of one object with reference to another as the position of the eye of the observer changes. In· Fig. 24 let A represent the thread of mercury in the *The avoidable* thermometer stem, and B the gradu- *able* ated scale. The eye of the observer *blunders.* placed at m will see the mercury thread at x on the scale, while the eye of the observer placed at n would see it at y. The correct position of the eye is such that the line of vision is at right angles to the stem of the thermometer. (3) The observer must avoid heating the thermometer by the presence of his own body. When the

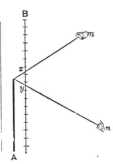

FIG. 24. — The Error of Parallax in reading a Thermometer.

difference in temperature between the indication of the thermometer and the observer's body is large, a considerable error may result if the temperature is not read quickly. On a cold zero morning in winter, an observer standing within one foot of an exposed thermometer for thirty seconds would probably increase its indication by at least two degrees. (4) A thermometer must be given time to indicate a new temperature.

F

It is necessary to allow at least five minutes for the thermometer to take up any considerable change in temperature.

The following are errors in the construction and use of the thermometer: (1) The fixed points may be slightly misplaced. (2) The bore may not be absolutely uniform. (3) Decomposition may have taken place in the fluid of the thermometer. (4) The bulb and stem may not be at the same temperature. (5) Pressure has a slight influence on the indication of a thermometer. The indication of a thermometer in vacuo and when exposed to the atmospheric pressure may change as much as one half a degree. The slight changes, therefore, in the pressure of the atmosphere which are continually taking place have a slight influence on the indications of a thermometer. (6) The thermometer must not be strained by exposing it to unusually high or low temperatures. If the temperature of a certain fluid were taken and the thermometer then exposed to a temperature of 100°, or more, and then placed in the same fluid again, it would probably not give the same indication. A thermometer should therefore not be exposed to direct sunlight any more than is necessary on account of the high temperature which it would indicate and on account of the possible decomposition of the fluid.

The errors of construction and use.

A standard thermometer costs from \$5 to \$25. This would indicate accurately the temperature within one fourth of a degree if the four blunders are avoided, but without taking account of the errors in the construction and use of the thermometer. A thermometer costing from \$.75 to \$3 would usually indicate the temperature correct to within 1° F. A cheaper thermometer may be in error anywhere from 1° to even 10° or 15°.

The accuracy of a thermometer.

64. The real air temperature. — It is not easy to determine the temperature of a gas, because the thermometer does not indicate the temperature of the surrounding medium, but the temperature of its own bulb. If a thermometer is surrounded by an opaque fluid or solid, it takes up, by conduction, the temperature of the surrounding medium, and thus indicates the temperature of that medium. If the bulb of a thermometer is exposed in a gas, however, its temperature is determined both by the conduction of heat to or from the surrounding medium and the difference between the radiant energy absorbed and emitted by the thermometer. A thermometer exposed in the open air gives neither during the day nor at night the real air temperature, even if it is an accurate, standard instrument. During the day it will indicate a temperature anywhere from

Why it is hard to get the real air temperature.

20 to 60° above the real air temperature, because the excess of the insolation received over the radiant energy emitted is large.   During the night the thermometer may indicate a temperature from *The indica-* 1 to 7° below the real air temperature, because it is radiat- *tions of a* ing more heat to the sky and earth than it is receiving.   A *thermom-* *eter in the* close approximation to the real air temperature can be deter- *open by day* mined in summer by exposing the thermometer in the shade *and at night.* of a small tree, and in winter by placing it on the inside of a north piazza post.   The indications of the instrument in these positions will give approximately the real air tempera-

ture, although they may differ in certain cases by as much as three or four degrees.

In meteorological work the real air temperature is determined in one of three ways; namely, by *The three* means of a thermometer *methods of* shelter, sling thermometer, *getting the* or ventilated thermometer. *real air* *tempera-*

65. Thermometer shel- *ture.* ter. — The thermometer shelter as used by the U. S. Weather Bureau consists of a cubical box *The ther-* about three feet on a side, *mometer* with a double sloping roof, *shelter of* *the U. S.* closed bottom, and latticed *Weather* sides.   Such a shelter is *Bureau.* represented in Fig. 25.   The purpose of the double roof is to shield the thermometers from the insolation of the sun.   The outer portion of the roof absorbs the insolation, and the

Fig. 25. — The Thermometer Shelter of the U. S. Weather Bureau.

circulation of air between the two portions of the roof prevents the lower layer from becoming heated.   The tight bottom excludes rising air currents and intercepts any radiations from the earth, while the latticed sides allow a free circulation of the air.   The shelter must be located at least five feet above the ground.   In the country or in a village the best location is in the open, over sod, although it may be attached to the north side of an unheated building.   A sufficient space must be left between the shelter and building to allow a free circulation of air. In a city, since the air stagnates in a narrow street, the best location is the roof of some high building. This does not, of course, give the temperature of the streets where the people must live, but it probably does give a close approximation to the real air temperature in the country immediately surrounding the city. It has been found as a result of many experiments that the indication of the thermometer in such a shelter

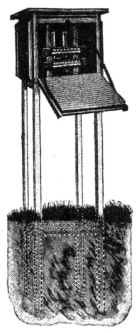

FIG. 26. — The French Thermometer Shelter.
(From ANGOT'S *Instructions Météorologiques*.)

FIG. 27. — The English Thermometer Shelter.

usually gives the real air temperature correct to within half a degree. It is most in error on still, clear, cold nights, and on still, hot, summer days. In both cases the indications of the sheltered thermometer are too conservative.

The thermometer shelter of other countries. The form of the thermometer shelter as used by the various countries is somewhat different, although the underlying principle is always the same. Figures 26, 27, and 28 illustrate the shelters used by the French, English, and Russian governments respectively.

66. Sling thermometer. — Another method of getting the real air temperature is by means of the sling thermometer, which was devised by Arago in 1830. Since a thermometer shelter is not portable, a sling thermometer is particularly advantageous for scientific expeditions and for those observations of air temperature which are not made constantly in the same place. The sling thermometer (see Fig. 29)

Description of the sling thermometer.

FIG. 28. — The Russian Thermometer Shelter.
(From WALDO'S *Modern Meteorology*.)

consists usually of two thermometers attached to a board which is provided with a chain or string so that it can be whirled rapidly. In some cases the bulbs of the thermometers are covered by wire netting in order to protect them from injury.

Whirling the thermometer brings a much larger quantity of air in contact with the bulb,

The principle of the sling thermometer.

and the exchange of heat with the surrounding medium by means of conduction is thus emphasized. The underlying principle in the use of the instrument in getting real air temperature is to so emphasize conduction that the effect of the radiant energy received or emitted becomes negligible in comparison. Several minor errors are introduced by whirling the thermometer. The centrifugal force causes a slight lowering in the indication of the instrument, and the friction of the bulb with the air would raise the temperature slightly, but these small errors counterbalance each other for the most part. It is

FIG. 29. — The Sling Thermometer.

supposed that the sling thermometer will give the real air temperature accurately to within a half a degree Fahrenheit under almost all circumstances.

67. Ventilated thermometer. – The ventilated thermometer, which is the best instrument for determining the real air temperature, was invented by Assmann at Berlin in 1887. Figure 30 gives a picture and a cross section of this instrument. It consists essentially of two sensitive, accurate thermometers (t and t' in the figure) whose bulbs are located within a double jacket. The instrument is provided with a motor A driven by a spring which draws

Description of the ventilated thermometer.

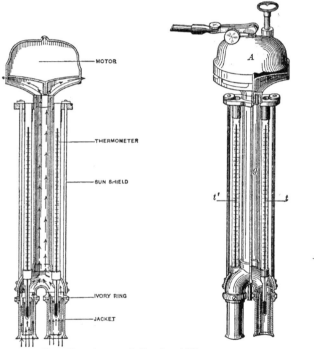

Fig. 30. — Assmann's Ventilated Thermometer.

(From Bornstein's *Leitfaden der Wetterkunde*.)

the air with a velocity of two or three feet per second past the bulbs of the thermometers up through a central column g and ejects it at the top. The instrument is made of burnished silver which reflects more than 90 per cent of the insolation, and thus is but little heated even in bright sunshine. It is provided with shields for keeping the insolation from the stems of the thermometers, and the two ivory rings prevent the conveyance of heat by conduction from the rest of the apparatus to the jackets which surround the thermometer bulbs. The jackets are double, so that any insolation which might be absorbed by the outer jacket would be communicated to the air between the two jackets and not directly to the air which surrounds the thermometer bulbs. Numerous tests of the accuracy of this instrument have been made, and it is stated that it will determine the real air temperature correctly to a tenth of a degree Fahrenheit under any conditions whatever.

Accuracy of the instrument.

FIG. 31. — The Draper Thermograph.

The instrument is extremely portable, and is by far the most convenient for determining the real air temperature. The only drawback is the expense, the cost being about $60.

68. **Thermographs.** — For many purposes continuous records of the temperature are desirable. These are obtained by means of the thermograph, of which there are two forms in ordinary use, the Draper and the Richard Frères.

The Draper thermograph (Fig. 31) is an American instrument and has a metallic

The Draper thermograph.

thermometer, one end of which is fixed while the other end is attached to a series of levers which magnify the small movements due to temperature changes, and transmit them to a pen which moves in and out across a dial. As the temperature rises the pen moves outward from the center, and as the temperature falls the pen moves inward. The pen contains a non-freezing glycerine ink, and rests against the dial, which is turned by clockwork. The dial is divided into days and hours by curved radial lines, and into degrees by concentric circles. The dial makes one complete revolution in a week, and a continuous record of the temperature is thus kept.

The Richard Frères thermograph (Fig. 32) is made in Paris and has **The Richard** been adopted by the U. S. Weather Bureau. The thermome-**Frères ther-** ter is here either a metallic thermometer or a bent tube of **mograph.** metal containing a non-freezing liquid. If a metallic thermometer is used, it consists of two curved strips of metal soldered to-

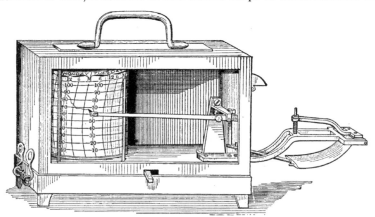

FIG. 32. — The Richard Frères Thermograph.

gether. In either case, as the temperature rises the bent thermometer tends to straighten; if the temperature falls the elasticity bends the thermometer into a sharper curve. These small motions are magnified and communicated by a system of levers to the pen, which moves up and down over the paper which is attached to the outside of the revolving drum. The drum contains the clockwork which drives it, and makes one complete revolution in a week or, if so ordered, in a day. The pen rises and falls with temperature changes, and thus the continuous record of temperature is kept. The thermograph is at best not an accurate instrument. In order to get the best results it must be standardized at least twice a day by comparing it with the indications of, preferably, a maximum and minimum thermometer kept near it.

Other forms of thermographs are in use for special purposes or for research work, but the two described above are the only ones commercially on the market. The Draper instrument costs about $20 and the Richard Frères costs about $50, although much less if it can be imported free of duty.

69. Maximum and minimum thermometers. — The purpose of maximum and minimum thermometers is to record the highest and lowest temperatures which have occurred during the interval since the thermometers were set. Figure 33 represents the maximum and minimum thermometers used by the U. S. Weather Bureau. The maximum thermometer, which is represented at the bottom of the figure, is a mercurial thermometer with a constriction in the stem just above the bulb. As the temperature rises the mercury is forced past the constriction; if the temperature falls, the mercury thread breaks at the constriction, leaving the stem filled with mercury. The highest temperature is thus

Description of the maximum and minimum thermometers of the U. S. Weather Bureau.

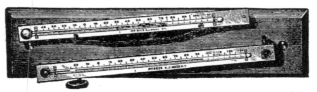

Fig. 33. — Weather Bureau Maximum and Minimum Thermometers.

indicated by the upper end of the thread of mercury in the thermometer stem. It is set by rapidly whirling the instrument, thus forcing back the mercury by means of centrifugal force.

The minimum thermometer is an alcohol thermometer and contains a small dumbbell-shaped glass index. As the temperature falls this index is dragged down the stem of the thermometer by surface tension. If the temperature rises, the fluid flows past the index, leaving it at its lowest position. The minimum temperature is thus indicated by the forward end of the glass index. The instrument is set by lifting the bulb end of the thermometer, thus allowing the index to slide down to the end of the column of fluid.

The Six maximum and minimum thermometer, which combines both thermometers in the same instrument, has come within the last few years into very general use. Although known for more than a century, it was not until within the last few years that sufficient accuracy has been attained to make its indications of value. Figure 34 gives an illustration and cross section of the instrument. It consists of a long bulb, containing a non-freezing fluid, a U-shaped stem containing a thread of mercury, and a bulb at the top containing compressed air. As the temperature rises the fluid in

The Six maximum and minimum thermometers.

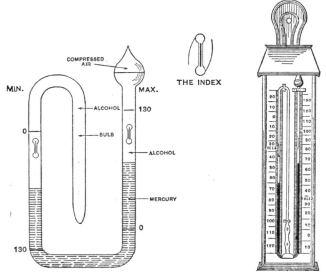

FIG. 34. — Six Maximum and Minimum Thermometer.

the bulb expands, forcing the mercury to rise on the right, thus raising the index and compressing the air in the bulb above. If the temperature falls, the fluid in the bulb contracts and the compressed air forces the mercury thread around, thus preventing the formation of any vacant space in the bulb. The index on the left is now raised, and it will thus indicate the lowest temperature. The lower ends of the two indices thus register the highest and lowest temperatures respectively. The glass index consists of a small piece of steel wire surrounded by a dumbbell-shaped glass covering provided with two hairlike appendages which prevent the index slipping in the stem of the thermometer. They are set by means of a magnet which acts on the steel wire in the center of each index.

Other forms of maximum and minimum thermometers have been devised, but have not come into general use either because they are hard to transport from place to place, or because they are liable to become deranged while in use.

70. Black bulb thermometer. — The black bulb thermometer as illustrated in Fig. 35 is a mercurial maximum thermometer, the bulb of which has been coated with lampblack or platinum black. The black The whole is inclosed in a glass jacket from which the air and bulb ther-moisture have been extracted. The transference of heat by mometer. conduction to or from surrounding objects is thus eliminated. The temperature of the bulb will rise to such a point that the energy given out in

FIG. 35. — Black Bulb Thermometer.

the form of radiation exactly balances the radiant energy absorbed. If such an instrument is surrounded by a case kept at a constant temperature, and if the insolation of the sun is allowed to fall upon it, the temperature recorded will give relative values of the insolation.   The instrument is thus an actinometer rather than a thermometer. Without the use of the case, which is kept at a constant temperature, the instrument will give only rough relative values of the insolation received.

71. Thermometers for special purposes. — The forms given to thermometers are many and varied, thus suiting them to a large variety of special purposes. As far as meteorology is concerned, but Thermome-three of these need be mentioned. These are thermometers ters for three for determining the temperature of the soil at various special depths beneath the surface, thermometers for determining purposes. the temperature of water at various depths and telethermometers. The first two thermometers will be briefly mentioned when the results of the observations of soil and water temperature are presented.   The purpose of a telethermometer is to indicate within a building the real air temperature outside. There are several forms of the instrument. It is very convenient but expensive, and is in use in only a few of the Weather Bureau stations.

THE RESULTS OF OBSERVATION

72. The Observations. — The stations of the U. S. Weather Bureau are of three kinds: regular stations, coöperative stations, and special stations. The regular weather bureau stations, of which
The three kinds of U. S. Weather Bureau stations. there are from 180 to 200, are located ordinarily in the larger cities, since they usually publish a daily weather map which must be distributed as soon as possible. There are from 3600 to 4000 coöperative stations, and from 300 to 500 special stations. The observations of temperature taken at a regular station are the real air temperature at 8 A.M. and 8 P.M., the highest and lowest temperatures during the preceding 12 hours, and a continu-
The instruments used and the temperature observations taken at each kind of station. ous thermograph record. The instruments for determining these temperatures are, a good mercury thermometer, maximum and minimum thermometers of the regular Weather Bureau form, and a Richard Frères thermograph. These instruments are located in a thermometer shelter which is ordinarily placed from 6 to 10 feet above the roof of some high building in the city. At a coöperative station the highest and lowest temperatures during a day are determined, and also the reading of the maximum thermometer just after it has been set. The purpose of taking this observation is to make sure that the maximum thermometer has been set and also to give the real air temperature at the time of observation. Maximum and minimum thermometers only are necessary for these observations, and the shelter which contains the thermometers is ordinarily located from 5 to 10 feet above sod in the open or on the north side of some building. Special stations take those observations for which the stations were established.

73. Normal hourly, daily, monthly, and yearly temperature. — From these observations of temperature which are made at the various Weather Bureau stations, certain normal temperatures and other tem-
The distinction between average, normal, and mean. perature data may be computed. At the very outset the three words " average," " normal," and " mean " must be carefully defined. By an average is meant simply the sum of a number of observations divided by the number of the observations. If the observations have been extended over a sufficient length of time so that the accidental irregularities are eliminated by taking the average, then the average value may be spoken of as a normal. Usually at least twenty years of observations are required

for a normal.   The word "mean" is used by various writers to cover both average and normal.

If a good continuous thermograph record for at least twenty years is available, the normal hourly temperatures for the various days of the year may be computed.   For example, the normal 9 A.M. **Normal** temperature for October 28 would be found by averaging the **hourly tem-** twenty or more values which had been recorded for this partic- **peratures.** ular hour on the date in question.   Similarly, the normal hourly temperatures for all the hours of all the days in the year may be computed. Usually this is not done for every day in the year, but the days are grouped by months.

The average temperature for a day is found by averaging the 24 values of hourly temperature observed during that day.   This requires a thermograph, and since thermographs have not been in general **The** use even at regular Weather Bureau stations for many years, **methods of** various combinations have been sought such that the average **determining** temperature for the day might be computed from the tem- **daily tem-** peratures observed at certain definite times during the day. **perature.** Some of the various combinations which have been used are the following: $\frac{1}{2}$ (8 A.M. + 8 P.M.) ; $\frac{1}{4}$ (7 A.M.+ 2 P.M. + 9 P.M. + 9 P.M.) ; $\frac{1}{2}$ (maximum + minimum).   Of these $\frac{1}{4}$ (7 A.M. + 2 P.M. + 9 P.M. + 9 P.M.) was formerly used by many Weather Bureau stations for computing the average daily temperature.   At the present time $\frac{1}{2}$ (maximum+ minimum) is used at all U. S. Weather Bureau stations for computing the average daily temperature.   The reasons are : the apparatus required is simple, the computation is very easy, and in the long run the average daily temperatures computed in this way approximate very closely the average daily temperatures found by taking one twenty-fourth of the sum of the twenty-four hourly values.   The normal daily temperature is found by averaging the average dailies for the date in question for a sufficient number of years to eliminate the accidental irregularities.   If the normals are based on twenty years of observations, it will be **Normal** found that there is not an even transition from day to day, **daily tem-** but jumps in temperature of even two or three degrees occur. **peratures.**

It might seem that the time were not sufficiently long to eliminate accidental irregularities.   It is found, however, that these abrupt changes do not disappear even if the normal is based on a hundred years of observations.   It may be that these abrupt changes are not due to the fact that the normal is based on too short a time, but to the fact that there are actual abrupt advances and recessions of temperature which take place on nearly the same date each year. Ordinarily, however, the normal daily temperatures are "adjusted" before they are published. That is, the irregularities are smoothed out and there are no jumps in temperature from day to day. The accompanying table gives the adjusted normal daily temperatures for every day in the year at Albany, N.Y. This station at Albany, N.Y., has been chosen for most of the data in this book because the record is a long one, the observations have been well taken, and it is typical of New England and the Middle Atlantic States.[1] The larger cities, as New York and Boston, are too near the ocean, which is always a disturbing factor, to be typical. In the case of an average

ADJUSTED NORMAL DAILY TEMPERATURES AT ALBANY, N.Y.

(Temp. F.)

DATE	JAN.	FEB.	MAR.	APR.	MAY	JUNE	JULY	AUG.	SEPT.	OCT.	NOV.	DEC.
1	24	22	27	38	53	64	71	72	66	57	44	32
2	24	22	27	39	53	65	71	72	66	57	44	32
3	24	22	28	40	54	65	71	72	66	56	43	31
4	24	22	28	40	54	65	71	72	66	56	43	31
5	23	22	28	41	55	66	71	71	66	55	42	30
6	23	22	29	41	55	66	71	71	65	55	42	30
7	23	22	29	42	55	66	71	71	65	54	42	30
8	23	22	29	42	56	66	72	71	65	54	41	30
9	23	22	30	43	56	66	72	71	64	53	41	29
10	23	23	30	43	57	67	72	71	64	53	40	29
11	23	23	30	44	57	67	72	71	64	52	40	29
12	23	23	30	44	58	67	72	70	64	52	40	28
13	22	23	31	45	58	68	72	70	63	52	39	28
14	22	23	31	45	58	68	72	70	63	51	39	28
15	22	23	32	46	59	68	72	70	63	50	39	28
16	22	24	32	46	59	68	72	70	62	50	38	27
17	22	24	32	47	60	68	72	69	62	50	38	27
18	22	24	33	47	60	69	72	69	62	49	37	27
19	22	24	33	48	60	69	72	69	61	49	37	27
20	22	24	33	48	61	69	72	69	61	48	37	26
21	22	25	34	48	61	69	73	69	61	48	36	26
22	22	25	34	49	61	69	73	68	60	48	36	26
23	22	25	34	49	62	70	73	68	60	48	35	26
24	22	26	35	50	62	70	73	68	60	47	35	25
25	22	26	35	50	62	70	73	68	59	47	35	25
26	22	26	36	51	63	70	73	68	59	46	34	25
27	22	26	36	51	63	70	73	68	58	46	34	25
28	22	27	36	52	63	70	72	67	58	45	34	24
29	22		37	52	64	70	72	67	58	45	33	24
30	22		38	52	64	71	72	67	57	44	33	24
31	22		38		64		72	67		44		24

[1] At the Albany station, Mr. George T. Todd is local forecaster and Mr. Herbert E. Vail is assistant. It is to the courtesy, kindly interest, and willing assistance of these gentlemen that the data for Albany in this book are due.

daily temperature, a departure from normal of 10° is common, of 20° is unusual, and of 30° is almost certainly record breaking.

74. The average monthly temperature is found by averaging the daily temperatures for the various days of the month. An average monthly temperature is practically independent of the way in which the average daily temperatures from which it is computed were found. A normal monthly temperature is found by averaging the average monthlies for a sufficient number of

Average and normal monthly temperatures.

AVERAGE AND NORMAL MONTHLY AND YEARLY TEMPERATURES AT ALBANY, N.Y.

(Temp. F.)

YEAR	JAN.	FEB.	MAR.	APR.	MAY	JUNE	JULY	AUG.	SEPT.	OCT.	NOV.	DEC.	AN.	
1874	28	22	30	*36*	56	66	70	66	63	49	36	26	46	
1875	*14*	*15*	25	38	56	65	*69*	69	*58*	47	*31*	36	*43*	
1876	29	24	29	42	55	70	73	72	60	47	40	*17*	47	46·6
1877	16	29	30	46	56	67	72	72	64	49	41	32	48	
1878	22	24	38	**52**	56	64	74	70	64	53	38	28	49	
1879	18	20	30	42	60	66	71	68	60	56	38	29	47	
1880	30	28	33	50	**66**	71	75	71	65	52	39	25	50	
1881	19	26	37	47	65	65	74	73	**71**	55	43	**39**	**51**	49.4
1882	26	31	38	46	55	69	74	72	65	56	41	30	50	
1883	21	28	29	46	58	71	73	74	61	51	**44**	30	49	
1884	23	**32**	36	47	59	71	71	72	67	51	39	28	50	
1885	23	*15*	*24*	46	57	67	73	68	61	51	41	31	46	
1886	21	23	33	51	59	66	72	71	64	53	40	24	48	
1887	21	25	28	43	65	69	**77**	69	60	50	38	27	48	47·8
1888	15	22	26	50	58	69	71	71	61	*46*	41	30	47	
1889	31	20	37	44	62	68	72	70	64	49	43	35	50	
1890	31	31	31	47	57	68	71	71	62	51	38	20	48	
1891	25	28	32	49	57	68	*69*	71	67	50	39	37	49	48·4
1892	24	26	30	46	56	71	73	72	62	51	38	26	48	
1893	17	22	31	44	58	70	72	72	59	54	39	26	47	
1894	27	21	40	48	60	70	75	69	67	53	36	29	50	
1895	23	19	30	47	62	**73**	70	72	67	47	41	32	48	
1896	20	25	27	50	64	68	74	73	62	48	**44**	26	48	49·0
1897	25	26	35	48	59	65	75	70	63	53	39	30	49	
1898	24	28	42	46	58	70	75	73	67	53	39	29	50	
1899	23	22	31	48	60	71	73	72	61	53	40	32	49	
1900	26	26	28	48	58	70	74	**75**	68	**58**	42	29	50	
1901	24	19	33	50	59	70	76	73	65	52	34	27	48	48·6
1902	23	24	40	48	57	64	70	68	63	51	43	23	48	
1903	24	27	**43**	48	62	*63*	71	*65*	64	53	36	23	48	
1904	15	17	31	44	63	69	72	69	61	49	35	20	45	
1905	21	18	33	46	59	67	74	69	63	52	38	32	48	
1906	**32**	22	28	47	59	69	73	73	66	52	39	24	49	47·6
1907	22	17	37	43	*53*	66	73	69	64	48	39	32	47	
1908	25	20	35	46	61	69	75	70	66	54	40	29	49	
Normal	23	24	33	46	59	68	73	71	64	51	39	29	48	

years to eliminate accidental irregularities.   It should also be equal to
the average of the normal daily temperatures for that month.

The average yearly temperature is found either by averaging the average monthly temperatures for the year, taking account of the number of
Average and
normal
yearly tem-
peratures.
days in each month, or by averaging the average daily temperatures for the year.   A normal yearly temperature is
found by averaging the average yearlies for a sufficient number of years.   It should also equal the average of the normal
monthlies, taking count of the number of days in each month, or the aver-

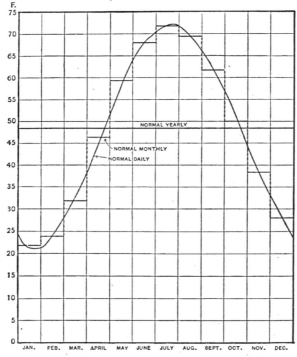

Fig. 36. — Graphical Representation of the Station Normals of Temperature
at Albany, N.Y.

age of the normal dailies for all the days in the year.   The table on
page 79 gives the average monthly temperatures and the average yearly
temperatures for several years, and also the normal monthly and yearly

temperatures for Albany, N.Y.   It will be seen in the case of the average monthly temperatures that a departure from normal of 3° is common, of 5° is unusual and of 7° is almost certainly record breaking   The departures in winter are usually larger than those in summer.   In the case of an average yearly temperature, a departure from normal of 0.5° is common, of 1.5° is unusual, and 2.5° almost certainly record breaking. In the table on pages 82 and 83 are given the normal monthly and annual temperatures for 20 cities in the United States and twenty-five foreign places together with certain facts in connection with their location.

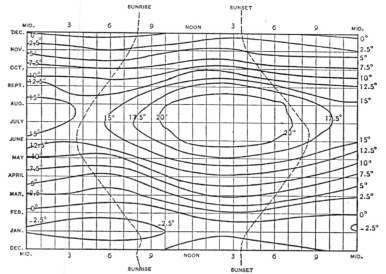

FIG. 37. — Thermo-isopleths, Centigrade, at Berlin, Germany.   (After BÖRNSTEIN.)

The data for the United States have been taken from the publications of the United States Weather Bureau while the data for the foreign places have been derived from HANN's *Lehrbuch der Meteorologie*.

75. The normal daily, monthly, and yearly temperatures are often spoken of as station normals of temperature, and these may be expressed graphically, as is shown by Fig. 36.

76. ·If the normal hourly temperatures for the various hours of each day in the year are known, these may also be expressed graphically, as is shown for the city of Berlin, Germany, in Fig. 37.   Months are here plotted along the Y-axis and hours

Graphical representation of station normals of temperature.

G

Normal Monthly and Annual Temperatures for Various Places
(Temp. F.)

Place	Latitude	Longitude	Elevation Ft.	Jan.	Feb.	Mar.	Apr.	May	June	July	Aug.	Sept.	Oct.	Nov.	Dec.	Year	Length of Record in Years
Werehojansk	67° 34' N	133° 51' E	459	-59.8	-49.5	-26.5	7.3	35.6	54.1	59.9	50.2	36.5	5.0	-36.0	-52.6	1.94	
Christiania	59° 55' N	10° 43' E	82	24.1	23.9	29.5	39.9	50.9	59.9	62.6	60.6	52.7	41.9	32.2	25.5	41.90	
Stockholm	59° 20' N	18° 5' E	148	26.6	25.7	28.9	37.8	47.4	57.4	61.9	59.5	52.7	43.2	34.7	28.4	42.08	
St. Petersburg	59° 56' N	30° 16' E	20	15.3	16.9	23.5	35.8	47.7	58.6	63.9	61.0	51.4	40.1	29.1	20.1	38.66	
London	51° 34' N	0° 8' W	118	38.1	39.7	42.1	48.0	53.8	60.3	63.1	62.1	57.6	49.8	43.0	39.2	49.82	
Paris (S. Maur)	48° 50' N	2° 20' E	164	36.1	38.5	42.6	49.8	55.4	61.7	64.9	63.9	58.5	50.2	42.4	36.9	50.00	
Madrid	40° 24' N	3° 42' W	2149	40.1	43.3	47.3	53.1	60.6	68.7	76.5	75.0	66.4	55.8	46.8	39.7	56.12	
Berlin	52° 33' N	13° 21' E	131	30.7	32.5	37.2	45.9	54.9	62.1	64.6	63.3	57.0	48.2	38.5	32.9	47.30	
Vienna	48° 15' N	16° 21' E	663	28.9	32.4	39.0	48.9	57.2	63.9	67.3	65.8	59.4	49.6	38.3	30.9	48.56	
Constantinople	41° 2' N	28° 58' E	246	41.4	41.4	46.2	53.4	62.4	70.3	74.3	74.5	68.4	62.2	53.2	45.7	57.74	
Athens	37° 50' N	23° 43' E	351	46.4	47.7	52.3	58.9	67.4	75.7	80.6	79.9	73.9	66.0	57.0	50.0	63.14	
Rome	41° 54' N	12° 28' E	102	44.2	46.9	50.7	56.7	64.0	70.9	76.3	75.6	70.0	61.5	52.1	45.7	59.54	
Peking	39° 57' N	116° 28' E	125	23.5	28.9	41.0	56.7	67.4	76.1	78.8	76.5	67.6	54.5	38.5	27.3	53.06	
Tokyo	35° 41' N	139° 45' E	69	36.7	38.1	44.4	54.3	61.7	69.1	76.3	77.7	71.6	60.1	49.6	41.2	56.66	
Manila	14° 34' N	127° 11' E	46	76.8	77.5	80.2	82.8	83.5	81.9	80.6	81.0	80.4	80.2	78.8	77.2	80.06	
Bombay	18° 54' N	72° 49' E	39	73.0	75.0	78.8	85.1	85.3	84.7	82.9	82.4	82.4	79.9	72.0	65.1	77.72	
Jerusalem	31° 48' N	35° 11' E	2460	44.6	48.4	50.9	59.4	66.0	70.3	73.2	73.2	70.7	66.6	56.1	47.8	60.62	
Cairo	30° 5' N	31° 17' E	108	54.8	57.6	62.4	69.4	75.9	81.1	83.3	81.9	77.5	73.8	64.6	57.9	69.98	
Capetown	33° 56' S	18° 27' E	39	69.6	69.3	66.6	63.1	58.3	55.4	54.6	55.8	57.6	61.2	64.6	68.0	62.06	
Sydney	33° 51' S	151° 11' E	154	71.2	70.5	69.1	64.6	58.5	54.7	52.2	54.5	58.5	63.5	66.4	69.6	62.78	

Station	Lat.	Long.	Elev.	Jan.	Feb.	Mar.	Apr.	May	June	July	Aug.	Sept.	Oct.	Nov.	Dec.	Year	Yrs.
Honolulu	21°18'N	157°50'W	49	70.3	70.5	70.9	72.7	74.1	76.1	77.2	77.7	77.4	76.3	73.8	71.8	73.9	32
Havana	23°9'N	82°21'W	56	70.3	72.0	73.2	76.1	79.2	81.3	81.9	81.9	80.4	77.9	74.7	71.6	76.6	31
Rio de Janeiro	22°54'S	43°10'W	217	77.4	77.7	77.0	74.5	71.2	68.5	67.8	68.8	69.4	71.1	73.2	76.5	72.9	33
Buenos Aires	34°37'S	58°21'W	72	76.4	74.3	69.8	63.1	56.1	51.4	50.6	52.9	57.0	61.9	67.8	73.0	62.8	31
Valparaiso	33°1'S	71°40'W	151	63.0	63.1	60.6	58.3	55.6	53.6	53.1	52.5	54.0	56.7	59.0	63.0	57.7	31
Portland, Me.	43°39'N	70°15'W	47	22	24	32	43	54	63	68	67	60	49	38	27	46	33
Boston	42°21'	71°4'	15	27	28	35	45	57	66	72	70	63	53	42	32	49	33
New York	40°43'	74°0'	37	30	31	38	48	60	69	74	73	66	56	44	34	52	33
Philadelphia	39°57'	75°9'	42	32	34	40	51	62	72	76	74	68	57	45	36	54	32
Baltimore	39°18'	76°37'	103	34	35	42	53	64	73	78	76	68	58	46	37	55	33
Washington	38°54'	77°3'	75	33	35	42	53	64	73	77	75	68	57	45	36	55	25
Norfolk	36°51'	76°17'	10	41	43	48	56	67	74	79	77	71	61	51	43	59	31
Charleston	32°47'	79°56'	11	50	52	58	65	73	79	82	81	76	67	58	51	66	33
Jacksonville	30°20'	81°39'	3	55	58	63	68	75	80	82	82	78	71	62	56	69	31
New Orleans	29°59'	90°4'	8	54	57	63	69	75	81	83	82	79	70	61	55	69	31
El Paso	31°47'	106°30'	3702	45	49	56	64	72	80	81	79	73	63	52	46	63	33
St. Paul	44°58'	93°3'	758	12	16	29	48	60	66	74	72	62	50	32	20	45	32
Omaha	41°16'	95°56'	1037	21	25	36	52	62	72	76	74	66	54	38	27	50	31
St. Louis	38°38'	90°12'	466	32	34	44	57	66	76	80	78	70	59	44	36	56	32
Chicago	41°53'	87°37'	595	24	26	34	46	57	66	72	71	64	53	39	29	48	32
Cleveland	41°30'	81°42'	659	27	27	34	46	58	67	72	70	64	53	40	31	49	33
Cincinnati	39°6'	84°30'	553	32	35	43	54	65	74	78	76	69	57	44	36	55	32
Denver	39°45'	105°0'	5219	29	32	39	48	57	67	72	71	63	51	39	33	50	31
Portland, Ore.	45°32'	122°43'	20	39	42	47	51	57	62	67	66	61	54	46	42	53	32
San Francisco	37°48'	122°26'	28	50	52	54	55	57	59	59	59	61	60	56	51	56	32

of the day along the X-axis. Lines joining points having the same values of temperature have also been drawn, and these are ordinarily known in Europe as thermo-isopleths. They are also sometimes called chronoisotherms. This method of graphical representation was first introduced by Lalanne about 1843. In order to note the relation between the time of minimum temperature and the time of sunrise, dotted curves representing the time of sunrise and sunset have been added. From this figure the normal temperature at any hour of any day may at once be found, also the time of the minimum temperature each day, the time of the maximum temperature, and the normal value of daily range. Similar charts for other cities will be found in various books and publications; for Greenwich in SCOTT, *Elementary Meteorology*, p. 48; for München in ARRHENIUS, *Cosmische* 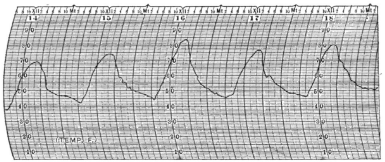 *Physik*, p. 556; for Aachen in *Meteorologische Zeitschrift*, April, 1904, p. 179; for Baltimore in FASSIG, *The Climate and Weather of Baltimore*, p. 62; for Chicago in WILLIS L. MOORE, *Descriptive Meteorology*, page 183.

Graphical representation of normal hourly temperatures.

77. **Diurnal, annual, and irregular variation.** — The graph which represents the daily variation in temperature is found by plotting to scale the normal hourly temperatures. If the values of the normal hourly temperatures are not known, an idea of the form of the curve may be obtained by noting the variation on some typical day, that is, some day when the other meteoro-

The daily variation in temperature.

FIG. 38. — Thermograph Record showing Typical Daily Variation of Temperature at Albany, N.Y., October 14–18, 1908. (U. S. Weather Bureau.)

logical elements have remained constant or followed as nearly normal courses as possible. Figure 38, which is a copy of the thermograph record at Albany, N.Y., for the five days ending October 18,

1908, shows a series of very typical daily variations.   Figure 4 also shows a typical daily variation in temperature at Williamstown, Mass. The lowest temperature usually comes at about the time of sunrise, and the highest temperature from 2.00 to 4.30 P.M., depending upon the season of the year.   The maximum occurs early in winter and later in summer.   The average temperature for a day occurs at about 9 A.M. and 8 P.M., the rise during the morning and early afternoon is sharp, the curve being convex; the drop during the afternoon and night is long and slow, giving a concave curve.   This curve, which represents the daily variation, varies slightly with the time of the year, the elevation, and the immediate surroundings.   It varies markedly with latitude and with nearness to the ocean.   The fact that the highest temperature does not occur at noon when receipts of energy from the sun are largest, but several hours later, needs explanation.   In Fig. 39 let A represent the insolation received from the sun on a horizontal surface dur- Why the ing some day in September.   The curve, which represents highest temthe receipts, begins at about 6 A.M., rises rapidly to a maxi- perature does not mum at noon, and drops down again to nothing at 6 P.M. occur at This represents the amount of energy received by a hori- noon. zontal surface provided the earth had no atmosphere, or provided the

atmosphere transmitted all of the insolation which falls upon it.   No question is here raised as to where the insolation is absorbed; whether in the atmosphere, at the surface of the ground, or in a body of water upon which it may fall.

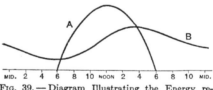

FIG. 39. — Diagram Illustrating the Energy received and given off by the Earth during a Day.

The amount of energy given off by the earth to space depends upon the temperature of the earth; thus the largest amount would be given off at the time of highest temperature and the least amount at the time of lowest temperature.   The graph which represents the energy given off is indicated by B.   During the early hours of the afternoon, although receipts have already commenced to fall off, it will be seen that they are still slightly in excess of expenditures, that is, a rise in temperature is still continuing and will continue until the receipts and expenditures become equal, which takes place at the time of highest temperature.

78.   The curve which represents the annual variation in temperature is found by plotting the normal daily temperatures to scale, and this is

pictured for Albany, N.Y., in Fig. 36. The time of lowest temperature is during the last part of January, and the time of highest temperature is
The annual variation in temperature. during the last part of July, in each case nearly forty days after the time of least and greatest receipts of energy from the sun. The rise from February until July is slow and regular, and the fall from the last of July until the first of February is of the same nature. The reason that the highest and lowest temperatures do not come at the time of greatest or least receipts of energy is the same as that

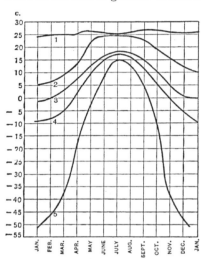

FIG. 40. — Annual Variation in Temperature at five different Places.— (1) St. Anns, Trinidad; (2) Palermo; (3) Berlin; (4) St. Petersburg; (5) Werchojansk, Siberia.

given for the time of occurrence of the maximum temperature during the day. The annual variation in temperature varies slightly with elevation and with the immediate surroundings of the station, and markedly with nearness to the ocean and with latitude. Figure 40 gives the annual variation at five different places on the earth's surface. The contrast between these five curves shows well the effect of latitude. At St. Anns, Trinidad, which has a tropical location, there are two maxima and two minima and an extremely small change during the year. With increasing latitude there is but one maximum and the change during the year increases steadily. Werchojansk with its high latitude and continental location has an immense change in temperature during the year. The graphs representing the annual variation in temperature can be constructed for many other places by plotting the data given in section 74.

The irregular changes of temperature are sometimes greater than the daily variation; that is, the temperature may fall steadily during the day
Irregular fluctuations in temperature. instead of rising, or it may rise during the night instead of falling. The irregular variations, however, are never larger than the annual variation.

79. Temperature data. — From the temperature observations which have been made at the different weather bureau stations, various temperature data in addition to the normals already fully discussed may be computed. Among these are the following: (1) *Average and normal daily range for the various months and for the year.* The daily range has been defined as the difference between the highest and lowest temperatures during the day. This range may be found, and normals may be computed, in exactly the same way as for observations of temperature. The accompanying table gives the average

Temperature data which may be computed from the observations taken at Weather Bureau stations.

AVERAGE AND NORMAL VALUES OF DAILY RANGE OF TEMPERATURES
AT ALBANY, N.Y.

(Temp. F.)

YEAR	JAN.	FEB.	MAR.	APR.	MAY	JUNE	JULY	AUG.	SEPT.	OCT.	NOV.	DEC.	ANNUAL
1874	20.3	17.8	17.4	15.2	21.7	18.9	18.4	20.0	19.7	16.8	15.3	16.5	18.2
1875	17.7	19.5	17.7	16.8	22.4	22.5	20.7	18.5	20.6	17.3	16.1	15.3	18.8
1876	18.5	17.1	17.5	18.5	20.8	19.2	20.1	23.3	15.6	15.5	11.5	15.2	17.7
1877	17.4	15.1	14.8	18.7	19.8	17.4	18.1	18.7	19.1	15.1	14.2	14.6	16.9
1878	17.3	17.9	17.2	14.2	17.8	18.9	18.1	17.6	17.8	18.4	11.2	12.5	16.5
1879	20.2	17.4	16.2	16.7	21.5	17.7	18.4	16.8	16.2	17.7	14.6	17.7	17.6
1880	16.6	18.6	14.6	18.3	18.0	17.2	15.8	16.8	14.8	15.7	13.0	10.8	15.8
1881	16.5	14.1	11.1	14.4	16.6	15.7	15.0	15.5	14.3	15.8	12.3	12.5	14.5
1882	15.5	16.4	13.8	16.4	15.9	16.8	16.0	16.7	13.8	15.4	12.7	10.8	15.0
1883	13.7	14.3	15.5	15.8	17.1	16.5	16.9	17.5	17.0	15.0	13.4	13.8	16.4
1884	16.3	13.3	11.5	13.9	16.1	20.0	15.6	16.9	16.5	17.8	14.9	14.7	15.6
1885	16.7	20.3	17.4	22.2	21.1	21.5	20.9	17.3	21.1	17.4	12.2	14.5	18.6
1886	15.2	17.6	14.7	20.1	20.6	18.8	21.3	22.1	20.1	19.3	16.9	16.5	18.6
1887	21.5	17.3	16.7	18.6	22.2	19.9	17.8	18.5	18.5	16.4	16.0	12.4	18.0
1888	15.9	19.6	16.3	20.0	18.3	20.4	22.3	19.5	16.8	14.5	14.3	14.0	17.6
1889	13.2	13.9	14.6	18.9	21.6	18.1	16.8	20.9	16.8	18.3	14.3	14.9	17.0
1890	14.9	15.2	14.9	21.8	19.4	20.9	20.7	17.1	15.7	13.1	14.1	14.1	16.8
1891	14.5	14.0	16.0	18.5	22.1	20.7	18.0	17.2	17.7	16.4	15.1	14.7	17.1
1892	13.9	14.4	13.8	18.4	15.6	17.9	20.9	17.5	18.5	16.6	11.1	10.9	15.8
1893	13.6	15.2	14.3	17.5	18.3	19.4	21.6	20.0	17.3	17.6	15.0	13.9	17.0
1894	14.5	16.5	16.7	17.4	19.1	19.4	20.9	19.8	18.6	15.3	12.4	13.2	17.0
1895	15.9	19.0	14.4	17.5	20.9	20.1	18.9	19.7	21.5	17.5	15.2	16.8	17.8
1896	13.5	15.7	16.2	20.8	20.9	20.0	19.2	19.6	19.5	15.2	14.7	14.4	17.5
1897	14.4	15.8	16.4	19.9	19.2	19.6	17.2	19.4	21.5	22.9	13.8	12.4	17.7
1898	15.5	14.6	17.5	16.6	15.9	19.2	20.7	20.7	20.2	15.7	14.2	14.9	17.0
1899	17.2	14.4	12.9	19.6	20.6	21.7	19.3	20.3	19.5	16.5	13.1	13.0	17.3
1900	17.7	16.2	17.4	19.1	24.3	21.7	22.0	20.5	19.5	18.0	13.0	14.5	18.7
1901	15.4	13.9	14.9	16.2	16.5	20.2	19.7	16.9	18.8	20.0	13.4	15.6	16.8
1902	15.1	14.0	15.5	17.1	19.7	18.6	17.2	19.7	18.1	17.0	16.5	18.1	17.2
1903	15.0	15.8	16.7	19.4	24.6	15.5	19.1	17.1	20.7	16.5	15.0	15.8	17.6
1904	17.8	18.3	14.7	17.6	21.2	19.8	18.9	19.3	18.0	17.5	14.2	15.6	17.7
1905	13.6	17.8	18.7	19.1	19.9	19.7	18.8	19.6	18.5	19.9	17.2	14.4	18.1
1906	14.9	19.9	14.0	19.7	21.4	20.4	19.2	20.0	21.1	18.4	13.5	15.0	18.1
1907	17.0	18.0	17.2	17.1	17.9	20.1	20.1	21.0	15.0	18.9	13.1	12.2	17.3
1908	17.6	16.7	16.2	18.8	18.6	22.2	20.6	19.6	22.4	21.5	13.8	14.5	18.5
Sums	564.5	575.6	545.4	630.8	687.6	676.6	665.2	661.6	640.8	600.9	491.3	500.7	603.8
Normal	16.1	16.4	15.6	18.0	19.6	19.3	19.0	18.9	18.3	17.2	14.0	14.3	17.3

value of the daily range for the various months, and for the year for several years, and also the normal values for Albany, N.Y. It will be seen that the range is greater in summer and smaller in winter, with a maximum in May and a minimum in November.

(2) *Monthly extremes of temperature.* By monthly extremes of temperature are meant the highest and lowest temperatures which have been observed during the month in question. (3) *Yearly extremes.* (4) *Absolute highest and absolute lowest for the various months and for the year.* By absolute highest and absolute lowest are meant the very highest and very lowest temperatures which have ever been observed. (5) *Variability.* By variability of temperature is meant the difference between successive daily averages. The average value of the variability for the various months and the year may be determined, and also normal values. The accompanying table gives the normal values for the various months and the year for several stations in the United States. The average variability for the various months and for the year and the normal values are also given for Albany, N.Y. The maximum of variability occurs in January and the minimum in August, one being more than twice the other. (6) *Freezing days.* By a freezing day is meant a day on which the temperature falls to 32° F. or below at some time during the day. (7) *Ice days.* By an ice day is meant a day on which the temperature remains below 32° throughout the whole day. (8) *Days above* 90°. (9) *Days above* 100°. (10) *Zero days.* By a zero day is meant a day when the temperature falls to zero or below at some time during the day. (11) *Temperature on special days.* The various normals of temperature for such special days as July 4, December 25, etc., may be computed. The table on page 91 gives for Albany, N.Y., the number of days above 90°, the number of days above 100°, and the number of zero days for a number of years.

At the regular stations of the U. S. Weather Bureau the following temperature records are kept constantly filled in and computed to date: monthly mean and departure from the normal; monthly mean maximum and minimum (to tenths); absolute maximum and date (each month); absolute minimum and date (each month); greatest daily range, mean daily range; absolute monthly range, mean variability (var. to tenths); lowest maximum, highest minimum; number of days with maximum 32° or below, 90° above; number of days with minimum 32° or below, zero or below; daily mean temperature (whole degrees); daily maximum temperature (whole degrees); daily minimum temperature (whole degrees); mean hourly temperature (to tenths).

NORMAL VARIABILITY OF TEMPERATURE
(Temp. F.)

	Jan.	Feb.	Mar.	Apr.	May	June	July	Aug.	Sept.	Oct.	Nov.	Dec.	Annual
Pacific Coast													
..., Oreg.	3.7	3.0	3.0	3.1	3.6	3.2	2.9	2.9	3.2	3.0	3.3	3.3	3.2
Sacramento, G.	2.3	2.4	2.3	2.7	3.0	2.9	2.9	2.9	2.9	2.5	2.2	2.7	2.6
San Diego, G.	2.3	2.3	1.8	1.8	1.4	1.3	1.1	1.2	1.5	1.9	2.2	2.3	1.8
Rocky Mountain and Plateau Regions													
Havre, M.	9.3	8.2	6.5	5.0	4.7	3.8	4.2	4.0	4.6	5.3	7.5	8.0	5.9
Salt Lake City, Utah	4.2	4.2	4.1	4.9	4.7	4.3	3.3	2.9	4.2	4.1	3.9	3.8	4.0
Denver, G.	8.1	6.6	6.7	5.7	5.2	4.1	3.5	3.2	4.6	5.3	6.3	6.7	5.5
Santa Fé, N. M.	4.2	4.1	4.2	4.7	3.9	3.2	2.2	2.1	2.7	3.3	3.8	4.0	3.5
Western Plains and Texas													
Bismarck, N. Dak.	9.0	8.4	7.1	5.6	4.9	4.4	4.4	5.0	5.7	5.7	6.9	7.7	6.2
Dodge City, Kls.	7.3	7.3	7.2	6.3	5.3	3.9	3.5	3.4	4.8	5.5	6.3	6.4	5.6
San ..., ...	6.2	6.4	5.8	3.7	2.8	1.9	1.4	1.5	2.2	3.4	5.3	5.5	3.8
... Valley													
St. ..., M.	8.2	8.5	6.0	5.1	4.4	3.6	3.5	3.7	4.6	5.2	6.3	7.0	5.5
St. Louis, M.	8.6	8.2	7.0	5.9	4.6	3.4	3.1	3.0	4.2	5.0	6.5	7.0	5.5
New ..., La.	6.0	5.0	4.5	2.9	2.0	1.7	1.6	1.6	1.7	2.8	4.6	5.6	3.3
Lake ... Mn.													
Detroit, Mh.	6.7	7.1	6.1	5.3	5.0	3.8	3.8	3.7	4.6	5.1	5.4	5.8	5.2
South Atlantic States													
..., G.	6.0	6.0	5.7	4.4	3.5	2.6	2.2	2.1	2.6	3.7	5.1	5.3	4.1
Jacksonville, ...	5.7	5.2	5.0	3.3	2.5	2.0	1.8	1.8	1.9	3.0	4.7	5.2	3.5
... States													
M... ..., III.	5.9	6.6	5.9	5.2	4.5	3.7	3.2	3.1	3.9	4.7	5.4	5.5	4.8
New England													
Boston, M.	7.3	7.3	5.3	5.5	5.5	5.0	4.0	3.9	4.9	5.3	6.2	7.4	5.6

AVERAGE AND NORMAL VALUES OF VARIABILITY OF TEMPERATURE
AT ALBANY, N.Y.

(Temp. F.)

YEAR	JAN.	FEB.	MAR.	APR.	MAY	JUNE	JULY	AUG.	SEPT.	OCT.	NOV.	DEC.	AN-NUAL
1874	6.6	6.2	7.5	5.2	4.7	5.3	3.6	3.4	4.0	3.0	5.0	9.3	5.3
1875	6.8	8.2	5.8	3.	3.5	4.5	.5	.6	.7		5.3	.8	5.0
1876	7.1	6.5	6.6	3.	5.3	3.1	.5	.7	.4		3.8	.8	4.9
1877	7.4	5.5	5.2	3.	3.5	4.8	.7	.7	.0		5.4	.9	4.4
1878	7.6	6.3	5.9	2.	4.1	3.1	.1	.4	.9		3.6	.5	4.4
1879	8.6	7.0	5.0	4.	5.1	4.2	.5	.9	.9		7.3	.8	5.5
1880	7.2	8.1	6.5	5.	5.3	3.3	.1	.2	.5		4.4	.2	5.1
1881	6.6	6.9	3.3	3.	5.0	3.3	.9	.6	.8		5.6	.3	4.7
1882	7.2	7.5	5.2	4.	4.5	3.3	.5	.8	.8		4.8	.3	4.5
1883	6.5	6.7	6.8	3.	4.5	3.3	.5	.3	.3		5.3	.5	4.8
1884	7.7	7.0	4.3	3.	4.3	3.5	.7	.1	.9		5.1	.1	5.0
1885	8.4	8.4	5.3	4.	3.9	4.9	.2	.2	.5		4.0	.5	5.4
1886	6.8	7.0	5.2	4.	4.0	3.3	.9	.6	.3		5.8	.2	4.7
1887	8.5	6.3	6.1	5.	3.7	3.6	.8	.6	.5		4.6	.6	4.7
1888	7.5	8.1	6.5	4.	4.2	4.0	.3	.7	.3		5.3	.9	5.2
1889	5.7	8.1	4.4	4.	4.8	3.6	.1	.9	.3		3.7	.0	4.6
1890	9.3	7.8	6.2	5.	4.5	4.6	.4	.5	.8		5.6	.6	5.6
1891	7.0	7.3	5.8	5.	5.4	4.9	.8	.7	.7		6.4	.0	5.3
1892	10.5	7.0	3.9	4.	4.1	4.2	.2	.6	.5		4.7	.5	4.8
1893	5.8	8.2	6.8	5.	4.6	4.7	.5	.5	.5		4.8	.6	5.5
1894	6.4	8.5	5.5	3.	4.5	3.8	.0	.2	.9		5.6	.6	5.0
1895	7.5	6.4	4.5	5.	4.9	3.2	.0	.0	.0		5.8	.0	5.1
1896	6.7	7.6	6.8	4.	4.7	3.7	.4	.5	.6		7.4	.0	5.1
1897	7.4	6.8	4.6	6.	4.4	4.6	.1	.8	.0		6.3	.1	5.2
1898	6.4	5.2	3.9	4.	3.4	3.4	.8	.5	.4		4.0	.0	4.4
1899	7.1	4.7	4.6	3.	3.9	5.0	.3	.9	.5		3.4	.8	4.7
1900	7.1	6.4	7.1	4.	6.6	3.7	.1	.1	.6		4.3	.2	5.2
1901	7.4	5.1	6.3	2.	3.7	3.3	.4	.6	.0		3.3	.0	4.6
1902	7.5	4.8	5.6	3.	5.2	3.9	.7	.3	.8		5.7	.9	5.0
1903	6.8	4.9	5.7	3.	3.9	3.2	.0	.6	.5		4.0	.5	4.4
1904	8.7	7.2	4.8	4.	3.5	4.4	.7	.7	.7		5.1	.4	5.4
1905	7.4	6.8	4.3	4.	4.6	3.7	.7	.4	.9		6.0	.2	4.9
1906	6.7	8.6	5.0	3.	5.5	3.7	.9	.6	.9		4.4	.7	5.2
1907	7.4	9.1	6.2	4.	5.7	3.4	.7	.9	.9		3.3	.9	5.2
1908	7.3	7.3	5.9	5.	4.1	4.7	.5	.7	.2		4.0	.4	5.1
Sums	256.6	243.5	193.1	150.0	157.6	137.2	120.1	110.8	157.5	162.3	173.1	227.1	173.9
Normal	7.3	7.0	5.5	4.3	4.5	3.9	3.4	.2	4.5	4.6	4.9	6.5	5.0

80. It has often been found desirable to compute a normal for a station at which observations have been taken for but a few
years, and the best method of procedure is the following:
Choose some near-by station which has a well-determined
normal and at which observations have been made during the
same interval. Assume that the normal at the station in
question will bear the same relation to the normal at the chosen station

The compu-
tation of a
normal
from insuffi-
cient obser-
vations.

THE NUMBER OF ZERO DAYS, DAYS ABOVE 90°, AND DAYS ABOVE 100°, FOR ALBANY, N.Y.

	ZERO OR BELOW	90° OR ABOVE	100° OR ABOVE		ZERO OR BELOW	99° OR ABOVE	100° OR ABOVE		ZERO OR BELOW	90° OR ABOVE	100° OR ABOVE
1874	13	2		1886	9	8		1898	7	14	1
1875	34	2		1887	8	13		1899	10	15	
1876	12	16		1888	16	8		1900	6	23	
1877	7	3		1889	1	1		1901	6	17	
1878	11	5		1890	5	7		1902	6	3	
1879	15	3		1891	3	9		1903	11	6	
1880	4	11		1892	6	15		1904	22	6	
1881	8	8		1893	7	12		1905	7	7	
1882	3	9		1894	7	16		1906	12	5	
1883	3	2		1895	10	15		1907	16	8	
1884	7	5		1896	12	19		1908	8	9	
1885	18	5		1897	4	7		1909	5	8	
								1910	11	5	

as the average for the few years at the station in question bears to the average for the same period at the chosen station. The normal determined in this way is usually much more reliable and correct than one determined from too short a period of observations.[1]

81. **Differences of Temperature with Altitude.** — The thermometer shelters at which the various temperature observations have been made all have definite locations; in a city usually on the roof of a high building, in the country usually a few feet above the sod or on the north side of some building. The natural question at once arises as to whether the observations would have been different if the elevation of the shelter above the ground had been different. In other words, what are the temperature differences in the small height of, say, 100 feet above the earth's surface. In the layer of air within five feet of the earth's surface marked differences in temperature will be found. During the day, when convection is operative and when wind velocities are large, the difference will be a comparatively small one, not more than a degree at most. The air in immediate contact with the ground is, of course, the warmer. At night the temperature differences are more

The differences in temperature in a thermometer shelter placed at different altitudes.

[1] *Monthly Weather Review*, April, 1910.

marked and will perhaps average as high as 2 or 3° Fahrenheit, with maximum values of even 5 or 6°. The layer of air in immediate contact with the ground is, of course, the colder. The layer of air from five to one hundred feet is so thoroughly mixed by the wind at night, as well as during the day, that a very small temperature difference will be found, probably not more than a degree at most, unless the air is held by natural or artificial barriers. Above 100 feet the regular vertical temperature gradient may be expected. As a general conclusion, then, a thermometer shelter should not be placed within five feet of the ground nor in a location where the air would be held stagnant by means of artificial or natural barriers. The temperatures observed at heights of from five to one hundred feet will probably be nearly the same.

82. **Temperature differences over a limited area.** — The question here arises as to whether the observations of temperature would be different

Temperature differences over a limited area.
if the thermometer shelter were placed at different points within a small area immediately surrounding the point in question. A small or limited area may be roughly defined as a square mile of surface in the form of a circle or square. During the daytime, on account of convection and the higher values of wind velocity, no appreciable difference in temperature over such a limited area will be found unless it is of unusual topography or the air is held stagnant by natural or artificial barriers. On some particularly favorable days, namely, those with plenty of sunshine and a low wind velocity, the lower points, particularly those in narrow valleys, may be a few tenths of a degree Fahrenheit warmer than the upper parts of the area. At night the layer of air next to the ground grows cold and denser, and drains like water into the valleys and places of small elevation. If the wind is unable to remove these pockets of cold air, a marked variation in temperature over a limited area will be found. For every limited area there will be a critical value of wind velocity, which for most areas is probably not far from three miles per hour. As long as the wind velocity remains larger than three miles these pockets of air will be removed and mixed with the air at other points, and no variation in temperature will be found. As soon as the wind velocity sinks below this critical value, a variation will begin to be manifest, and it is the valley station and those of low elevation which are ordinarily the coldest. Since the question of the variation in temperature depends upon the interplay between the drainage of colder air and the ability of the wind to remove those pockets of cold air, the variation will depend not only upon the elevation, but also upon the openness of the valleys,

their direction, the roughness of the surface, and the direction from which the wind comes.   The average difference of temperature between the warmest and coldest places in the limited area will probably average about 4° throughout the year, and will show at times differences as much as from 10° to 15°.   The location of the shelter in a limited area will thus make a great difference with the temperature observations ; and if records are to be of value, the limited area surrounding each station should be critically investigated for several years.

The chief factor which determines the variation in temperature over a limited area, particularly at night, is without doubt the interplay between the drainage of colder air into the valleys, and the ability of the wind to remove these pockets of cold air.   A second factor is the nature of the surface.   That this plays an important part has been well shown by the research work of Professor Henry J. Cox, of the U. S. Weather Bureau and others in connection with the minimum temperatures and frosts observed over the cranberry marshes of Wisconsin.[1] It has been found that cultivation, drainage, and sanding are very efficacious in preventing destructive frosts.   The minimum temperatures observed have sometimes been 10° F. and even more, higher than over near-by untreated marshes.   Here there is essentially no difference in elevation and the wind velocity can be assumed to have been the same.   The differences in temperature were brought about entirely by the nature of the surface. In causing temperature differences the surface acts in two ways.   Different surfaces heat unequally during the day and at night they cool off at very different rates.   Since on frosty nights the air is always particularly quiet, its temperature is determined almost entirely by the temperature of the surface upon which it rests.

The Distribution of Temperature over the Earth

83. Construction of Isothermal Charts. — Observations of temperature have been made at many stations in all parts of the world, and some of the records are long ones.   From these observations the normal monthly and annual temperatures may be computed.   The stations are, however, at different elevations above mean sea level, and in order to make the observations comparable with one another it is necessary to reduce them all to mean sea level. *The reduction of the station normals for the land stations.*
Since the average vertical temperature gradient is 1° F. for 300 feet, the reduction could well be made by using this factor and adding 1° F. for

[1] See Bulletin T, U. S. Weather Bureau.

each 300 feet of elevation. As a matter of fact the reduction factors used by different investigators in preparing charts are very different. Buchan used 1° F., for 270 feet, while others have used values all the way to 1° F. for 500 feet.

The normal temperatures of the air over the ocean have been computed from temperature observations made by vessels while at sea. At

The method of determining the air temperature over the ocean.

present nearly all vessels take meteorological observations, and these are reported to the weather bureaus of the various countries as soon as port is reached. These observations are grouped by months and by squares, each square being generally 5° on a side. For some squares of the Atlantic Ocean thousands of observations are made during a single month. Although these observations are not always simultaneous, yet, since the daily variation in temperature over the ocean is so small, the normals are nearly as correct as those for the land stations.

These normal temperatures may be charted on a map and lines drawn through those places which have the same temperature. These lines

The construction of isothermal charts.

are called ordinarily isothermal lines or simply isotherms.[1] The chart is often spoken of as an isothermal chart. The first isotherms for the world were drawn by Humbolt in 1817. These isotherms were improved by Kämtz in 1831 and Mohlmann in 1841 as new data were gathered. Dove in 1852 was the first to chart the normal monthly temperatures.

84. Isothermal lines for the year. — The normal yearly temperatures have been used in the construction of chart I,[2] and this chart thus repre-

Isotherms for the year.

sents the isotherms for the year. There are several characteristics of these isothermal lines which deserve careful consideration and explanation.

(1) *Highest temperature at the equator and lowest temperatures at the poles.* The first and most obvious fact in connection with the isothermal

Highest temperature at the equator and lowest at the poles.

lines for the year is the existence of a hot belt near the equator and low temperatures at the two poles. The northern part of South America, most of Africa, India, and a portion of Australia are surrounded by a line marked 80° F. This means that all points within this closed curve have a normal annual temperature of 80° F. or more. This is the hot belt, and a line passing through its center is often spoken of as the heat equator or the thermal equator. The temperature at the north pole is 0° F., while the

---

[1] From the Greek : ἴσος = equal ; θέρμη = heat.
[2] See end of book for the 50 charts.

isothermal lines near the south pole have been omitted on account of insufficient data. The explanation of the existence of this hot equatorial belt and the low polar temperatures is the well-known fact that the equator receives much more insolation from the sun than the polar regions. The ratio between the equator and pole is 347 to 143, and this is sufficient to account for the temperature differences.

The reason for the temperature difference.

(2) *The deflection of isothermal lines from parallels of latitude.* All places with the same latitude receive the same amount of insolation from the sun. It would thus be expected that all places on the same parallel of latitude would have the same temperature, and that the isothermal lines would be parallel to the parallels of latitude. This is, however, by no means the case, and the chief cause of the deflection is the existence of ocean currents.

The deflection of isothermal lines.

It is beyond the scope of this book to discuss fully the causes for the existence and the direction of ocean currents. The factors which cause ocean currents and determine their direction are: temperature differences between different parts of the earth; the permanent wind system of the earth; the evaporation which is greater at some parts of the ocean than at others; the inflow, which is also greater at certain places than at others; varying degrees of saltiness and thus density; the earth's rotation on its axis; the configuration of the coast line.

The factors which determine the existence and direction of ocean currents.

Chart II presents the general scheme of the ocean currents. In the Atlantic Ocean the Gulf Stream, consisting of warm water which has made the circuit of the Gulf of Mexico and emerged between Florida and Cuba, together with a considerable quantity of water which has passed northward outside of the West Indies, sets diagonally across the Atlantic Ocean towards England and Scandinavia. The return current flows southward along the coast of Europe and along the northern part of Africa and back again along the equator. Another return current flows southward along Greenland and Newfoundland. The northern Pacific Ocean possesses a similar set of ocean currents. In the South Atlantic and South Indian and South Pacific oceans there is an oval circulation turning in a counterclockwise direction.

The general scheme of ocean currents.

On the European side of the Atlantic Ocean the isothermal lines over England and Scandinavia are carried far to the north by the Gulf Stream and even bend backward on themselves. The cool return current along the coast of Spain and northern Africa carries the isothermal lines towards the equator. The

The effect of ocean currents on isotherms.

result is a fan-shaped spreading out of the isothermal lines over Europe. On the North American side of the Atlantic the cold return current from Greenland carries the isothermal lines southward, while the warm water coming up from the Gulf carries them northward, and the result is a crowding together of the isotherms. The same difference in temperature will be found in one half the distance on the North American side of the Atlantic as on the European. Numerous other illustrations of the deflection of isotherms by ocean currents can be noticed by comparing the scheme of ocean currents with the isothermal lines for the year. Although the ocean currents are the chief cause of deflection, they are not the only causes, as diversity of surface, whether land or water, vegetation covered or barren, and the general wind system, also play a part.

(3) *Regularity in the southern hemisphere.* It will be noticed that the isothermal lines are much more regular in the southern hemisphere than

Regularity in the southern hemisphere.

in the northern. The reason for this is that the southern hemisphere is largely a water hemisphere, while the northern hemisphere has a diversified surface consisting of both land and water. A water surface tends to even out temperature irregularities and also to equalize the temperature between equator and pole.

(4) *The heat equator north of the geographical equator.* It will be noticed that the central line of the hot belt lies, on the whole, north of the

The heat equator north of the geographical equator.

geographical equator. The reason for this is not far to seek. Since the southern hemisphere is largely a water hemisphere the equatorial portion has been cooled and the polar regions have been warmed by the exchange of water between the equator and poles. In the case of the northern hemisphere this is not so easily possible, since it is largely a land surface. For this reason the equatorial belt of high temperature lies north of the geographical equator.

(5) *The hot belt not of the same width and temperature throughout.* It will be noticed that the hot belt inclosed by the isothermal line 80° is

The hot belt not of the same width throughout.

widest over South America and Africa. It narrows markedly in crossing the Atlantic Ocean and disappears entirely over the Pacific Ocean. The reason for this is again the tendency of the ocean to equalize equatorial and polar temperatures.

85. Chart III represents the isothermal lines for the year for the United States.

86. **Isotherms for January and July.** — In the construction of Charts IV and V the normal January and July temperatures have been used, and

these charts thus present the isothermal lines for January and July. The first three characteristics noted in connection with the isothermal lines for the year are the same for the isothermal lines for January and July. These three characteristics are : the hot belt at the equator and the low temperatures at the poles ; the deflection of isothermal lines from parallels of latitude ; the regularity in the southern hemisphere. In addition there are two new characteristics which deserve attention and explanation. *The iso-therms for January and July.*

(1) *The hot belt and all isothermal lines migrate north and south in the course of the year.* On the January chart it will be seen that the highest temperatures occur in South Africa and in Australia, while on the July chart it will be seen that the highest temperatures occur over the southern part of North America, Northern Africa, and Australia. There has thus been a decided migra- *The migra-tion of iso-thermal lines.* tion of the hot belt between January and July. If any other isothermal line is considered it will be found that it too has migrated. On the Pacific the hot belt migrates from 15 to 20° of latitude and on the Atlantic it migrates still less, and only on the western portion of this ocean does it cross the geographical equator to the southern hemisphere. On the con- tinents it shifts over a somewhat greater distance. In Africa it moves from about 23° north to 20° south latitude. In America the migration is from about 35° north to 15° south. The average, however, is less than 47°, which is the amount the sun migrates in the course of the year. Three things may be noted in connection with this migration : (*a*) the migration is less than the migration of the sun ; (*b*) the hot belt lags behind the sun in its migration ; (*c*) the migration is greatest on land and least on the ocean.

(2) *The highest and lowest temperatures on land.* On the January chart the highest temperatures are in the northern part of Africa and in Australia, while the lowest temperature is in north central Siberia. On the July chart it will be seen that the highest temperatures are in central North America, North Africa, and Arabia. The truth of the statement that the highest and lowest temperatures occur on land is thus demonstrated. *The highest and lowest tempera-tures always on land.*

The reason for this is not far to seek. The land, as compared with the ocean, is always radical in its behavior as regards temperature changes. During the day and during the summer it heats to a high temperature, during the night and during the winter it cools correspondingly low. Thus the highest and lowest temperatures are always to be expected on land.

H

87. Charts VI and VII represent the July and January isotherms for the United States.

88. Poleward temperature gradient. — The diminution in temperature in going from equator to pole is often spoken of as the poleward temperature gradient. By examining the isothermal charts for the year, and for January and July, the three following generalizations may be formed: (1) the poleward temperature gradient is larger in winter than in summer; (2) the poleward temperature gradient is larger in the northern hemisphere than in the southern hemisphere; (3) the poleward temperature gradient is larger on land than over the ocean. It will

The characteristics of the poleward temperature gradient.

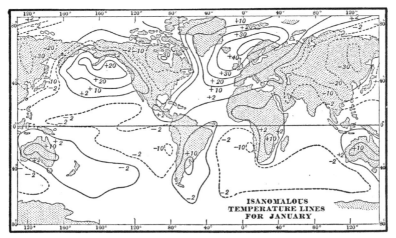

Fig. 41. — Isanomalous Temperature Lines for January (Temp. F.). (After BATCHELDER.)

be seen later that to the first of these, namely, the larger value of poleward temperature gradient in winter, is due the increased wind velocity and the greater violence of storms during the winter.

It is also an interesting fact that the poleward temperature gradient is about 800 times smaller than the vertical temperature gradient. That is, it would be necessary to travel poleward 800 miles to get the same diminution in temperature which would be gained by ascending one mile.

89. Thermal anomalies. — The average temperature of a given parallel of latitude may be found by knowing the actual temperature at equally distant intervals along this parallel and finding

Thermal anomalies.

their average.   The difference between the temperature of a place and
the average for its parallel of latitude is called the thermal anomaly.
These thermal anomalies for January and July are pictured in Figs.
41 and 42.   On the January chart it will be seen that the north Atlantic

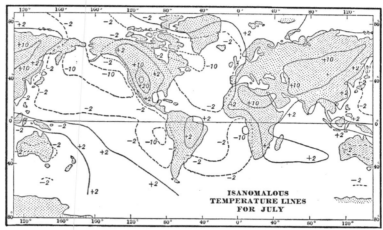

FIG. 42. — Isanomalous Temperature Lines for July (Temp. F.).   (After BATCHELDER.)

is 40° above the average of this latitude, the north Pacific is 20° above
the average, while the central part of Asia and the central part of North
America are 30° below the average for their latitude.   If the July chart is
examined, it will be seen that the center of North America and of Asia
are above their latitudes
in temperature, while
the Pacific and the
Atlantic Oceans average
below their latitudes.
These statements may
be summarized as a
general law: in summer
the continents are above
the average in tempera-
ture, while the oceans are below ; during the winter the continents
are below the average in temperature, while the oceans are above.

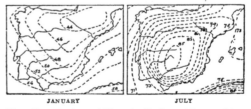

FIG. 43. — Isothermal Lines for Spain and Portugal for
January and July (Temp. F.).

Figure 43, which represents the isothermal lines for Spain and Portugal for January and July, illustrates this general law for a small area.

90. Annual range of temperature. — Figure 44 represents the annual range of temperature for the earth's surface. Many different kinds of
The char- annual range may be computed. The kind represented here
acteristics is the difference between the January and July normals. The
of the greatest value of range is 120° F. in north central Siberia.
annual
range of The northern part of North America comes next with 80° F.,
temperature. while South America, South Africa, and Australia have 30°
each. It will be noticed that the greatest value of range always occurs on land, and that it is roughly proportional to the amount of

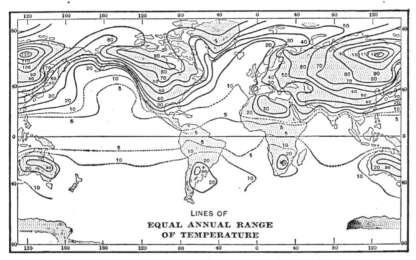

Fig. 44. — Annual Range of Temperature (Temp. F.).

(After Connolly — from Davis's *Elementary Meteorology.*)

land surrounding the place in question. This is but another illustration of the well-known principle that a land surface is radical in its temperature behavior as compared with an ocean surface. A land surface becomes excessively warm in summer and correspondingly cold
The highest during the winter, while a water surface is more conserva-
and lowest tive in its behavior.
tempera-
tures ever **91. Extremes of temperature.** — The lowest temperature
observed in ever observed on the earth's surface is − 90.4° F. at
the world. Werchojansk (or Verkhoyansk) in north central Siberia.
This temperature was observed on January 15, 1885. The highest

temperature ever observed was 127.4° F. at Ouargla in Algeria.   It was observed July 17, 1879.   North central Siberia and the northern portion of Africa are thus the coldest and hottest places in the world respectively.

The lowest temperature ever observed in the United States is −65° F. at Poplar River in Montana.   It was observed January 1, 1885.   The highest temperature ever observed in the United States is 119° F., and this was observed at Phoenix, Arizona.   The isothermal lines for the United States, for January and July, indicate the portions of the country which are hottest and coldest respectively.

The highest and lowest temperature ever observed in the United States.

In making these statements only those temperatures which have been observed in regular series at weather bureau stations have been taken into account.   A temperature of 122° F. is reported for Death Valley, Cal., during the summer of 1891, and a temperature of 130° F. is said to have been observed at Mammoth Tank, Cal., on August 17, 1885.   Since the thermometers may not have been properly sheltered, these temperatures are not usually considered as authentic.   Temperatures as high as 154° F. have been reported from parts of the Sahara, and a temperature as low as −96° F. is reported from the Arctic regions of North America.

At every place abnormally high and abnormally low temperatures have occurred if long periods of time are considered.   Unusually cold winters and unusually hot summers have also been described.   At many European cities where the records are long ones, these abnormalities make very interesting reading.   Space does not permit, however, a full treatment of this subject.

92. **Other temperature charts.** — All the temperature data mentioned in section 79 may be charted, provided the data are available for many stations in all parts of the world or in a given country.   Figures 45, 46, and 47 represent the highest temperatures ever observed in the United States, the lowest temperatures ever observed in the United States, and the variability of temperature for January in the United States.

Other temperature charts.

93. **Polar temperatures.** — Charts VIII and IX represent the normal temperatures for the north polar regions[1] for January and for July. .It will be seen that during July the north pole is the coldest part of the northern hemisphere.   During January north central Siberia is the coldest part of the northern hemisphere.   Both

North polar temperatures.

[1] For isotherms for the north polar regions see *Meteorologische Zeitschrift*, 1906, p. 111.

of these facts are somewhat anomalous, since for a short time during the summer the North Pole receives more insolation than any other place in

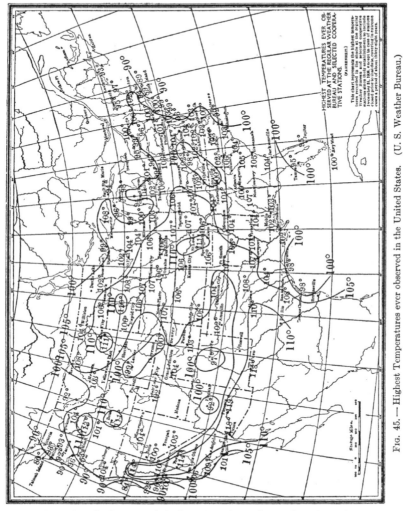

FIG. 45. — Highest Temperatures ever observed in the United States. (U. S. Weather Bureau.)

the northern hemisphere, and during the winter it receives less than any other place in the northern hemisphere. These seeming anomalies can,

however, be easily explained.   During the summer the North Pole is the
coldest part of the northern hemisphere, in spite of its value
of insolation, for three reasons: first, these large values of **The North**
Pole coldest
insolation last for a very short time ; secondly, the polar **in summer.**

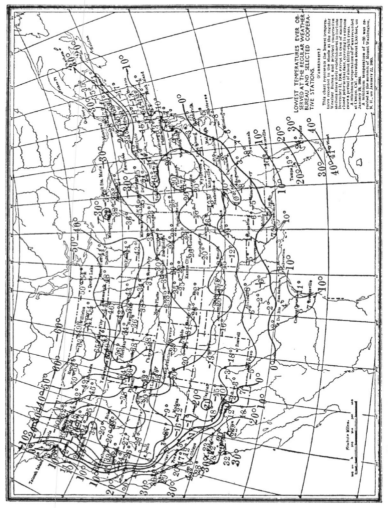

FIG. 46. — Lowest Temperatures ever observed in the United States.   (U. S. Weather Bureau.)

regions are covered with snow and ice, which reflect about 30 to 40 per cent of the insolation, which is thus lost as far as heating is concerned;

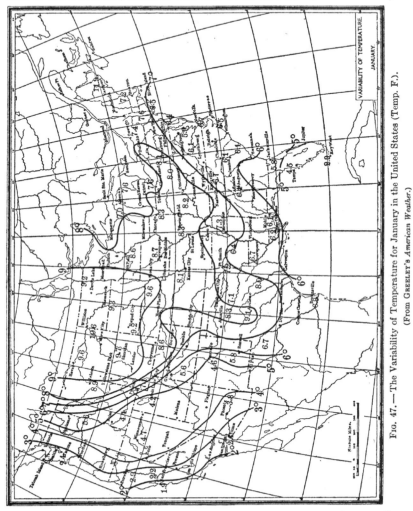

FIG. 47. — The Variability of Temperature for January in the United States (Temp. F.).

(From Greeley's *American Weather*.)

thirdly, the temperature cannot be raised markedly above 32° F. until all the snow and ice is melted, and this never occurs.

There are two reasons why north central Siberia surpasses the polar regions for cold during the winter; first, the exchange of water between the equator and pole warms the polar regions. This is not possible in the case of a land surface. Secondly, north central Siberia is a land surface, while the polar regions is largely a water surface, or a water surface covered with ice and snow. Land is always radical in its temperature behavior, and thus cools to a greater extent than a water surface. *North central Siberia coldest in winter.*

The Temperature of Land and Water

94. Ocean temperatures.— The normal annual temperature of the surface of the ocean varies from between 80° and 90° F. at the equator to about 28° F. in the polar regions. The temperature of the bottom of the ocean varies from 36° in the equatorial regions to about 28° in the polar regions. Thus, at the equator the temperature change between the surface and bottom is from between 80° and 90° to 28°; in the polar region the temperature is 28° throughout. The change in temperature between day and night is extremely small, not amounting to more than a degree or two at most. The temperature change between winter and summer is also small at the equator and in the polar regions, but greater in middle latitudes. At New York the temperature change between summer and winter is from about 70° to 30° F., and this is perhaps the greatest yearly change anywhere in the Atlantic Ocean. On account of the ocean currents the difference in temperature between places but a few hundred miles apart may be considerable. Salt water freezes at 27° F. Thus all harbors north of 50° north latitude are frozen shut in winter, and ice forms all over the polar seas, seldom, however, to a depth of more than 5 or 6 feet. *The temperature of the surface and bottom of the ocean.* *The annual change in the temperature of surface.the*

Special thermometers must be used for determining the temperature of the ocean water, particularly at considerable depths. The thermometer must not be influenced by pressure and must indicate the temperature at the required depth, regardless of the temperature of the layers of water through which the thermometer must be raised in drawing it to the surface. *Characteristics of the special thermometers used.*
For a more detailed treatment of this subject the reader must be referred to special works on ocean temperatures.

95. Lake temperatures. — In the case of a deep lake in the middle latitudes, the temperature of the surface water in summer will be between

60° and 80°, while at considerable depths the temperature will be 39° or above. With the coming of winter the surface water cools and thus

The temperature of a lake from top to bottom at different times of year. becomes heavier and sinks to the bottom. This process continues until the temperature of the lake becomes 39° throughout. 39° F. is the temperature of water at its maximum density. As the temperature falls below this, the water again expands and becomes lighter. Thus, as the surface water cools below 39°, it is now lighter than the water below and remains at the top. It cools finally to 32° and ice forms and the thickness of the ice grows greater and greater during the winter. Thus, in winter the temperature of the surface of a lake next the ice will be 32° F., while the bottom will have a temperature of 39°. With the coming of spring the ice melts and the surface layers become warmed until the temperature again becomes 39° throughout. From this point on the warmer layers remain at the top until finally in late summer the surface may have heated up to 60° or even 80° F. The temperature of the bottom, if a lake is deep, will remain 39°; if shallow, the lake may have become heated throughout, and the bottom temperature may be somewhat above 39°.

96. River temperatures. — If a river is deep and slow flowing, its

River temperatures. temperature behavior will be the same as that of a lake. In winter the temperature of the water underneath the ice will be 32°, and the temperature of the bottom 39° or slightly lower.

If the river is shallow and rapidly flowing, the water will be so thoroughly mixed that the temperature will be the same throughout.

97. Temperature below the surface of the land. — The ground is a very poor conductor of heat, and for this reason the daily variation in

The annual variation does not penetrate more than fifty feet. temperature does not penetrate to a greater depth than two or three feet, and requires many hours to reach even that depth. The annual variation does not penetrate to a depth of more than fifty feet and requires nearly six months to reach that depth. Thus, at a depth of thirty or forty feet in the ground the highest temperatures will be experienced in winter and the lowest temperatures in summer.

Below fifty feet comes the layer of invariable temperature. This will have a thickness from perhaps fifty feet to several hundred feet, and this

The layer of invariable temperature. temperature is always the normal annual temperature for the place in question. Caves are often located in this layer of invariable temperature, and the temperature of the air in such caves is always the normal yearly temperature for the place. Some cellars are also sufficiently deep to be located in this layer. They will have a constant temperature throughout the year, and its value will be the normal annual temperature for the place. If the normal annual temperature is below 32°, there will be a layer which remains constantly frozen throughout the year.

Below the layer of invariable temperature, the temperature increases at the rate of 1° for every 52 feet (although this varies from 40 to 100 feet). These observations have been obtained from tunnels and from deep mines. These extend, however, not more than a mile below the surface of the earth, which is a mere scratch compared with the radius of the earth. It would be *Temperature increase with depth.* impossible from astronomical and geological considerations for the temperature to increase at this rate all the way down to the earth's center.

The temperature of spring water gives an idea of the depth of the layer through which the water percolates and finally emerges as a spring. If the water comes through the layer of invariable temperature, the temperature of spring water would remain the same throughout the year and would have the normal annual temperature for the place. If the layer through which the water *The temperature of spring water.* passes is below this, the temperature would remain the same throughout the year, but have a higher value than the normal yearly temperature of the place. If the spring is a shallow one, there would be a change in temperature between summer and winter, but the average would again be the normal yearly temperature for the place in question.

QUESTIONS

(1) Define thermometry. (2) Name the three systems of thermometry. (3) How are the fixed points numbered in each system? (4) State the interrelation between the three scales. (5) What is the best method of making a mental computation of the Centigrade temperature corresponding to a Fahrenheit? (6) When and where was the thermometer invented? (7) Describe the early form of the thermometer. (8) Describe the origin of the Fahrenheit, Centigrade, and Réaumur thermometers. (9) Describe the appearance and state the working principle of a thermometer. (10) Describe the various steps in the construction of a thermometer. (11) Name the essentials in a good thermometer. (12) What are the advantages and disadvantages of having a large bulb? (13) What is the ordinary form of the bulb and the reason for it? (14) Why should the graduation be placed on the stem of a thermometer itself? (15) Name the inaccuracies in determining a temperature. (16) What is meant by the error of parallax? (17) What is the effect of atmospheric pressure on a thermometer? (18) What is the cost and accuracy of a thermometer? (19) Why is it hard to determine the temperature of a gas? (20) How should a thermometer be placed to give approximately the real air temperature? (21) What are the three methods of determining the real air

temperature? (22) Describe the thermometer shelter of the U. S. Weather Bureau. (23) Describe the other types of thermometer shelter. (24) Describe the sling thermometer. (25) Describe in full the ventilated thermometer. (26) What are the two kinds of thermographs in ordinary use? (27) Describe the Draper thermograph. (28) Describe the Richard Frères thermograph. (29) What is the working principle in each case? (30) What is the purpose of maximum and minimum thermometers? (31) Describe the maximum and minimum thermometers of the regular Weather Bureau form. (32) Describe the Six maximum and minimum thermometer. (33) How are maximum and minimum thermometers set? (34) Describe the black bulb thermometer. (35) What is the purpose of the black bulb thermometer? (36) Name some special purposes for which thermometers of various forms have been adopted. (37) State the three kinds of U. S. Weather Bureau stations. (38) What temperature observations are made at each? (39) What instruments are used? (40) Where are these instruments located? (41) Where is the shelter located? (42) Distinguish between average, mean, and normal. (43) How are normal hourly temperatures computed? (44) How is the average daily temperature computed? (45) How are the average and normal yearly temperatures computed? (46) What is meant by station normals of temperature? (47) How may these be represented graphically? (48) What are thermo-isopleths. (49) How is the graph which represents the daily variation found? (50) Explain the time of occurrence of the highest temperature. (51) Describe the annual variation in temperature. (52) Describe the irregular variation in temperature. (53) What temperature data beside normals may be computed? (54) Define variability of temperature. (55) How are normal temperatures computed for stations where the record is a short one? (56) State the differences in temperature in a thermometer shelter placed at different altitudes. (57) Describe the temperature differences over a limited area during the day. (58) Describe the temperature differences over a limited area at night. (59) What observations are used for the construction of isothermal charts? (60) How are isothermal charts constructed? (61) State the characteristics of the isothermal lines for the year. (62) What are the causes of ocean currents and their direction? (63) Illustrate the effect of ocean currents on isothermal lines. (64) Explain the characteristics of the annual isothermal lines. (65) State the characteristics of the isotherms for January and July. (66) Describe the migration of the hot belt. (67) What is meant by poleward temperature gradient? (68) What are its characteristics? (69) Define thermal anomaly. (70) What are its characteristics? (71) What are the characteristics of the annual range of temperature? (72) In what regions of the world have the highest and lowest temperatures been observed? (73) In what parts of the United States have the highest and lowest temperatures been observed? (74) Name other temperature charts which can be constructed. (75) Describe polar temperatures during the summer and during the winter. (76) Why is the north pole the coldest place in the northern hemisphere during the summer? (77) Why is north central Siberia colder than the north pole in winter? (78) Describe the temperature of the ocean. (79) State the diurnal variation of this temperature. (80) Describe the annual variation of this temperature. (81) Does the ocean water freeze? (82) Describe the temperature changes which take place in the water of a lake between summer and winter. (83) Describe the condition of a river as regards temperature during the summer and in winter. (84) To what depths does the daily variation of temperature penetrate? (85) To what depths does the annual variation of temperature penetrate? (86) What is meant by the layer of invariable temperature? (87) What evidences are there of its existence?   (88) What is the temperature of the ground below the invariable layer of temperature?   (89) To what depths have observations been made? (90) Upon what does the temperature of the water of a spring depend?

TOPICS FOR INVESTIGATION

(1)   The early history of the thermometer.

(2)   The details of the methods used in constructing thermometers in large numbers.

(3)   The inaccuracies in determining a temperature and the methods of eliminating them.

(4)   Thermometer shelters — their construction, use, and accuracy.

(5)   Thermometers for determining the temperature of the earth.

(6)   Thermometers for determining the temperature of deep water.

(7)   The methods of computing the average daily temperature and their accuracy.

(8)   Temperature differences over a limited area.

(9)   Ocean currents.

(10) Abnormalities of temperature and seasons at various places.

(11) Ocean temperatures and the thermometers used in determining them.

(12) Lake temperatures.

(13) The te perature below the earth's surface.

PRACTICAL EXERCISES

(1) Draw the three thermometer scales side by side and of such a length that the fixed points fall together.   Number the points of division so that corresponding temperatures to the nearest degree can be read off.

(2) If physical apparatus is available, study critically one or two good thermometers.   Determine their fixed points; test the uniformity of the bore ; determine the sluggishness; determine the effect of sudden changes in temperature, etc.

(3) Compare the indications of thermometers of several different forms in the open, thermometers in a thermometer shelter, a sling thermometer, and a ventilated thermometer under the most varied conditions.

(4)   Determine the transmission coefficient of the atmosphere on several different days by means of the black bulb thermometer in vacuo.

(5) Determine the inaccuracies and behavior of a thermograph by checking its indications by means of a thermometer.

(6) Compute the station normals of temperature for some station and represent them graphically.

(7) Determine how well the different methods of computing an average daily temperature agree.

(8) Plot the daily and annual variation in temperature for several stations in different parts of the world and in each case explain the characteristics of the graph.

(9) Work up all or some of the temperature data mentioned in section 79 for several stations.   Those stations may be chosen in which the student has a particular interest.

(10) Contrast two near-by stations.   One should be chosen in a large city and the other in the country near by.

(11) Compare the observations made at the base and top of some tower.

(12) Investigate the limited area surrounding a given station for temperature differences.

(13) If the observations can be obtained, construct temperature charts showing for a certain country or for the world some of the temperature data mentioned in section 79.

(14) Determine the temperature behavior of certain springs during the year and then determine the depth of the layer through which the water must come.

(In the case of nearly half of the above problems, if the results were carefully worked out, they would be worthy of publication in some meteorological magazine.)

REFERENCES

For the history, description, illustration, and use of apparatus for determining temperature, see:

ABBE, *Meteorological Apparatus and Methods*, Washington, 1888. Pages 11 to 107 treat of thermometers and thermometry.

BOLTON, HENRY C., *Evolution of the Thermometer*.

The apparatus catalogues of such firms as:

Henry J. Green, 1191 Bedford Ave., Brooklyn, N. Y.
Julien P. Friez, Belfort Observatory, Baltimore, Md.
Queen & Co., 8th and Arch Sts., Philadelphia, Pa.
Negretti & Zambra, 38 Holborn Viaduct, London.
James J. Hicks, 8-10 Hatton Garden, London, E. C., England.
C. F. Casella & Co., 11-15 Rochester Row, Victoria St., London, S. W.
R. Fuess, Düntherstrasse 8, Steglitz bei Berlin, Germany.
Wilh. Lambrecht, Göttingen, Germany.
Max Kohl, Chemnitz, Germany.

For the exposure and care of thermometers, see:

HAZEN, HENRY A., Thermometer Exposure, Professional Papers of the Signal Service, No. XVIII, 1885.

Instructions for coöperative observers, U. S. Weather Bureau.

See also the various guides for observers mentioned in Appendix IX in group (2) B.

For meteorological observations, usually somewhat summarized, consult:

(1) The serial publications of the U. S. Weather Bureau.
　　(*A*) Daily Weather Map.
　　(*B*) National Weather Bulletin, weekly during the summer, monthly during the winter.
　　(*C*) Climatological Reports. These were issued monthly at 44 section centers until July, 1909. Since then they have been combined with Monthly Weather Review.
　　(*D*) Weather Bulletins, weekly during the summer at 44 section centers. (Discontinued in 1908.)
　　(*E*) Snow and Ice Bulletins, weekly during the winter.
　　(*F*) Monthly Weather Review with annual summary and index.
　　(*G*) Mount Weather Bulletin.
　　(*H*) Annual Report of the Chief of the Weather Bureau.
(2) The periodical publications of the weather bureaus of the various countries. For a list of these see BARTHOLOMEW'S *Physical Atlas*, Vol. III (Atlas of Meteorology).
(3) The serial publications of many private stations and observatories. For example: Meteorological Observations of the Massachusetts Agri-

cultural Experiment Station, published monthly since January, 1889; Observations at Blue Hill Observatory by Professor A. L. Rotch, published since 1886 in the *Annals of the Harvard College Observatory*.
(4) Scientific magazines and special publications. The best way to locate these is to consult some digest of meteorological literature; as, *Fortschritte der Physik* (part three).

For temperature normals for various places, see:

BUCHAN, ALEXANDER, *Report on Atmospheric Circulation*.

HANN, *Lehrbuch der Meteorologie*.

HANN, *Handbuch der Klimatologie*.

VAN BEBBER, *Handbuch der Meteorologie*.

Report of the Chief of the Weather Bureau, particularly for 1891–1892, 1896–1897, 1897–1898, 1900–1901, 1901–1902.

Temperature and Relative Humidity Data, Bulletin O, U. S. Weather Bureau, WILLIAM B. STOCKMAN.

Climatology of the United States, Bulletin Q, U. S. Weather Bureau, ALFRED JUDSON HENRY.

The Daily Normal Temperature and the Daily Normal Precipitation in the United States, Bulletin R, U. S. Weather Bureau, FRANK H. BIGELOW.

Report on the Temperatures and Vapor Tensions in the United States, Bulletin S, U. S. Weather Bureau, FRANK H. BIGELOW.

Summary of the Climatological Data for the United States by Sections (106 are to be issued).

For isothermal and climatological charts, see:

BARTHOLOMEW, *Physical Atlas*, Vol. III (Atlas of Meteorology), Prepared by Bartholomew and Herbertson and edited by Alexander Buchan, 1899.

BUCHAN, ALEXANDER, *Report on Atmospheric Circulation* (Report on the scientific results of the voyage of H. M. S. *Challenger*).

HANN, *Atlas der Meteorologie*, 1887, A section of the Berghaus Atlas, but can be bought separately.

HILDEBRANDSSON, H. H., ET TEISSERENC DE BORT, *Les bases de la méteorologie dynamique* (2 vols. have appeared), Paris, 1900–1907.

Summary of International Meteorological Observations, Bulletin A of the U. S. Weather Bureau.

ELIOT, SIR JOHN, *Climatological Atlas of India*, Edinburgh, 1906.

Russia, *Atlas climatologique de l'empire de Russie*, St. Petersburg, 1900.

BLODGET, LORIN, *Climatology of the United States*, Philadelphia, 1857.

Isothermal Lines for the United States, 1871–1880, by A. W. GREELY.

Professional Papers of the Signal Service, No. II, 1881.

GREELY, GEN. A. W., *American Weather*, New York, 1888.

Report of the Chief of the Weather Bureau, particularly for 1896–1897, 1897–1898, 1900–1901, 1901–1902.

Climatic Charts of the United States, U. S. Weather Bureau, Washington, D. C., 1904 (W. B. 301).

Climatology of the United States, Bulletin Q, by A. J. HENRY, 1906 (W. B. 361).

For temperature differences over a limited area, see:

Monthly Weather Review:
   July, 1905, XXXIII, p. 305.
   August, 1906, XXXIV, p. 370.
   August, 1908, XXXVI, p. 250.

CHAPTER IV

THE PRESSURE AND CIRCULATION OF THE ATMOSPHERE

A. THE OBSERVATION AND DISTRIBUTION OF PRESSURE

B. THE OBSERVATION AND DISTRIBUTION OF THE WINDS

112

A. THE OBSERVATION AND DISTRIBUTION OF PRESSURE

The Determination of Atmospheric Pressure

98. **Atmospheric pressure.** — The second meteorological element to be considered is the pressure of the atmosphere. We are probably less conscious of atmospheric pressure and its changes than of any of the other meteorological elements. It is true that great changes in atmospheric pressure do produce marked physiological effects, but we are absolutely unconscious of the ordinary changes in atmospheric pressure from day to day. In meteorological work and weather forecasting, however, the pressure, with the possible exception of temperature, is the most important of the elements.

Pressure is an important meteorological element.

The atmosphere has mass and is acted upon by gravity, and thus possesses weight and exerts a downward pressure. The pressure of the atmosphere is simply the weight of the column of air above the station in question, extending to the limits of the atmosphere. Atmospheric pressure thus diminishes with elevation above the earth's surface because there is a less quantity of air to exert a downward pressure. Since the atmosphere is a gas, the pressure, as in the case of all fluids, is exerted in every direction. If there were no temperature differences and thus no winds, the pressure would be the same at all points on a level surface, as for example, at all points on the hydrosphere. This is not the case, however, and the pressure is different at different points at the same level and is also constantly changing at the same station. It is desirable, therefore, to have instruments for determining the pressure of the atmosphere.

The atmospheric pressure is the weight of the atmosphere.

Atmospheric pressure not constant.

The instrument for determining atmospheric pressure is called a barometer;[1] and there are two kinds of barometers, those employing a fluid and those without fluid. Since mercury is the fluid ordinarily used, such barometers are usually called mercurial barometers. The two kinds of barometers. Barometers which do not use a fluid are called aneroid[2] barometers. The pressure of the atmosphere might be expressed in poundals per square foot, or in dynes per square centimeter. As a matter of fact, however, it is expressed in terms of The pressure is expressed in terms of inches of mercury. the length of an equivalent or balancing mercury column. A pressure of thirty inches thus means that the pressure of the atmosphere is the same as the pressure exerted by a column of mercury thirty inches long.

99. **Mercurial barometer: history, construction, corrections.** — The history of the mercurial barometer[3] dates from a series of experiments made by Torricelli in 1643. It was a remark of Galileo Torricelli's experiment in 1643. Galilei of Pisa, the father of experimental science, when it was called to his attention that water would not rise in a pump more than eighteen cubits above the level of a well, that nature probably did not abhor a vacuum above that height, which attracted the attention of Torricelli, who was then his pupil and later his successor in the chair of Philosophy and Mathematics at Florence. Not being satisfied with this explanation, he instituted a series of experiments which led to the invention of the barometer in 1643. His most famous experiment, as pictured in Fig. 48, consisted in filling a glass tube The invention of the barometer. more than thirty inches long with mercury, covering the open end, and then inverting it over a vessel containing mercury. When the open end was uncovered, the mercury immediately fell to a height of about thirty inches regardless of the length of the glass tube. Torricelli's explanation was that it was the pressure of the atmosphere which was supporting the column of mercury. This explanation received full confirmation a few years later, in 1648, when Pascal persuaded his brother-in-law Perrier to ascend the Puy de Dôme near Clermont with a Torricellian barometer. The diminution in length of

Fig. 48. — Torricelli's Experiment.

[1] βάρος = weight; μέτρον = measure.   [2] ά = without; νεφός = fluid.

[3] For an historical account of the barometer, see *Quarterly Journal of the Royal Meteorological Society*, No. 59, July, 1886, p. 131; or *Meteorologische Zeitschrift*, 1894, p. 445.

the mercury column during the ascent supplied final proof that it was the pressure of the atmosphere which was supporting the column of mercury.

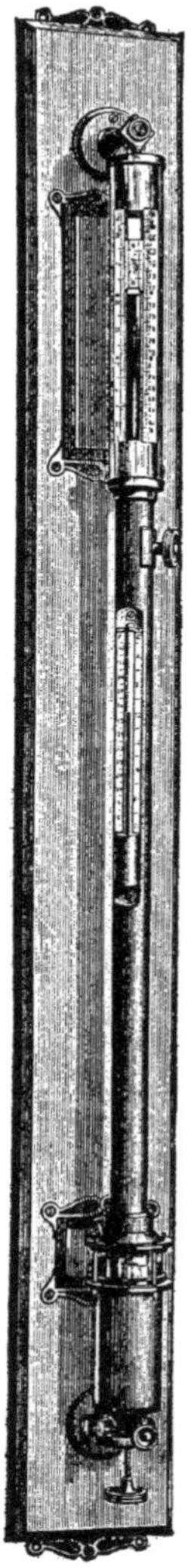

100. A mercurial barometer of the Fortin form as used at the present time consists essentially of a glass tube more than thirty-four inches long, filled with mercury, and inverted over a vessel containing mercury and called the cistern. The mercury used in the tube must be pure, and it must have been previously boiled in order to extract all air and moisture. An air trap is often inserted in the tube to prevent the ascent of any air or moisture into the Torricellian vacuum above the mercury column. The whole is inclosed in a brass case for protection, in order to make it somewhat portable, and to enable it to be suspended vertically. The brass case is cut away at the top, exposing the glass tube containing mercury, and is provided with a scale and vernier for reading the height of the mercury in the tube.

Description of a mercurial barometer.

A barometer is usually suspended from the top and held vertically by means of a ring with three set screws at the bottom. The back board is provided with two translucent windows at the bottom and top to illuminate the cistern and top of the mercury column, particularly at night, when a light is placed back of them. Figures 49 and 50 represent the barometer as a whole and a sectional view.

The cistern is made up of a glass cylinder F, which allows the surface of the mercury q to be seen, and a top plate G, through the neck of which the barometer tube t passes, and to which it is fastened by a piece of kid leather, making a strong but flexible joint. To this plate, also is attached a small ivory point h, the extremity of which marks the commencement or zero of the scale above. The lower part, containing the mercury, in which the end of the barometer tube t is plunged, is formed of two parts i, j, held together by four screws and two divided rings. To the lower piece j is fastened the flexible bag N, made of kid leather, furnished in the middle with a socket k, which rests on the end of the adjusting screw O. These parts, with the glass cylinder F, are clamped to the flange B by

Detailed description of the cistern of a barometer.

FIG. 49.
Standard
Barometer.

means of four long screws P and the ring R; on the ring R screws the
cap S, which covers the lower parts of the
cistern, and supports at the end the adjusting screw O. G, i, j, and k are of boxwood;
the other parts of brass or German silver.
The screw O serves to adjust the mercury
to the ivory point, and also, by raising the
bag, so as to completely fill the cistern
and the tube with mercury, to put the instrument in condition for transportation.

A thermometer is attached to the middle
of the barometer to indicate the temperature of the instrument. A good barometer
costs from $30 to $200.

101. There are two steps in reading a
barometer. In the first place the screw at
the bottom of the instrument must be
turned until the surface of the **The two**
mercury in the cistern has been **steps in**
brought to the end of the scale or **reading a**
barometer.
to the end of the ivory point
which serves as the end of the scale. This
can be done very exactly because the ivory
point is reflected in the mercury in the
cistern. The surface is raised until the
point and its image appear just to touch.
The vernier is then set tangential to the
top of the mercury column and the scale
reading observed. There are three corrections to be applied to this reading of the
length of the mercury column: (1) *the
meniscus correction;* (2) *the tem-* **The three**
perature correction; (3) *the grav-* **corrections**
ity correction. Since mercury **to be**
applied.
does not wet glass, capillary action will depress the mercury column and
give it a rounded top, called the meniscus,
as shown in Fig. 51. In reading a barometer the zero of the vernier is placed
tangential to the upper surface of the

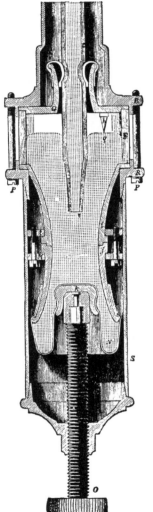

FIG. 50. — Cross-section of the
Cistern of a Barometer.

meniscus. A correction must thus be applied in order to determine what the reading would be if capillary depression did not exist, and the mercury column were cut square across instead of rounded. This correction is usually applied by the maker by moving the scale the proper amount. If this has not been done, the correction is usually combined with the temperature correction and furnished by the maker of the instrument as a table of corrections.

The meniscus correction.

The pressure inside a warm room is the same as out of doors, whatever the temperature may be there. Ordinary buildings are not sufficiently air-tight to permit differences of pressure between the inside and the outside to exist for more than a few moments. If a barometer were taken from a warm room into the cold outside air, both the mercury and scale would contract and the reading would become different, although the pressure would be the same. It is necessary, therefore, to reduce all readings of a barometer to a standard temperature. 32° F. or 0° C. are considered standard temperatures, and the accompanying table gives the corrections to be applied to a mercury barometer with a brass scale for various temperatures and pressures.

The temperature correction.

Fig. 51. — The Meniscus.

Temp. F. Pressure in Inches	0°	10°	20°	30°	40°	50°	60°	70°	80°	90°	100°
26	+ 0.068	+ 0.044	+ 0.020	− 0.003	− 0.027	− 0.050	− 0.074	− 0.097	− 0.121	− 0.144	− 0.167
27	+ 0.070	+ 0.046	+ 0.021	− 0.003	− 0.028	− 0.052	− 0.077	− 0.101	− 0.125	− 0.150	− 0.174
28	+ 0.073	+ 0.047	+ 0.022	− 0.003	− 0.029	− 0.054	− 0.080	− 0.105	− 0.130	− 0.155	− 0.180
29	+ 0.076	+ 0.049	+ 0.023	− 0.004	− 0.030	−0.056	− 0.082	− 0.109	− 0.135	− 0.161	− 0.187
30	+ 0.078	+ 0.051	+ 0.024	− 0.004	− 0.031	− 0.058	− 0.085	− 0.112	− 0.139	− 0.166	− 0.193
31	+ 0.081	+ 0.053	+ 0.024	− 0.004	− 0.032	− 0.060	− 0.088	− 0.116	− 0.144	− 0.172	− 0.200

The value of gravity is not the same at all points on the same level surface. Thus a column of mercury of the same length would The gravity not give the same pressure at all points on a level surface. It correction. is necessary, therefore, to reduce the readings to a standard value of gravity. The value of gravity at 45° north latitude is considered standard, and the accompanying table gives the corrections to be applied for various latitudes.

Latitude	90°	80°	70°	60°	50°	40°	30°	20°	10°	0°
		+ 0.07		+ 0.04		− 0.01		− 0.06		− 0.08 inches
Correction	+ 0.08		+ 0.06		+ 0.01		− 0.04		− 0.07	

A good mercury barometer will indicate the pressure accurately to the hundredth of an inch, and will give a fair approximation to the thousandth. There are, however, several sources of error which Accuracy depend upon the accuracy of construction. Some of these and sources are the accuracy of the scale, the correct adjusting of the ivory of error. pointer, the purity of the mercury, the excellence of the vacuum above the mercury column, etc.

There are various forms of mercury barometers. A modified form is sometimes used on shipboard; and other special forms, such as the siphon barometer, and others, have been devised. For a full treat- Modified ment of these the reader must be referred, however, to special forms of the treatises on this subject.[1] barometer.

102. The aneroid barometer. — The aneroid barometer, as the name implies, is a fluidless barometer, and is thus much more portable than the mercury barometer. It was invented by Vidi in 1848. It Description consists essentially of a so-called vacuum box about an inch of an aneroid barom-and a half in diameter and one quarter inch thick, made of oid barom-German silver with corrugated top and bottom. The air has eter. been exhausted and it is hermetically sealed. It is kept from collapsing by a strong leaf spring which extends over the vacuum box. If the pressure increases, the box is pressed together against the action of the spring; and conversely, if the pressure decreases, the elasticity of the spring causes the box to expand slightly. These small motions are magnified by a system of levers and communicated to a pointer which moves over a dial which is graduated to inches to correspond to a mercury barometer. Figures 52 and 53 represent an aneroid barometer and its internal construction.

[1] Both the English and metric barometric scales are used. Sometimes both are put on the same barometer. In Appendix III a graphical comparison of the two scales is given.

FIG. 52. — An Aneroid Barometer.

FIG. 53. — The Internal Construction of an Aneroid Barometer.

The meniscus and gravity corrections do not exist in connection with the aneroid barometer.   The instrument is, however, slightly affected by temperature.   In order to compensate for temperature changes a small amount of air is often left by the maker in the vacuum box, and such an instrument is usually marked with the word "compensated" on its face. How the three corrections are taken account of.

An aneroid barometer is at best an inaccurate instrument as compared with the mercurial barometer.   The great advantage lies in its portability.   If it is not jarred unduly and is frequently compared with a mercury standard, its indications can be trusted to a tenth of an inch and may give a fair approximation to the hundredth.   The cost of a good aneroid barometer varies from $5 to $40.   The words fair, storm, change, rain, very dry, and the like, often found on the face of an instrument, are meaningless. The accuracy of an aneroid.

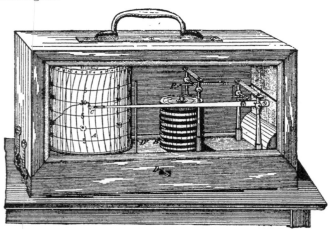

FIG. 54. — The Richard Frères Barograph.

103. Barographs. — For many purposes it is desirable to have a continuous record of barometric pressure and the instrument for keeping this continuous record is called a barograph.   The Richard Frères form of barograph (Fig. 54) is the one ordinarily used by the U. S. Weather Bureau and the one commercially on the market.   It consists of a battery of from six to ten of the vacuum boxes of the aneroid barometer placed one above the other.   The Description of the Richard Frères barograph.

reason for the large number of vacuum boxes is to lessen the effect of the irregularities in any one, and to make the instrument more sensitive. The motion of these vacuum boxes is communicated by a system of levers to an arm which carries the V-shaped trough which contains the non-freezing glycerine ink. As the pressure changes this pen moves up and down. The details in the construction of the pen and recording mechanism have been fully described in connection with the thermograph in section 68.

There are other more accurate forms of barographs in use for the purpose of research and in special observatories. The reader must again be referred to special treatises on the subject for the full description of these.

104. Other so-called barometers. In addition to the barometers just described, there are two instruments, usually called barometers, which deserve a passing mention. One is hardly more than a scientific toy and the other merely masquerades under the name of barometer. The one, to translate its German name, is called a "mouth-barometer" and, as represented in Fig. 55, consists of a bulb usually filled with a colored liquid which does not readily evaporate. The bulb is attached to a graduated stem with a moderately small bore. The fluid does not entirely fill the bulb, but an air space is left and the stem protrudes into the bulb in such a way that the air which constitutes this bubble cannot enter the stem in any position of the instrument. The size of the air bubble and thus the height of the fluid in the stem depends upon the temperature and pressure. If the temperature could be kept constant, the height of the fluid in the stem would depend upon the pressure alone and the instrument would thus be a barometer. Now the temperature of the human body is remarkably constant, and if the bulb of the instrument is held in the mouth, it can be assumed that the temperature is always the same. Such an instrument is cheap and very portable, and, when calibrated in terms of a mercurial barometer, it will give results comparable with those obtained with an aneroid barometer.

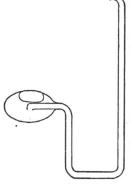

FIG. 55. — The Mouth-Barometer.

The instrument which merely masquerades under the name of barometer is usually designated by its makers as "the cottage barometer" or " the signal service barometer " or sometimes as " the chemical weather glass." It is a sealed glass tube about six or eight inches long and half an inch in diameter filled with a clear liquid which has in it a flocculent, sometimes partly crystalline, substance. The amount of this substance, its appearance, and position in the tube are supposed to indicate the coming weather; and it is, of course, advertised as an infallible guide to the coming weather. It is usually mounted in the same case with a thermometer and sold for 50 cents or much less.   The actual composition and action of the instrument is this : The clear fluid is nearly always alcohol and the substance in it consists of equal parts of nitrate of potash, camphor, and ammonium chlorid. More of this mixture has been added than the alcohol can dissolve and the excess appears in the tube as the solid substance in the clear liquid. It is a thick tube and is hermetically sealed so that it is not affected in the least by changes in pressure and thus is no barometer at all.   It is affected simply by temperature changes.   As the temperature rises, more of the substance goes into solution.   As the temperature falls, more of the substance must come out of solution and appear as solid.   Now the rapidity of temperature changes, unequal temperatures on different sides of the tube, the direction and the amount of the light falling upon it, may make a difference in the form, amount, and position of the substance which comes out of solution as the temperature drops, and this probably accounts for the varied appearauces of the instrument.   Now, temperature changes alone are no indicator of the kind of weather that is coming or of the characteristics of a coming storm.   Thus this instrument is neither a barometer nor an indicator of the coming weather.

The description of the composition and action of the so-called chemical weather glass.

THE RESULTS OF OBSERVATION

105. **The observations.** — At the regular stations of the U. S. Weather Bureau the atmospheric pressure is determined at 8 A.M. and 8 P.M. by means of a mercurial barometer, and a continuous barograph record is also kept.   The instruments used for taking these observations are a good mercurial barometer and a Richard Frères barograph.   These instruments are located ordinarily in the office part of the Weather Bureau Station and not in the thermometer shelter or in the open. They are more conveniently and safely located within the building, and the atmospheric pressure is the same within the build-

The observations taken at the Weather Bureau stations and the instruments used.

ing as out in the open. No building is sufficiently air-tight to permit differences of pressure to exist for more than a few moments except perhaps in the case of a tornado. No pressure observations are required of the coöperative and special stations.

106. Normal hourly, daily, monthly, and yearly pressure. — These normals of pressure are computed in exactly the same way as the corresponding temperature normals. In order to obtain a good normal hourly pressure, a barograph record for at least twenty years is necessary. If, then, for any given date the pressures at any given hour for the last twenty years are averaged, the result would be the normal pressure at that hour for the given day. In this way the normal pressure for every hour of every day in the year might be determined. As a matter of fact, the days in a month are usually grouped together, and thus the normal hourly pressures for a January day or a February day, etc., are determined. If the average pressure for a day is to be determined, the barograph record is almost always used. One half of the pressure observed at 8 A.M. and 8 P.M., one third of the pressure observed at 7 A.M., 2 P.M., and 9 P.M., would give a rough approximation to the average daily pressure. Normals of pressure are not ordinarily computed for any station where a barograph record has not been kept for several years.

The method of computing pressure normals.

107. Diurnal, annual, and irregular variation. — If the normal hourly pressures are plotted to scale, the resulting graph represents the diurnal variation in pressure. If these normals are not available, a fairly good idea of the characteristics of the diurnal variation can be formed by considering the change in pressure on some day when the other meteorological elements have been as normal as possible. Figure 56 illustrates a series of days during which the remaining elements remained unusually normal. The general characteristics of the daily variation are these: the chief maximum usually occurs at about 10 in the morning, the chief minimum at 4 in the afternoon, a secondary maximum at 10 in the evening, and a secondary minimum at 4 in the morning. The pressure is thus subject to a double oscillation in the course of a day. The amplitude or amount of the daily variation is always small, never amounting to more than 0.2 of an inch, and in many places being often less than a tenth of this. It varies somewhat with the time of year, being usually greater in summer and somewhat less in winter. It varies also with latitude, its greatest values being found in the equatorial regions, while the amount grows steadily less with the higher latitudes. It is

Description of the daily variation in pressure.

also somewhat less on cloudy days than on days with plenty of sunshine. It is also somewhat greater for interior stations than for coast or island stations.

The amount varies with the season, latitude, location, etc.

For interior stations the secondary maximum and minimum become less prominent, while for coast and island stations the two are of equal prominence, or the night maximum and minimum may even become the most important. Elevation also plays an important part in the time of occurrence of the maximum and minimum.

Approximate values of the amplitude or amount of the daily variation for various places are here given: Calcutta, India, lat. 24°, 0.116; Greenwich, lat. 52°, 0.020; Dublin, 0.020; St. Petersburg, 0.012; Fort Conger, lat. 83°, 0.010; Yuma, Arizona, 0.129; San Antonio, 0.117; Denver, 0.079; Albany, 0.074; St. Louis, 0.068; Philadelphia, 0.061; San Francisco, 0.052; Bismarck, 0.038; Sitka, 0.014. Figure 57 represents graphically the daily variation at Mexico City, San Francisco, St. Louis, New York, and Sitka,

The amount and characteristics of the variation at different places.

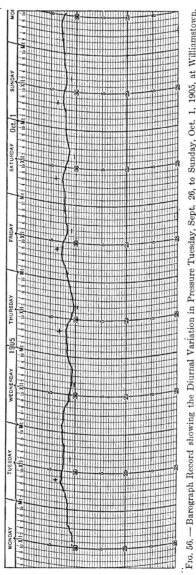

Fig. 56. — Barograph Record showing the Diurnal Variation in Pressure Tuesday, Sept. 26, to Sunday, Oct. 1, 1905, at Williamstown, Mass. The secondary maxima and minima at 10 P.M. and 3 A.M. are not discernible. The chief maxima and minima are marked with a plus and minus sign respectively.

Alaska. By contrasting the values here given for various places, the truth of the foregoing statements as to the amount of the oscillation may be tested.

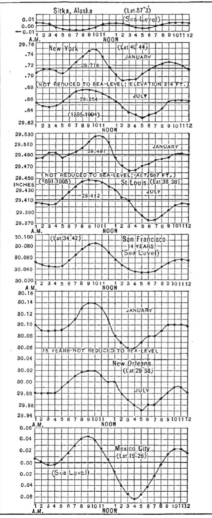

FIG. 57. — The Diurnal Variation in Pressure at Sitka, New York, St. Louis, San Francisco, New Orleans, and Mexico City.

Since the local time of occurrence of the maxima and minima is approximately the same for all places, the diurnal changes in barometric pressure may be thought of as produced by waves of higher and lower pressure which move westward from the Atlantic Ocean, cross the continent, and pass off into the Pacific Ocean. The location and height of these waves have been computed by Dr. Oliver O. Fassig for each hour of the day for the western hemisphere. The results were found by using the daily variation in pressure, which had been determined from observations at many stations in both North and South America, and drawing lines through those places which at the hour in question showed the same departure from the normal for the day. These charts will be found in Bulletin No. 31 of the U. S. Weather Bureau

The change in pressure considered as caused by moving waves of higher and lower pressure.

or in the Monthly Weather Review for November, 1901. Two of them, for 10 A.M. and 4 P.M., the times of greatest and least pressure, are reproduced as Figs. 58 and 59. The numbers here represent departure from the normal for the day expressed as thousandths of an inch.

108. The cause of this diurnal variation in pressure is not a tide in the atmosphere caused by either the sun or the moon, for if it were both the solar and lunar influence would be noticed and the corresponding periods detected.

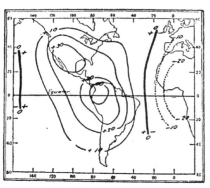

Fig. 58. — Diurnal Barometric Wave at 10 A.M., 75th Meridian Time.

(Fassig, U. S. Weather Bureau).

Temperature and the formation of dew without doubt play a large part in causing this diurnal variation, but the exact way

Temperature change and dew as possible causes.

in which these two operate to produce the result cannot be satisfactorily stated. Due to the temperature change alone, the maximum would be expected at the time of least temperature or a little later, that is, at sunrise or a few hours after. The minimum would be expected at the time of highest temperature, that is, from two to four in the afternoon, or somewhat later. Due to the formation of dew the largest amount of moisture is present in the atmosphere in the late afternoon and the least amount at the time of sunrise. The interaction of these two influences of temperature and dew, however, cannot account for all the characteristics which have been observed in connection with the daily variation.

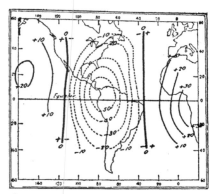

Fig. 59. — Diurnal Barometric Wave at 4 P.M., 75th Meridian Time.

(Fassig, U. S. Weather Bureau.)

By means of harmonic analysis [1] this daily oscillation or variation in pressure may be separated into two components, one with a daily or twenty-four hour period, and another with a half-daily or

Its separation into two components with different periods. twelve-hour period. If this is done for many stations, choosing those in equatorial regions as well as those in higher latitudes, those in the interior of the continents as well as on the seashore or on islands, those located in valleys as well as on mountain sides and on mountain tops, it will be found that the twelve-hour periodic oscillation is remarkably regular and shows essentially the same characteristics everywhere, while the twenty-four hour oscillation proves to be due almost entirely to local causes and is very different at different stations. Some have even gone so far as to ascribe a cosmic cause to this twelve-hour oscillation, that is, to ascribe it to some influence outside of the earth itself. It has also been thought that this twelve-

One period corresponds to nothing in nature. hour oscillation may be simply a free oscillation of the atmosphere as a whole, considering it as an elastic body. This twelve-hour periodic oscillation is nearly as large as the twenty-four hour oscillation, and whenever this is the case and no reason in nature for a twelve-hour period can be found, it is questionable whether this separation into two components can be considered as corresponding to anything in nature, or may not be merely a device of the mathematician.

109. The annual variation in pressure may be found by plotting to scale the normal monthly pressures. If this is done, it will be noted for

The annual variation in pressure and its cause. the interior of continents that the pressure is somewhat higher than the yearly normal in winter and less in summer ; for the oceans the opposite is true, the pressure being somewhat higher in summer and lower in winter. The cause for this is not far to seek. During the winter, as was seen in connection with temperature anomalies, the continents are colder than the surrounding oceans. That means that the air is colder, denser, and heavier, and thus an increase in pressure during the winter is to be expected. During the summer the contrary is true. The continents are warmer than the surrounding oceans, the air is expanded and light, and the atmospheric pressure is correspondingly lower. This variation, again, does not amount ordinarily to more than a few tenths of an inch.

The irregular variations in pressure, particularly in the temperate zones, are far larger than any of the periodic variations. Variations in

[1] For an illustration see *Monthly Weather Review*, November, 1906.

pressure of an inch or two follow each other in rapid succession, and at irregular intervals.

110. **Barometric data.**—In addition to the normals described above, but few results are computed for the various stations from the observations of pressure. Among those sometimes computed, the following may be mentioned: (1) *The daily range.* From the barograph record the highest and the lowest pressure for the day may be determined. The difference between these gives the daily range. From these values of range, normal values for the day, for the month, and for the year may be computed in the regular way. (2) *Absolute range for the months and the year.* By the absolute range in pressure for the months and for the year is meant the difference between the very highest and very lowest pressures observed during the period of time in question. (3) *Frequency of irregular variations.* The interval of time between successive, marked, irregular variations in pressure may be determined from the records of the barograph, and the normal frequency of these irregular fluctuations can thus be determined for the various months and for the year as a whole. (4) *Magnitude of irregular fluctuations.* The magnitude of each irregular fluctuation can be determined. If these are averaged in the regular way, the normal amount of the irregular fluctuations for the various months and for the year as a whole may be determined.

Barometric data which may be computed from the observations of pressure.

None of these results is of any great interest or of far-reaching practical importance.

At the regular stations of the U. S. Weather Bureau there are three tables for pressure which are kept up to date. These contain :

(1) Highest and lowest in inches and hundredths (reduced to sea level) (the data are given for each month and the year as a whole).

(2) Mean station and absolute monthly range (inches and hundredths) (the data are again given for each month and the year as a whole).

(3) Mean hourly pressure (inches and hundredths).

The Variation with Altitude

111. **Reduction to sea level.** — Since the pressure of the atmosphere decreases with elevation, in order to compare pressures observed at different stations, it is necessary to take account of the elevation. *The old way.* This was formerly done by determining the difference between the observed pressure and the normal yearly pressure for the sta-

K

tion in question. These differences could then be compared and thus the conclusion reached as to which place had the higher or lower pressure.

At the present time all pressure observations are reduced to the same level; that is, to what the pressure would be if the observation had been made at sea level. In order to make this correc-

The modern way is to add the weight of the column of air reaching down to sea level. tion, the weight of a column of air reaching from the station in question to sea level must be added to the observed pressure. Now the weight of this column of air is not a constant, but varies with the pressure, with the temperature, and slightly with the moisture in it. If the pressure is high, then the air is dense and heavy. If the temperature is high, it is correspondingly expanded and light. Moist air is lighter than dry air, when other conditions are the same. In order to determine the value of this correction, elaborate tables are ordinarily used. The best set of tables is probably the Smithsonian Meteorological Tables, published at Washington and carrying the number 1032 in the Smithsonian Miscellaneous Collections. In this volume are contained the necessary tables for reducing barometric pressure to sea level, together with the explanation and derivation of the formulas used. In Appendix IV a short table is given, but this is intended not to serve for the reduction of well-taken observations, but simply to give a rough idea of the amount of the correction in a few instances.

In a certain sense this reduction to sea level is fictitious. The mass of a column of air reaching down to sea level is added. The mass of this column of air is quite different from what it would be if there were no mountain or plateau and the station were actually located at sea level. For this reason, the older method by means of differences from normal might be better, particularly in the construction of weather maps.

112. Barometric determination of altitude. — Since barometric pressure depends upon elevation, it is possible, by observing the barometric pressure at two different, near-by stations to determine their difference of elevation. In order to make a precise determina-

Simultaneous observations of pressure and temperature at the two stations must be obtained. tion, simultaneous observations of pressure and temperature are necessary. The moisture at the two stations is also sometimes determined. The practical details in securing these simultaneous observations will at once suggest themselves.

If two observers and a sufficient number of instruments are available, it may be easily arranged to take the observations at some definitely appointed hour. For the pressure observations at the higher elevation, particularly if the station is located on a

mountain top, the aneroid barometer is ordinarily used.   This is not as accurate as a mercurial barometer, but it is much more portable and convenient to carry.   The temperature is ordinarily deter- mined by means of a sling thermometer.   If a ventilated thermometer is available, it will give far better results.   It is ordinarily too troublesome to construct or to carry up a thermometer shelter for protecting an ordinary thermometer. Practical de- tails in se- curing the observa- tions.
If two observers and the proper apparatus are not available, fair results can be obtained by taking observations at the base station, both before and after the ascent, and then determining by interpolation the proper value of pressure, temperature, and possibly moisture, at the time when the observations were taken at the summit.   If the aneroid barometer is used, it should be compared both before and after the ascent with a mer- curial standard, in order to check its accuracy and to be sure that it has not become deranged by the jars of transportation.

113. Let B_0 and T_0 indicate the pressure and temperature at the base station, that is, at the station whose elevation is known, and let B_1 and T_1 indicate the pressure and temperature at the summit or at the station for which the elevation is to be determined.

The first and simplest method of computing the difference in eleva- tion is to take no account of the temperature and moisture, and allow 90 feet for each tenth of an inch of pressure difference.   It has been found that the general average for all conditions of pressure and of temperature and of moisture is about 90 feet to a tenth of an inch.   Thus, if the difference in pres- sure between two stations is two inches, it means a differ- ence in elevation of approximately 1800 feet.   Whenever aneroid barometers are provided with a separate scale for indicating elevations, it is always divided so that a tenth of an inch corresponds to 90 feet. There are four methods of computing a difference in elevation from the ob- servations made.

	TEMP. F.										
Pressure	0°	10°	20°	30°	40°	50°	60°	70°	80°	90°	100°
22 inches .	111	114	116	119	122	124	127	130	132	135	138
24 . . .	101	104	106	109	111	114	116	119	121	124	126
26 . . .	94	96	98	101	103	105	107	110	112	114	116
28 . . .	87	89	91	93	95	98	100	102	104	106	108
29 . . .	84	86	88	90	92	94	96	98	100	102	104
29.5 . .	83	85	87	89	91	93	95	97	99	101	103
30.0 . .	81	83	85	87	89	91	93	95	97	99	101
30.5 . . .	80	82	84	86	88	90	92	94	96	98	100
31.0 . .	78	80	82	84	86	88	90	92	94	96	98

A better method of computing the difference in elevation is to tak
from the accompanying table the actual number of feet which correspond
to a tenth of an inch difference in pressure for the average pressure and
the average temperature at the two stations.

The third way of computing the difference in elevation, which is perhaps a little more accurate than the use of the table given above is by
means of the following formula : —

$$\text{Difference in elevation} = \frac{B_0 - B_1}{B_0 + B_1} \left\{ 55,761 + 117 \left(\frac{t_0 + t_1}{2} - 60 \right) \right\}$$

This formula, as will be seen, contains nothing but the observations of
pressure and temperature at the two stations.

If the most exact possible use of the observations is to be made in
determining the difference in elevation, the elaborate Smithsonian
Meteorological Tables referred to above must be used. The formula is
there derived, the necessary tables are given and their use explained.

114. Variation with altitude. — The accompanying table shows approximately, for average conditions, the barometric pressure which cor-

The baro-
metric pres-
sure which
corresponds
to various
altitudes.

responds to various elevations. These values, however, correspond to average conditions, and are not exact enough for
the determination of elevation, particularly as the barometric pressure at any given station is constantly changing.

BAROMETRIC PRESSURE	ALTITUDE	BAROMETRIC PRESSURE	ALTITUDE
30 inches	0 feet	21 inches	9,300 feet
29 inches	910 feet	20 inches	10,600 feet
28 inches	1,950 feet	18 inches	13,200 feet
27 inches	2,820 feet	16 inches	16,000 feet
26 inches	3,800 feet	15 inches	3.6 miles
25 inches	4,800 feet	$7\frac{1}{2}$ inches	6.8 miles
24 inches	5,900 feet	$3\frac{3}{4}$ inches	10.2 miles
23 inches	7,000 feet	$\frac{4}{100}$ inches	24.1 miles
22 inches	8,200 feet		

THE DISTRIBUTION OF PRESSURE OVER THE EARTH

115. Construction of isobaric charts. — Isobaric charts are con-

The method
of construct-
ing an iso-
baric chart.

structed by making use of the normals of pressure which
have been determined for various stations in all parts of
the world. These normals of pressure must first be reduced
to sea level, and they are then charted on a map, and lines

are drawn connecting those stations which have the same value. These lines are called isobaric lines, or simply isobars.[1]

116. Isobars for the year. — Chart X shows the isobars for the world for the year. It is the normal annual pressures which have been here used for the construction of the chart. The following charac- *The three* teristics are at once evident: (1) *The equatorial belt of low* *character-* *pressure and the belts of high-pressure at 35° N. and 30° S.* *istics of iso-* *latitude.* It will be noticed that along the equator there is a *year for the* belt of low-pressure, the least pressure being 29.80 inches. *world.* This belt is irregular, of varying width, and not of the same pressure throughout, and it lies somewhat on the north side of the equator. The belts of high pressure are also not of the same pressure throughout. The one at 35° N. has its areas of highest pressure over the Pacific Ocean, the Atlantic Ocean, and central Siberia. In passing over Siberia it lies far north of the equator. The southern belt of high pressure has its peaks of pressure over the Pacific Ocean, the South Atlantic Ocean, and the Indian Ocean. From these two belts of high pressure, the pressure diminishes rapidly toward the poles; in the southern hemisphere quite regularly; in the northern, however, with less regularity. (2) *Regularity in the southern hemisphere.* It will be noticed that the low pressure near the north pole consists of two depressions, one in the North Pacific near Alaska, and the other in the North Atlantic near Iceland, each with a central pressure of 29.70 inches. In the southern hemisphere, however, the pressure drops much more uniformly and rapidly from the belt of high pressure toward the pole. (3) *Lower in southern hemisphere* *than in the northern hemisphere.* It will also be seen that the diminution in pressure is much greater in the southern hemisphere than it is in the northern.

117. Vertical section along a meridian. — A study of the distribution of pressure in a vertical section along a meridian, as shown in Fig. 60, will prove instructive. The north and south poles of the *How the 30-* earth and the equator are indicated by N., S., and E., respec- *inch line is* tively. At the equator, in order to reach a barometric pres- *determined.* sure of 30 inches, it is necessary to go a certain distance below sea level, and the same is true at the north and south poles. At 35° N. and 30° S., it is necessary to ascend a certain distance above the earth's surface in order to find the 30-inch line. The heavy line marked 30 in the diagram thus indicates in the vertical section those places which would have a pressure of 30 inches. In order to locate the 29-inch line, it is necessary

[1] From the Greek: ἴσος = equal; βάρος = pressure.

to ascend a certain distance above the 30-inch line, on the average about
900 feet. At the equator, the air is warm and moist, while at the pole

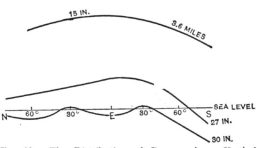

**How the
other lines
are located.** it is cold
and ' com-
paratively
dry. As a result, it
is necessary to ascend
a larger number of
feet at the equator to
obtain the same dimi-
nution in pressure
than at the pole. A
glance at the table in
section 113 will show
approximately the

FIG. 60. — The Distribution of Pressure in a Vertical
Section along a Meridian.

amount. The figures would be about 810 feet at the pole and 990
feet at the equator. As a result, the isobaric lines will show a greater
and greater upward bulging at the equator and drooping at the
poles. By the time the 27-inch line has been reached, the equatorial
depression in the 30-inch line will have practically disappeared, and,
with increasing elevation, the upward bending or convexity of the
isobaric lines becomes more and more pronounced, as is shown in the
figure.

118. **Isobaric surfaces.** — If there were no temperature differences,
the pressure at sea level would be everywhere the same. As it is,
**Tempera-
ture differ-
ences cause
difference in
pressure.** the temperature differences cause the air to be light and
expanded at one place and dense and contracted in
another. As a result, movements of air take place, increasing the pressure at one point and lessening it at another.

By an isobaric surface is meant a surface at every point of which the
pressure is the same. If there were no temperature differences, the
**Definition of
an isobaric
surface.** surface of the hydrosphere would be an isobaric surface,
and all other isobaric surfaces would be concentric with
it and at a certain distance above it. The 29-inch
surface, for example, would be concentric with the hydrosphere and
**The normal
form.** approximately 900 feet above it everywhere. Temperature
differences, however, warp and twist these isobaric surfaces
so that they are no longer parallel to the hydrosphere. Whenever they
are so much warped as to intersect the hydrosphere, their intersection
forms isobaric lines or isobars. In the center of the area of high pressure,

shown in Fig. 61, it is necessary to ascend a certain distance in order to find the 29.9-inch surface, and a slightly greater ascent is necessary in order to reach the 29.8-inch surface. Over such an area of high pressure, then, the isobaric surfaces are warped upward

Their form over areas of high or low pressure.

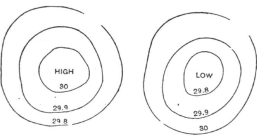

FIG. 61. — Areas of High and Low Pressure.

and have the form of an inverted bowl. Conversely, over an area of low pressure the isobaric surfaces are warped downward and have the appearance of a saucer. If the isobars for the year are interpreted as being the intersections of isobaric surfaces with the hydrosphere, it will be seen that over the equator and at the poles they are warped downward, whereas at 35° N. and 30° S. latitude they are warped upward. From the isobaric lines the isobaric surfaces can always be constructed.

119. Isobars for January and July. — Charts XI and XII show the isobars for the world for January and July. It is the normal pressure for January and for July, corrected for elevation, which have been used in the construction of these charts. The same three characteristics which were noted above in connection with the lines for the year are again evident. (1) *The equatorial belt of low pressure and high pressure belts at 35° N. and 30° S.* (2) *Regularity in the southern hemisphere.* (3) *Lower in the southern than in the northern hemisphere.* In addition, two new characteristics are to be noted. (1) *The belts of high and low pressure no longer remain continuous belts, but break up into peaks and depressions.* Thus, the equatorial belt of low pressure shows depressions over South America, South Africa, and Australia on the January chart, and one very pronounced depression over India on the July chart. This depression over India is so pronounced that it persists in the annual averages and appears on the chart of the isobaric lines for the year. On the January chart the belt of high pressure in the northern hemisphere shows two immense peaks of pressure over North America and Siberia. In each case these is a lip of high pressure extending towards the west and forming really a small area of high pres-

The observations used for the charts.

Three old and two new characteristics are to be noted.

sure over the ocean to the west of the continent. On the July chart
the highs over the continents have disappeared, and two peaks of
pressure over the Pacific and Atlantic oceans have built up
where these small areas were located. The southern belt
of high pressure shows three peaks of pressure over the
South Atlantic, South Pacific, and Indian oceans, and these
peaks are the same on both the January and July charts.
In addition, a small peak of high pressure appears over South Africa
on the July chart. In the north polar regions a marked depression
appears near Iceland on both the January and July charts. A·decided
low appears on the January chart near Alaska, but this does not exist
on the July chart. There are thus eight highs and six lows to be considered. It will be noticed in some cases that the highs are located on
the land in winter and over the oceans in summer. The reason for
these highs and lows will be considered in a later section.

The belts break up into peaks and depressions.

(2) *The belts of pressure migrate somewhat with the sun.* The migration, however, is not as great as the migration of the sun, 47°, or of
the temperature belts, which have been considered before.
The pressure belts also lag behind the sun in their
migration.

The pressure belts migrate.

120. **Other pressure charts.** — In addition to the three charts which
have been given and which represent the isobars for the year, for January, and for July for the whole world, the isobars for the
various months and for the year for various countries, or
for the world as a whole, may be represented. In addition, all of the
barometric data mentioned in section 110, such as the absolute range
for the months and for the year, the frequency of irregular variations,
the magnitude of the irregular variations, etc., may be charted.

Other pressure charts.

B. THE OBSERVATION AND DISTRIBUTION OF THE WINDS

THE DETERMINATION OF THE DIRECTION, FORCE, AND VELOCITY OF THE WIND

121. **Wind direction, force, and velocity.** — Air in motion near the
earth's surface and nearly parallel to it, is called wind. All other motions of masses of air should be spoken of as air currents,
although this distinction is not always recognized. In connection with wind there are three things to be determined
or measured; namely, the direction, the velocity, and the
force or pressure.

Wind defined. Three things to be measured.

The wind is named from the direction from which it comes; thus if air moves from the north toward the south, it is spoken of as a north wind. The direction from which the wind comes is called windward, and the direction toward which it goes is called leeward. Whenever the wind direction changes steadily in a clockwise direction, as, for example, from southeast through the south and southwest to west, it is said to veer; when the direction changes steadily in a counterclockwise direction, as, for example, from east through the northeast and north to northwest, it is said to back. In noting wind direction, eight points of the compass are used; namely, the four cardinal points, north, south, east, and west; and the four intermediary points, northwest, southwest, south-east, and northeast.

Windward and leeward, veering and backing, defined.

Eight wind directions are recognized.

The velocity of the wind and the force or pressure of the wind are so related that when one has been observed the other can at once be determined. The relation between them is such that the pressure, in pounds per square foot, is equal to a constant multiplied by the square of the velocity in miles per hour. Experimental determinations of the value of the constant have given results which vary from .004 to .007. If an average value of the constant is used, the relation may be expressed by the simple formula, $P = .005V^2$. The accompanying table computed from the above formula gives the pressure in pounds per square foot for several wind velocities.

The relation between wind velocity and the pressure of the wind.

Velocity in Miles per Hour	Pressure in Lbs. per Sq. Foot	Velocity in Miles per Hour	Pressure in Lbs. per Sq. Foot
0	0	50	12.50
5	0.12	60	18.00
10	0.50	70	24.50
15	1.12	80	32.00
20	2.00	100	50.00
25	3.12	125	78.12
30	4.50	150	112.50
40	8.00		

The value of the constant depends to a slight extent on temperature, pressure, and moisture. When the temperature is high, or the pressure is low, or the moisture content is large, the air is less dense, and thus the pressure exerted by the moving air would be smaller. Conversely, if the temperature is low, or the pressure is high, or the moisture content is small, the air is denser, and thus the pressure exerted would be greater.

122. Wind vane. — The direction of the wind is determined by means of a wind vane, and wind vanes are too well known to need any further description here. A wind vane is often spoken of popularly as a weather vane, but this is a misnomer, since the wind vane indicates wind direction only, and not the present or future condition of the weather. The Weather Bureau form of wind vane has a tail made up of two thin pieces of wood, making an angle of about 22° with each other. There are two advantages of this form, since the wind vane is more sensitive in a light wind and steadier in a gusty wind. The wind vane is sometimes attached to a vertical rod which extends down into a building below and is there attached, perhaps by means of a series of cog wheels, to a pointer which moves over a dial so that the wind direction can be read within the building.

Wind vane, not weather vane.

The Weather Bureau form of wind vane.

There are instruments for getting the direction of air currents as well as of the wind. These consist ordinarily of a large wind vane which indicates the horizontal direction. To this is attached a second wind vane pivoted about a horizontal axis so as to be free to indicate the vertical component of the air motion.

The direction of air currents.

123. Anemoscope. — The purpose of an anemoscope is to give a continuous record of wind direction. The vertical rod to which the wind vane is attached sometimes reaches down into a room below, and a cylinder is attached to the lower end of the rod. This cylinder turns with the rod and is covered with a piece of paper for obtaining the record. A pen presses lightly against this paper and is carried downward at a slow and regular rate by clock-work, so as to descend through the length of the cylinder in a day or, if so constructed, in a week. With every change in wind direction the cylinder turns underneath the pen, and thus a continuous record of wind direction is made.

The cylinder form of anemoscope.

At regular Weather Bureau stations the wind direction is recorded automatically by attaching a contact maker to the wind vane and its direction is then recorded electrically in the office. The contact maker, as shown diagrammatically in Fig. 62, consists of four blocks lettered *N*, *S*, *E*, and *W*, from which wires extend to four electro magnets connected with the recording drum in the office below. An arm is attached to the vertical rod of the wind vane, and at the end of this arm is a contact shoe which runs over the four contact blocks. The circuit is closed automatically by a clock every minute. If the wind direction is north, the contact shoe

The Weather Bureau form of anemoscope.

will be located over the block marked N, and the corresponding circuit will be closed, and the armature in the office below will print N, or make a dot in the proper place on the revolving drum.   The contact shoe is of such form and size that if the wind direction is northeast, it covers a portion of both the N and E blocks, and thus when the circuit is closed, both N and E are printed on the record or two appropriately placed dots are made.   In this way, an automatic record of the wind direction every minute is kept at Weather Bureau stations. In the actual instrument the contact blocks are not placed horizontally or out in the open. The whole is made compact, usually of cylindrical form, and is carefully protected.   The principle, however, is as described.

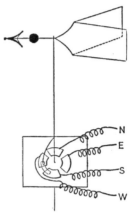

Fɪɢ. 62. — The Electrically Recording Wind Vane of the Weather Bureau.

124. **Velocity estimations.** — Before instruments were in use for the measurement of wind velocity, its value was deter- The Beaumined by the estimation of its fort wind effects.   The first attempt to make scale. definite these estimations was made by Admiral Beaufort of the English Navy in 1805.   He devised the so-called twelve-point wind scale, in which twelve names for different winds were introduced, and these were defined in terms of the amount of sail which a vessel could carry under the different conditions.   Velocities corresponding to those various winds have since been determined experimentally.   From 1805 on, many wind scales were proposed in which the number of winds named and defined varied all the way from twelve to four.   An The tenattempt was made recently by international agreement to point wind adopt a ten-point wind scale.   The accompanying table gives scale. the names of the winds, the average velocity in miles per hour and in meters per second, for the Beaufort twelve-point and this ten-point wind scale.   In order to change from the metric to the English system, it should be remembered that the velocity in miles per hour multiplied by .447 equals the velocity in meters per second; or, conversely, the velocity in meters per second multiplied by 2.237 equals the velocity in miles per hour.

THE TEN-POINT WIND SCALE

SCALE NUMBER	NAME OF WIND	MILES PER HOUR	METERS PER SECOND
0	Calm	0	0
1	Very light breeze	0 to 4.5	0 to 2.0
2	Gentle breeze	4.6 to 9.0	2.1 to 4.0
3	Fresh breeze	9.1 to 13.5	4.1 to 6.0
4	Strong wind	13.6 to 22.5	6.1 to 10.1
5	High wind	22.6 to 31.5	10.1 to 14.1
6	Gale	31.6 to 40.5	14.2 to 18.1
7	Strong gale	40.6 to 49.5	18.2 to 22.1
8	Violent gale	49.6 to 67.5	22.2 to 30.2
9	Hurricane	67.6 to 85.5	30.3 to 38.2
10	Most violent hurricane	85.6 ———	38.3 ———

THE BEAUFORT TWELVE-POINT WIND SCALE

SCALE NUMBER	NAME OF WIND	MILES PER HOUR	SCALE NUMBER	NAME OF WIND	MILES PER HOUR
0	Calm	0	7	Moderate gale	40
1	Light air	3	8	Fresh gale	48
2	Light breeze	13	9	Strong gale	56
3	Gentle breeze	18	10	Whole gale	65
4	Moderate breeze	23	11	Storm	75
5	Fresh breeze	28	12	Hurricane	90
6	Strong breeze	34			

At the present time wind scales are of little advantage, since instruments for determining wind velocity have become so common. **The effects produced by winds of different velocities.** In estimating wind velocities, however, it is of advantage to know about what effects are produced by a wind of a certain velocity. The following table may be of use in guiding one's estimation of wind velocity:

MILES PER HOUR

0   No perceptible movement of anything.

0 to 4   This just moves the leaves of a tree without moving or swaying the branches.

4 to 12   This moves the branches of a tree, and blows up dry leaves and paper from the ground.

12 to 22   This sways the branches of trees, blows up dust from the ground, and drives leaves and paper rapidly before it.

22 to 32   This sways whole trees, blows twigs and small branches along the ground, raises clouds of dust, and hinders walking somewhat.

32 to 72   This breaks small branches, loosens bricks on chimneys, etc., litters the ground with twigs and branches from trees, and hinders walking decidedly.

72 on   This brings about more or less complete destruction of everything in its path.

The adjectives used in describing wind in order are: light, feeble, gentle, mild, fresh, strong, high, heavy, violent. The names used for various winds are often calm, breeze, wind, storm, **Terms used.** gale, hurricane. A weak adjective should not, of course, be used with a strong wind or a strong adjective with a weak wind.

125. Anemometers. — The anemometer,[1] as the name implies, is an instrument for determining the velocity of the wind. The first anemometer was devised in 1667, and since that time anemometers of such varied form and in such **There are many kinds of anemometers, but ordinary ones may be divided into three groups.** large numbers have been devised that it is well-nigh impossible to classify them, to say nothing of describing them here. Those in ordinary use may be divided into three groups; namely, deflection anemometers, pressure anemometers, and rotation anemometers.

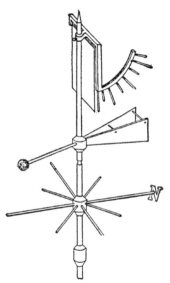

FIG. 63. — A Simple Deflection Anemometer.

(ABBE, U. S. Weather Bureau.)

The simplest deflection anemometer consists of a square board or sheet of metal hinged along its upper **A simple deflection anemometer.** edge and always turned to face the wind by means of a wind vane. As the wind velocity increases, the deflection from the vertical position increases and gives a measure of the pressure, and thus the velocity of the wind. A device for recording the maximum wind velocity can easily be added to this instrument.

A simple pressure anemometer consists of a small board directed against the wind by means of a wind vane and held in position by a

[1] ἄνεμος = wind; μέτρον = measure.

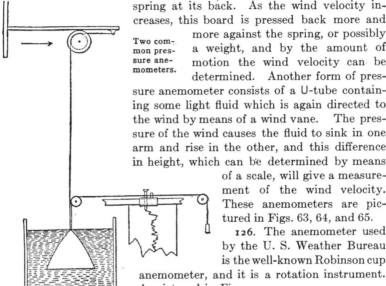

spring at its back. As the wind velocity increases, this board is pressed back more and more against the spring, or possibly a weight, and by the amount of motion the wind velocity can be determined. Another form of pressure anemometer consists of a U-tube containing some light fluid which is again directed to the wind by means of a wind vane. The pressure of the wind causes the fluid to sink in one arm and rise in the other, and this difference in height, which can be determined by means of a scale, will give a measurement of the wind velocity. These anemometers are pictured in Figs. 63, 64, and 65.

Two common pressure anemometers.

126. The anemometer used by the U. S. Weather Bureau is the well-known Robinson cup anemometer, and it is a rotation instrument. As pictured in Fig. 66, it consists essentially of two horizontal arms at right angles to each other, which carry hemispherical cups at the ends of the arms. The pressure of the wind is greater on the concave side of the cup than on the convex side, and thus the cups rotate with the convex side forward. The cross arms which carry the cups are attached to a vertical rod which is connected with cog wheels for recording the number of revolutions of the cups. It is also usually arranged so that contact is made at the end of each mile of wind that passes. It has been found experimentally that the velocity of the motion of the centers of the cups must be multiplied by three in order to obtain the wind velocity,

Robinson's cup anemometer.

Fig. 64. — A Simple Pressure Anemometer.
(From J. W. Moore's *Meteorology, Practical and Applied.*)

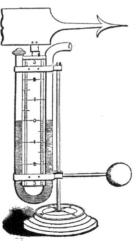

Fig. 65. — Lind's Pressure Anemometer.
(From J. W. Moore's *Meteorology, Practical and Applied.*)

and this factor is used in the construction of the instruments. This is unfortunately, however, not a constant, but it varies with the size of the cups, with the length of the horizontal arms, and with the friction in the instrument. It has been found to vary all the way from 2.2 to 3.1. This constant is determined experimentally by attaching the anemometer to a long arm and whirling it at an even rate of speed. There are several sources of error in this form of anemometer; one of the large ones is the fact that due to inertia, the instrument is apt to run past lulls in the wind, and to gain speed slowly if the velocity of the wind suddenly increases.

The relation of the velocity of the cups to the wind velocity.

Fig. 66. — Robinson's Cup Anemometer.

The constant used by the U. S. Weather Bureau for the anemometer of the size employed is three. All recorded velocities are thus based on this assumption. It has been found that the recorded velocities are nearly correct for wind velocities below 15 miles an hour, but that corrections must be applied for higher velocities. In the accompanying table the corrected velocities are given for all recorded velocities from zero to 90.

	+0	+1	+2	+3	+4	+5	+6	+7	+8	+9
0	..	..	..	..	..	5.1	6.0	6.9	7.8	8.7
10	9.6	10.4	11.3	12.1	12.9	13.8	14.6	15.4	16.2	17.0
20	17.8	18.6	19.4	20.2	21.0	21.8	22.6	23.4	24.2	24.9
30	25.7	26.5	27.3	28.0	28.8	29.6	30.3	31.1	31.8	32.6
40	33.3	34.1	34.8	35.6	36.3	37.1	37.8	38.5	39.3	40.0
50	40.8	41.5	42.2	43.0	43.7	44.4	45.1	45.9	46.6	47.3
60	48.0	48.7	49.4	50.2	50.9	51.6	52.3	53.0	53.8	54.5
70	55.2	55.9	56.6	57.3	58.0	58.7	59.4	60.1	60.8	61.5
80	62.2	62.9	63.6	64.3	65.0	65.7	66.4	67.1	67.8	68.5
90	69.2	..	..	..	..	..	..	..	..	..

127. In addition to the four anemómeters described above, there are many forms of instruments belonging to the deflection, pressure, and rota-
Other forms of anemom- eters.
tion type, and there are also several other types of anemometers. The rate of evaporation of a fluid has been used to determine wind velocity. Sand and mercury have been allowed to fall vertically and to be caught on a tray with a series of compartments. The compartment receiving the largest quantity of the falling substance would give a measure of the wind velocity. Arrangements have also been used in which the wind causes a musical sound, and the velocity has been determined by means of the pitch of the sound.[1]

Fig. 67. — Pocket Anemometer.
(Cut furnished by KEUFFEL and ESSER Co., New York.)

For determining small wind velocities or the velocity of motion of air currents, so-called
Anemometers for determining small wind velocities.
pocket anemometers like the one pictured in Fig. 67 are sometimes used. Here a fan with a series of plates takes the place of the cups. An instrument of this kind is best calibrated, and the wind velocity determined by walking at a known rate of speed both toward and away from the wind. Great care must be, of course, taken not to influence the instrument by the presence of the observer's body. Very light wind velocities may also be determined by watching the velocity of motion of light substances in the air. For this purpose thistledown or very light balls of cotton may be used, or an artificial cloud may be formed by combining the fumes from hydrochloric acid and ammonia.

[1] See DINES, "Anemometer Comparisons," *Quarterly Journal of the Royal Meteorological Society*, Vol. XVIII, 1892, p. 165.

The Location of Observatories

128. Effect of surroundings. — Wind, both as regards direction and velocity, is probably more affected by the immediate surroundings of the station at which the observations are made than any other one of the meteorological elements. There are four things to be especially considered; namely, valleys, buildings, nature of the surface, and altitude. Valleys influence wind direction markedly and velocity to a slight extent. The wind ordinarily blows harder on a mountain top than in a near-by valley. A good illustration of the effect of the valley on wind direction may be seen in considering the prevailing wind direction in the Hudson River Valley and in the Mohawk Valley in New York State. The Mohawk Valley runs nearly east and west, and the prevailing wind direction is also west. The Hudson River Valley runs nearly north and south, and the prevailing wind direction, only a few hundred miles from the Mohawk Valley, is nearly north. Valleys, therefore, have a tendency to cause the wind to blow along their length. Buildings increase the wind velocity near them and also make the wind gusty. In fact, one result of all unevennesses in the surface over which air passes is to cause gusts. The nature of the surface, also, has a marked influence on wind velocity. On land the wind velocity is very much reduced near the earth's surface. This is brought about not only by friction, but also by the intermingling of air masses, and by the formation of eddies which result from the uneven surface. Wind velocity increases markedly with altitude. The increase is very rapid in the first hundred or two hundred feet, particularly over the land. On the top of the Eiffel Tower at Paris, at an altitude of 990 feet, the average wind velocity is 3.1 times the velocity at an altitude of 60 feet. The accompanying table gives the wind velocity at various altitudes as determined by cloud observations at Blue Hill, near Boston.

(margin notes: Valleys, buildings, nature of surface, and altitude must be considered. The influence of a valley. The effect of the nature of the surface. The effect of altitude.)

Altitude, meters	200 to 1000	1000 to 3000	3000 to 5000	5000 to 7000	7000 to 9000	9000 to 11,000	11,000 to 13,000
Mean velocity in meters per second { Summer	7.5	8.2	10.6	19.1	23.5	31.1	35.2
{ Winter	8.8	14.7	21.6	49.3	54.0	..	..

The immediate surroundings of a station, then, influence markedly wind as regards both direction and velocity, and a station should be very carefully chosen in order to give values which shall be characteristic for the section in question.

129. Hill and mountain observatories. — In order to get observations of the meteorological elements at considerable altitudes above the earth's surface, many stations have been established on hills and moun-

The advantage of mountain observatories. tains. Such observations are also freer from the influence of the immediate surroundings of the station. Every hill and mountain, however, influence to a certain extent the meteorological elements, and the values are not the same as would be obtained at the same altitude in the open air. For this reason kite and balloon observations in recent times have supplanted the observations taken at hill and mountain stations. In this country among the more important mountain observatories may be mentioned the one on Blue Hill, near Boston, at an altitude of 675 feet, which has been main-

Some well-known mountain observatories in this country. tained many years by Professor A. L. Rotch, and the research observatory of the U. S. Weather Bureau at Mount Weather in Virginia. In 1871 an observatory was established on Mt. Washington, New Hampshire, at an altitude of 6279 feet, and in 1877 one was established on Pike's Peak, Colorado, at an altitude of 14,134 feet. These last two observatories were given up in 1887 and 1889 respectively. The Lick observatory on Mt. Hamilton, California, at an altitude of 4400 feet, maintains a full meteorological record. Among the many important mountain observatories in Europe

European mountain observatories. may be mentioned Ben Nevis, Scotland, 4407 feet; Brocken, North Germany, 3743; Hoch Obir, Austria, 7047; Pic du Midi, France, 9381; Puy de Dôme, France, 4800; Schneekoppe, Germany, 5246; Sentis, Switzerland, 8215; Sonnblick, Austria, 10,155; Wendelstein, Germany, 5669. The Eiffel Tower in Paris at an altitude of 990 feet has given very interesting meteorological records. These records are of especial value because the slender form of the tower causes no disturbing influence in the condition of the air around it.

THE RESULTS OF OBSERVATION

130. The observations. — At all regular stations of the U. S. Weather Bureau continuous records of the wind direction and wind velocity are maintained. The instrument used for determining the wind direction is the wind vane with an electric contact maker. It will be remembered that by means of a sliding contact one or two of the four circuits N, S, E, and W, are constantly kept closed. The current is sent through the instrument every minute by means of a clock, and on the revolving drum in the office below the wind direction is thus recorded to eight points of the compass. The instrument used for determining wind velocity is the Robinson cup anemometer. The contact is made after every mile of wind has gone by, and a spur is made on the revolving drum to indicate that fact. Wind direction, wind velocity, amount of rainfall, and the duration of sunshine[1] are all recorded on the same revolving drum, which is usually *The observations of wind at a regular station; the instruments used and their location.*

spoken of as a triple register or meteograph and is located in the office part of the Weather Bureau. The wind vane and anemometer are usually exposed at the top of some high building at a considerable distance above the ground. The purpose of this is to give them as free an exposure as possible and thus prevent local influences from disturbing the wind direction or velocity.

At the coöperative stations of the U. S. Weather Bureau the wind direction for the day is the only thing recorded. If the wind has shifted during the day, the middle of the arc through which it has shifted during the twenty-four hours is determined and the corresponding direction recorded. *Observations at a coöperative station.*

131. Prevailing wind direction; wind roses. — Since wind direction is not a mere number, normals cannot be computed in the usual way. For this reason, instead of the three words mean, average, and normal, the single word "prevailing" is used. This distinction, however, is not always recognized. The prevailing wind direction may be expressed in two *Prevailing wind direction; two ways of expressing it.*

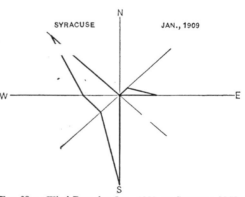

FIG. 68. — Wind Rose for Jan., 1909, at Syracuse, N.Y.

ways: either by means of a table, or graphically by means of what is

[1] One circuit is used for the double purpose of recording rainfall and sunshine. But little confusion results, because it seldom rains while the sun is shining.

called a wind rose. If the prevailing wind direction, be it for a month or a year, or an indefinite time, is expressed by means of a table, this table contains simply the number of times each wind direction was observed. The number of calms must also be noted. The graphic representation of the table is known as the wind rose. The four directions — north, south, east, and west — are first drawn from a central point, and then the four intermediary directions. On these eight lines distances are laid off proportionately to the number of times each of these wind directions was observed. If these points are connected by straight lines the resulting figure is called a wind rose. The number of calms may be expressed by a circle described about the center with a radius proportionate to the number of calms. In the following table is given the number of times the wind blew from each direction at Syracuse in the Mohawk Valley, and Albany in the Hudson River Valley during January, 1909. Figures 68 and 69 are the corresponding wind roses. The reason why east and west winds occurred

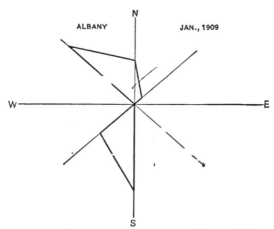

FIG. 69. — Wind Rose for Jan., 1909, at Albany, N.Y.

The method of constructing a wind rose.

	SYRACUSE	ALBANY
N	0	5
NE	1	1
E	4	0
SE	0	0
S	10	10
SW	3	5
W	3	0
NW	10	10
Calms	0	0
	31	31

at Syracuse and not at Albany is probably because the Mohawk Valley runs east and west, while the Hudson Valley runs north and south. In other respects the wind roses are quite similar.

132. Normal hourly, daily, monthly, and yearly velocity. — Since the wind velocity is a mere number, the various normals may be computed in the usual way. In determining a normal hourly wind velocity, however, it is not customary to determine it for every hour of every day in the year, but for the various hours of the months as a whole. For example, it would be the 8 A.M., 9 A.M., 10 A.M., etc., wind velocities for January which would be determined.

The method of computing the normal velocities.

133. Diurnal, annual, and irregular variation. — The graph which represents the diurnal variation in wind velocity may be determined by plotting to scale the normal hourly wind velocities. For New England and the larger part of the United States the higher values of wind velocity occur during the day and the lower velocities at night. The maximum usually occurs between twelve and four in the afternoon, and the minimum at about the time of sunrise. The reason for this daily variation is to be found in convection. During the day the layers of air near the ground become heated; and as they rise, due to convection, the colder air must come down to take the place of the rising air and bring with it the higher wind velocities of the upper atmosphere. The diurnal variation

The daily variation in wind velocity and its cause.

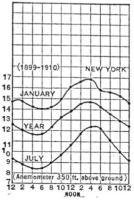

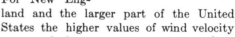

FIG. 70. — The Daily Variation in Wind Velocity at New York, St. Louis, and San Francisco for January and July.

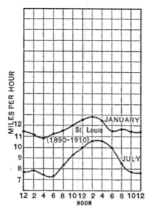

FIG. 70 a. — The Daily Variation in Wind Velocity at New York, St. Louis, and San Francisco for January and July.

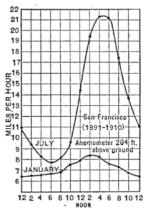

FIG. 70 b. — The Daily Variation in Wind Velocity at New York, St. Louis, and San Francisco for January and July.

of wind velocity is greatest on land and practically disappears over the ocean. It is less in winter than in summer and less on cloudy days than on clear days. The reason for these facts is evident, since convection on land, during the summer and on clear days, is far greater than under the opposite conditions.

Figure 70 shows graphically the change in wind velocity during the day at New York, St. Louis, and San Francisco for both January and July and illustrates well the truth of the statements which have just been made.

There is also a slight diurnal variation in wind direction. This, however, is so masked and changed by local conditions, particularly if the station is located near the seashore, where the land and sea breezes blow, or near mountains, where the mountain and valley breeze is felt, that the typical diurnal variation is hardly noticeable. In order to observe this diurnal variation in all its simplicity, it would be necessary to have a station surrounded on every side by practically identical conditions. If this were the case it would be found that the wind shifts slightly in a clockwise direction during the daytime and shifts back again in a counterclockwise direction during the night. The reason for this is again to be found in convection. The upper air currents always blow in a direction turned somewhat clockwise as contrasted with the surface winds. Thus, during the day, as these upper air masses come down, they will bring with them a wind direction turned slightly in a clockwise manner.

The daily variation in wind direction and its cause.

The diurnal variation in wind velocity and in wind direction as described above occurs only at low altitudes. At high altitudes, at the level of the clouds for example, exactly the reverse is true; that is, the higher values of wind velocity occur during the night and the wind direction shifts in the opposite direction.

The effect of altitude on the diurnal variation.

If the prevailing wind direction for the various hours of the day and the normal hourly wind velocities for any station are known, they may be expressed graphically, as in Fig. 71, for the top of the Eiffel Tower at Paris. The wind direction during any hour of the day may be found by connecting the corresponding hour with the point O, and the wind velocity by noting the length of the line connecting the hour in question with the point O. It will be seen that at this slight elevation, 990 feet, a modification of the surface conditions is already apparent.

Graphic representation of the diurnal variation in wind direction and velocity.

134. If the normal daily or the normal monthly velocities are plotted to scale, the graph will indicate the annual variation in wind velocity. In general, the wind blows harder in winter than in summer. The maximum usually comes in the very late winter, February, March, or April, and the minimum during the summer, July or August. The reason for this is twofold. In the first place, when the trees are covered with leaves and vegetation is most luxuriant, wind velocities are lessened much more by friction than during the winter, when the trees are bare and the ground is snow-covered and frozen. In the second place, the poleward

The annual variation in wind velocity; its characteristics and causes.

temperature gradient is much greater in winter than in summer (see section 88), and it will be seen later that it is this difference in temperature between equator and pole which drives the convectional circulation which is at the foundation of the general wind system of the globe. The greater the difference in temperature between the equator and pole, the more energetic will be this circulation, and thus the higher will

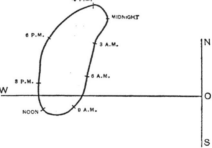

Fig. 71. — Diurnal Variation in Wind Direction and Velocity at the Top of the Eiffel Tower during June, July, and August.

(After Angot.)

be the wind velocities. The following table gives the normal monthly wind velocity for several stations in the United States. In Fig. 72 the results are shown graphically for four stations, Philadelphia, Chicago, Phœnix, and San Francisco. The graphs for Philadelphia and Chicago are typical for the northeastern and central portion of the country and illustrate the description of the annual variation which has just been given. In the Southern States and on the Pacific coast the characteristics are very different because the whole type of weather and weather control are different.

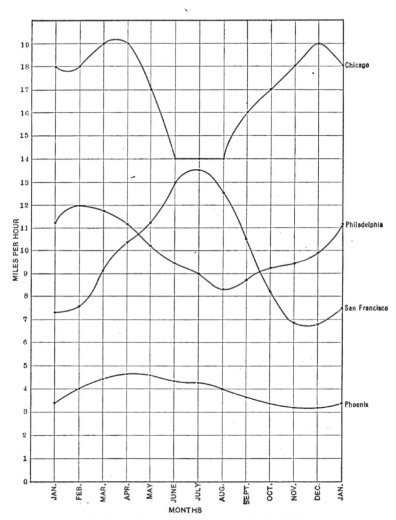

FIG. 72. — The Annual Variation in Wind Velocity.

THE NORMAL WIND VELOCITY IN MILES PER HOUR FOR THE VARIOUS MONTHS AND FOR THE YEAR

Station	Height of Anemometer above Ground	Length of Record	Jan.	Feb.	Mar.	Apr.	May	June	July	Aug.	Sept.	Oct.	Nov.	Dec.	Annual
Bismarck, N. Dak.	35	34	8	9	10	12	11	10	9	9	10	10	9	8	10
Charleston, S.C.	92	38	9	12	11	12	11	10	10	9	9	9	9	9	10
Chicago, Ill.	274	15	18	18	19	19	17	14	14	14	16	17	18	19	17
Columbus, Ohio	222		9	10	10	9	8	7	7	6	7	8	9	9	8
El Paso, Tex.	133		10	12	14	13	12	10	7	9	9	9	9	10	10
Indianapolis, Ind.	164	12	11.7	11.5	12.1	11.3	9.9	8.9	8.2	7.4	8.3	9.4	10.4	11.5	10.0
Key West, Fla.	53	38	11	11	11	11	9	8	8	7	8	11	11	11	10
New Orleans, La.	121	39	8.8	9.3	9.4	8.9	7.6	6.1	6.1	6.1	7.4	7.8	8.3	8.8	7.9
Omaha, Neb.	121	37	9	9	10	11	9	8	7	7	8	8	9	9	9
Philadelphia, Pa.	184		11.1	12.0	11.6	11.3	10.2	9.4	9.0	8.4	8.9	10.3	10.4	10.9	10.3
Phœnix, Ariz.	56	15	3.4	4.0	4.5	4.6	4.6	4.4	4.5	4.0	3.8	3.5	3.3	3.2	4.0
Portland, Ore.	106	37	6	7	6	6	6	6	5	6	6	6	6	6	6
St. Louis, Mo.	317	10	12.0	12.1	12.4	12.	10.9	9.0	8.8	8.0	9.2	10	11.8	11.8	10.7
St. Paul, Minn.	124	36	7.8	8.3	8.8	9.	8.7	7.7	7.1	7.1	8.0	8	8.1	7.8	8.1
San Francisco, Cal.	204	21	7.3	7.5	9.1	10.4	11.1	13.0	13.5	12.6	10.4	8.1	6.9	6.9	9.7
Seattle, Wash.	151		7.1	7.3	7.4	6.8	6.4	6.2	5.5	4.8	5.4	5.6	7.0	7.0	6.4

The year 1908 is the last year included in these normals.

There is also an annual variation in wind direction. This, however, depends so much upon the location of the place on the earth's surface that no general rules can be laid down. For New England the prevailing winds of winter are more northwest and north than during the summer, when they shift more to the south and southwest. The reason in every case is the building up of areas of high pressure over the continents during the winter and of low pressure over the continents during the summer. It is to these changing areas of high and low pressure that the annual change in wind direction is due.

The annual variation in wind direction and its cause.

135. In addition to these regular diurnal and annual changes of wind direction and velocity, there are many irregular variations. In the first place, both wind direction and wind velocity are constantly changed slightly from moment to moment. We characterize this by saying that the wind is usually gusty. This is due entirely to the irregularities in the surface of the earth over which the air is moving. In addition, wind direction and

Irregular variations in wind direction and velocity.

wind velocity are very different on different days. The reason for this change in direction and velocity is to be found in the various storms, which are to be discussed later (Chapter VI).

136. Wind data. — In addition to the prevailing wind direction and the various normals in connection with the velocity which have been
Extremes   described above, very few results are computed from the ob-
of velocity.   servations of wind. Practically the only one of any great interest is the extremes of velocity which have occurred at various places. At nearly all stations in the United States wind velocities have gone above fifty miles an hour for a period of five minutes and at nearly every sea-coast station velocities above seventy miles for the same period have been recorded. The accompanying table gives the maximum wind velocity ever recorded at a number of stations in the United States. At the regular stations of the U. S. Weather Bureau there are five tables of wind data which are kept constantly up to date. These contain

(1) Total movement in miles. (The data are given for the various months and for the year as a whole.)

(2) Prevailing direction and average hourly velocity (velocity to tenths).

(3) Maximum velocity, direction, and date.

(4) Mean hourly wind velocity (miles and tenths per hour).

(5) Prevailing wind direction (hourly).

STATION	MAXIMUM VELOCITY, MILES	DIRECTION	DATE OF OCCURRENCE	LENGTH OF RECORD IN YEARS	LAST YEAR INCLUDED IN RECORD
Abilene, Texas . . .	66	SW	May 8, 1892	22	1908
		SW	June 8, 1892	22	1908
Albany, N.Y. . . .	70	W	Feb. 2, 1876	34	1907
		E	Oct. 23, 1878	34	1908
Atlanta, Ga. . . .	60	NW	Feb. 16, 1903	29	1908
		NW	April 8, 1907	29	1908
Atlantic City, N.J. .	72	NE	Sept. 10, 1889	34	1908
Augusta, Ga. . .	52	NE	July 28, 1893	37	1908
Baltimore, Md. . .	70	W	June 20, 1902	37	1908
Binghamton, N.Y. .	44	W	Jan. 20, 1907	17	1907
Bismarck, N.D. . .	74	NW	Mar. 10, 1878	33	1908
Block Island, R.I. .	90	NE	Oct. 27, 1898	28	1908
Boise, Idaho . . .	55	SW	May 13, 1900	9	1908
Boston, Mass. . . .	60	NE	Mar. 3, 1891	36	1908
		N	Oct. 27, 1898	37	1908
		NE	Oct. 24, 1901	37	1908
		E	Nov. 5, 1900	37	1908
Buffalo, N.Y. . . .	90	SW	Jan. 13, 1890	37	1907
Burlington, Vt. . .	60	SE	Jan. 20, 1907	22	1907

STATION	MAXIMUM VELOCITY, MILES	DIREC- TION	DATE OF OCCURRENCE	LENGTH OF RECORD IN YEARS	LAST YEAR INCLUDED IN RECORD
Cairo, Ill.	84	W	June 21, 1891	36	1908
Charleston, S.C. . .	96	E	Aug. 28, 1893	37	1908
Chattanooga, Tenn.	60	W	May 12, 1895	29	1908
Chicago, Ill. . . .	84	NE	Feb. 12, 1894	37	1908
Cincinnati, O. . . .	52	NW	May 23, 1901	37	1908
Cleveland, O. . . .	73	S	Nov. 26, 1895	37	1908
Columbus, O. . . .	66	NW	Jan. 20, 1907	29	1907
Davenport, Iowa . .	72	SW	Aug. 7, 1872	37	1908
Denver, Col. . . .	68	NW	May 1, 1902	36	1908
Des Moines, Iowa. .	64	SW	April 1, 1892	29	1908
Detroit, Mich. . . .	76	SW	Oct. 26, 1895	37	1908
Duluth, Minn. . . .	78	NE	Aug. 16, 1881	37	1908
El Paso, Texas . . .	78	W	Mar. 5, 1895	29	1908
Galveston, Texas * .	84	NE	Sept. 8, 1900	37	1908
Hatteras, N.C. . .	105	N	July 17, 1899	33	1908
Havre, Mont. . . .	76	NW	June 9, 1890	26	1908
Helena, Mont. . . .	60	W	Feb. 6, 1890	27	1908
		W	Dec. 25, 1890	27	1908
Honolulu	55	SE	Dec. 31, 1906	33	1908
Huron, S.D. . . .	72	NE	Jan. 6, 1903	26	1907
Indianapolis, Ind. .	60	W	June 25, 1882	37	1908
Jacksonville, Fla.. .	75	SW	Feb. 16, 1903	36	1908
Kansas City, Mo. .	55	NW	July 10, 1902	20	1908
Key West, Fla. . .	88	SW	Oct. 19, 1876	37	1908
Lexington, Ky. . .	68	W	Sept. 8, 1899	23	1908
Los Angeles, Cal. .	48	NE	Jan. 28, 1882	30	1907
Memphis, Tenn. . .	75	SW	Mar. 9, 1901	37	1908
Milwaukee, Wis. . .	60	SW	July 24, 1874	37	1908
		SW	Oct. 16, 1880	37	1908
Minneapolis, Minn. .	84	NW	July 20, 1904	17	1908
New Haven, Conn. .	62	SE	Oct. 21, 1904	35	1908
New Orleans, La.. .	60	E	Aug. 19, 1888	37	1908
New York, N.Y. . .	80	N	Mar. 20, 1899	37	1908
Omaha, Neb. . . .	64	NE	July 13, 1905	37	1908
Philadelphia, Pa. . .	75	SE	Oct. 23, 1878	37	1908
Phœnix, Ariz. . . .	48	SE	July 25, 1903	12	1908
Portland, Me. . . .	60	SE	Mar. 21, 1876	36	1908
		SW	Dec. 2, 10, 1878	36	1908
Portland, Oregon . .	55	S	Mar. 25, 1897	37	1908
St. Louis, Missouri .	80	NW	May 27, 1896	37	1908
St. Paul, Minn. . .	102	NW	Aug. 20, 1904	37	1908
Salt Lake City, Utah	66	NW	Nov. 15, 1906	34	1908
San Francisco, Cal. .	64	NE	Nov. 30, 1906	37	1908
Savannah, Ga. . . .	76	NW	Aug. 21, 1898	37	1908
Springfield, Mo. . .	64	W	May 29, 1905	20	1908
Syracuse, N.Y. . .	66	S	Mar. 24, 1907	5	1908
Washington, D.C. .	66	SE	Sept. 29, 1896	37 ′	1908
Yuma, Arizona . .	54	NW	Mar. 17, 1894	30	1908

* Anemometer blew away — 120 miles by estimation.

137. **Prevailing winds of the world.** — Charts XI and XII, in addition to the pressure, indicate the prevailing wind direction for the world for

January and July. In preparing these charts, that wind direction
which predominated during the month in question was the
one charted. If these charts are carefully compared, it

The characteristics of the prevailing winds of the world.

will be found that the following generalizations can be
made. In the northern hemisphere the wind blows spirally
inward toward areas or belts of low pressure, turning in a
counterclockwise direction. From areas or belts of high pressure the
wind blows spirally outward, turning in a clockwise direction. The
results are the same for the southern hemisphere with the exception
that the direction of rotation is the converse. In the centers of the
areas or belts of high and low pressure calms occur.

138. Other wind charts. — In addition to the two charts described above
many other wind charts might be prepared. The prevailing wind direc-

Other wind charts.

tion for all the months in the year for the world as a whole
or for any separate country might be charted. Also, the
normal wind velocity or the extremes of velocity for all the months in
the year or for the year as a whole might be charted for the whole
world or for any given country.

C. THE CONVECTIONAL THEORY AND ITS COMPARISON WITH OBSERVED FACTS

THE CONVECTIONAL THEORY

139. General convectional motion. — The general convectional motion of the atmosphere can best be illustrated by means of a fluid analogy.

FIG. 73. — Diagram illustrating Convection in a Long Tank of Fluid.

In Fig. 73 let *NES* represent a long
tank filled with fluid and

Convectional circulation in a fluid.

heated at the bottom and in
the center. The layer of
fluid next the bottom will be
heated by conduction and will expand,
thus raising the layers of fluid above it
and causing a slight bulging of the upper surface. Gravity acting
upon this will cause the fluid to flow towards the ends of the tank,
thus decreasing the pressure in the center and increasing it at the ends.
This increased pressure at the ends will drive in the colder fluid toward
the center, forcing the lighter expanded fluid to rise and thus starting
the convectional circulation.

The atmosphere, as was fully discussed in Chapter II, is heated primarily at the bottom, and the amount of heat supplied to the equator is

nearly three times the amount applied at the poles, the exact ratio being 347 to 143. One would thus expect a convectional circulation to take place between the equator and poles, the warm air rising at the equator, flowing poleward on the outside of the atmosphere, descending in the polar regions, and returning toward the equator along the earth's surface. The continents are warmer than the adjoining oceans in summer and colder in winter; one would thus expect a convectional circulation between continents and adjoining oceans. Along the seashore the land is warmer than the water during the day and colder than the water at night; one would thus expect a convectional circulation between the land and the near-by water. For one reason or another a certain locality might be warmer than nearby places; small local convectional circulations would thus be expected. As a matter of fact, as will be seen in later sections, all of these convectional circulations actually exist. The first gives rise to the most important of the general winds of the world; the second is the cause of the monsoons; the third is the reason for land and sea breezes; while certain cloud formations, thunder showers, and dust whirlwinds are evidences of the last.

The convectional circulation between equator and pole.

Other convectional circulations.

140. Arrangement of isobaric surfaces in a general convectional circulation. — If there were no temperature differences and no winds, the isobaric surfaces would be parallel to the hydrosphere and at the same distance above it.

A vertical section along the meridian of these isobaric surfaces would show a series of lines parallel to the earth's surface. These are represented in Fig. 74 by heavy continuous lines. Suppose now that the at-

The meridional section before convection started.

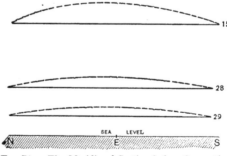

FIG. 74. — The Meridional Section before Convection started.

mosphere is suddenly heated at the bottom and at the equator more than elsewhere. The air will expand and the isobaric surfaces will be warped upward, the upper surfaces showing an increasing amount of bulge. The vertical section along the meridian of these warped surfaces is represented by the dashed lines in Fig. 74. The excess of air at the equator will now, under the influence of gravity, flow off toward the poles, decreasing the equatorial pressure and increasing the pressure at the poles. This increased pressure at the poles will force in the colder air along the earth's surface and cause the warm equatorial air to rise, thus starting the convectional circulation. Figure 75 shows a vertical section along a meridian of the isobaric surfaces, and the

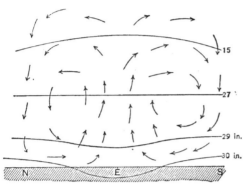

FIG. 75.— The Meridional Section after Convection was permanently established.

The meridional section after convection was permanently established.

air circulation after the convectional circulation has become permanently established. A meridional view of the air circulation is shown more in detail in Fig. 76. One half of the atmosphere is below three miles of elevation, and 30° N. latitude marks the dividing line on the earth's surface, one half of the atmosphere being between it and the equator and one half between it and the pole. Thus, between 30° N. latitude and the equator the air would be rising, flowing poleward outside of the three-mile limit, dropping down to lower levels between 30° N. latitude and the pole, and returning equatorward below the three-mile limit. These last two

A meridional view of air circulation.

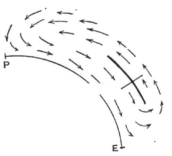

FIG. 76. — A Meridional View of the Air Circulation.

diagrams illustrate the convectional circulation between equator and pole which would be expected on a non-rotating earth heated at the equator and colder at the poles. In all three of the above diagrams the vertical scale is much exaggerated as compared with the horizontal.

141. Condition of steady motion. — The condition of steady motion can be stated in a single sentence. As long as in the atmosphere differences of temperature are maintained, a steady convectional circulation will continue.   In the case of the earth, the equatorial portions are constantly maintained at a higher temperature than other regions, and the atmosphere is heated at the bottom.   Thus a permanent convectional circulation between equator and pole is to be expected.

The condition of steady motion.

142. **Barometric gradients.** — The barometric gradient is defined in exactly the same way as the two temperature gradients which have already been described (sections 53 and 88).   The vertical temperature gradient was defined as the change in temperature with elevation above the earth's surface ; the poleward temperature gradient was defined as the change in temperature with distance in going from equator to pole.   Similarly, the barometric or pressure gradient may be defined as the change in barometric pressure with distance.   This is ordinarily expressed as a change of so many hundredths of an inch of pressure in 500 miles, or so many millimeters per latitude degree (69.5 statute miles).   It can be found by noting the difference in pressure between two places and the distance between them, and then reducing this to the difference per 500 miles or per latitude degree.

Definition of the barometric gradient.

How expressed.

143. **Relation of wind direction to pressure gradient.** — The relation of wind direction to the barometric or pressure gradient can be expressed best by means of a diagram.   In Fig. 77, let A, B, and C represent respectively three isobaric lines, with a pressure of 30.3, 30.0, and 29.7 inches respectively.   Suppose that a mass of air M is located on the 30-inch line.   The pressure on the K face of this mass of air is slightly greater than 30 inches, because it lies between the 30-inch line and the 30.3-inch line.   In the same way, the pressure of the L face of this mass of air is less than 30 inches because it lies between the 30-inch line and the

Relation of wind direction to pressure gradients and isobaric lines.

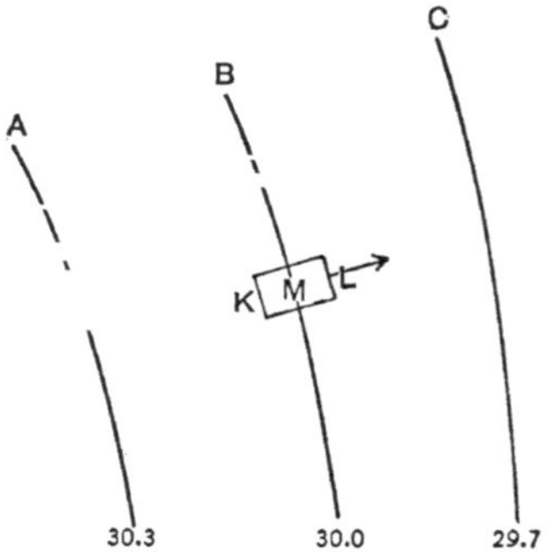

FIG. 77. — The Relation of Wind Direction to Pressure Gradients.

29.7-inch line.   There is thus an unbalanced pressure on this mass of air tending to push it from the 30-inch line to the 29.7-inch line. This may be generalized by saying that air tends to move along

pressure gradients and at right angles to isobaric lines. If this princi-
Air motion about "highs" and "lows." ple is applied to the air in an area of high and low pressure, as illustrated in Fig. 78, it will be seen that the air should move directly outward from areas of high pressure and directly inward toward areas of low pressure.

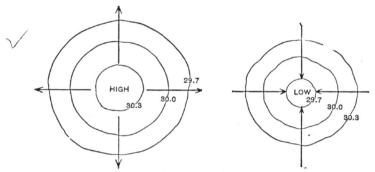

FIG. 78. — Air Motion about "Highs" and "Lows" on a Non-rotating Earth.

144. Effects of the earth's rotation on wind direction and pressure. — The introduction into meteorology of the idea that the rotation
The first recognition of the effect of the earth's rotation. of the earth might influence the direction of moving air masses was slow. The first use of this principle to explain observed wind directions was made by Hadley in 1735. It had been known, for a considerable time previous to this, that the trade winds did not blow directly towards the equator, but had an oblique movement. This had been observed especially in the case of the trade winds from 30° N. latitude to the equator which blow from the northeast. His explanation of the
Hadley's explanation of the direction of the trade winds. oblique movement was essentially as follows: A mass of air starting from 30° N. latitude directly toward the equator would be passing over regions which would have a greater easterly velocity of motion due to the earth's rotation, than the places from which it had come. As a result it would lag behind the rotating earth and thus, instead of moving due south, it
The mistakes in Hadley's explanation. would deviate to the right and become a northeast wind. Hadley's explanation contains the germ of the truth, but is erroneous in at least two directions. In the first place, it tacitly assumes that the effect of the earth's rotation would be felt only by air masses moving north or south, and not by masses

moving in an easterly or westerly direction; and in the second place, it is a natural corollary of Hadley's explanation that the wind velocity must increase as the equator is approached, for with the motion from the north must be compounded the motion from the east due to lag.   Both of these are, however, mistakes, since the earth's rotation influences air moving in any direction, and the velocity does not increase as the equator is approached.

The deflective effect of the earth's rotation on moving bodies on its surface was worked out later by various mathematicians, but its complete application to the motion of air masses, and the first **Ferrel's** rational explanation of the general wind system of the globe, **work.** was begun by Ferrel in 1856   He was a school teacher at Nashville, Tenn., and was a self-taught mathematician of remarkable originality and ability.   It is probably not too much to say that Ferrel's work caused a revolution in the science of meteorology.

145. The law which expresses the effect of the earth's rotation on moving air can be briefly stated as follows: If a mass of air starts to move on the earth's surface, it deviates to the right in the **The law** northern hemisphere (to the left in the southern hemisphere), **which ex-** and tends to move in a circle the radius of which depends **presses the effect of the** upon its velocity and the latitude of the place.   The accom- **earth's ro-** panying table indicates, for several velocities and latitudes, **tation.** the radius of this circle.   It will be seen that the earth exercises a deviating influence on air moving east or west of exactly the same kind as if the air moved north or south.

RADIUS OF CURVATURE (IN MILES) FOR FRICTIONLESS MOTION ON THE EARTH'S SURFACE

Latitude	0°	5°	10°	20°	30°	40°	50°	60°	70°	80°	90°
20 miles an hour . .	∞	880	150	150	153	150	150	88	82	78	77
10 miles an hour . .	∞	440	150	150	76	150	50	44	41	39	38
5 miles an hour . .	∞	220	110	56	38	30	25	22	20	19	19

146. The effect of this deviation to the right on the air masses which, due to convection, are moving from the equator toward the poles on the outside of the atmosphere must now be considered.   Instead of moving directly from the equator poleward, the air masses will be deviated to the right in the northern hemisphere and become more and more a

M

west air current encircling the pole in a great whirl. It is a principle of mechanics that whenever a rotating body is not acted upon by out-side forces, the moment of momentum must remain a con-

The effect of the earth's ro-tation on the air masses ap-proaching the poles. stant. The formula for the moment of momentum is $\Sigma M V R$, where M represents the mass of each particle of the rotating body, V the velocity of the particle, and R the radius, that is, the distance of the particle from the center of rotation. This product $M V R$ must be summed up for all the particle

of the rotating body, and thus $\Sigma M V R$ represents the moment of momentum of the body. As this ring of whirling air about the pole approaches it, the mass remains constant, the radius is decreasing,

Low polar pressures explained. and thus the velocity must steadily increase, and it has been computed that, if the velocities were not held down by fric-tion, they would amount to hundreds of thousands of miles

per hour. A whirl of air with these high velocities must cause cen-trifugal force, and this centrifugal force will hold air away from the pole, thus causing a diminution in the barometric pressure. The amount of land at the north pole is much greater than at the south pole. One would thus expect wind velocities and the diminution in pressure to be larger at the south pole than at the north pole. An exact analogy to this process can be seen in the escape of water from a washbowl through

An analogy. a central vent. Due to some cause, the escaping water usually takes up a motion of rotation. As the water ap-proaches the vent to escape, the velocity of rotation becomes greater and greater, and the centrifugal force developed is often sufficient to hold the water away from the center and cause an empty core above the vent.

FIG. 79. — The Meridional Section of the Isobaric Surfaces in a Convectional Circulation on a Rotating Earth.

The vertical section along a meridian of the isobaric surfaces on a rotating earth should thus have the appearance of Fig. 79.

The verti-cal section along a meridian of the isobaric surfaces on a rotating earth. On a non-rotating earth, the pressure at the equator would be low and at the poles high. This was illustrated in Fig. 75. It is the centrifugal force due to the circumpolar whirls which has held the air away from the poles and turned the high pressure into the low pressure. The larger velocities at the south pole are responsible for the lower barometric

pressure there.

147. The effect of the earth's rotation on the motion of air in connection with areas of high and low pressure must also be considered. In Fig. 80 the dashed lines represent the air motion on a non-rotating earth (see Fig. 78); the full lines represent the air motion on a rotating earth. The deviation to the right in each case is evident. These diagrams apply to the northern hemisphere. In the southern hemisphere the direction of deviation is the opposite. *The effect of the earth's rotation on air motion about highs and lows.*

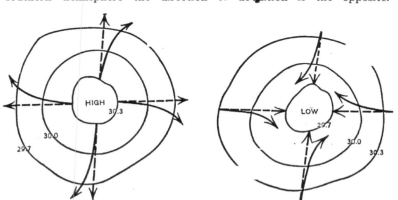

Fig. 80. — Air Motion about "Highs" and "Lows" on a Rotating Earth.

148. Buys Ballot's law. — About 1850 Buys Ballot, after a careful study of the air circulation about storms, generalized a law which was of great practical value. The usual statement of the law is that if one stands with his back to the wind, the pressure on his left hand is lower than on his right. This law derived from observations found its way into text books on meteorology, was used by the captains of vessels, and had a very wide practical application. This is its chief importance, for considered in connection with the preceding sections, it is simply the inevitable result of the rotation of the earth on air moving in towards areas of low pressure. In Fig. 80, if one applies Buys Ballot's law, its truth becomes at once evident. *Buys Ballot's law.* *The reason for its truth.*

Comparison of the Consequences of the Convectional Theory with the Observations of Pressure and Wind

149. In the first two subdivisions of the present chapter, the material was presented from the inductive standpoint (see section 30). The instruments for making various observations were described, the various observations taken were then stated, and from these observations several generalizations were made. It was found, for instance, that there was a belt of low pressure at the equator, belts of high pressure at 35° N. and 30° S. latitude, and low pressure at the poles, the pressure at the south pole being much lower than at the north pole. It was furthermore seen that air tended to move spirally outward from areas of high pressure and spirally inward toward areas of low pressure, turning clockwise about areas of high pressure, counterclockwise about areas of low pressure in the northern hemisphere. These conclusions were simple generalizations from observations and serve as a good example of the inductive method of gaining information.

The inductive method of gaining information about the pressure distribution and air motion.

In the third subdivision of this chapter, the deductive method of reasoning was followed. Two general principles were used; namely, that the atmosphere was heated at the bottom and more at the equator than elsewhere, and secondly, that the earth turned eastward on its axis. From these two general principles it was determined what the pressure distribution over the earth ought to be and what the air circulation about areas of high and low pressure ought to be. If these deductions from the two principles are tested, by comparing them with the generalizations derived from the observations, exact agreement will be found. This stamps the whole process of reasoning, as well as the conclusions, as correct. The present illustration of the inductive and deductive method of reasoning is the best one to be found in the realm of meteorology.

The deductive method.

The comparison.

D. A GENERAL CLASSIFICATION OF THE WINDS

The Classification of the Winds

150. The chief characteristics of a general convectional circulation between equator and pole on a rotating earth have just been stated; it now remains to treat in detail the various components of this circulation. These, together with certain other winds, due in most cases to convection on a smaller scale, are usually spoken of as the general winds of the globe. These were first classified

The general winds of the globe.

by Dové, a German meteorologist, nearly one hundred years ago.
He divided them into permanent, periodic, and irregular; the permanent being those whose direction remained the same   The Dové
throughout the year, the periodic being those whose direc-   classifica-
tion changed during a certain interval of time, the irregular   tion.
being those which showed marked irregular changes in either direction
or velocity.   In the *American Meteorological Journal* for March, 1888,
Professor W. M. Davis published a classification based upon another
principle.   This classification is also treated in the seventh   The Davis
chapter of his book, *Elementary Meteorology*, Ginn and   classifica-
Company, 1894.   In this classification, the winds are divided   tion.
into sun-caused, moon-caused, and earth-caused winds.   In neither of
these classifications is any account taken of the importance, or intensity,
or the amount of the earth's surface covered by these winds.   In the
following sections these general winds of the globe will be treated roughly
in the order of their importance, although the treatment is based largely
upon the Davis classification.   In each case the position of the wind in
both the Dové and Davis classification will be given.

PLANETARY WINDS

151. Typical system. — By planetary winds are meant those winds
which would occur on any planet heated at the equator and turning
eastward on its axis.   Any migration of the point of appli-   Definition
cation of the greatest heat and all irregularities of surface   of planetary
are here left out of consideration.   The typical system of   winds.
planetary winds can be best illustrated by means of a series of diagrams.
In section 146 it was shown that a planet heated at the   The pres-
equator and turning eastward on its axis would develop a   sure belts.
belt of low pressure at the equator, belts of high pressure in the tropical
regions, and cups of low pressure at the poles.   These deductions are
also in exact agreement with the results of the pressure observations
made in different parts of the earth.   One would thus expect calms at
the equator and in the tropical regions, and winds blowing outward from
the belts of high pressure, turning to the right in the northern   The air cir-
hemisphere and to the left in the southern.   These are   culation.
illustrated in Fig. 81.   The equatorial belt of calms is called the doldrums; the tropical belts of calms are called the horse   The   nes
latitudes; the northeast winds blowing from the tropical   of th
belt of high pressure in the northern hemisphere and the   wind
southeast winds blowing from the tropical belt of high pressure n the

southern hemisphere toward the equator in each case are called the trade winds; the southwest wind, blowing in the northern hemisphere from the belt of high pressure poleward, and the northwest wind blowing from the high pressure belt in the southern hemisphere poleward are called the prevailing westerlies. This surface distribution of the winds extends to an altitude of about 3000 feet. At this height, as was seen in connection with the meridional section of the isobaric surfaces (Fig. 60) the equatorial belt of low pressure disappears and the isobaric lines become practically straight with a slight droop at the poles.

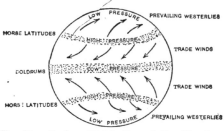

FIG. 81. — Surface Distribution of the Planetary Winds; 0 to 3000 ft.

The extent of this surface distribution.

152. On the outside of the atmosphere, that is, from a height of about 13,000 feet on, the air currents move from the equator toward the pole in both hemispheres, turning to the right in the northern hemisphere so as to become a southwest air current, and to the left in the southern hemisphere so as to become a northwest air current. These air currents approaching the poles in spiral paths give the impression of circumpolar whirls in each hemisphere. These are indicated in Fig. 82. Right at the equator the air moves from east to west, and the reason for this can be readily stated. The air, rising at the equator, in order to start its circumpolar journey, goes into regions which have a larger eastward velocity of motion, due to the earth's rotation, than the earth's surface itself. As a result, these rising air masses lag behind and give the impression of an east wind.

The air currents in the outer layer of the atmosphere

The explanation of the east air current at the equator.

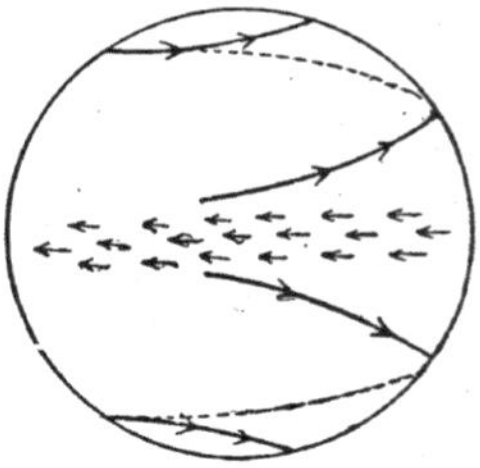

FIG. 82. — The Air Currents in the Outer Layer of the Atmosphere; 13,000 ft. on.

153. From the tropical belt of high pressure to the pole in the northern hemisphere, both the upper air currents and the surface winds blow from the southwest. If the air is not to become massed at the pole, a return current

in an intermediate layer is necessary. The direction of the air motion in this intermediate layer from, say, 3000 feet to 13,000 is illustrated in Fig. 83. The air would start from the poles with a slight velocity, and would tend to be deviated toward the right in the northern hemisphere and thus become a northeast wind, but it is imprisoned between two layers of air, both of which have a high velocity from some westerly point. As a result, it is carried eastward and becomes a northwest return current in spite of the deviation to the right caused by the earth's rotation. After passing the tropical belt of high pressure the surface wind becomes a north-

The air motion in the intermediate layer.

The explanation of its direction.

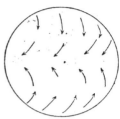

Fig. 83. — The Air Circulation in the Intermediate Layer; 3000 to 13,000 ft.

east wind and the intermediate layer then also obeys the deviating force of the earth's rotation and becomes a northeast wind.

154. The north-south component of the circulation of the atmosphere can be illustrated by projecting it upon the plane of the meridian. This is shown in Fig. 84. On the outside of the atmosphere, that is, from 13,000 feet on, the motion is northward, while in the intermediate layer, from 3000 to 13,000 feet, the motion is southward. The surface winds move southward from the tropical belt of high pressure toward the equator, and northward from this belt toward the pole. At the equator there are rising air currents and in the tropical belt of high pressure descending air currents. The circulation in the southern hemisphere is exactly the same.

The atmospheric circulation in cross section.

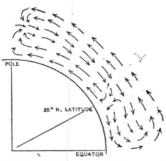

Fig. 84. — The North-south Component of the Circulation of the Atmosphere.

155. This general circulation of the atmosphere must be modified in two ways, as will be seen later, to adapt it to the actual earth, and it is also invaded by storms, principally in the region of the prevailing westerlies which cause much confusion and reduce wind velocities to a marked extent.

The typical system must be modified.

If the earth did not rotate on its axis, the air motion would be a simpler direct north and south motion, as illustrated in Fig. 78, and not an oblique circulation as at present. The effect of the deflec-

The result on the winds if the earth stopped rotating.

tive force due to the earth's rotation is to increase some-what the wind velocities; but by causing the circulation to be oblique, it also retards the exchange of air between equator and pole and thus causes greater temperature dif-ferences to exist between equator and pole than otherwise would. In the Dové system all these winds would be classified as per-manent, in the Davis classification they would be considered sun-caused.

156. Trade winds. — The trade winds blow from the tropical belts of high pressure toward the equator; as a northeast wind in the northern

Description of the trade winds.

hemisphere and a southeast wind in the southern hemisphere. They cover nearly half of the earth's surface and their name is derived, not from their importance to commerce, but from the steadiness with which they blow. They blow with moderate to brisk velocity and particularly over the ocean have an unchanged direc-tion for perhaps weeks at a time. They carry but few clouds by day, and by night the sky is usually cloudless. As they approach the equator the velocity steadily increases, and the amount of moisture contained in them also becomes greater. The trade winds were first described by Halley in 1686, and since that time many observations have been made of them. Their thickness is about 13,000 feet. This information is

Thickness of the trades and how found.

gained by noting the motion of the cirrus and cirro-cumulus clouds which occur at this level or higher, from kite and balloon observations, and also from the observations made on mountain tops which exceed this elevation. In some cases the smoke coming from volcanoes has been observed to drift from the southwest in the air currents of the outer layer of the atmosphere, while the lower clouds drifting in the trade winds moved from the northeast.

157. Doldrums. — The doldrums are the equatorial belt of calms. The air, laden with moisture and at a high temperature, brought in by

Description of the doldrums.

the trade winds, here loiters and finally rises to commence its poleward journey on the outside of the atmosphere. This rising air cools, reaches the dew point, and then yields cloud and precipitation. The doldrums are thus characterized by light, baffling breezes, frequent calms, overcast sky and heavy rains, often in the form of thunder storms and squalls.

158. Horse latitudes. — In the tropical belts of high pressure, at 35° N. and 30° S. latitude, are located the horse latitudes. These stand

in marked contrast to the doldrums, although the wind is light and variable and the velocity small.   Calms are frequent.   The sky, however, is nearly always clear and the amount of moisture The horse in the atmosphere is small.   This is due chiefly to the fact latitudes. that here we have to do with descending instead of ascending air currents.

159. **Prevailing westerly winds.** — In the northern hemisphere, from the tropical belt of high pressure poleward, the wind blows from the south-west or some westerly quarter.   In the southern hemisphere The pre-the wind direction is from the northwest.   Both these vailing west-winds are spoken of as the prevailing westerly winds.   In the erly winds. southern hemisphere the velocity is larger, sometimes amounting to a gale, and for this reason, particularly among sailors, these winds are often spoken of as the " brave west winds," and the region of their occur-rence in the southern hemisphere called the " roaring forties."   Sailing vessels, in going from Europe to Australia often make the outward jour-ney around the Cape of Good Hope and the return journey around Cape Horn, thus using the prevailing westerlies which encircle the south pole. The prevailing westerlies are more invaded by storms than any other of the permanent winds, and in many parts of the United States the succession of storms is so rapid that the prevailing westerly wind only makes itself manifest in the general averages.

160. **Upper currents.** — The upper currents move from the equator as a southwest air current in the northern hemisphere The direc-and a northwest air current in the southern hemisphere, tion of mo-spirally poleward, and are often spoken of as the anti-trades. tion and The velocities are high and much greater in winter than in istics of the summer.   The motion of these air currents has been fre-upper quently determined by the drifting of smoke from lofty vol-currents. canoes, by the cirrus clouds which occur at these levels, by observa-tions on high mountains, and by kite and balloon observations.

TERRESTRIAL WINDS

161. **Definition.** — The typical system of planetary winds must be modified in two ways.   In the first place, the axis of the The planet-earth is inclined $23\frac{1}{2}°$ to the plane of the ecliptic, which ary system causes a change in the presentation of the northern hemi-of winds sphere to the rays of the sun.   On account of this the sun modified in migrates 47° in the course of a year, being farthest north on two ways. the 21st of June and farthest south on the 21st of December.   As will be

seen later, this migration modifies the wind system. In the second place the earth's surface is not uniform, but is very diversified, being composed of both land and water and with numerous mountain ranges.

Definition of terrestrial winds. This diversity of surface also affects the typical system of planetary winds. By terrestrial winds are meant the typical planetary system modified by taking account of the first condition; namely, that the sun migrates 47° in the course of a year.

162. **Annual migration of the winds.** — This migration of the sun through 47° causes a migration of the heat belt. The heat belt migrates less than the sun and lags behind it from four to six weeks.

The migration of the sun, heat belt, equatorial low pressure belt, and wind system. The migration of the equatorial belt of high temperature causes the equatorial belt of low pressure, of which it is the cause, to migrate. This migration in turn lags behind the heat belt and migrates over a smaller distance. The migration of the equatorial belt of low pressure causes the permanent wind system of the globe to migrate. The migration of the wind system lags some two months behind the migration of the sun, and on the average covers a distance of

The amount of the migration. 5 or 6°. The following table, which gives for the Atlantic and Pacific oceans the boundaries of the trade winds and doldrums for March and September, makes clear this migration of the wind system.

	ATLANTIC OCEAN		PACIFIC OCEAN	
	March	September	March	September
N.E. Trades	26° N. – 3° N.	35° N. – 11° N.	25° N. – 5° N.	30° N. – 10° N.
Doldrums	3° N. – 0°	11° N. – 3° N.	5° N. – 3° N.	10° N. – 7° N.
S.E. Trades	0° N. – 25° S.	3° N. – 25° S.	3° N. – 28° S.	7° N. – 20° S.

163. **Subequatorial and subtropical wind belts.** — This migration of the wind system over 5 or 6° in the course of a year gives rise to

The subequatorial and subtropical wind belts. three belts on the earth's surface which require particular consideration. These are called the subequatorial and the subtropical belts, and they are illustrated in Fig. 85. In the case of the northern subtropical belt, when the horse latitudes are farthest north, most of the belt is covered by the northeast trade winds. When the horse latitudes are farthest

The kind of winds found in each belt. south, most of the belt is covered by the prevailing westerlies. Thus places in this belt in the course of a year will experience calms, southwest winds, and northeast winds. The corresponding subtropical belt in the southern hemisphere will experience calms, southeast trade winds, and northwest winds. In the case of the subequatorial belt a further complication arises. When the doldrums are farthest north, the southeast trade winds from the southern hemisphere cross the equator. After crossing the equator, they come under the deflective influence of the earth's rotation and become a southwest wind. Thus, places in the subequatorial belt north of the equator experience calms, northeast winds, and southwest winds in the course of a year. A place south of the equator in this belt will experience calms, southeast trade winds, and northwest winds.

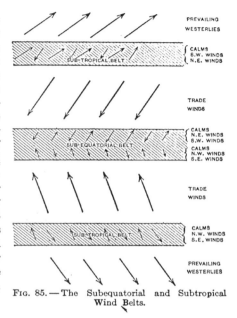

FIG. 85. — The Subequatorial and Subtropical Wind Belts.

CONTINENTAL WINDS

164. Definition. — The typical system of planetary winds which must exist on every planet heated at the equator and turning eastward on its axis must be modified in two directions in order to fit it more exactly to our earth as it actually exists. The first one of these two modifications has already been discussed. The sun migrates 47° in the course of a year and, as a result of this migration, the equatorial belt of high temperature, the equatorial belt of low pressure, and the wind system of the world migrate north and south, lagging behind the sun and each migrating over a less distance. This migration of the wind system gives rise to three belts known as the subequatorial and the two subtropical belts, where calms and winds blowing from two opposite directions in the course of a year exist. The planetary winds thus modified are called the terrestrial winds. The second modification is due to the fact that the earth's surface is not uniform. It is not an all land surface, or an

The two modifications of the typical system of planetary winds.

Definition of terrestrial and continental winds.

all water surface, but is highly diversified, consisting of part land and part water. In addition the surface of the ocean has not the same temperature along a parallel of latitude due to the presence of ocean currents. The terrestrial winds modified by taking account of this diversity of surface are called continental winds.

In the chapter on temperature, the fact was emphasized that the continents during the winter are colder than the adjoining oceans under the same latitude, while in summer they are warmer than the adjacent oceans. In view of the convectional theory, one would expect the continents to become areas of low pressure during the summer and of high pressure during the winter.

The effect of diversities in the surface.

This tendency must strongly modify and distort the belts of pressure running round the world. The oceans, due to the presence of hot or cold ocean currents, have not the same temperature along a parallel of latitude. As a result one would expect highs to build over the colder parts of an ocean while lows would form over the warmer parts. As a result of this the belts of pressure would again be modified or broken up. The general tendency, then, of diversities in the surface would be to form peaks and depressions and to break up continuous belts.

In section 119, when the isobars for January and July were being considered, it was found that there were eight areas of high pressure and six areas of low pressure to be considered. Six of these highs and three of these lows either persist throughout the whole year or are so prominent during a large portion of it that they make themselves felt in the annual averages and appear on the chart of isobars for the year. These peaks and depressions are often spoken of as the " permanent highs and lows " or as " the centers of action " because they exert such a marked influence, as will be seen later, on the wandering areas of high and low pressure which constitute our storms and determine the weather to such a marked extent. This tendency of diversities in the surface to break up the belts of pressure modifies to such an extent the typical planetary wind system that it is sometimes claimed that it does not really exist on the poleward side of the tropics, but is replaced by an air circulation about these highs and lows.

The eight highs may be divided into two groups. There are five which are located over the ocean and in the region of the high pressure belts at 35° N. and 30° S. latitude. They are also nearly permanent in position and persist throughout the year. They are also located on the eastern side of the ocean.

A study of the eight highs and six lows.

They are located in the North Pacific opposite California, in the North

Atlantic west of Spain, in the South Pacific, in the South Atlantic, and in the Indian Ocean.   In each case, they are located near the place where a cold ocean current flowing towards the equator crosses the belts of high pressure and to this is without doubt due the breaking up of the belt and the building of the peak of pressure.   These facts have just (1911) been brought out by Humphreys in a valuable article in the Mount Weather Bulletin and for further details the reader must be referred to this article.   The other three highs are located over land surfaces, namely, over Siberia, over North America, and over South Africa.   In each case they appear during the winter and disappear during the summer.   The Siberian high builds to such an immense height during the winter that its influence is felt in the yearly averages and it also appears on the chart showing the isobars for the year.   These are without doubt due to the low temperatures to which the continents fall during the winter.   During the summer, when the land surface heats up, they disappear.

The six lows may also be divided into two classes.   Four of them are located in the equatorial belt of low pressure.   They are in South Africa, South America, Australia, and India.   In each case they appear during the summer and disappear during the winter. They are due without doubt to the excessive heating of the land surface during the summer.   The other two lows are located near Iceland and just south of Alaska, in each case over the ocean.   The Iceland low persists throughout the year, while the Aleutian Islands low disappears during the summer.   These are the two depressions into which the polar cap of low pressure breaks up.   These are without doubt due, as Humphreys has also pointed out, to the Gulf Stream and the Japan current.   The warm water brought to the far north by the Gulf Stream stands in sharp temperature contrast to the perpetual ice and snow of Greenland and Iceland.   As a result a low persists over the ocean throughout the year.   In the case of the ocean near the Aleutian Islands the same thing is true during the winter.   During the summer, both Alaska and the coast of Asia heat markedly.   As a result the temperature contrast and also the low disappear.

Charts XIII and XIV represent the air circulation of the Atlantic Ocean for January–February, and July–August.   The various components of the planetary wind system are clearly evident.   The trade winds, the prevailing westerlies, the doldrums, and the horse latitudes are all easily recognized. They are not separated, however, by lines that run parallel to latitude

The air circulation of the Atlantic Ocean.

lines ; but a distinct tendency is shown to blow in a spiral direction around these areas of high and low pressure.

165. Monsoons. — This tendency of the continents to become the centers of high pressure during the winter with spirally outflowing winds, Definition of and the centers of areas of low pressure during the summer a monsoon. with spirally inflowing winds, is not usually sufficient to cause a complete reversal of the wind direction between winter and summer. The temperature differences and the resulting pressure differences are too slight to bring about this result. The usual result is an increase or decrease in wind velocity, and a slight change in wind direction between summer and winter. There are several instances, however, when a complete reversal is brought about, and these winds which change direction with the seasons are spoken of as monsoons. The word is of Arabic origin and means " season." Monsoons are particularly notice-Where mon- able in connection with those places which are located in the soons occur. subtropical or subequatorial wind belts. The monsoonlike changes in the wind direction brought about by the migration of the wind system when aided by continental influences readily develop into pronounced monsoons.

The most typical monsoon in the world occurs in India. During the winter an immense peak of high pressure forms over southern Siberia, with a central pressure of 30.50 inches. The wind blowing The cause and de-scription of the India monsoon. spirally outward from this area of high pressure causes north-east winds over India. The air coming over southern Siberia and across the Himalaya Mountains is dry, and this is spoken of as the dry or winter monsoon. During the summer the equatorial belt of low pressure migrates to the north of India, and the wind blowing spirally inward towards this belt of low pressure causes southwest winds to blow over India. The air coming from the ocean is moisture-laden, and when forced to rise in this area of low pressure and also by the Himalaya Mountains, it causes drenching rains over India. This is called the southwest or wet monsoon. The coming of the south-west monsoon has been vividly described by many authors. For several weeks previous, the air has been nearly calm and the breezes have been light. Then the wind begins to blow from the southwest, at first fit-fully, but gradually attaining greater velocity and steadiness. The dark clouds on the horizon and the breaking of the waves on the shore announce the coming of the steady southwest wind. Soon, with light-ning flashes and thunder, the rain, which descends in torrents, perhaps lasts for two or three weeks and marks the beginning of the monsoon. The

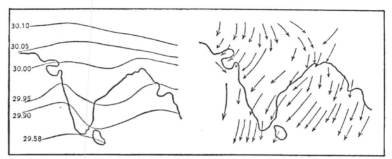

FIG. 86. — Pressure and Wind Distribution over India during January.
(From SALISBURY'S *Physiography*.)

clouds then break away, and wind blows steadily and freshly from the
southwest, and rains are frequent during the whole of the monsoon
period.   Figures 86 and 87 indicate the pressure and wind distribution
over India during January and July.

Monsoon winds are also noticeable in the peninsula of Spain and
Portugal.   Figures 88 and 89 illustrate the pressure and wind direction
during January and July.   It will be seen that the penin- The mon-
sula is the seat of low pressure with inflowing winds during soons of
the summer and the seat of high pressure with outflowing Spain and
winds during the winter.   Australia also exhibits a well- and of
marked monsoon, as illustrated by Fig. 90.   Other portions Australia.
of the world also exhibit monsoon tendencies, but not to such
marked extent as those countries just mentioned.

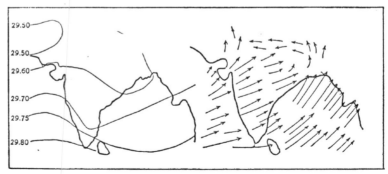

FIG. 87. — Pressure and Wind Distribution over India during July.
(From SALISBURY'S *Physiography*.)

166. Other land effects. — The various results due to the fact that
the earth is not a water surface throughout, but has a diversified surface
The seven   made up of part land and water, may be briefly brought
land effects. together and summarized here: (1) *The belts of pressure
are broken up into peaks and depressions which are called the permanent
highs and lows.* (2) *The continents are the cause of the monsoons which*

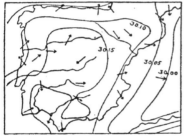

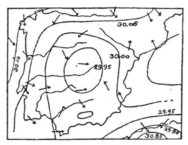

Fig. 88. — Pressure and Wind Direction   Fig. 89. — Pressure and Wind Direction
in Spain and Portugal during January.   in Spain and Portugal during July.

have just been described. (3) *The diurnal change in wind velocity and wind
direction is brought about by the presence of land.* During the day the

SUMMER   WINTER
Fig. 90. — Wind Direction in Australia during January and July.

wind velocities are greater than during the night. The wind direction
also changes slightly during the day and shifts back again during the
night. The cause of this is convection, which causes the upper air during
the day to come down to the earth's surface, bringing with it its higher
velocity and changed direction. Over the ocean the diurnal change in
either wind velocity or direction is barely noticeable. (4) *The conti-*
N

nents also cause an annual change in wind direction. This is brought about by the fact that they become centers of high pressure during the winter and low pressure during the summer. (5) *Mountain chains cause changes in wind direction and velocity.* This is not only true on a small scale in connection with local barriers, but the longer and higher mountain chains of the world influence the general winds of the world both as regards direction and velocity. (6) *Wind velocity is much reduced by friction when the air is passing over land.* This is proven by the fact that the normal wind velocity over the ocean is nearly twice what it is over the land. (7) *The Arctic winds, if they exist, are without doubt the result of a land surface.* Only few observations have been made in the polar regions, but they would seem to show that the barometric pressure does not decrease steadily as the pole is approached, but that it reaches a minimum in about 70° or 80° north latitude, and then increases slightly toward the pole. In the polar regions, winds blowing from the north instead of the southwest have also been observed. These would be the inevitable result of an increase of pressure toward the pole. If these observations are trustworthy, the explanation is without doubt to be found in the increasing percentage of land near the north pole. This would decrease the wind velocity so that the highest values of wind velocity and thus the least pressure would be found, not at the north pole itself, but perhaps 20° or 30° from it. The observations, however, are not sufficiently numerous or trustworthy to make an elaborate explanation or treatment of the subject necessary.

LAND AND SEA BREEZES

167. At the seashore the land is heated more than the adjacent water during the day and at night by rapid cooling becomes colder than the adjacent water. This gives rise to a small convectional circulation known as land and sea breezes. The sea breeze ordinarily begins about ten or eleven o'clock in the morning and blows gently but with increasing velocity from the ocean toward the land. It does not usually penetrate more than ten or fifteen miles inland, and it reaches its maximum velocity ordinarily at two or three in the afternoon. At about sunset the sea breeze dies out, and it is replaced during the night by the land breeze, which blows from the land toward the ocean. The velocity of the sea breeze and its regularity are usually more marked than in the case of the land breeze. The land and sea breezes, as has been fully shown by observations made

The description of land and sea breezes.

by kites and captive balloons, do not ordinarily extend to a greater height than, say, a thousand feet. The phenomenon, therefore, of land and sea breezes is confined to a space of fifteen miles either side of the coast line, and to the lower thousand feet of atmosphere.

The explanation of the land and see breezes may be thus stated: During the early hours of the morning the land heats rapidly under

The explanation of land and sea breezes. sunshine and becomes warmer than the adjacent water. The air expands, flows off aloft over the ocean, thus increasing the pressure slightly over the ocean, and decreasing it over the land. This increased pressure over the ocean drives in the air in the form of the sea breeze. At night the land and the layer of air next to it cool rapidly, and the converse process takes place, the air being forced to move by the increased pressure from the land toward the ocean.

It has been often observed that, when the sea breeze first makes its appearance, it starts some distance from the land, as is shown by the

Why the sea breeze starts some distance from the land. rippled surface of the water, and then slowly beats its way in toward the shore. This can be readily explained. As the air over the land becomes heated during the early morning hours, it tends to expand laterally as well as to flow off above over the ocean. This lateral expansion would tend to hinder the coming of the sea breeze near the shore. Furthermore it is the increased pressure over the ocean which drives in the sea breeze, and, as this is first felt at a considerable distance from the shore, it is there that one ought to expect the first beginnings of the sea breeze.

In many places the sea breeze is not felt as a breeze blowing directly from the ocean during the day, and the land breeze at night does not

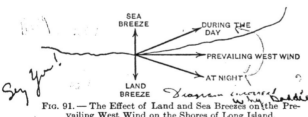

FIG. 91. — The Effect of Land and Sea Breezes on the Prevailing West Wind on the Shores of Long Island.

Land and sea breezes combine with the prevailing wind. blow directly from the land, but they are compounded with the prevailing wind and thus merely cause changes in the direction or velocity of the prevailing wind. On the shores of Long Island the prevailing wind direction is west, and as is shown in Fig. 91, the sea breeze changes this to a southwest wind by day and the land breeze is combined with it and becomes a northwest wind during the night. On the coast of California and Chile, where the prevailing wind direction is west, that is, from the ocean toward the land, the sea breeze during the day causes a marked increase in the wind velocity, while the land breeze during the night reduces it to practically a calm. This is said to be particularly pronounced in the portions of Chile near Valparaiso. Here the wind velocity during the day may amount almost to a decided gale, hindering walking, making the transaction of business unpleasant, and almost cutting off intercourse between vessels in the harbor and the shore. At night an almost complete calm reigns.

MOUNTAIN AND VALLEY BREEZES

168. The so-called mountain and valley breezes are well known in all mountainous countries, but they are particularly noticeable in long, narrow valleys which emerge on a large open plain below. They are periodic, sun-caused winds which are particularly well developed on still clear days. During the night, from a few hours after sunset until early morning, the mountain breeze flows down from the mountains through the valley to the plain below. It is not only easily detected, but at times it attains a moderate velocity; and those who camp at night in the open in mountainous countries soon learn to build camp fires on the downhill side, so that the mountain breeze may blow the smoke away from their tents. During the day the valley breeze is felt as a gentle breeze blowing up through the valley and up the mountain slopes. It is usually hardly noticeable and is never as well developed as the mountain breeze.

Description of the mountain and valley breeze.

During the night, the layer of air next the ground becomes cooled by radiating its heat to the colder ground and to the sky, and also by conduction to the cold ground. This layer of cold, and thus dense, air drains, just as water would, into the valleys and places of less elevation than surrounding regions. As a result, depressions are filled at night with pockets of colder air which have drained into them from the surrounding slopes. In a long, narrow valley this drainage of colder air makes itself felt as a mountain breeze. If a glacier is located in the valley, the layer of air next it may be cooled sufficiently to cause a mountain breeze even during the daytime. As the mountain breeze moves down the valley the air is compressed and thus is heated to the extent of 1.6° for every 300 feet of descent. If the motion down the valley is slow so that there is sufficient time for the air

The cause of the mountain breeze.

to radiate its heat to the ground and sky, or to lose it by conduction, it may reach the open plain below as cold or even colder than the air over the plain. If, however, the descent has been rapid, the air will probably reach the plain with a higher temperature than the air over the plain, and to this fact may be due the often observed result that frosts are less severe on a plain opposite the point of entering of a long valley.

During the day the air at the bottom of a valley is heated more than the air along the mountain slopes, particularly as one side of a valley is nearly always in shade. As a result of this heating, the isobaric surfaces are warped upward.

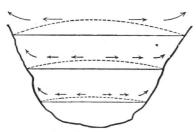

FIG. 92. — Cross Section of a Valley, showing the Isobaric Surfaces.

The cause of the valley breeze. In Fig. 92, which shows a cross section of a valley, the normal isobaric surfaces are represented by full lines, while the upward bulging of these surfaces is indicated by dotted lines. As a result of this upward bulging of the isobaric surfaces, the air flows off toward the side; and, when it meets the sloping sides of the valley, it is deflected upward, thus causing valley breezes which flow up the mountain slopes during the day.

ECLIPSE, LANDSLIDE, TIDAL, AND VOLCANIC WINDS

169. Eclipse winds. — When, during an eclipse of the sun, the shadow of the moon falls upon the earth, the sun's radiant energy is withheld from a belt which represents the path of the eclipse. As a result of this, the air ought to be colder here than in the surrounding regions, and thus a belt of high pressure should be developed along the path of an eclipse of the sun. The air would gently descend in this belt of high pressure and flow outward in both directions from it along the earth's surface. The few observations which have been made in connection with eclipses would seem to show that these conditions are actually realized. But few observations have, however, been made; and the winds due to an eclipse are of such seldom occurrence, of such little importance, and of such slight intensity that they are hardly worth considering. They would be classified as irregular sun-caused winds.

The characteristics and cause of eclipse winds.

170. **Landslide and avalanche winds.** — Masses of air are often pushed ahead of landslides and avalanches in sufficient quantity to cause destruction even at considerable distances. The rush of air ahead of an avalanche is sometimes felt at a distance of a mile, and destruction has been caused at a distance of a quarter of a mile. These winds are ordinarily classified as *Landslide and avalanche winds.* irregular earth-caused winds. In the case of an avalanche wind, however, it is indirectly of solar origin, for it is the energy of the sun which has evaporated the water and caused the circulation of the air which has deposited the snow on the mountain side to build the avalanche.

171. **Tidal winds.** — In some gulfs and bays, where the rise and fall of the tide is considerable, a slight motion of the air away from the gulf or bay when the tide is coming in and toward the gulf or bay when the tide is going out has been observed. This would seem to be caused by the actual raising of the atmosphere by the incoming water as the tide comes in and by the *The characteristics and cause of tidal winds.* falling of the atmosphere to take the place of the out-going water as it recedes. These tidal winds, if they exist, would be compounded with the land and sea breeze in such a way that it would require careful observations at many stations surrounding the gulf or bay to be sure of their presence. Such breezes, if they exist, would be classified as periodic moon-caused winds.

172. **Volcanic winds.** — It has been observed in connection with certain volcanoes while in eruption that there seems to be a rising air column above the volcano accompanied by inflowing breezes from all directions. There are two causes for the rising of air over *Volcanic winds.* the volcano. In the first place it may be caused by the explosive action of the eruption; and secondly, the heating of the air by the hot lava and other material ejected from the volcano may cause convectional motion. A volcano in violent eruption is sometimes capped by a thundershower with violent thunder and lightning. These volcanic winds are of telluric origin and are irregular as to time of occurrence.

CYCLONIC STORMS

173. The general winds of the earth have now been considered in detail; and, as a result of what has been said, it ought to be possible to state for any portion of the earth exactly what changes in wind direction and wind velocity ought to take place in the course of a day or a year. In the northeastern portion of the United States, for example,

the prevailing wind direction should be from some westerly quarter, since this portion of the country is located in the region of the prevailing westerlies. The wind ought to blow more from the southwest in summer and from the northwest in winter, when the extensive area of high pressure builds over North America due to the low temperatures. The wind velocity ought to be higher in winter than in summer, reaching its maximum in February or March and its minimum in August or September. The velocity ought to be somewhat greater during the daytime than at night, and there should also be a slightly diurnal change in wind direction. If a place is located near the seashore or in a mountainous region, the land and sea breezes and the mountain and valley breezes ought to make themselves felt. If careful observations of the changes in wind direction or in wind velocity are, however, made for any period of time, it will be at once noticed that the irregularities both as regards direction and velocity are numerous. For example, during the winter, when the wind ought to be blowing from the northwest with moderate velocity, the wind will perhaps be blowing from the east with high velocity accompanied by driving snow. This may continue during a whole night and, instead of increasing, die down as the day comes. All of these irregular and unlooked-for wind directions and wind velocities are due to the presence of so-called storms, of which there are four different kinds: extratropical cyclones or the lows of our weather map; tropical cyclones, which are sometimes called hurricanes in the West Indies or typhoons in the China Sea; thunder showers; tornadoes. These four storms have been arranged in order of size, for the extratropical cyclones sometimes cover an area a thousand or more miles in diameter, while the tornadoes, although more violent than any of the others, cover but an extremely small area. These are usually spoken of as cyclonic storms, because the air in each case is usually in motion in a spiral direction. These will be fully considered in one of the following chapters.

The characteristics as regards direction and velocity of the winds in the northeastern part of the United States.

There are numerous irregularities.

The irregularities are due to storms, of which there are four kinds.

WINDS OF OTHER PLANETS

174. A careful consideration of those changes in the general winds of our own earth which would result from a change in the conditions upon which they depend will serve as an introduction to the consideration of the wind system which ought to exist on the various planets.

If the earth turned more rapidly on its axis, the deflective force would be greater, and the circulation between equator and pole would be more oblique than at present. As an example, the trade winds, in the northern hemisphere, instead of blowing from the northeast would blow from a still more easterly point. The effect of having a more oblique circulation would be to cause higher wind velocity, to retard the exchange of air between equator and pole, and thus to accentuate the differences in temperature which already exist. *The effect on the wind system of a more rapid rotation of the earth.*

If the earth's axis were inclined to the plane of its orbit at a greater angle than 23.5°, the migration of the wind belts would be more pronounced, the subtropical and subequatorial belts would be wider, and the tendency to monsoons would be felt over a larger portion of the earth's surface. *The effect on the wind system of a larger inclination.*

If the earth's surface were entirely a water surface instead of being a diversified land and water surface, land and sea breezes, and mountain and valley breezes would be absent; the belts of pressure would be uniform around the world; monsoons would be unknown; the daily change in wind direction and wind velocity would no longer exist, and the yearly change in wind velocity would be very much less. *The effect on the wind system of a uniform surface.*

The factors, then, upon which the characteristics of the wind system depend are the rotation of the earth on its axis, the location of the belt of highest temperature, the migration of this belt of highest temperature during the year, and the character- istics and diversity of the surface. If these facts were known in detail for any planet, it ought to be possible to determine by deduction the characteristics of the wind system which should be present. *The factors which de- termine the wind system.*

175. Mercury and probably Venus turn the same face toward the sun always, and thus rotate on their axes very slowly, Mercury making a complete rotation in eighty-eight days and Venus in two hundred and twenty-five days. As a result the center of the illuminated hemisphere ought to become an area of low pressure with air blowing spirally inward along the surface of the planet, rising in this area of low pressure, *The proba- ble wind systems of the various planets.* flowing outward on the outside of the atmosphere, and descending again in an extensive area of high pressure which would form on the dark hemisphere of the planet. In the case of Mars, since the surface features are but seldom obscured by haze or cloud, no observations are available for determining the characteristics of the air circulation on that planet. Since it turns on its axis in about the same time as the earth, and has an axis inclined at about the same angle to the plane of its orbit, it ought to have a system very similar to ours.   In the case of Jupiter, Saturn, Uranus, and Neptune, the planets are probably continually cloud-covered, and, particularly in the case of Jupiter and Saturn, there are belts parallel to the equator which are easily observable.   These are probably due to the atmospheric circulation, but since no facts are known concerning the surface of these planets, it is impossible to form any definite deductions as to what wind system ought to exist.

QUESTIONS

(1) Why is pressure an important meteorological element?   (2) Why does the atmosphere exert a pressure?   (3) Is the atmospheric pressure constant? (4) How is the pressure of the atmosphere measured and expressed?   (5) Give a detailed history of the invention of the mercurial barometer.   (6) Describe a mercurial barometer as used at present.   (7) Describe in detail the cistern of a mercurial barometer.   (8) What are the two steps in reading a mercurial barometer.   (9) Name and treat fully the three corrections to be applied to the reading of a mercurial barometer.   (10) Describe an aneroid barometer. (11) What corrections must be made to the reading of an aneroid barometer?   (12) Compare the accuracy of an aneroid and mercurial barometer.   (13) What is a barograph?   (14) Describe the Richard Frères barograph.   (15) Describe the construction and action of the so-called " mouth barometer."   (16) Describe the construction and action of the " chemical weather glass."   (17) What causes the changes in its appearance?   (18) What observations of pressure are taken at the various Weather Bureau stations? (19) What instruments are used and where are they located?   (20) How are the pressure normals computed? (21) Describe the diurnal variation in pressure.   (22) With what does the diurnal variation in pressure change?   (23) Treat fully the various explanations of this diurnal variation.   (24) Describe the annual variation in pressure and state its cause.   (25) What other barometric data may be computed from the observations of pressure?   (26) Describe the old method of reducing pressure observations in order to compare them.   (27) Describe the modern way of reducing pressure observations to sea level.   (28) What observations are necessary to determine altitude by means of the barometer?   (29) How are these observations secured in practice?   (30) Describe in detail the four methods of computing the difference in elevation from the observations made.   (31) How are isobaric charts constructed?   (32) Describe the chief characteristics of the isobars for the year.   (33) How is the vertical section along a meridian constructed?   (34) Describe the course of the thirty-inch line.   (35) Why do differences in pressure exist over the earth's surface?   (36) What is meant by an isobaric surface?   (37) What is the normal form of an isobaric surface? (38) What is the form of isobaric surfaces over areas of high and low pressure? (39) Describe the chief characteristics of the isobars for January and July. (40) What other pressure charts might be constructed?   (41) Define wind and state what three things can be measured.   (42) Define windward, leeward, veering, and backing.   (43) To how many points of the compass is the wind

direction determined? (44) What is the relation between wind velocity and the pressure of the wind? (45) Describe the wind vane of the Weather Bureau form. (46) What are the advantages of this form of wind vane? (47) What is the anemoscope? (48) Describe the rotating cylinder form of anemoscope. (49) Describe the electrical contact form of anemoscope. (50) What is meant by a wind scale? (51) Describe the ten-point and the Beaufort wind scale. (52) State the three groups of anemometers. (53) Describe a simple deflection anemometer. (54) Describe two pressure anemometers. (55) Describe the Robinson cup anemometer. (56) How is this instrument tested and its constant determined? (57) Describe other forms of anemometers. (58) How can small wind velocities be determined? (59) What are the chief effects of surroundings on wind direction and velocity? (60) Describe the influence of a valley. (61) Describe the effect of buildings. (62) Describe the effect of the nature of the surface. (63) What is the effect of altitude? (64) What are the advantages of mountain observatories? (65) What observations of wind are taken at regular Weather Bureau stations? (66) State the instruments used and their location. (67) Define prevailing wind direction. (68) How may prevailing wind direction be expressed? (69) Describe the construction of a wind rose. (70) How are normal wind velocities computed? (71) Describe the daily variation in wind velocity and state its cause. (72) Describe the daily variation in wind direction and state its cause. (73) How may the diurnal variations in wind direction and velocity be expressed graphically? (74) Describe the annual variation in wind velocity and state its cause. (75) Describe the annual variation in wind direction and state its cause. (76) What is the cause of irregular variations in wind direction and velocity? (77) What other wind data may be determined from the observations? (78) State the different characteristics of the prevailing winds of the world. (79) What other wind charts may be prepared? (80) Describe the general convectional motion in a long tank heated at the bottom. (81) Why may a convectional circulation between equator and pole be expected? (82) What other convectional circulations would be expected? (83) Illustrate by means of two diagrams the arrangement of the isobaric surfaces in the general convectional circulation just before its beginning and after it has become permanently established. (84) What is the meridional view of the air circulation? (85) State the condition of steady motion. (86) What is meant by a barometric gradient? (87) How is it expressed? (88) What is the relation of wind direction to pressure gradient and isobaric lines? (89) What is the reason for the relation? (90) Describe the air motion about highs and lows. (91) State the historical rise of the recognition of the effect of the earth's rotation on air motion. (92) What was Halley's explanation of the direction of the trade winds? (93) Describe Ferrel's work. (94) State the effect of the earth's rotation on the air motion. (95) State the effect of the earth's rotation on the air masses moving poleward on the outside of the atmosphere. (96) Explain in full the cause of the polar caps of low pressure. (97) Illustrate by means of two diagrams the meridional section of pressure on a rotating and on a non-rotating earth. (98) Illustrate by means of two diagrams the air motion about highs and lows on a rotating and non-rotating earth. (99) What is Buys Ballot's law? (100) What is the reason for its truth? (101) Contrast the inductive and the deductive method of gaining information as illustrated in this chapter. (102) What is meant by the general winds of the globe? (103) What is the basis of the Dové classification? (104) What is the basis of the Davis classification? (105) Define planetary winds. (106) Illustrate by means of a diagram the air circulation near the earth's surface. (107) Illustrate by means of a diagram the air circulation in the upper layer of the atmosphere. (108) Illustrate by means of a diagram the air motion in the intermediate layer. (109) Illustrate by means of a diagram the atmospheric circulation in cross section. (110) What are the effects of the earth's rotation? (111) Describe the trade winds. (112) Describe the doldrums. (113) Describe the horse latitudes. (114) Describe the prevailing westerlies. (115) Describe the upper currents. (116) In what two ways must the planetary system of winds be modified? (117) Define terrestial winds. (118) What causes the yearly migration of the wind system? (119) What are the characteristics and amount of this migration? (120) Describe the subequatorial and subtropical wind belts. (121) Illustrate by means of a diagram the kind of winds found in each belt. (122) Define continental winds. (123) What is the effect of a diversified surface on the belts of pressure? (124) What effect does the building of peaks of pressure and depressions have on the air circulation? (125) What is a monsoon? (126) Describe the monsoons of India. (127) State in full the cause. (128) At what other place on the earth's surface are monsoons felt? (129) State the various effects of land on wind direction and wind velocity. (130) Treat fully the so-called Arctic winds. (131) Describe the land and sea breezes. (132) What is the explanation of land and sea breeze? (133) Why does the sea breeze start over the ocean. (134) In what way is land and sea breeze combined with other winds? (135) Describe the mountain and valley breezes. (136) Explain in detail the valley breeze during the day. (137) Explain in detail the mountain breeze during the night. (138) Describe in detail the air motion caused by eclipses, landslides, tides, and volcanoes. (139) How are these various winds classified in each system of classification? (140) What is the cause of irregularities in the general winds of the world. (141) Name the four storms. (142) What would be the effect on the wind system if the earth turned more rapidly on its axis? (143) What would be the effect on the wind system if the earth's axis were inclined at a greater angle? (144) Name the factors upon which the characteristics of the wind system of a planet depend. (145) Describe the probable wind systems of the various planets.

TOPICS FOR INVESTIGATION

(1) The history of the mercurial barometer.
(2) Modified forms of mercurial barometers.
(3) The various forms of barographs.
(4) The construction and action of chemical weather glasses.
(5) The diurnal variation in barometric pressure.
(6) The barometric determination of altitude.
(7) Wind scales.
(8) Anemometers.
(9) Mountain observatories and their work.
(10) Ferrel's contributions to meteorology.
(11) The monsoons of India.

PRACTICAL EXERCISES

(1) Test carefully an aneroid barometer. This might include its accuracy for a given range of pressures, the effects of temperature changes, and the effects of jars due to transportation.
(2) Study a chemical weather glass.

(3) Learn to read a mercury barometer and to reduce the observations by applying all the corrections.

(4) Plot the graphs showing the annual variation in pressure at several stations and explain the peculiarities of each graph.

(5) Determine a difference in elevation by means of a barometer.

(6) Work up some or all of the barometric or pressure data mentioned in section 110 for several stations. Those stations may be chosen in which the student has a particular interest. A barograph record of considerable length is almost essential. Extremes of pressure and the frequency and magnitude of the irregular variations would be the most interesting. The attempt should be made to draw general conclusions by contrasting different places, summer and winter, etc.

(7) Determine the wind velocity on the stillest nights of winter.

(8) Construct several wind roses for different places for the same month or year and for the same place for different months.

(9) Work up some or all of the wind data mentioned in section 136.

(10) Perform the experiment illustrated in Fig. 70.

(11) Investigate some valley for mountain and valley breezes.

REFERENCES

For the description, illustration, construction, and use of apparatus to measure pressure and wind, see:

ABBE, *Meteorological Apparatus and Methods* (Washington, 1888); pp. 111 to 179 and 183 to 336 treat of pressure measuring and wind measuring instruments respectively.

MOORE, JOHN W., *Meteorology*, 2d ed., pp. 120 to 138 and 265 to 292.

PLYMPTON, *The Aneroid Barometer* (D. Van Nostrand Company, New York).

WALDO, *Modern Meteorology*, pp. 59 to 135.

WILD, H., *Ueber die Bestimmung des Luftdruckes*, Rep. für Met. III, N. 1, pp. 1–145.

Report on the Barometry of the United States, Canada, and the West Indies, Report of the Chief of the Weather Bureau, 1900–1901, Vol. II.

MARVIN, C. F., Circulars A, D, F, Instrument Division of the U. S. Weather Bureau.

Instructions for Coöperative Observers (U. S. Weather Bureau, Washington).

Apparatus catalogues of various firms. See p. 110.

(See also the various guides to observers mentioned in Appendix IX, in group 2(B).)

For observations of pressure and wind consult the publications mentioned on p. 110.

For normal values of pressure and wind, see:

BUCHAN, ALEXANDER, *Report on Atmospheric Circulation*.

HANN, *Klimatologie*.

Report of the Chief of the Weather Bureau particularly for 1891–1892, 1896–1897.

Climatology of the United States (Bulletin Q of U. S. Weather Bureau by A. J. HENRY, 1906).

Summary of the Climatological Data for the United States, by Sections (106 are to be published).

For charts of pressure and wind, see:

BARTHOLOMEW, *Physical Atlas, Vol. III*.

Buchan, Alexander, *Report on Atmospheric Circulation* (Report on the Scientific Results of the Voyage of H.M.S. *Challenger*).

Hann, *Atlas der Meteorologie*, 1887.

Hildebrandsson, H. H., et Teisserenc de Bort, *Les bases de la météorologie dynamique*, Paris, 1900–1907.

Segelhandbuch der Deutschen Seewarte; Hierzu ein Atlas (for the Atlantic, Pacific, and Indian Oceans).

Summary of International Meteorological Observations (Bulletin A of the U. S. Weather Bureau).

Eliot, Sir John, *Climatological Atlas of India*, Edinburgh, 1906.

Russia, *Atlas climatologique de l'empire de Russie*, St. Petersburg, 1900.

Blodget, Lorin, *Climatology of the United States*, Philadelphia, 1857.

Greely, *American Weather*.

Report of the Chief of the Weather Bureau for 1900–1901, Vol. II.

Climatic Charts of the United States (U. S. Weather Bureau).

Climatology of the United States (Bulletin Q of the U. S. Weather Bureau).

For the daily change in barometric pressure, see:

Alt, Eugen, *Die Doppeloszillation des Barometers insbesondere im Arktischen Gebiete*. (Inaug. Diss.), 22 pp., 1909.

Angot, H., "Etude sur la Marche Diurne du Barometre," *Ann. Bu. Cent. Met.*, Paris, 1889.

Cole, Frank N., The Daily Variation of Barometric Pressure, W. B. Bulletin No. 6, 1892.

Fassig, Oliver O., The Westward Movement of the Daily Barometric Wave, Bulletin No. 31 of U. S. Weather Bureau or Monthly Weather Review for November, 1901.

Hann, Julius, " Theory of the Daily Barometric Oscillation," *Quart. Journ.*, Vol. 20, p. 40.

Hann, Julius, *Untersuchungen über die tägliche Oscillation des Barometers*, Wien, 1889.

Wagner, Arthur, "Die Temperatur verhältnisse in der freien Atmosphäre," *Beiträge zur Physik der freien Atmosphäre*, Band III, Heft 2/3.

Wild, Heinrich, Repetorium für Meteorologie, Vol. VI, No. 10 (La marche diurne du baromètre en Russie par M. Rykatchew).

For the barometric determination of altitude, see:

Smithsonian Meteorological Tables. See section 111 of this book.

Whymper, *How to Use the Aneroid Barometer* (John Murray), London.

Wilson, *Topographic Surveying* (Wiley and Sons).

For a classification, description, and explanation of the winds and the general circulation of the atmosphere, see:

Abbe, Cleveland, *The Mechanics of the Earth's Atmosphere, a Collection of Translations.* Second Collection; Washington, 1891. Third Collection; Washington, 1910.

Bigelow, Frank H., *Report on the International Cloud Observations* (Washington, 1900).

Brillouin, Marcel, *Mémoires originaux sur la circulation générale de l'atmosphere* (Paris, 1900).

Coffin, J. H., *The Winds of the Globe*.

Davis, *American Meteorological Journal*, March, 1888.

Ferrel, *A Popular Treatise on the Winds.* The preface of this book contains a list of his previous articles on the same and kindred subjects.

CHAPTER V

THE MOISTURE IN THE ATMOSPHERE

A. THE WATER VAPOR OF THE ATMOSPHERE

EVAPORATION

THE CONDITION OF THE ATMOSPHERE AS REGARDS MOISTURE

THE DETERMINATION OF THE MOISTURE OF THE ATMOSPHERE

THE RESULTS OF OBSERVATION

B. DEW, FROST, FOG

DEW

A. THE WATER VAPOR OF THE ATMOSPHERE

EVAPORATION

176. Water vapor. — The two terms, the moisture of the atmosphere and the water vapor in the atmosphere, are synonymous, and both refer to the water in invisible gaseous form which is always present in the atmosphere, but in amounts varying at the earth's surface from almost nothing to 4 per cent as a maximum. The change from the solid or liquid state to this invisible gaseous state is called evaporation; its opposite is condensation. Water vapor is supplied to the atmosphere from a variety of sources. Nearly three fourths of the earth's surface is a water surface, and the oceans, lakes, and other bodies of water supply by evaporation the larger portion of the water vapor to the atmosphere. Other sources of water vapor are the surface of the ground which is nearly always moist, the leaves of plants and vegetation in general, and the air exhaled from the lungs of animals. *Definition of water vapor. Evaporation and condensation. The sources of the water vapor of the atmosphere.*

Water vapor is lighter than air, the ratio being 100 : 62. That is a cubic foot of water vapor at a certain temperature and under a certain pressure has .62 of the mass of a cubic foot of air at the same temperature and under the same pressure. Moist air is thus lighter than dry air because a portion of the dry air has been replaced by water vapor, which is lighter. *Relative lightness of water vapor.*

177. Latent heat. — When evaporation takes place from a water surface, the molecules near the surface break away from the attractions of their neighbors and make their way into the air above. ⟨This breaking of the bond of adhesion between the molecules requires an expenditure of energy; and the energy used up in *Definition of latent heat.*

evaporation, that is, in separating the molecules, is called latent heat. This energy may be supplied from two sources. The molecular energy

The two sources of latent heat. of the body itself may be used up and in this case the body becomes cold; or radiant energy in the form of ether waves may be used up directly to supply the latent heat. The amount of latent heat involved in the change from the solid to the liquid, or from the liquid to the gaseous state, is large. If the Centigrade

Values of latent heat. system of measuring temperatures is used and the unit of heat is defined as the amount of heat required to raise the temperature of a gram of water 1° C., then the amount of latent heat required to change a gram of ice to a gram of water at the same temperature is 79, and the number of heat units required to change the gram of water to a gram of water vapor at the temperature of boiling water is 536. If the Fahrenheit system of measuring temperature is employed and the heat unit is defined as the amount of heat required to raise one gram of water 1° F., then the latent heat in changing from the solid to the liquid state is 143, and from the liquid to the gaseous state 966. If evaporation takes place at a lower temperature than the boiling point of water, a larger amount of latent heat is required. For example, in the Fahrenheit system of thermometry it requires 1092 heat units to change a gram of water at 32° to water vapor at the same temperature.

This latent heat makes itself felt in many different ways. One reason why the ocean rises so little in temperature under the direct rays of the

Illustrations of the part played by latent heat. sun is because so large an amount of the radiant energy of the sun is used up in causing evaporation instead of in heating the water. One reason why the air temperature at the north pole is so low in summer in spite of the large amount of insolation received is because the surface is a snow and ice surface and the insolation is used up as latent heat in melting the snow and ice instead of in raising the temperature of the air. After it has rained in summer, the air usually remains cooler for a considerable time. The reason is that after a rainfall the ground and all surfaces are wet, and the drying of these surfaces uses up the radiant energy of the sun, and this prevents a rapid rise in the air temperature.

178. **Amount of evaporation.** — The amount of evapora-

The amount of evaporation depends upon many things. tion depends upon a great variety of things: such as the nature of the surface from which the evaporation is taking place, the amount of water vapor already in the atmosphere, the temperature, the velocity of the wind, and the barometric pressure. Each surface has its own rate of evaporation. Areas

covered with vegetation under the same conditions evaporate about
one third more water than a free water surface.   In measuring evaporation, a free water surface is usually taken as the standard.   The more
water vapor there is already in the atmosphere, the slower the evaporation, while high temperature and high wind velocity favor evaporation.
Evaporation is least when the barometric pressure is largest.   If the
barometric pressure is higher, the number of molecules in each cubic
foot of air resting upon the evaporating surface is larger
and the molecules of water find more hindrances and
greater difficulty in making their way into the air
above.   Thus the higher the pressure, the slower the
evaporation.

The amount of evaporation is usually determined by
exposing large pans of water in the open and measuring
the amount evaporated in a given time.
These evaporating pans should have a large   The method
area and considerable depth, so that the   of determin-
temperature will be approximately the same   ing the
                                              amount
as the temperature of the surrounding areas.   evaporated.
Values of the amount evaporated have also been determined by measuring the evaporation from inclosed
tanks, reservoirs, etc.   Relative values of the amount
evaporated may be determined by means of an ingenious little instrument pictured in Fig. 93, and   The Piche
called a Piche evaporimeter.   It consists   evaporim-
essentially of a long glass tube graduated   eter.
to cubic centimeters or cubic inches.   A piece of rough
paper is held against the open end of this glass tube by
means of a brass spring and plate.   The rough paper
is kept wet, and as the water evaporates from it, it is
replaced by water from the glass tube.   The amount

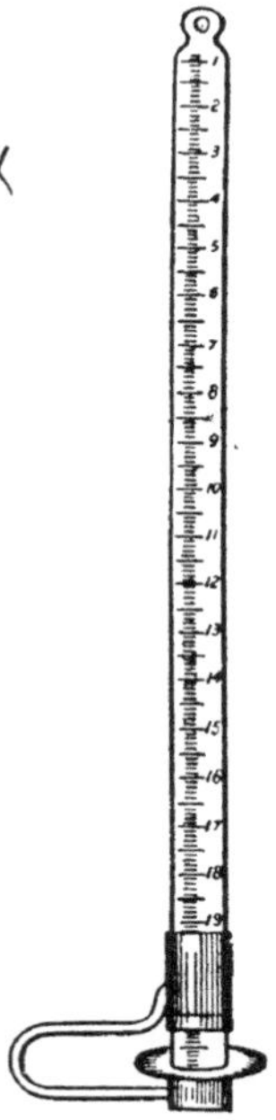

FIG. 93. — Piche
Evaporimeter.

evaporated from the rough paper in a given time
can thus be determined by simply reading the amount of water
remaining in the glass tube.   By means of this piece of apparatus,
which is easily managed and quickly read, relative values of the
amount of evaporation on days of different types and at different
times ·of the year, and in different places, may be readily determined.   In order to determine by means of this instrument
the absolute amount evaporated, it must be standardized in
terms of a free water surface near it, under the same conditions.

o

Automatically recording evaporimeters have also been recently devised.[1]

The amount of evaporation varies all the way from a few inches to several hundred inches per year in dry, hot countries. The amount of The amount evaporation is greater during the day than at night and evaporated. greater during the summer than in winter, except under very special conditions. Unless the atmosphere is to become unduly filled with water vapor, the amount which condenses from it in the form of precipitation must equal the amount which evaporates in it. Thus for the earth as a whole the amount of evaporation must equal the amount of precipitation.

179. **Distribution of the water vapor.** — The water vapor which is supplied to the atmosphere by evaporation is distributed by three pro- Water vapor cesses, — diffusion, convection, and wind. Diffusion is is distri- the process whereby the molecules of water vapor, due to buted by diffusion, their own motion, make their way slowly from point to convection, point between the molecules of the air. It is at best a slow and wind. process, and due to this cause alone, water vapor would be transported but a few feet in many hours. Air rising due to convection carries the water vapor along with it, and in this way it is distributed throughout the lower three to five miles of the atmosphere. The wind and air currents complete the distribution by transporting the water vapor immense distances and scattering it widely.

THE CONDITION OF THE ATMOSPHERE AS REGARDS MOISTURE

180. **Capacity of air for water vapor.** — By the capacity of air for Definition water vapor is meant the amount of water vapor which a of capacity. given quantity of air can hold. The capacity depends It depends upon the temperature only. With increasing temperature on temper- the amount of water vapor which the air can hold in- ature only. creases rapidly; in fact, it increases at an increasing rate.

181. **Saturation.** — If a given quantity of air contains all the water vapor which it can hold; in other words, if its capacity is entirely Definition of satisfied, then it is said to be in a state of saturation. The saturation. amount of water vapor which saturates a given quantity of air may be expressed in grains per cubic foot or grams per cubic meter, or it may be expressed in terms of the pressure which it exerts in milli-

[1] For recent works on evaporation, see articles by BIGELOW in the *Monthly Weather Review* since 1905.

meters or inches. The following table, which applies to saturated air, gives for the Fahrenheit scale the amount of moisture in grains per cubic foot, the pressure in inches, and the mass of the satu- Table of rated air in grains per cubic foot. It also gives for the values. Centigrade scale the amount of moisture in grams per cubic meter, the pressure in millimeters, and the mass of the saturated air in kilograms per cubic meter.

Temperature Degrees F.	Vapor Pressure Inches	Amount of Water Vapor in Grains per Cu. Ft.	Mass of Saturated Air in Grains per Cu. Ft.
−30°	0.010	0.12	650
−20	0.017	0.21	634
−10	0.028	0.35	620
0	0.045	0.54	606
+10	0.071	0.84	593
20	0.110	1.30	580
30	0.166	1.97	568
40	0.246	2.86	556
50	0.360	4.09	544
60	0.517	5.76	533
70	0.732	7.99	521
80	1.022	10.95	509
90	1.408	14.81	497
+100	1.916	19.79	487

Temperature Degrees C.	Vapor Pressure MM.	Amount of Water Vapor in Grams per Cu. Meter	Mass of Saturated Air in Kilograms per Cu. Meter
−30°	0.38	0.44	1.45
−20	0.94	1.04	1.40
−10	2.15	2.28	1.35
0	4.57	4.87	1.30
+10	9.14	9.36	1.25
20	17.36	17.15	1.20
30	31.51	30.08	1.15
+40	54.87	50.67	1.11

In order to compute the amount of water vapor or saturated air which may be present in any given space, it should be held in mind that 7008 grains constitute one pound avoirdupois. Sup- Illustration. pose a room 15 feet square and 10 feet high were filled with saturated air at a temperature of 70° F. Its volume would be 2250 cubic feet and it would thus contain 2.6 pounds of water vapor and 167 pounds of moisture-laden air.

182. Humidity; absolute and relative humidity. — Humidity is defined as the state of the atmosphere as regards moisture. If the air were

absolutely dry, its humidity would be spoken of as zero. It is the humidity, as much as the temperature, which adds to the uncomfort-

Definition of Humidity. ableness of a sultry day in summer. In fact, the human body feels three things: temperature, wind, and moisture, and not, as is sometimes popularly supposed, temperature only. The cold on a windy day in winter is more penetrating than still cold,

The human body feels temperature, moisture, and wind. because the wind drives the cold air through the clothing into contact with the skin. The cold is also more penetrating on a damp day than on a dry day. The reason is because the moisture makes the clothing a better conductor and thus lessens the heat of the body. A moist, hot day in summer is much more oppressive than a dry, hot day, because the moisture in the atmosphere prevents that free evaporation of the perspiration from the human body which cools it. If an instrument could be invented which could indicate the feelings of the human body, it would be necessary to take account of temperature, wind, and moisture in its construction.

Absolute humidity is defined as the actual quantity of moisture present in a given quantity of air. It may be expressed as a certain

Definition of absolute and relative humidity. number of grains per cubic foot or a certain number of grams per cubic meter. By relative humidity is meant the ratio of the actual amount of water vapor present in the atmosphere to the quantity which could be there if it were saturated. Relative humidity is always expressed in per cent.

183. Dew point. — If the temperature of a quantity of air containing moisture is lowered, a temperature will be finally reached when the

Definition of dew point. given quantity of air is saturated with moisture and is containing all the moisture that it can hold. This temperature is spoken of as the dew point, and any further reduction in temperature must result in the condensation of some of the moisture in the form of dew, frost, fog, cloud, or precipitation.

184. Problems. — The four quantities, absolute humidity, relative

The interrelation of temperature, dew point, absolute humidity, and relative humidity. humidity, dew point, and temperature, are so linked together by the table given in section 181, that if any two of these four quantities have been observed or determined, the other two may be obtained by computation. The following examples will make clear the meaning of these four terms and their interrelations.

Example I. If the temperature is 70° F. and the dew point is 50° F., find the relative and absolute humidity.

From the table it is seen that air at 50° F. can hold 4.09 grains per cubic foot.

This same amount of moisture was present at 70° F. The absolute humidity is thus 4.09 grains per cubic foot. This simply says that if air containing 4.09 grains at a temperature of 70° F. is cooled down to 50° F., it is containing all the moisture it can, it is saturated, and has reached the dew point.

Air at 70° F. can contain 7.99 grains per cubic foot. The relative humidity is thus $\frac{4.09}{7.99}$ or 51 %.

Example II. If the temperature is 40° F. and the relative humidity (R. H.) is 70 %, find the absolute humidity (A. H.) and the dew point.

Air at 40° F. can hold 2.86 grains per cubic foot. Thus

$$.70 = \frac{A.\,H.}{2.86}. \qquad A.\,H. = 2.00.$$

From the table it is seen that 2.00 grains saturates air with a temperature a little above 30° F. The dew point thus lies between 30° and 31° F.

Example III. If the absolute humidity is 4.09 grains and the relative humidity is 80 %, find the dew point and temperature. From the table it is seen that 4.09 grains saturates air at 50° F. 50° F. is thus the dew point.

$$.80 = \frac{4.09}{\text{Possible amount}}. \qquad \text{Possible amount} = 5.11 \text{ grains.}$$

The air must have a temperature between 50° F. and 60° F. to hold this amount. Interpolation would give a value of about 56° F.

Since there are six possible combinations of four things taken two at a time, there are six possible examples. Only three have been solved, but the solution of the other three would be along the same lines as indicated above. The formula R. H. $= \dfrac{A.\,H.}{\text{Possible Amount}}$ and the table are sufficient to solve all six.

THE DETERMINATION OF THE MOISTURE OF THE ATMOSPHERE

185. Hygrometers for determining absolute humidity. — It has just been shown that the four quantities, temperature, absolute humidity, relative humidity, and dew point, are so linked together by means of a table that if any two of the four are known, the other two may be determined by computation. The methods for determining the real air temperature have already been fully discussed in Chapter III. The necessary apparatus is either a thermometer in a thermometer shelter, a sling thermometer, or a ventilated thermometer. The experimental methods and the necessary apparatus for determining the other three quantities must now be carefully considered. These instruments for determining the moisture

The four quantities, temperature, relative humidity, absolute humidity and dew point may all be determined by means of suitable apparatus.

of the atmosphere are called, in general, hygrometers [1] or moisture measurers.

Absolute humidity is determined ordinarily by means of the so-called chemical hygrometer. This consists usually of two U-tubes containing calcium chlorid ($CaCl_2$) and sulfuric acid (H_2SO_4). Anhy-

The chemical hygrometer for determining absolute humidity. drous phosphoric acid is perhaps better than sulfuric acid. A known quantity of the moisture-laden air is drawn through these U-tubes so slowly that the moisture is absorbed and thus by weighing the tubes both before and after the given quantity of air was drawn through, the amount of moisture in it, and thus the absolute humidity, may be determined. In performing the experiment care must be taken to have the air pass so slowly through the U-tubes that all of the moisture will be absorbed, and furthermore, certain precautions are necessary in order to prevent any moisture from gaining access to the tubes except from the air which has passed through them.

Other methods of determining absolute humidity. Another method of determining absolute humidity is to inclose a given quantity of the moisture-laden air in a glass vessel and then to extract the moisture by chemical means. By measuring the diminution in pressure, or if the pressure is kept the same, by noting the diminution in volume, the absolute humidity may be determined.

186. **Hygrometers for determining relative humidity.** — Relative humidity is determined ordinarily by means of the hair hygrometer.

The construction of a hair hygrometer. This consists essentially of a long human hair, from which the oil has been extracted by soaking it in alcohol or a weak alkali solution. A weak solution of caustic potash (KOH) or caustic soda (NaOH) is ordinarily used. The hair thus treated changes its length with changes in moisture, and it has been found by experiment that these changes in length are nearly proportional to changes in relative humidity. As shown in Fig. 94, one end of the hair is fastened rigidly to the frame, while the other passes over a cylinder and is held taut by a weight. An index is attached to the

How the instrument may be standardized. cylinder, which moves over a dial graduated from 0 to 100 per cent, and this indicates the relative humidity. The instrument may be standardized by comparing it with some other accurate hygrometer or by determining the 0 and 100 per cent points. The 0 point is verified by exposing the hygrometer to air which has been entirely desiccated by chemical

[1] ὑγρό s = moist; μέτρον = measure.

means. The instrument must be left a sufficient time in this moisture-free air to take up a constant reading. The 100 per cent point may be verified by exposing the hygrometer to completely saturated air. Air may be completely saturated by blowing live steam into it and then cooling it down to an ordinary temperature, or by spraying moisture into it by means of an atomizer. The hair hygrometer is also affected somewhat by temperature changes, but as these are slight they are ordinarily neglected. At best, the Its hair hygrometer is an inaccurate accuracy. instrument. If recently standardized, its indications may be trusted perhaps to 2 or 5 per cent. If this is not the case, its indications are not reliable to within 10 or 15 per cent.

FIG. 94. — The Hair Hygrometer.

Another form of hygrometer which works on the same principle is in very common use. The outer case is usually circular Another and has a diameter of an inch or form with two. The working part consists the same of a fine piece of spring copper principle. which has been bent into the form of a spiral spring. This is coated with some hygrometric material. A bamboo preparation is often used for this purpose and it is generally colored red. This hygrometric material changes length with moisture changes, and thus causes the spiral spring to wind up or unroll. These motions are communicated to a pointer which moves over a dial graduated to indicate relative humidity.

187. There are two other instruments for determining relative humidity, but they are scientific toys rather than accurate meteorological instruments. One is the so-called weather house. Two scien-This consists of a little box usually in the shape of a house tific toys. and ordinarily provided with two openings in which are figures of a man and of a woman. It is so arranged that if the air is dry, the woman appears; while if the air is damp, the man appears. These two figures are held by a twisted strand of hygrometric material which winds up or unwinds with moisture changes, and this causes the motion of the figures.

It has been found that a solution of cobalt chlorid of the proper strength will turn pink if the air is damp, and bluish in color if the air is dry. A piece of filter paper or cheese cloth saturated with a solution of the proper strength thus becomes a rough indicator of the relative humidity. An attempt has been made by standardizing the strength of the solution and arranging a scale of color, to make the instrument give quantitative results.

188. Dew point hygrometer. — The simplest possible instrument for determining the dew point consists of a bright tin cup, provided with

The construction and action of the dew point hygrometer. a thermometer as a stirring rod and partly filled with water. If ice water or cold water is added to the water in the cup, its temperature will be steadily reduced and there will come a moment when the outside of the cup will become coated with moisture. This means that the layer of air in contact with the cup has been cooled below the dew point, and the moisture has been deposited in the form of dew. The reading of the thermometer will thus indicate the dew point. A more accurate determination can be made if the water in the cup is now allowed to rise in temperature until the moisture on the outside disappears. The average of the two temperatures will give a very close approximation to the dew point. This simple method of determining the dew point has been modified in various ways. The cooling may be produced by causing ether in the cup to rapidly evaporate, and black glass has also been often used instead of the bright metal. In all the instruments, however, the fundamental principle remains the same.

189. Psychrometer. — The psychrometer [1] is an instrument which indicates directly neither absolute humidity, relative humidity, nor dew

Description of a psychrometer. point, but all three of these may be indirectly determined by means of its indications. It consists of two identical thermometers attached to a frame. One remains in its ordinary condition, while the bulb of the other is covered with a linen jacket to which is attached a wick which extends down to a vessel of water. The instrument is pictured in Fig. 95. The evaporation from the wet bulb thermometer cools it and thus causes a difference in the readings of the two thermometers. The underlying principle is this: the larger the amount of moisture in the atmosphere the less the evap-

The underlying principle. oration, and thus the smaller the difference between the indications of the two thermometers. This difference, however, is affected by pressure and by wind, as well as by the moisture present in the atmosphere. The higher the barometric pres-

[1] $\psi\upsilon\chi\rho\acute{o}s$ = cold ; $\mu\acute{\epsilon}\tau\rho\sigma\nu$ = measure.

sure, the larger the number of molecules in each given volume, and, as a result, the smaller the evaporation. The effect, however, of the pressure on the rate of evaporation is so slight that it is ordinarily neglected. If the air in contact with the wet bulb thermometer stagnates, it will soon become filled with moisture and further evaporation will cease. It is therefore necessary to maintain a constant supply of fresh air in contact with the wet bulb ther- *It is also affected by pressure and wind.*

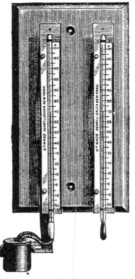

Fig. 95. — The Psychrometer.   Fig. 96. — The Whirled Psychrometer.

mometer. This may be done by rapidly whirling the instrument, or by blowing air against it by means of a fan, or by placing it in a tube and drawing air rapidly through the tube. Ordinarily, the thermometer is whirled rapidly, and such an instrument is illustrated in Fig. 96. The instrument is practically useless below 32° F. When the water in the wick and linen jacket freezes, evaporation still takes place, but the supply can no longer be kept good. Furthermore, the compression caused by the formation of the ice around the bulb of the thermometer often causes it to indicate a temperature from one-half to a degree too high. Psychrometer readings are for this reason often discontinued after the tempera- *How the effect of wind is obviated.* *Difficulties below 32° F.*

ture goes below 32° F. The accompanying table gives for various temperatures and various differences between the wet and dry bulb thermometer, the relative humidity and the dew point. For the reduction of a long series of observations, a more elaborate table will be found more convenient.

Difference of Readings of Dry and Wet Bulbs		Temperature of Air — Fahrenheit											
		−10°	0°	10°	20°	30°	40°	50°	60°	70°	80°	90°	100°
1	D.P.	−22	−7	5	16	27	38	48	58	69	79	89	99
	R.H.	55	71	80	86	90	92	93	94	95	96	96	97
2	D.P.	−76	−18	−1	12	24	35	46	57	67	77	87	98
	R.H.	10	42	60	72	79	84	87	89	90	92	92	93
3	D.P.	—	−39	−9	7	21	33	44	55	66	76	86	96
	R.H.	—	13	41	58	68	76	80	84	86	87	88	90
4	D.P.	—	—	−22	1	17	30	42	53	64	74	85	95
	R.H.	—	—	21	44	58	68	74	78	81	83	85	86
6	D.P.	—	—	—	−18	7	24	37	49	61	72	82	93
	R.H.	—	—	—	16	38	52	61	68	72	75	78	80
8	D.P.	—	—	—	—	−8	16	31	45	57	68	79	90
	R.H.	—	—	—	—	18	37	49	58	64	68	71	74
10	D.P.	—	—	—	—	—	4	25	40	53	65	77	87
	R.H.	—	—	—	—	—	22	37	48	55	61	65	68
12	D.P.	—	—	—	—	—	−16	17	35	49	62	74	85
	R.H.	—	—	—	—	—	8	26	39	48	54	59	62
14	D.P.	—	—	—	—	—	—	5	28	45	58	70	82
	R.H.	—	—	—	—	—	—	16	30	40	47	53	57
16	D.P.	—	—	—	—	—	—	−20	20	39	54	67	79
	R.H.	—	—	—	—	—	—	5	21	33	41	47	51
18	D.P.	—	—	—	—	—	—	—	8	33	50	63	76
	R.H.	—	—	—	—	—	—	—	13	26	35	41	47
20	D.P.	—	—	—	—	—	—	—	−13	25	45	60	73
	R.H.	—	—	—	—	—	—	—	5	19	29	36	42
22	D.P.	—	—	—	—	—	—	—	—	15	39	56	69
	R.H.	—	—	—	—	—	—	—	—	12	23	32	37
24	D.P.	—	—	—	—	—	—	—	—	0	32	51	66
	R.H.	—	—	—	—	—	—	—	—	6	18	26	33

190. Recording hygrometers. — It is just as desirable to keep a continuous record of the moisture of the atmosphere as of the other meteorological elements. It is the relative humidity of which the continuous record is ordinarily kept, and it is accomplished by means of a recording Richard Frères hygrom-

The recording hygrometer for relative humidity.

eter. As pictured in Fig. 97, this consists of a strand of hair fastened at each end and held taut in the middle by means of a hook. The

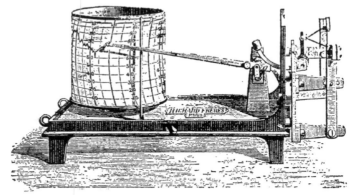

, Fig. 97. — The Recording Hygrometer. (The case has been removed.)

changes in length of this strand are magnified by a system of levers and communicate to a pen which rises and falls over the revolving drum.

The Results of Observation

191. The observations. — At the regular stations of the U. S. Weather Bureau, psychrometer observations are made at 8 A.M. and 8 P.M., the hours at which observations of the other meteorological elements are made. In addition, a continuous record of relative humidity, by means of the Richard Frères recording hygrometer, is usually kept. At the coöperative stations no observations of moisture are required. At a great many foreign stations the hair hygrometer is used instead of the psychrometer for temperatures below the freezing point.

The observations taken.

192. Normal hourly, daily, monthly, and yearly values of absolute and relative humidity. — Since absolute humidity and relative humidity are expressed by mere numbers, the various average and normal values may be computed in exactly the same way as the corresponding averages and normals for the other meteorological elements. In computing the normal hourly values it is customary not to compute the value for each hour of every day in the year, but to compute the value for each hour of the day for the various months of the year.

The averages and normals can be computed in the regular way.

193. Daily and annual variation in absolute humidity. — The daily variation in absolute humidity may be shown graphically by plotting

The characteristics of the daily variation in absolute humidity.

to scale the normal hourly values. If these are not known, a fairly good idea of the characteristics of the variation may be obtained by noting the actual change in the absolute humidity on some day when the other meteorological elements have followed as closely as possible a normal course. The maximum usually occurs in the late afternoon and the minimum at the time of sunrise. During the day evaporation has gone on rapidly from all water surfaces and from the damp ground, and increased activity on the part of the plants and animals during the day has also added a large amount of moisture to the atmosphere. As a result the absolute humidity is greatest in the late afternoon or early evening. During the night, a large quantity of water vapor comes out of the atmosphere in the form of dew, and evaporation is also less. As a result, the minimum is reached about the time of sunrise.

During the summer months in some warm, moist, low-lying countries, there is a secondary minimum in the middle of the afternoon which causes two maxima, one during the morning and the other in the early evening. The reason for this is to be found in convection, which becomes very energetic during the afternoon and carries up the warm, moisture-laden air, replacing it with dryer air from above. The result is a decrease in the absolute humidity which gives rise to the secondary minimum and the two maxima.

194. The annual variation in absolute humidity is found by plotting to scale the normal monthly values. The maximum usually occurs in

The characteristics of the annual variation in absolute humidity.

the late summer and the minimum during the winter. The reason for this is the increase in evaporation during the summer due to the higher temperatures and the fact that the ground is not frozen. Plant life is also much more luxuriant. In the accompanying table will be found

Illustrations.

a series of values of absolute humidity for the various months of the year and for the year as a whole for various stations in the United States. In Fig. 98 some of these results are shown graphically and illustrate the points just mentioned.

195. Geographical variation in absolute humidity. — The

Correlation of the general wind system and absolute humidity.

geographical variation in absolute humidity is closely correlated with the general wind system. The largest values occur at the equator, because temperatures there are highest and the air is relatively calm. The amount decreases through

NORMAL ABSOLUTE HUMIDITY IN GRAINS PER CUBIC FOOT FOR THE VARIOUS MONTHS AND FOR THE YEAR

Station		Length of Record	Jan.	Feb.	Mar.	Apr.	May	June	July	Aug.	Sept.	Oct.	Nov.	Dec.	Annual
Albany, N.Y.	A.M.	15	1.07	1.15	1.55	2.35	3.66	5.26	5.87	5.43	4.5	3.08	2.12	1.41	3.13
	P.M.	15	1.2	1.2	1.7	2.47	3.86	5.51	6.26	5.7	6	3.28	2.2	1.30	3.38
Bismarck, N. Dak.	A.M.		0.4	0.4	0.7	1.82	2.79	4.20	4.91	3.8		1.92	1.0	0.	2.1
	P.M.		0.5	0.5	1.0	2.28	3.25	4.86	5.28	4.0	85	2.39	1.3	0.	2.44
Boise, Idaho	A.M.		1.2	1.5	1.8	2.21	2.77	3.09	2.80	2.9	66	2.08	1.8	1.	2.22
	P.M.		1.4	1.6	1.4	2.07	2.61	3.09	2.93	3.2	88	2.58	2.1	1.	2.38
Boston, Mass.	A.M.	1	1.1	1.1	1.5	2.21	3.39	4.88	5.67	5.5	.	3.14	2.1	1.	3.0
	P.M.		1.3	1.3	1.7	2.42	3.53	4.98	5.93	5.7	.	3.19	2.2	1.60	3.2
Buffalo, N.Y.	A.M.		1.1	1.0	1.4	2.18	3.26	4.86	5.79	5.4	.	2.91	1.9	1.39	2.9
	P.M.		1.2	1.2	1.5	2.24	3.34	4.91	5.96	5.4	.	3.06	2.0	1.83	3.0
Charleston, S.C.	A.M.		2.8	3.00	3.6	4.61	6.18	7.67	8.38	8.4	.	5.08	3.7	2.35	5.29
	P.M.		3.0	3.28	3.9	4.70	6.38	8.27	8.75	8.8	.	5.33	4.1	3.25	5.64
Chicago, Ill.	A.M.	15	1.14	1.14	1.5	2.30	3.51	5.15	5.82	5.3	.	2.99	1.8	1.5	3.04
	P.M.	15	1.31	1.48	1.8	2.57	3.53	5.14	6.07	5.9	.	3.14	2.1	1.0	3.30
Columbus, Ohio	A.M.		1.4	1.35	1.8	2.62	3.89	5.40	5.98	5.6	.54	3.04	2.1	1.1	3.29
	P.M.		1.5	1.56	2.0	2.86	4.07	5.53	5.96	5.8	.56	3.21	2.2	1.0	3.44
Denver, Col.	A.M.		0.8	0.91	1.1	1.64	2.38	3.16	3.75	3.3	.4	1.65	1.1	0.5	1.94
	P.M.		1.1	1.16	1.3	1.75	2.49	2.99	3.82	3.2	.4	2.04	1.3	1.8	2.0
El Paso, Tex.	A.M.		1.4	1.4	1.4	1.52	2.01	2.97	4.94	4.8	.8	2.62	1.7	1.2	2.5
	P.M.		1.4	1.4	1.1	1.14	1.51	2.23	4.08	4.2	.6	2.39	1.7	1.9	2.2
Galveston, Tex.	A.M.		3.7	4.0	4.7	6.16	7.20	8.69	9.92	8.9	.1	6.30	4.8	4.7	6.4
	P.M.		3.9	4.3	5.1	6.26	7.20	8.42	8.87	8.9	.9	6.41	5.1	4.5	6.4
Havre, Mont.	A.M.		0.7	0.6	0.9	1.67	2.63	3.37	4.00	3.3	.5	1.77	1.2	0.4	1.9
	P.M.		0.8	0.8	1.3	1.86	2.76	3.54	3.60	3.5	.9	2.20	1.5	1.1	2.1
Helena, Mont.	A.M.		0.7	0.8	1.0	1.52	2.17	2.74	3.06	2.8	.3	1.69	1.2	0.9	1.7
	P.M.		0.8	0.9	1.2	1.82	2.22	2.92	2.90	2.4	.3	1.81	1.4	1.4	1.8
Indianapolis, Ind.	A.M.		1.3	1.3	1.7	2.70	4.00	5.36	5.69	5.6	.5	2.96	2.0	1.2	3.2
	P.M.		1.5	1.5	2.0	2.91	4.28	5.71	6.23	5.8	.7	3.17	2.2	1.9	3.5
Key West, Fla.	A.M.		5.9	6.2	6.3	6.62	7.50	8.84	8.87	8.8	.9	8.38	7.0	6.4	7.4
	P.M.		6.1	6.2	6.3	6.62	7.60	8.57	8.72	8.9	.7	8.02	7.1	6.0	7.4
Los Angeles, Cal.	A.M.		2.4	2.7	3.1	3.46	4.08	4.52	5.17	5.3	.7	3.91	2.9	2.9	4.3
	P.M.		3.3	3.3	3.7	3.87	4.27	4.71	5.28	5.5	.2	4.77	3.9	3.6	4.3
New Orleans, La.	A.M.		3.42	3.7	4.3	5.21	6.59	8.07	8.69	8.5	.4	5.43	4.2	3.5	5.7
	P.M.		3.59	3.9	4.5	5.35	6.47	7.87	8.60	8.3	.4	5.60	4.5	3.84	5.8
New York, N.Y.	A.M.		1.42	1.3	1.6	2.48	3.67	5.13	6.14	6.0	.1	3.38	2.2	1.65	3.3
	P.M.		1.59	1.5	1.8	2.66	3.94	5.45	6.39	6.3	.4	3.57	2.4	1.85	3.5
Omaha, Neb.	A.M.		0.88	0.9	1.4	2.45	3.77	5.36	6.07	5.9	.1	2.67	1.6	1.5	3.0
	P.M.		1.16	1.2	1.7	2.71	4.06	5.52	6.36	6.1	.4	2.83	1.9	1.2	3.3
Philadelphia, Pa.	A.M.		1.49	1.4	1.7	2.58	4.02	5.56	6.41	6.1	.2	3.32	2.3	1.0	3.5
	P.M.		1.54	1.6	1.9	2.72	4.18	5.38	6.46	6.3	.3	3.52	2.4	1.8	3.6
Phœnix, Ariz.	A.M.		1.96	2.1	2.0	2.11	2.26	2.55	5.19	5.3	.0	2.83	2.1	1.8	2.8
	P.M.		2.20	1.9	1.9	1.91	2.09	2.18	4.27	4.8	.0	2.48	2.4	1.90	2.7
Portland, Me.	A.M.		1.01	1.0	1.2	2.04	3.31	4.61	5.43	5.3	.3	2.78	1.8	1.21	2.8
	P.M.		1.01	1.1	1.5	2.18	3.44	4.73	5.65	5.6	.4	2.93	1.8	1.28	2.9
Portland, Ore.	A.M.	15	2.18	2.3	2.5	2.80	3.39	3.80	4.21	4.6	.2	3.58	2.8	2.44	3.2
	P.M.	15	2.31	2.4	2.6	2.91	3.43	4.17	4.21	4.4	.1	3.78	3.1	2.61	3.3
St. Louis, Mo.	A.M.	6	1.46	1.4	2.0	2.95	4.66	6.24	6.65	6.3	.1	3.34	2.1	1.70	3.6
	P.M.	6	1.62	1.7	2.3	3.41	4.79	6.20	6.76	6.4	.5	3.51	2.3	1.90	3.8
St. Paul, Minn.	A.M.	5	0.65	0.6	1.0	2.06	3.12	4.63	5.47	5.1	.8	2.4	1.3	0.95	2.6
	P.M.	5	0.75	0.9	1.3	2.28	3.24	4.76	5.55	5.1	.8	2.6	1.5	1.00	2.7
Salt Lake City, Utah	A.M.	5	1.19	1.3	1.5	1.75	2.41	2.67	2.98	2.9	.3	1.9	1.6	1.	2.0
	P.M.	5	1.48	1.5	1.8	2.01	2.51	2.62	2.96	3.1	.5	2.3	1.9	1.	2.4
San Francisco, Cal.	A.M.	5	3.04	3.1	3.3	3.46	3.80	4.03	4.26	4.5	.4	4.2	3.8	3.	3.7
	P.M.	5	3.28	3.2	3.4	3.56	3.87	4.00	4.42	4.6	.4	4.1	3.8	3.	3.8
Seattle, Wash.	A.M.	3½	2.42	2.4	2.5	2.77	3.23	3.63	4.07	4.3	.9	3.4	2.8	2.	3.1
	P.M.	3½	2.48	2.4	2.4	2.72	3.33	3.85	4.34	4.3	.1	3.5	2.9	2.	3.2
Washington, D.C.	A.M.	5	1.49	1.5	1.9	2.76	4.31	5.87	6.76	6.8	.3	3.5	2.3	1.	3.6
	P.M.	5	1.63	1.6	2.1	2.96	4.61	6.24	6.95	6.9	.6	3.6	2.4	1.	3.8

The last year included in these normals is 1903. The time used is Eastern Standard Time.

the trade wind belts both on account of the lower temperatures and on account of the mixing caused by the larger wind velocities. In the horse latitudes the moisture is slightly less, both because of

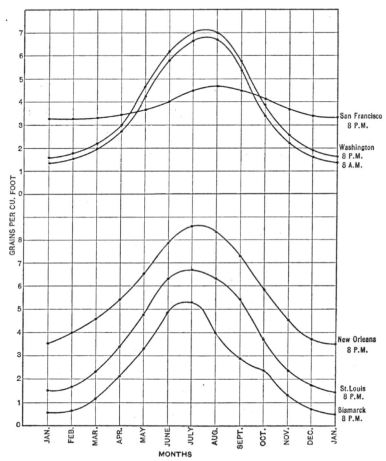

FIG. 98. — The Annual Variation in Absolute Humidity.

lower temperatures and because we have to do here with descending air currents which are always relatively dry. In the region of the prevailing westerlies the amount of moisture decreases steadily both on account of the still lower temperatures and the rapid motion of the air. It must not be supposed that the absolute humidity is the same for all places which have the same latitude or lie in the same wind belt. The various factors which determine the **The factors upon which** amount of absolute humidity are the temperature, distance **the absolute** from the ocean, the inclosure by mountains, and the altitude. **humidity depends.** On the whole, the higher the temperature, the larger the amount of moisture in the atmosphere. The amount decreases with distance from the ocean and is markedly less if the place is surrounded by mountains. It also decreases markedly with altitude and practically disappears at a height of ten miles.

196. Daily and annual variation in relative humidity. — The daily variation in relative humidity is found by plotting to scale the normal hourly values. The minimum usually occurs during the **The char-** early hours of the afternoon and the maximum just before **acteristics** sunrise. During the morning the amount of moisture in the **of the daily** atmosphere rapidly increases, but with the rising tempera- **relative** ture the capacity of the air for moisture increases so much **humidity.** more rapidly that the relative humidity decreases and reaches a

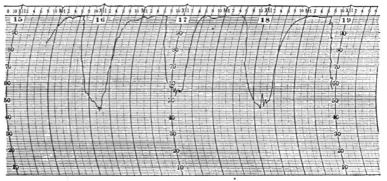

FIG. 99. — Relative Humidity at St. Louis, Mo., September 16–19, 1908.
(U. S. Weather Bureau.)

minimum in the early afternoon. Soon after sunset the air cools rapidly and thus the capacity of the air for moisture is rapidly lessening. In most well-watered countries the saturation point is reached in the early evening. That means that the relative humidity has obtained a value of 100 per cent and it usually remains at this value all during the night until the rise of temperature occurs again on the following morning. In Fig. 99 the trace of a recording hygrometer for September 16 to 19, 1908, at St. Louis, Mo., is given. Here the characteristics which have just been mentioned are particularly pronounced.

Illustration.

197. The annual variation in relative humidity may be found by plotting to scale the normal monthly values. The maximum usually occurs in the autumn, when the falling temperatures are causing a great decrease in the capacity of the air for water vapor or during the winter itself. The minimum usually occurs during the spring, when the rapidly rising temperatures are causing a great increase in the capacity of the air for water vapor or during the summer itself.

The characteristics of the annual variation in relative humidity.

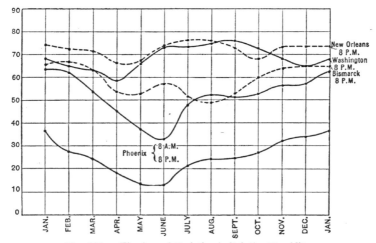

FIG. 100. — The Annual Variation in Relative Humidity.

The following table, which gives the normal values of relative humidity for the various months and for the year for several stations in the United States, illustrates these characteristics. In Fig. 100 some of these results are shown graphically.

Illustrations.

198. Geographical variation in relative humidity. — The geographical variation in relative humidity is again closely correlated with the general wind system. The normal yearly value in the equatorial regions is more than 80 per cent, it

The characteristics of the geographical variation in relative humidity.

NORMAL VALUES OF RELATIVE HUMIDITY IN PERCENTAGES FOR THE
VARIOUS MONTHS AND FOR THE YEAR

STATION		LENGTH OF RECORD	JAN.	FEB.	MAR.	APR.	MAY	JUNE	JULY	AUG.	SEPT.	OCT.	NOV.	DEC.	ANNUAL
*Albany, N.Y.	8 A.M.	1	83	81	80	74	73	75	76	75	81	84	83	83	79
	8 P.M.	15	79	78	75	65	65	69	69	67	74	75	78	80	73
Bismarck,	8 A.M.	1	77	78	78	76	76	81	80	80	79	81	80	79	79
N. Dak.	8 P.M.	21	66	67	65	55	54	59	52	49	52	60	63	67	59
*Boise, Idaho	8 A.M.		84	82	74	72	73	66	54	56	63	73	76	82	72
	8 P.M.	5	72	63	46	36	36	32	21	26	31	45	63	73	45
Boston, Mass.		14	72	71	68	66	71	72	71	75	77	75	75	71	72
*Buffalo, N.Y.	8 A.M.	1	79	79	76	71	72	74	75	75	75	74	75	72	75
	8 P.M.	15	77	77	73	68	69	70	70	68	70	70	73	70	71
Charleston,	8 A.M.		81	78	79	76	77	78	79	82	83	79	80	80	79
S.C.	8 P.M.	38	77	76	75	75	77	79	80	81	80	76	77	79	78
Chicago, Ill.	8 A.M.	20	85	84	81	76	76	76	74	74	75	76	80	83	78
	8 P.M.	20	80	80	77	70	69	71	67	69	68	68	73	79	72
Columbus,	8 A.M.		84	82	80	74	75	77	76	79	80	81	82	83	79
Ohio	8 P.M.		77	74	69	61	61	63	59	61	62	64	70	76	66
*Denver, Col.	8 A.M.	15	58	64	61	62	65	63	63	63	59	58	55	56	61
	8 P.M.	15	49	49	42	36	38	34	36	33	30	34	41	48	39
El Paso, Tex.	8 A.M.	16	62	55	43	36	55	41	60	63	63	60	58	58	53
	8 P.M.	16	34	27	18	13	13	16	29	32	33	30	31	34	26
Galveston,	8 A.M.		86	87	86	85	82	82	81	82	82	79	82	84	83
Texas	8 P.M.		82	83	83	81	77	77	74	75	73	73	78	81	78
Havre, Mont.		22	80	81	77	62	63	63	57	56	62	68	75	79	69
Helena, Mont.	8 A.M.		72	73	71	65	66	67	60	59	64	66	68	71	67
	8 P.M.		66	63	56	42	42	41	31	30	39	48	56	65	48
Indianapolis, Ind.		21	79	77	73	66	67	68	65	67	68	68	73	77	71
Key West, Fla.	8 A.M.		82	81	78	73	74	76	73	74	77	79	80	82	77
	8 P.M.		80	78	76	74	75	77	75	75	78	78	79	80	77
Los Angeles,	8 A.M.	22	67	75	80	83	87	88	90	88	83	79	71	60	79
Cal.	8 P.M.	22	64	65	65	63	67	64	62	64	64	68	71	61	64
New Orleans,	8 A.M.	21	84	84	84	85	83	81	81	82	84	83	80	84	83
La.	8 P.M.	21	73	.2	71	68	68	71	74	75	73	68	73	74	72
*New York,	8 A.M.	15	76	75	74	70	73	76	77	78	79	77	77	75	76
N.Y.	8 P.M.	15	72	71	69	65	69	71	70	73	73	71	72	71	71
Omaha, Neb.	8 A.M.	21	81	81	76	73	75	79	78	80	80	76	77	81	78
	8 P.M.	21	71	70	64	53	55	57	57	59	58	55	63	70	61
Philadelphia, Pa.			73	72	69	64	68	68	70	72	74	71	71	72	70
Phœnix, Ariz.	8 A.M.	8	64	62	53	45	38	32	49	54	51	51	57	57	51
	8 P.M.	8	37	28	24	18	15	12	21	25	25	26	32	32	25
Portland, Me.		14	75	74	72	69	76	76	76	80	81	79	77	75	75
Portland, Ore.		21	84	79	74	70	69	69	64	66	71	79	84	85	74
St. Louis, Mo.		21	75	72	72	66	69	69	67	69	70	67	70	76	70
St. Paul, Minn.	8 A.M.	21	84	84	81	75	75	79	79	83	83	81	81	83	81
	8 P.M.	21	76	75	68	55	54	58	55	56	60	63	69	76	64
Salt Lake City, Utah		22	74	68	58	48	48	39	34	34	40	51	52	71	52
San Francisco,	8 A.M.	19	87	86	85	85	87	89	92	93	88	86	85	84	88
Cal.	8 P.M.	19	75	72	70	69	72	72	77	79	73	71	71	73	73
Seattle, Wash.	8 A.M.		88	86	85	85	86	85	85	87	89	88	88		87
	8 P.M.		81	74	66	58	59	56	51	54	63	74	81	82	67
*Washington,	8 A.M.	15	77	75	75	70	75	76	77	80	81	80	79	76	77
D.C.	8 P.M.	15	69	66	65	59	68	71	72	74	76	72	69	67	69

The time is expressed in Eastern Standard Time. Ordinarily the last year included in the normals is 1908. If the station has a *, the last year is 1903.

P

is less in the trade wind belts, and reaches a minimum of about 70 per cent in the horse latitudes. It then increases again in value, and in the polar regions is between 80 and 90 per cent, thus surpassing the value at the equator. Locally, it is influenced by the same four factors which determine the absolute humidity.

199. **Other moisture data and charts.** — In connection with moisture, the average and normal values for the various months and for the

The use made of the observations.

year are the only results which are computed from the observations taken. In the case of the U. S. Weather Bureau, the relative humidity at 8 A.M. and 8 P.M. are the only data which are kept constantly computed to date. If these normal values are known for a country or for the world, corresponding charts may be prepared. Charts XV and XVI give the relative humidity for the United States for January and July.

THE EFFECT OF WATER VAPOR ON THE GENERAL CIRCULATION

200. As was stated in section 176, water vapor is lighter than air under the same conditions of pressure and temperature, the ratio

The moisture acts with the temperature in causing pressure differences and wind.

of mass being 62 to 100. Thus, moisture-laden air is lighter than dry air. Now the amount of moisture in each cubic foot of air at the equator is normally six times the amount in the same volume of air at the pole. This excess of moisture at the equator will thus cause the air at the equator to be lighter than at the pole, and will thus operate in the same way as the higher equatorial temperatures to accentuate the pressure differences between the equator and pole, and to cause a general circulation of the atmosphere.

B. DEW, FROST, AND FOG

DEW

201. **Condensation.** — Condensation is the opposite of evaporation; it is the passage of the moisture of the atmosphere from

Definition.

The seven forms of condensation.

the invisible gaseous form, water vapor, into some visible, solid or liquid form. When condensation takes place, the water vapor takes the form of dew, frost, fog, cloud, rain, snow, or hail. There are thus seven forms of condensation, and each of these forms must receive careful consideration.

There are two ways in which the condensation of some of the water vapor in a given quantity of air may be brought about: by compression or by cooling. Of these, the first can be performed in the laboratory, and never occurs in nature. Condensation by cooling is the only process which occurs in nature. Suppose a vessel contains two cubic feet of air at a temperature of 70° F., and suppose, furthermore, that it contains 7 grains of moisture per cubic foot. Let this air be compressed until its volume is only one cubic foot, and suppose that the temperature has been held at 70° by cooling the containing vessel. This cubic foot of air will now contain 14 grains of moisture, and by referring to the table in section 181, it will be seen that a cubic foot of air at 70° can contain only 7.99 grains. Therefore the excess of moisture, namely, 6.01 grains, must come out of the air in some form. In nature this can never occur, because there is no means by which the compressed air may be kept at a constant temperature. Descending air currents are so warmed by compression that the capacity for water vapor increases much more rapidly than the increase in the amount of moisture in each cubic foot. Thus, descending air currents soon become dry.

The water vapor in a given quantity of air may be condensed by compression or cooling.

Description of condensation by compression.

If a given quantity of air is cooled, it will soon reach a temperature at which the amount of moisture in it saturates it, and any further cooling will cause moisture to come out of it in some one of the seven forms of condensation. This cooling of the air in nature may be brought about by three processes. It may be caused by expansion, and in this case the air grows colder without the addition or subtraction of heat from any outside objects; secondly, it may be cooled by mixture with colder air; thirdly, it may be cooled by conducting its heat to surrounding objects which are colder or by radiating its heat to space or surrounding objects.

The three ways in which air may be cooled in nature.

202. **Dew.** — In the early morning, more particularly in summer, shortly after sunrise, the temperature rises rapidly and the amount of moisture added to the atmosphere by evaporation from bodies of water and the moist ground also increases rapidly. Activity on the part of plants and animals also adds large quantities of water vapor to the atmosphere. As a result, all during the day, the amount of moisture in the atmosphere, that is, the number of grains in each cubic foot, is steadily increasing. In the early evening, just after sunset, the leaves of trees and plants, the grassy covering of the ground, the ground itself, and in short all material objects lose their

Description of how dew is formed.

heat rapidly by radiation to space and become colder than the overlying layer of air. The layer of air loses its heat by conduction, and to a certain extent by radiation, and soon reaches the saturation point. Any further reduction in temperature will cause the excess moisture to come out on any solid object as dew. Dew has been studied for a little more than a century, but the first correct explanation was given by Wells in 1814. It was essentially the explanation just given above.

The formation of water drops on the outside of an ice pitcher in a warm room on a summer's day is also an illustration of this process.

Ice pitcher illustration.
The ice pitcher corresponds to the ground which has become cold, due to radiation. The layer of air next it corresponds to the layer of air near the surface of the ground, and the water which collects on the outside of the pitcher corresponds to the dew.

203. There are three sources of the moisture which go to make up dew. A large part of it comes from the lower layer of the atmosphere itself. Some of it comes from the leaves of trees and plants.

The three sources of the moisture.
The moisture exudes from these, and, since evaporation is not possible, it collects on them and adds to the supply of dew. Furthermore, a certain amount of moisture comes up from the subsoil by capillary action, and as soon as it evaporates it again condenses on the blades of grass and the leaves of plants above.

Various estimates and measurements have been made of the amount of dew. The figures usually given are these: in a well-watered region,

The amount of dew.
during a single night, the dew would, perhaps, amount to 0.01 of an inch of water, and the total amount of dew which occurs in the course of a year might amount to an inch. In desert regions, of course, the amount is practically nothing.

204. The part played by latent heat. — When evaporation takes place, energy is required to separate the molecules, and this energy is called latent heat. Conversely, when condensation takes

Condensation liberates latent heat.
place, this latent heat again makes itself felt. It may either heat the surface upon which the condensation takes place, or retard its rate of cooling by adding to its supply of heat.

The small daily range of temperature in a well-watered region is an illustration of the part played by latent heat. During the day, high

Latent heat lessens daily range.
temperatures are prevented because so much energy is used up as latent heat in evaporation and not in heating the air. At night, cooling is retarded by the liberation of so much latent heat.

205. Conditions for the formation of dew. — There are two condi_
tions for the formation of dew; a clear sky and absence of wind.   The
reason for this is because a large drop in temperature during
the night is essential for the formation of dew, and both   The two
clouds and wind prevent this.   Clouds prevent it by hinder-   for the for-
ing that free radiation of heat to space which is essential for   mation of
a large drop in temperature.   Wind prevents it by mixing   clear sky
the layer of air in contact with the ground which has be_   and absence
come cold by conduction with the warmer layers above.
When dew is deposited, an inversion of temperature nearly always occurs.

More dew forms in valleys than on hilltops, and there are two reasons
for this.   It is in the valleys that the ponds, lakes, streams, and water
courses are usually located, and these add a large amount
of moisture by evaporation to the atmosphere.   Further-   More dew
more, at night the cold and thus denser air drains into the   valleys than
valleys from the surrounding hilltops.   In riding over a   on hilltops.
hilly road at night, the increase in moisture and the lower temperatures
on descending into the valleys is always plainly evident.

FROST

206. Frost. — Frost is dew which has formed with the temperature
below 32° F.   It consists of small, translucent, frozen drops placed
side by side, and also of feathery, spinelike forms.   It is   The appear-
sometimes thought that the translucent, frozen drops are   ance of frost.
due to the moisture which has exuded from the leaves of plants and
frozen there.   It may be the dew which had formed before the temperature went below 32° F.   The moisture which comes from the atmosphere and from the ground usually has a feathery, spinelike form.   In
winter, the moisture which comes up from the subsoil usually forms ice
crystals just beneath the surface of the ground, raising the surface and
giving to the ground a spongy character.

The word frost is really used in two different senses.   It is used to
designate the deposit of frozen dew which may form at any time of year,
provided the temperature is below 32° F. and the dew point   Light and
has been passed.   It is also used to designate the occurrence   killing
of those temperatures, particularly in the late spring and   frosts.
early autumn, which are destructive to vegetation.   When used in this
last sense, two kinds of frosts may be distinguished; light frosts or
white frosts, and killing frosts.   If the temperature does not fall much

lower than 4° below the freezing point, only the very tenderest vegetation is killed and the frost is spoken of as a light or white frost. If the temperature goes below 28° F., even the hardier forms of vegetation will be killed, and the frost is then spoken of as a killing frost.

207. Two processes operate to produce the cooling which may result in the destructive frosts of late spring and early autumn. These are, first, the importation of colder air, and secondly, the radiation of heat from the ground and the cooling of the air next to it by conduction and radiation. On some occasions, a strong northwest wind will import cold, dry air, thus holding down the maximum temperature during the day, and causing a low temperature. If the wind dies down during the night and the sky becomes clear, it takes but little radiation to cause sufficient cooling to produce a frost. On other occasions, there seems to be but little importation of cool, dry air. The sky is quite clear and there is almost no cloud. Radiation is excessive, and the resulting large drop in temperature may cause a frost. While both of the above processes are usually active, it is generally easy to see which predominates. As in the case of dew, a clear, still night is necessary for frost formation. These conditions are most likely to be fulfilled on the first or second night following the passage of an area of low pressure and the transition of the weather control to an area of high pressure. As will be seen later, this facilitates both the importation of colder air during the day and the radiation at night, — the two processes which cause the low temperatures required for a frost.

The causes of the destructive frosts of late spring and early autumn.

208. **Prediction of frost.** — A frost is predicted by the U. S. Weather Bureau in exactly the same way as any other temperature. The observations of the various meteorological elements are made at many stations at 8 A.M. and 8 P.M., and these are distributed by telegraph to all stations where predictions are made. From the weather map which is prepared from these observations, and from the sky appearance, the probable clearness of the sky during the following night, and the probable wind velocity must be estimated. From these, chiefly, the probable drop in temperature is determined. In this way, the probable minimum temperature of the following morning and thus the likelihood of the frost is determined.

The U. S. Weather Bureau method of predicting frost.

209. There is a widespread opinion that if the frost prediction is delayed until the maximum temperature of the preceding day and the dew point at the time of the maximum have been determined, that

a much more accurate estimation of the probable minimum temperature of the following morning can be made. It is generally supposed that the temperature will drop rapidly until the dew point is reached and then any further drop will be much retarded by the liberation of latent heat. In other words, the dew point will serve as a guide in determining the probable amount of drop. As a matter of fact, in the northeast portion of the United States the air is so dry and the dew point lies so low on all those days when a frost is imminent, that the dew point is seldom reached and plays practically no part in determining the minimum temperature. The drop is, also, far from constant, and must be carefully estimated for each individual case, taking into account the probable characteristics of the afternoon and night.

Frost prediction from the maximum temperature and dew point of the day before.

210. If, after the probable minimum temperature in the thermometer shelter has been estimated, it is desired to determine the probable temperature of low-growing vegetation in the open, at various points in a limited area surrounding the station in question, three things must be taken into account : First, that plant temperatures go below the real air temperature because the plants are in the open, without such a hindrance to radiation as is the shelter about a thermometer. On the average, vegetation in the open will go about 2° F. below the real air temperature and in extreme cases 3° or 4° below. Second, vegetation is located near the ground and not at the height of the instruments in the thermometer shelter. The difference in temperature between the surface of the ground and the thermometer shelter, say from 5½ to 6 feet above the surface of the ground, will average perhaps 3° or 4°, and in extreme cases might amount to 6° or 7°. Third, the variation in temperature over a limited area may amount to several degrees. In fact, for a limited area it may easily amount to 6° or 7° and in certain cases may be much more than that. Thus, the temperature of vegetation in the open, near the ground, in the coldest portion of a limited area, may be expected to average from 10° to 14° lower than the estimated minimum in the thermometer shelter, and under extreme circumstances may differ by from 5° to 10° more.

The difference between the minimum temperature in a thermometer shelter and the temperature of vegetation.

211. **Protection from frost.** — Tender vegetation is usually protected from frost by covering it with cloth or paper or some such covering. The radiating surface is thus transferred from the ground or the leaves of the plants to the covering, and it is thus protected against frost.

Ordinary protection against frost.

It has often been proposed to cover a whole village or even a portion of a state with a smoke layer by building smudge fires. If concentrated

The use of a smoke layer. action of this kind could be secured, and a smoke layer of sufficient thickness could be produced, the radiation would be transferred from the surface of the ground to the surface of the smoke layer, and temperatures from 5° to 6° higher than would otherwise occur might be maintained underneath the smoke layer. Experiments are just beginning to be made . on such a large scale as to make possible definite conclusions.

212. **Frost observations, frost data, and charts.** — The observations made of frosts are the dates of the first light frost and the first killing

The observations taken and the results computed. frost in the autumn, and the last light and the last killing frost in the spring. From these observations, the normal dates of the first and last light and killing frost may be determined. If these dates are known for a whole country or for the whole world, they may be indicated on charts. The

departure from normal, however, in the case of these dates, may amount to three weeks in either direction. Charts XVII and XVIII depict, for the United States, the normal dates of the first killing frost of the autumn and the last killing frost of the spring.

FOG

213. **The nature of fog.** — If a layer of air of considerable thickness falls below the temperature of saturation, the moisture is unable to come

The nature of fog. out as dew or frost at the bottom of the layer, but condenses as fog. This condensation takes place on the dust and other particles in the air. A fog particle, therefore, is a minute water drop, about 0.001 inch in diameter, with a dust particle or some other particle for its nucleus (see section 227). Whenever fog occurs, the relative humidity is usually very high, at least 90 per cent. There are a few cases on record, however, when a dense fog has existed with a relative humidity even below 60 per cent. In these very exceptional cases there must have been something which prevented the evaporation of the fog particles, as a coating of oil or the like.

The time of year when the maximum number of foggy days occurs is very different in different parts of the world. On the Atlantic coast of

The time of maximum fogginess. North America, from Maine northward, fogs occur chiefly during the summer. In the southern part of the United States, particularly inland, fogs occur chiefly during the winter. In New England and the middle Atlantic States, particularly inland, fogs are of two entirely different kinds.   One is most prevalent during the late winter and early spring, while the other occurs chiefly during the late summer and early autumn.   One is caused by the transportation of air and the other by radiation.   In fact, all fogs are caused in one or the other of these two ways.   In the late winter, warm moisture-laden air may come up from the south over the cold snow-covered regions at the north.   As a result, a layer of air of considerable thickness is cooled below its dew point, and condensation in the form of fog takes place. Again, due to rain or a continued thaw, the air may have become warmed and moisture-laden.   The wind suddenly changes to the northwest, the temperature drops rapidly, and a layer of air of considerable thickness may go below the dew point and the moisture come out in the form of fog.   In both cases, the fog has been caused by the transportation of air, and this transportation is brought about by the presence of storms.   The formation of fog in the autumn is entirely different.   During the day, the temperature rises to considerable heights and evaporation is considerable.   The nights are growing longer and the drop in temperature is considerable.   At night the temperature often goes below the dew point for a considerable thickness of air, and the moisture comes out in the form of fog.   These fogs always commence during the night and are thickest at sunrise.   They usually disappear during the forenoon.   The disappearance of these fogs, and in fact all fogs, is brought about by the wind or a rise in temperature.   The wind may mix the fog-laden layer of air with a dryer layer, thus causing its disappearance. A rise in temperature may so increase the capacity of the air for vapor that the fog particles evaporate and become invisible water vapor.

The transportation fogs of spring and the radiation fogs of autumn.

.It has been sometimes stated that the city fogs are entirely different from the country fogs.   This is only true in the sense that combustion products play a larger part in their formation, thickness, and continuance.   The number of fogs in London has steadily increased with the growth of the city and the increase in the combustion products.   This is shown in the accompanying table.   The increase

City fogs.

Year	Normal Annual Number of Foggy Days
1871–1875	50.8
1876–1880	58.4
1881–1885	62.2
1886–1890	74.2

Very recent data would seem to show that the number of foggy days has reached its maximum and is now decreasing. [1]

[1] See *Meteorologische Zeitschrift*, March, 1911, p. 135.

in the dust and smoke, due to condensation, furnishes more nuclei for the condensation of water vapor and also facilitates the radiation of heat during the night, and thus the formation of fog. Furthermore, after the fog particle has once formed, it becomes coated with oil, and it is thus much harder for it to be evaporated when the temperature rises. In fact, the fog forms to such a thickness that the heating during the day is no longer sufficient to disperse the fog entirely, and it continues to grow thicker and thicker on successive days and only disperses when a marked change in the weather occurs.

214. Fog observations, fog data, and charts. — The only observations made of fog are the day on which it occurs. From these observations, the normal number of foggy days for the various months and for the year may be determined. If these normals are known for many stations, they may be charted for a given country or the world.

The observations taken and the results computed.

C. CLOUDS

The Classification of Clouds

215. Early History. — Although the varying forms of the clouds must have been observed by the earliest peoples, no cloud names or cloud classifications have come down to us from antiquity. The first classification was proposed by Lamarck, a French naturalist, in 1801. He distinguished six kinds of clouds, to which French names were given. Luke Howard in 1803 proposed the scheme from which the one in general use at the present time was destined to develop. His classification included seven forms in all, four type forms and three combinations of the type forms to indicate intermediate varieties. To the four type forms of cloud he gave the Latin names *Cirrus*, *Cumulus*, *Stratus*, and *Nimbus*, meaning respectively lock of hair, pile, layer, and storm cloud. The name in each case indicates the most striking characteristic of the cloud. The *cirrus* is the delicate, fibrous, hairlike cloud. The *cumulus* is the lumpy, piled-up form of cloud. The *stratus* is the level sheet or layer. The *nimbus* is the cloud from which precipitation is falling. After Howard's time many other systems of classification and nomenclature were proposed, some being mere modifications or extensions of the Howard system. By 1891 nearly twenty systems had been proposed and had been used more or less, so that there was great uncertainty, inexactness, and indefiniteness in the use of names.

Howard's classification.

216. The international system. — In 1891, Hildebrandsson of Upsala and Abercromby, an English meteorologist, presented a report on cloud classification to the International Meteorological Conference which was then in session at Munich. This report was adopted, and the system thus inaugurated has since found almost universal acceptance. In this system thirteen names are used to designate the different kinds of clouds. Howard's four type forms were retained, four names for intermediate varieties were formed by combining the names for the type forms, two names were formed by prefixing *alto*, meaning high, and three by prefixing *fracto*, meaning broken or windblown, to the names for type forms, thus making in all thirteen names. The following is an arrangement of the four type forms in order of height above the earth's surface:

The adoption of the international system.

The thirteen cloud names.

Cirrus

Cumulus   *Stratus* [1]

Nimbus   *Stratus*

The thirteen cloud names are:

Type forms:	*cirrus, stratus, cumulus, nimbus*
Intermediate varieties:	*cirro-stratus, cirro-cumulus, strato-cumulus, cumulo-nimbus*
Alto forms:	*alto-stratus, alto-cumulus*
Fracto forms:	*fracto-stratus, fracto-cumulus, fracto-nimbus*

The underlying principle of the classification is evident from the arrangement as regards height. Other possible names would have been alto-cirrus, alto-nimbus, fracto-cirrus, cumulo-cirrus, strato-cirrus, cumulo-stratus, strato-nimbus, nimbo-stratus, cirro-nimbus, nimbo-cirrus, nimbo-cumulus. But these are all impossible combinations, redundant, or unnecessary. For example, alto-cirrus is unnecessary because cirrus is always high; fracto-cirrus is unnecessary because cirrus is always a detached broken cloud. Strato-nimbus is unnecessary because a nimbus cloud is in the form of a layer. Strato-cirrus is considered the same as cirro-stratus, cumulo-stratus the same as strato-cumulus, etc. Cirro-nimbus is impossible because the two kinds of clouds never occur at the same altitude, and alto-nimbus is impossible because a nimbus cloud is always low.

The underlying principle of the classification.

[1] Stratus is here considered simply as a cloud in a horizontal sheet or layer. When not near the earth's surface it is always spoken of as alto-stratus.

These cloud names may also be grouped as follows:

Other methods of grouping the thirteen cloud forms.

Cirrus type:	*cirrus, cirro-stratus, cirro-cumulus*
Stratus type:	*cirro-stratus, alto-stratus, stratus, fracto-stratus, strato-cumulus*
Cumulus type:	*cirro-cumulus, alto-cumulus, cumulus, fracto-cumulus, strato-cumulus, cumulo-nimbus.*
Nimbus type:	*cumulo-nimbus, nimbus, fracto-nimbus*

The International Committee grouped them as follows:

Upper clouds:	*cirrus, cirro-stratus*
Intermediate clouds:	*cirro-cumulus, alto-stratus, alto-cumulus*
Lower clouds:	*strato-cumulus, nimbus, fracto-nimbus*
Clouds formed by diurnal ascending currents:	*cumulus, fracto-cumulus, cumulo-nimbus*
High fogs:	*stratus, fracto-stratus*

In order to give more exact names to the cloud forms the international system is sometimes subdivided. This is usually done by adding · Latin adjectives or nouns to the thirteen names of the international system. An exact description of the variety to which the name is to apply should then be given or a reference given to some place in the literature on the subject where such a description can be found.

Subdivision of the system.

217. The thirteen cloud forms.—

Cirrus (abbreviation Ci.)

Description: The name Cirrus is applied to detached clouds generally of white color made up of slender, delicate, irregularly curling fibers or wisps of cloud. This kind of cloud is variously described as made up of wavy sprays of cloud, irregularly branching structures, bundles of drawn-out filaments, long curving threads of clouds, or fluffy heads with long streamers. It sometimes takes the form of belts which cross the sky and converge in perspective towards one or two points of the horizon. It also takes the form of feathers or ribbons or delicate fibrous bands and is sometimes striated or rippled. It is often called popularly " cats' whiskers " or " mares' tails." In its less typical form it may be composed of gauzy, clotted masses, or almost

Cirrus.

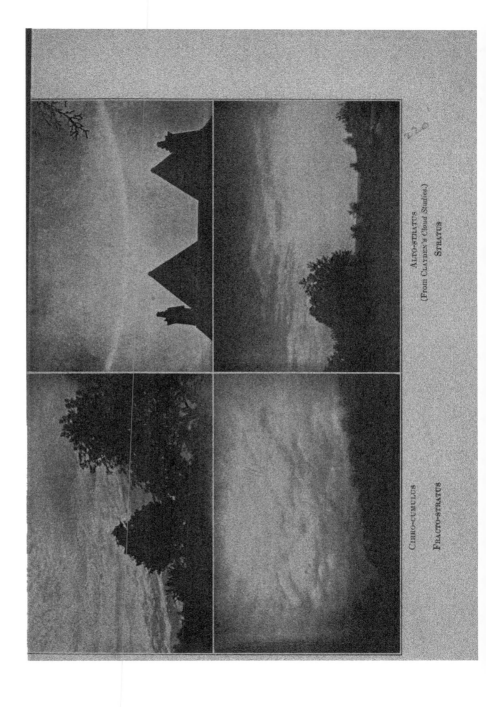

CIRRO-CUMULUS

ALTO-STRATUS
(From Clayden's Cloud Studies.)

FRACTO-STRATUS

STRATUS

NIMBUS
(From BÖRNSTEIN's *Leitfaden der Wetterkunde.*)

FRACTO-NIMBUS
(From BÖRNSTEIN's *Leitfaden der Wetterkunde.*)

CUMULO-NIMBUS
(From HANN's *Lehrbuch der Meteorologie.*)

ALTO-CUMULUS

CIRRUS
CIRRO-CUMULUS

CIRRUS
CIRRO-STRATUS

CUMULUS   FRACTO-CUMULUS

CUMULO-NIMBUS   STRATO-CUMULUS

CUMULUS
CUMULUS .

with absence of structure. It very seldom causes a halo. The direction of motion is usually from the west, and no shadows are caused by it.

Transition forms: Thickening: Ci. Cu. or Ci. St.

Thinning: Disappears.

Height: Mean 28,000 ft., max. 46,000 ft., min. 9000 ft.

Methods of formation: (4), (5); perhaps (6), (3), (8); possibly (7), (9).[1]

Stratus (abbreviation St.)

Description: Stratus was originally defined as lifted fog in a horizontal stratum. It is now applied to any low-lying horizontal cloud sheet of approximately uniform thickness. It is a flat, structureless cloud, usually of wide extent and causes what is called "gray weather." It is sometimes thin in places so that the under surface appears as parallel lines or rolls of clouds all around the horizon. In breaking up it sometimes appears in lenticular patches.

Transition forms: Thickening: Fr. St. or Nb.

Thinning: St. Cu. or A. St.

Height: Mean 2100 ft., max. 6000 ft., min. 400 ft.

Methods of formation: (1), (3); perhaps (7).

Cumulus (abbreviation Cu.)

Description: Cumulus clouds are thick, rounded lumps, whose summits are domes or turrets with protuberances and whose bases are flat. When viewed opposite the sun, they are white with dark centers. When viewed near the sun, they are dark with brilliant dazzling white edges. They cast dense shadows, appear in greatest abundance during the warm part of the day, and look like exploded cotton bales. They often appear to be arranged in rows parallel to the horizon, in which case the flatness of the base is particularly noticeable.

Transition forms: Thickening: St. Cu. or Cu. Nb.

Thinning: A. Cu.

Height: Top: Mean 6000 ft., max. 15,000 ft., min. 2400 ft.

Bottom: Mean 4400 ft., max. 12,000 ft., min. 1600 ft.

Methods of formation: (2), (3).

[1] These refer to the nine methods of cloud formation to be discussed later (sections 231–239).

Nimbus (abbreviation Nb.)

Description: The Nimbus is the rain cloud and is a dense, dark sheet of formless cloud from which precipitation is falling. It is widely extended, and when breaks occur an upper cloud area is usually seen.

Nimbus.

Transition forms: Thickening: Heavier nimbus
Thinning: St., Fr. St., or St. Cu.
Height: Mean 3600 ft., max. 18,000 ft., min. 200 ft.
Methods of formation: (1), (2), (3).

Cirro-stratus (abbreviation Ci. St.)

Description: Cirro-stratus is a thin, whitish sheet or veil of cloud which gives the whole sky a milky appearance. It is widely extended, always distinctly fibrous and streaky in appearance, and looks like a tangled web of short, curling fibers matted together. It often produces halos around the sun and moon, is grayish or white in color, and may be flocculent, granular, or banded at times.

Cirro-stratus.

Transition forms: Thickening: A. St.
Thinning: Ci.
Height: Mean 22,000 ft., max. 42,000 ft., min. 7000 ft.
Methods of formation: (1), (3), (8); perhaps (6), (7), (9).

Cirro-cumulus (abbreviation Ci. Cu.)

Description: Cirro-cumulus consists of small white balls or flakes of semi-transparent clouds without shadows or with very faint ones. These are arranged in groups and often in rows. These have been likened to a thin sheet broken in many places and curled at the edges, to a flock of sheep lying down, to the foam in the eddying wake of a steamer, or to floating ice cakes on the surface of a river. The popular name is "mackerel sky."

Cirro-cumulus.

Transition forms: Thickening: A. Cu.
Thinning: Disappears or Ci.
Height: Mean 20,000 ft., max. 35,000 ft., min. 7000 ft.
Methods of formation: (2), (3), (4); possibly (8), (9).

Alto-stratus (abbreviation A. St.)

Description: Alto-stratus is a thick veil of gray or bluish color, brighter near the sun, without fibrous structure. The sun and moon are sometimes faintly visible through it, and coronas but not halos are produced by it. When breaking up it sometimes appears in lenticular patches. It appears most often in the early morning and in winter. Alto-stratus.

Transition forms: Thickening: St.

Thinning: Disappears or Ci.

Height: Mean 15,000 ft., max. 36,000 ft., min. 3800 ft.

Methods of formation: (1), (3), (4), (6); possibly (9).

Alto-cumulus (abbreviation A. Cu.)

Description: Alto-cumulus is the name given to white or grayish balls of dense fleecy cloud with shaded portions which are detached, but frequently lie close together. The cloud balls are often grouped in flocks or arranged in rows, sometimes in two directions, and the balls in the center of a group are generally the largest. Alto-cumulus is more flattened and disklike than the typical cumulus. Alto-cumulus.

Transition forms: Thickening: St. Cu.

Thinning: Ci. Cu.

Height: Mean 12,000 ft., max. 27,000 ft., min. 2700 ft.

Methods of formation: (2), (3), (5).

Fracto-stratus (abbreviation Fr. St.)

Description: When a widely extended cloud sheet which would ordinarily be called stratus is torn by the wind or mountain summits into irregular fragments, it is called fracto-stratus. Fracto-stratus.

(The remaining facts are the same as for stratus.)

Fracto-cumulus (abbreviation Fr. Cu.)

Description: Fracto-cumulus is the name given to a cumulus cloud which has been torn by high winds so that its margins are ragged and irregular. It is a flat, tattered, broken cumulus. Fracto-cumulus.
Sometimes the cloud appears to have been rolled instead of torn.

(The remaining facts are the same as for cumulus.)

Strato-cumulus (abbreviation St. Cu.)

Description : Strato-cumulus consists of large balls or rolls of dark cloud, flat at the base, often covering the whole sky and leaving only a little blue sky here and there in the breaks. ˙It is some-times defined as stratus thickened here and there into cumulus or cumulus joined together with a common, flat base to make a layer.

Strato-cumulus.

Transition forms : Thickening: Fr. S., N., or Fr. N.
Thinning: Fr. Cu., or Cu.
Height : Mean 7000 ft., max. 14,000 ft., min. 1000 ft.
Methods of formation : (2), (3) ; possibly (8).

Cumulo-nimbus (abbreviation Cu. Nb.)

Description : Cumulo-nimbus is the thundercloud. It consists of heavy masses of cloud rising like mountains, towers, or anvils. Below is the dark, formless nimbus, or fracto-nimbus, cloud from which rain is falling. At the top a veil or cap of fibrous texture called false cirrus is often seen. It is a detached cloud, but usually covers large areas.

Cumulo-nimbus.

Transition forms : Thickening: Large Cu. Nb.
Thinning: Cu.
Height : Top: Mean 28,000 ft., max. 42,000 ft., min. 16,000 ft.
Bottom: Mean 4400 ft., max. 8000 ft., min. 600 ft.
Methods of formation : (2), (3).

Fracto-nimbus (abbreviation Fr. Nb.)

Description : When nimbus is torn into small patches, or if low, ragged, detached fragments of cloud (scud) move rapidly below the main mass, it is called fracto-nimbus.

Fracto-nimbus.

(The remaining facts are the same as for nimbus.)

218. The sequence of cloud forms. — Many observations have been made to determine the probability that fair weather or precipitation will follow the appearance of a certain cloud form. With the exception of nimbus, however, which by definition is a cloud from which precipitation is falling, the preponderance of probability is so small that it is of no value in forecasting.

The usual sequence of cloud forms.

There is, however, a fairly regular sequence of cloud forms in pass-ing from a clear sky to stormy weather. First comes the cirrus ; this thickens to cirro-stratus or cirro-cumulus and then becomes alto-stratus or alto-cumulus; next the nimbus appears which often becomes fracto-nimbus. In clearing off, the sequence is generally this: the nimbus usually breaks up into fracto-nimbus, disclosing often an upper cloud area of cirrus or cirro-cumulus. The upper cloud layer gradually disappears and the lower fracto-nimbus becomes strato-cumulus, fracto-cumulus, and finally cumulus.

THE OBSERVATIONS OF CLOUDS AND CLOUDINESS AND THE RESULTS OF OBSERVATION

219. Height of clouds. — Simultaneous observations of the altitude (angular distance above the horizon) and the azimuth (direction) of a cloud obtained from two stations a mile or two apart and connected by telephone, permit the determination of the height of the cloud above the earth's surface and its distance from the two stations. It is necessary that the two stations be connected by a telephone in order to agree upon the portion of the cloud which is to be observed at a given moment. Cloud heights may also be determined photographically by taking simultaneous photographs from two stations, as before. Cloud heights may also be determined by a single observer if he is provided with an accurate map of the surrounding country. The necessary observations in this case are the altitude and azimuth of the cloud, the location of its shadow on the surrounding country, and the date and time of observation. The student of trigonometry will readily see how these observations can be utilized in computing the height and distance of the cloud. These three do not exhaust the possible methods of determining a cloud height, for nearly a dozen have been proposed.

The methods of determining the height of clouds.

Long series of observations of cloud heights have been made at the Blue Hill observatory, near Boston, at Upsala in Sweden, and at Berlin. The results of these measurements show that the height of clouds varies all the way from zero, which is actual contact with the earth's surface, to about ten miles in the case of the lofty cirrus clouds. The greatest, least, and mean height of the various kinds of clouds have already been given in section 217.

Where many observations have been made, and the results.

It has also been found from these observations that the average height of clouds in summer is larger than that in winter.

Clouds are higher in summer than in winter.

Q

220. Direction and velocity of motion. — The direction and velocity of motion of a cloud is usually determined by estimation. A cloud is spoken of, for example, as moving rapidly or slowly from the northwest. If the cloud is near the horizon, however, reliable estimations cannot be made.

Obtained ordinarily by estimation.

Exact determinations of the direction of motion and the velocity of motion of a cloud may be made at the same time that its height is determined, if the same cloud is observed twice at an interval of twenty or thirty minutes.

Ordinarily, an instrument which is called a nephoscope,[1] and is pictured in Fig. 101, is used for determining the direction and velocity of motion of clouds. This consists essentially of a horizontal mirror of black glass, provided with a divided circle. A small compass box is usually attached to the mirror so that the zero of the scale may be set at either magnetic or true north. This

A nephoscope and its use.

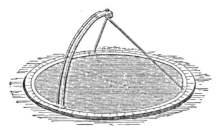

FIG. 101. — A Nephoscope.

is provided with an eyepiece held in a curved framework so that it can be moved up and down and at the same time remain at an unchanging distance from the center of the mirror. The mirror is usually provided with a series of concentric circles at equal distances from each other and with a scale etched across its face. When a cloud is to be observed, the mirror is turned and the eyepiece placed at such a height that the image of the cloud is observed to follow the divided scale across the mirror. The direction of the cloud motion can then be determined by reading the horizontal divided scale, and if its height is estimated, its velocity of motion may be determined by observing the distance on the mirror passed over in a given interval of time.

The results of such observations show that the clouds move much more rapidly than the surface winds. When surface winds are moving with a velocity of 50 miles an hour, it is not at all unusual for the cloud velocity to be 150 or even 200 miles per hour.

Results.

Definition of cloudiness.

221. Cloudiness. — The term cloudiness refers to the amount of the sky covered by clouds, and has nothing to do with the kind of clouds. It is usually determined by a naked-eye

---

[1] νέφος = cloud; σκοπέω = I look at.

estimation, and for this estimation the sky is divided into two portions by a small circle at an altitude of 45°. The amount of the upper clouds and of the lower clouds is estimated separately and the *How esti-* average determined. It is customary to estimate the *mated.* amount in fifths, since the sum of these estimations will in this case give at once the cloudiness in tenths for the whole sky.

There are five adjectives used to express the degree of cloudiness. Cloudless is used when no clouds at all are visible. The sky is said to be clear if it is from 0 to $\frac{3}{10}$ covered with cloud. The term partly cloudy is used to designate an amount of cloudiness *The five adjectives* between $\frac{3}{10}$ and $\frac{7}{10}$ inclusive. The word cloudy designates *used to* a sky from $\frac{7}{10}$ to all covered with cloud. If the sky is com- *designate* pletely cloud-covered, it is usually spoken of as overcast. *cloudiness.* Very often but three of these terms are used in connection with cloudiness: clear, partly cloudy, and cloudy. In this case, cloudless is included in clear, and overcast in cloudy. The term fair is also sometimes used. It is synonymous with partly cloudy and allows $\frac{1}{100}$ of an inch of precipitation, but no more.

222. Sunshine records. — There are various instruments for *The three kinds of* determining *sunshine* the amount of *recorders.* sunshine, which is the converse of cloudiness. The three in most general use are called the burnt paper, the photographic, and the electrical contact sunshine recorder.

The burnt paper sunshine recorder, often

Fig. 102. — The Burnt Paper Sunshine Recorder.

called the Campbell-Stokes sunshine recorder, is illustrated in Fig. 102, and consists of a sphere of glass which focuses the *The burnt* rays of the sun on a piece of paper held in a curved *paper sun-* framework at the back. This faces south and is exposed *recorder.* so that the sun may shine upon it from sunrise until sunset. Whenever the sun is shining, the glass bulb focuses its rays on the paper and chars a track. If the sun goes under a cloud, it is evident on the chart by the absence of charring. The strip of paper is renewed every day, and in this way an accurate record of the duration of sunshine is kept.

Fig. 103. — The Photographic Sunshine Recorder.

The photographic sunshine recorder, often called the Jordan recorder, is illustrated in Fig. 103. It consists of a light-tight cylindrical box (sometimes two) containing a piece of sensitized photographic paper. The

The photographic sunshine recorder.

sun's rays are allowed to fall upon this paper through a very small opening, and as the sun passes from rising-to setting, a track is left on this photographic paper. Any interruption in the sunshine makes itself manifest as an interruption in this track. By changing the position of the small opening, records for a whole month may be kept on a single piece of paper.

223. The sunshine recorder used by the U. S. Weather Bureau is the electrical contact recorder.

The Weather Bureau sunshine recorder.

It is pictured in Fig. 104, and consists essentially of a black bulb thermometer surrounded by a glass jacket from which the air has been extracted. Two wires have been sealed in the stem of the thermometer. When the sun shines, the black bulb quickly absorbs its radiant energy, the mercury rises in the stem, contact is made, and in the office below the pen makes a step every minute in the line it is tracing on a revolving drum. If

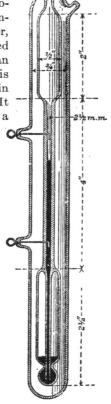

Fig. 104. — The U. S. Weather Bureau Sunshine Recorder.

the sun ceases to shine, the black bulb thermometer radiates its heat, the mercury drops, contact is broken, and the pen makes a straight line upon the revolving drum. In this way a continuous record of sunshine is kept.

The amount of sunshine may be expressed in two ways: either as the absolute duration in hours and decimals of an hour, or as a percentage. The percentage is the ratio of the actual duration of the sunshine to the total possible. *How the amount is expressed.*

224. Observations of clouds, cloudiness, and sunshine. — At the regular stations of the U. S. Weather Bureau, the kind of cloud and the cloudiness are observed at eight in the morning and eight in the evening. A continuous record of the sunshine is also kept by means of an electrical contact sunshine recorder. The black bulb thermometer, with its vacuum jacket, is exposed on the roof of the building. The record is made on a revolving drum in the office, on which is recorded wind direction, wind velocity, and rainfall, as well as sunshine. At the coöperative stations, the average cloudiness of the day as a whole is the only thing which is recorded. *The observations made.*

225. Normal values, data, and charts. — From these observations of cloudiness and sunshine, the following average and normal values may be computed: the normal cloudiness at 8 A.M. and 8 P.M. for the various months and for the year, the normal hourly sunshine, the normal monthly sunshine, and the normal yearly sunshine. For these normals the annual variation in cloudiness and the diurnal and annual variation in sunshine may be determined in the regular way. In the case of the regular stations of the U. S. Weather Bureau, the records which are kept constantly computed to date are: hours of sunshine and percentage of possible; average cloudiness at 8 A.M. and 8 P.M.; number of days clear, partly cloudy, and cloudy; hourly sunshine in percentages. *The normal values which are computed.*

226. For the northeastern part of the United States and for the larger part of the earth's surface, the amount of cloudiness is a maximum during the day and a minimum during the night. The reason for this is convection, which causes an increase in the amount of cumulus clouds by day. The annual variation in cloudiness depends more upon the locality. In general, the summer time is freer from clouds than the winter time. The following table, which gives the normal values of cloudiness for the various months and for the year for various stations in the United States, will make clear the characteristics of the variation. *The characteristics of the daily and annual variation in cloudiness.*

NORMAL VALUES OF SUNSHINE (PERCENTAGE OF POSSIBLE) FOR THE
VARIOUS MONTHS AND FOR THE YEAR

STATION	LENGTH OF RECORD	JAN.	FEB.	MAR.	APR.	MAY	JUNE	JULY	AUG.	SEPT.	OCT.	NOV.	DEC.	ANNUAL
Albany, N.Y. . . .	7	43	59	55	60	60	65	61	68	64	55	39	34	54
Bismarck, N. Dak. .	10	51	56	53	56	59	60	69	66	63	60	49	46	57
Boise, Idaho . . .	5	33	45	53	55	64	76	87	82	79	63	45	45	61
Boston, Mass. . .	10	51	57	53	53	57	60	60	60	62	54	45	52	55
Buffalo, N.Y. . .	14	31	42	49	55	55	66	65	66	60	48	27	24	49
Charleston, S.C. . .		51	48	53	63	66	61	54	51	57	55	58	50	56
Chicago, Ill. . . .	10	46	53	51	62	63	68	71	68	63	61	42	38	57
Columbus, Ohio . .	10	35	44	44	59	61	65	71	70	65	60	39	32	54
Denver, Col. . . .	14	73	67	67	68	61	69	67	66	75	76	71	68	69
Galveston, Tex. . .	14	47	44	47	58	66	72	70	67	68	73	59	51	60
Helena, Mont. . .	10	43	48	53	57	54	60	73	74	61	62	42	41	56
Indianapolis, Ind. .	7	43	47	40	52	57	61	72	64	67	63	46	39	54
Los Angeles, Cal. .	7	68	73	69	70	60	67	78	76	76	75	79	80	73
New Orleans, La. .	14	46	44	50	54	65	55	50	51	64	64	51	47	53
New York, N.Y. .	10	53	58	54	57	56	60	57	60	60	52	51	52	56
Omaha, Neb. . . .	7	60	56	54	60	60	65	74	68	67	59	54	49	60
Philadelphia, Pa. .		50	53	51	55	58	61	63	63	66	60	53	55	57
Phœnix, Ariz. . .	8	72	79	80	85	89	94	83	82	87	87	84	85	84
Portland, Me. . .		57	60	57	60	68	61	65	66	64	54	46	53	58
Portland, Ore. . .	14	31	36	41	43	49	69	61	49	44	23	20	22	41
St. Louis, Mo. . .	13	53	51	52	58	63	67	70	71	71	70	52	49	60
St. Paul, Minn. . .	8	49	55	49	58	55	59	66	59	61	52	44	44	54
Salt Lake City, Utah	14	43	44	52	59	64	79	81	77	79	69	55	44	62
San Francisco, Cal.	9	53	56	60	70	67	76	71	60	67	67	57	56	63
Seattle, Wash. . .	9	23	39	46	50	49	52	62	57	48	35	16	18	41
Topeka, Kan. . . .		60	56	57	59	61	63	72	73	64	66	51	54	61
Washington, D.C. .	13	48	50	48	53	55	63	64	55	68	60	51	54	57

The last year included in these normals is 1903.

If these normal values of cloudiness and of sunshine have been determined for many stations, over the earth's surface, charts may be prepared which will show these values. Charts XIX and XX show the normal sunshine in per cent for January and for July for the United States. Chart XXI gives the normal annual cloudiness for the whole earth. Lines connecting those places which have the same amount of cloudiness are called isonephs, and these were first drawn for the world by Teisserenc de Bort in 1884. It will be seen that the geographical variation of cloudiness is somewhat correlated with the general wind system. The average cloudiness at the equator is between 50 per cent and 60 per cent. This drops down to a minimum of about 40 per cent at latitude 25° or 30° N. It then increases with increasing latitude to a maximum at about 65° N.

The cloudiness of the United States and the world. [side note]

with a value of between 60 per cent and 70 per cent, and then drops somewhat as the pole is approached. Cloudiness varies from place to place because of altitude, temperature, inclosure by mountains, nearness to bodies of water, and the character of the storms.

The Nature of Clouds

227. Nuclei of condensation. — There are several laboratory experiments which would seem to show that the condensation of water vapor always takes place on some material object or on the dust or water particles which are present in the atmosphere. If a vessel contains a quantity of air from which all dust particles have been removed, the temperature may be lowered many degrees below the dew point without causing any condensation in the form of fog or cloud. The condensation takes place slowly and entirely on the sides and bottom of the containing vessel. The reduction in temperature is ordinarily brought about by quickly rarefying the air and thus cooling it by expansion. If, however, ordinary air which contains a large amount of dust is used instead of the dust-free air, as soon as the temperature is lowered below the dew point, it immediately becomes filled with a foglike condensation. It is accordingly supposed that a dust or moisture particle serves as the nucleus or center of condensation of the water vapor which forms clouds or fog.

In laboratory experiments nuclei of condensation are necessary.

There are certain facts, however, which are not in accord with this supposition. The amount of sediment in rain water is extremely slight, yet each raindrop is composed of myriads of cloud particles, each one of which is supposed to have a dust particle for its nucleus. This seeming contradiction is explained by two considerations. In the first place, the dust particles in the atmosphere are probably extremely minute. It has also been found that ionized air permits condensation as well as dusty air. It has been furthermore found that the air is always in a more or less ionized condition. This condition can be brought about by electrical discharges, by ultra-violet light, by cathode rays, or even by the impact of the air against obstacles. Condensation, however, does not readily take place on ions. It has been found that the air must contain several times as much moisture as will saturate it, before condensation on the ions can be forced. It would thus seem that dust particles must play by far the larger part in serving as nuclei of condensation in the atmosphere.

There is but little sediment in rain water.

Ionized air acts the same as dusty air.

228. Size and constitution of cloud particles. — Cloud particles vary in size from $\frac{1}{4000}$ to $\frac{1}{1000}$ of an inch in diameter. These results have

Size of cloud particles.
been obtained by direct measurement and also by deductions from certain optical phenomena in which the cloud particles play a part. It was formerly supposed that the cloud particles were hollow spheres. The reasons for this assumption

Cloud particles are solid, not hollow.
were the fact that cloud particles remained suspended for such long times in the atmosphere, and furthermore certain optical phenomena were better explained on this assumption. Recent observations, however, have shown conclusively that cloud particles are solid throughout. The reason for the suspension in the atmosphere lies in their small size. It can be shown

Cloud particles held in suspension by rising air currents.
that a water drop $\frac{1}{1000}$ of an inch in diameter would ordinarily fall at the rate of less than two inches per second in air at all ordinary pressures. It would thus require but a very feeble ascending air current to keep a cloud particle in suspension or even to cause it to rise.

229. Haze. — Haze occurs both high up in the atmosphere and near

Two kinds of haze.
the earth's surface. The high haze pales out the blue color of the sky by day and gives to it a whitish or washed-out appearance. At night it makes itself manifest by dimming the light

Description and explanation of the high haze.
of the stars and rendering the fainter ones invisible. In this it acts as a very thin cirrus cloud, and thus thick haze is often spoken of as cirrus haze. It is caused by minute ice particles in the upper atmosphere, and is thus a condensation product.

Haze near the earth's surface makes itself manifest by obscuring and rendering indistinct the outlines of distant objects. It is well known

Description and explanation of the haze near the earth's surface.
that dust plays a large part in this, and some authorities credit the cause of haze to the presence of an abnormal amount of dust in the atmosphere. The fact that haze occurs chiefly in September and October, when the dry leaves are falling from the trees and being blown about by the increasing winds, would seem to lend color to this explanation. It has also been thought that haze is due to the condensation of water vapor on these large dust particles near the earth's surface, but as haze often occurs when the air is far above the dew point, it would hardly seem that condensation on any such scale is possible. A good part of the indistinctness of distant objects is probably optical and not mechanical. That is, it is due to the

mixing of layers of air, of different temperature or containing very different amounts of invisible water vapor, and thus having different refractive indices.

THE FORMATION OF CLOUDS

230. Introduction. — The clouds are formed by the condensation of water vapor in a quantity of air which has become supersaturated; that is, which has gone below its dew point and is containing more moisture than the given quantity of air at the tempera- ture in question can contain.   This condition may be produced by the addition of water vapor to a quantity of air already near the point of saturation, or by lowering the temperature so that the air can no longer hold the water vapor which it formerly contained.   It must be held in mind that when condensation occurs in the free atmosphere, the conditions are very different than they are in the usual laboratory experiments.   When the amount of water vapor which saturates a given quantity of air is being determined in a laboratory, large objects and particles are present upon which the condensation may take place.   In 'the free atmosphere the condensation must take place upon the extremely minute particles or ions.   This explains why air has often been found which contains more than the saturating amount of moisture, and yet without condensation.   Valuable information could be gained by determining the amount of water vapor which saturates air when condensation must take place on minute particles.   The amount would be much greater than that given in ordinary tables, and would probably be different for particles of different sizes.

How clouds are formed.

There are nine processes which may lead to cloud formation.   Eight of these produce condensation by cooling the air, and one by adding water vapor.   These nine processes may be divided into three groups.   The first three are the major processes in cloud formation.   The second group of three are those processes which are common, but yet are not such powerful cloud producers as those in the first group.   The three processes which form the third group are ordinarily unimportant and insignificant.   These nine processes are numbered one to nine inclusive, and in the previous sections they have been referred to by these numbers.

The nine processes of cloud formation may be divided into three groups.

231. Condensation in warm winds blowing over cold surfaces (Method 1). — If warm, moisture-laden air blows over a cold surface, its temperature will be reduced, and it may be lowered below the dew point, in which case condensation must take place. There are four minor illustrations of this method of cloud or fog formation. The first one is the spring fogs while it is thawing. The ground is perhaps snow-covered and frozen, and the temperatures are well below the freezing point. The wind begins to blow from the south, and warm air is brought by transportation and is carried over these snow-covered, frozen surfaces. The snow begins to melt and the air goes below its dew point, and a fog is the result. A second example of this process is the formation of fog near Newfoundland. A warm, moisture-laden wind from the south blows over the Gulf Stream against the icebergs and cold water brought down from the arctic regions. The result is a cooling of the warmer air below the dew point and the formation of the many fogs which are prevalent in this region. This is especially true in summer, when the temperature contrasts are particularly marked. A third example is the so-called mountain cloud. This is particularly noticeable if the mountain is snow-capped. The air moving over the summit is cooled below the dew point, and a cloud banner streams out from the mountain top in a direction opposite to the wind direction. This cloud usually has a certain definite length. It disappears, due to the fact that the air which had become cooled in passing the mountain top has become warmed again by mixture or by absorbing the sun's radiant energy sufficiently to go above its dew point and contain the moisture in the form of invisible water vapor. A fourth example of this process is the formation of the so-called frost work in winter. A warm wind blows against objects which are colder than the air. The air in contact with them is cooled below the dew point, and the moisture is deposited on these objects in the form of frost work. This frost work grows out to windward against the air current. These forms are particularly prevalent on the iron work of towers or buildings which are located on high mountains.

How the condensation is produced.

The four minor illustrations of this process.

The major example of this cloud-forming process is the formation of clouds when the wind is south. A south wind transports large quantities of warm, moisture-laden air from southern countries and carries it over colder surfaces farther north. As a result, immense quantities of air are cooled below the dew point, and clouds result. Due to irregularities in the surface and friction, there is probably a certain amount of mixing on the part of the air, and also rising and falling. Other cloud-forming processes

The major example of this cloud-forming process.

would thus probably be operative at the same time. These clouds are chiefly of the stratus and nimbus variety, and occur ordinarily, about one half mile to two miles above the earth's surface. The Kind and height of the cloud. reason for this particular height is twofold: the air near the earth's surface is retarded by friction so that the greater the height above the earth's surface, the more vigorous the transportation. Due to this cause, the most vigorous transportation, and thus condensation, would take place at the greatest heights. But the amount of moisture in the atmosphere decreases with altitude. Thus, two causes are counterbalanced to produce a maximum of moisture transportation, and thus cloudiness at the height stated above.

232. **Condensation in ascending currents due to convection** (Method 2). — When air rises, due to convection, it comes into regions of less barometric pressure, and consequently it expands. How the condensation is produced. The expansion causes cooling at the rate of 1.6° F. for 300 ft. (0.993° C. for 100 meters). As a result the dew point is often passed, and the excess moisture must then condense in the form of cloud. Since convection occurs usually during the day, clouds formed by this process are generally daytime clouds, and the kind is some one of the cumulus forms.

Two constants make possible the computation of the height at which a convection-formed cloud ought to appear. These two constants are the adiabatic rate of cooling, that is, the cooling due to the expansion as the air rises, and the lowering of the dew point due to expansion. The value of the first constant is 1.6° F. for 300 feet. The other constant is introduced here for the first time. Suppose that a cubic foot of air contains a certain The computation of the height at which the cloud will form. amount of moisture. If this air rises a certain height, it will have expanded and become two cubic feet; and since the amount of moisture has remained unchanged, each cubic foot will now contain one half the moisture which it contained before. As a result, the dew point will have become lowered. This dropping back of the dew point, due to expansion, amounts to .33° F. for 300 feet, or .2° C. for 100 meters. Thus, as air rises, its temperature lessens 1.6° F. for each 300 feet, and the dew point is lowered .33° F. for each 300 feet. The result is that the temperature of the air approaches the dew point at the rate of the difference, namely 1.27° F. for each 300 feet. The height at which a convection-caused cloud should be formed can thus be computed by dividing the difference between the temperature and the dew point by 1.27, and multiplying it by 300. For example, suppose on a summer afternoon the temperature is 85° F. and the dew point has been found to be 70° F. · The height of a convection-caused cloud would be $\dfrac{85-70}{1.27}$ 300 or 3543 feet.

After a cloud begins to be formed, the air continues to rise, and it will continue to rise until its temperature has fallen to the temperature of its surroundings. In order to determine this, the vertical temperature gradient must be known. There are three things which favor further rise of the air after it has commenced to become cloudy. These are the absorption of radiant energy, the latent heat of condensation, and the latent heat of fusion. As soon as the rising air becomes cloudy, it absorbs the radiant energy of the sun and earth. Furthermore, the latent heat of condensation is supplied to the rising air, thus lessening the rate of cooling, and causing further rise. As soon as the freezing point is passed, the latent heat of fusion is added to the latent heat of condensation. As a result, all of these convection-caused clouds are usually extremely thick, because the rising air supplied with heat in these three ways must rise to considerable heights before it has cooled by expansion to the temperature of the surrounding air. In the case of a thundercloud, the thickness may amount to as much as five or even six miles.

Convection-caused clouds are usually thick.

There are several illustrations of clouds formed in this way. The cumulus clouds, which occur particularly in summer, and generally during the hotter parts of the day, are usually convection-formed. This can be decided by noting whether they disappear in the late afternoon, as soon as convection stops.

Illustrations of convection-formed clouds.

A thunderstorm, as will be seen later, is convection-caused, and is simply an overgrown cumulus cloud. The clouds formed over certain islands are also good illustrations of this process. The ground warms more rapidly during the day than the ocean. This causes convection, the warm air being forced to rise by the cool air which comes in from the surrounding ocean. Thus certain islands are cloud-covered during the day, and the sky becomes clear again at night when convection ceases. In some cases this process becomes vigorous enough to cause a thundershower each day.

233. Condensation in forced ascending currents (Method 3). — When air is forced to rise, it will expand and become cooler, and perhaps cloudy, in just the same way as in convection. There are two ways in which air may be forced to ascend; one is by passing over a mountain or hill or other barrier

Forced ascent may be caused in two ways.

which forces it to rise, and the second is by being forced to rise in a storm center, that is, in an area of low barometric pressure.

If air is forced to rise by a barrier, a bank of cloud parallel to the barrier ordinarily forms on the leeward side. This is a fairly common cloud in mountainous regions, and the position of the cloud is determined by the wind direction. Air forced to rise by storm centers is one of the chief causes of the stratus and nimbus clouds which often cover large sections of the country. The formation of these immense cloud areas will be fully discussed in the chapter on storms. *The characteristics of the clouds formed.*

234. Condensation caused by diminishing barometric pressure (Method 4). — If, for any reason, the barometric pressure diminishes, the air will at once expand and become cooler. If the air is nearly saturated with moisture, that is, if it is near its dew point, it will require but little diminution in pressure to cause the air to become supersaturated, and condensation in the form of cloud will then take place. This is a process *Clouds formed by diminishing barometric pressure.* of intermediate importance in cloud formations, and gives rise chiefly to clouds of the cirrus or stratus variety.

235. Condensation in atmospheric waves (Method 5). — Whenever a fluid flows over obstacles, waves are usually formed in it. There are many illustrations of waves formed in this way in nature. The surface of a field of drifting snow usually becomes wavy. Sea sand underneath the water also becomes wavy, due to the passing backward and forward of the water. A stream *How the atmospheric waves are formed.* flowing over an irregular bed usually has waves formed on its surface. In just the same way, when air flows over a rough surface, waves are formed in it. These waves may have a height of only a few hundred feet from trough to crest, or at times they may attain a height of half a mile or even more. When the air rises, due to the passage of one of these waves, it expands, becomes cooler, and may go below its dew point. When it falls it is compressed, rises in tempera- *The characteristics of the clouds formed by atmospheric waves.* ture, and usually is cloudless. The waves in the atmosphere would thus give rise to a series of clouds in the form of parallel bars. This is of very common occurrence, and all the rippled or striated clouds are due to atmospheric waves. This appearance is particularly noticeable in connection with cirrus or stratus clouds.

236. Condensation caused by radiation (Method 6). — A layer of air near its dew point may radiate its heat to space or the cold ground, and go below its dew point and become cloudy. This is often noticed in the northeast portion of the United States

**The charac-
teristics of
a radiation-
caused
cloud.** on the still, clear, cold winter mornings. Excessive radiation has taken place from the upper layers of the atmosphere, they have cooled below the dew point, and a thin overcasting in the form of an alto-stratus cloud is the result. It is visible in the early morning, and with the rise in temperature it soon disappears.

237. Condensation due to conduction (Method 7). — This is a very minor factor in cloud production. A layer of air could lose suffi-

**How con-
duction
could pro-
duce a cloud.** cient heat by conduction to adjoining layers to go below its dew point and become cloudy. Suppose a layer of air is at a temperature of 40° and is saturated with moisture, and suppose, furthermore, that the layer of air above or below it has a temperature of 30°. By conduction, this layer of air might lose its heat to the adjoining layer, go below the dew point, and become cloudy. This process of cloud formation would probably produce some cirrus or stratus cloud form.

238. Condensation by mixing air (Method 8). — If two quantities of saturated air of different temperatures are mixed, the result is always

**How mixing
saturated air
at different
tempera-
tures pro-
duces cloud.** condensation. A numerical example will illustrate the truth of this. Suppose a cubic foot of saturated air with a temperature of 70° F., and another cubic foot of saturated air with a temperature of 50° F. are mixed. The cubic foot of air with the temperature of 70° F. contains 7.99 grains of water vapor per cubic foot, while the cubic foot of air with the temperature of 50° F. contains 4.09 grains. These values are obtained from the table in section 181. The resulting air must contain the average of these two values, namely 6.04. The resulting temperature will be 60° F., and it will be seen from the table that air at 60° F. can contain but 5.76 grains per cubic foot. The excess must condense in the form of cloud. As a matter of fact, only a fraction of this excess would actually condense in the free atmosphere, for as soon as condensation started, latent heat would cause a rise of temperature and thus enable the air to hold more moisture. The variety of cloud formed in this way would be of the stratus, perhaps cirrus, type.

Historically this process of cloud formation was one of the first to be

**Historical
importance.** considered, and in fact it was thought that it was one of the major processes in cloud formation. It is now known, however, that it plays a very minor part in the formation of cloud.

239. Condensation by diffusion of water vapor (Method 9). — Suppose a given quantity of air saturated with water vapor is situated between two layers of air containing a larger amount of moisture. By diffusion some of this moisture may pass to the layer of air in question, thus supersaturating it and causing cloudy condensation. Cirrus or stratus would probably be the cloud form produced. *How diffusion may cause condensation.*

240. Conditions that favor a clear sky. — The converse or opposite working of all the methods of cloud formation just mentioned would favor a clear sky. One of the nine processes, however, has no opposite or converse, and this is convection. The omission of this one would leave two major processes of cloud formation. There are thus two major processes which favor a clear sky. These would be cold winds blowing over a warm surface and forced descending air currents. It is a well known fact of observation that clear skies are very prevalent when the wind is north or northwest, and when the station is covered by an area of high barometric pressure, which causes descending and outflowing air currents. *The conditions that favor a clear sky.*

D. PRECIPITATION

THE KINDS OF PRECIPITATION

241. Rain. — Four of the seven forms of condensation have already been fully considered. These are dew, frost, and fog, which occur near the earth's surface, and clouds, which form high up in the atmosphere: The remaining three, rain, snow, and hail, are all included under the general term of precipitation, and they require the most vigorous condensation for their production. *Precipitation.*

Raindrops are formed from the cloud particles. These are not all of the same size, and the larger ones will fall faster than the smaller ones, or, if they are being carried up by ascending air currents, it will be the larger particles which are carried up less rapidly. Due to these motions many collisions must occur, and whenever two cloud particles collide, they coalesce into a diminutive raindrop. As soon as a raindrop begins to fall at all rapidly, it will soon come into layers of air warmer than itself, and condensation of water vapor will then take place on the cold drop. A raindrop thus increases in size, due to collision and condensation, until it reaches the base of the cloud and begins its final fall to the surface of the earth. *The formation of a raindrop.*

But for two circumstances, all clouds would yield precipitation. In the first place, the velocity of the ascending air currents is often sufficient to hold stationary or even carry up drops of considerable size. Then again, after a raindrop leaves the base of a cloud, it at once begins to evaporate, and often disappears long before the earth's surface is reached. It is not at all uncommon to see a dark trail of rain depending from a cloud while no trace of precipitation reaches the earth below. It was formerly thought that electricity played a large part in preventing the formation of raindrops by a cloud. It was known that all clouds are highly electrified, and it was supposed that the small cloud particles, being charged with like electricity, would be kept from coalescing by electrical repulsion. It was accordingly supposed that raindrops could form only when the cloud particles had been discharged by a lightning flash or in some other way. Later experiments seem to show that in most cases electricity helps rather than hinders, if it plays any part at all.

Why every cloud does not yield rain.

In size, raindrops vary from very small to perhaps $\frac{1}{10}$ of an inch in the case of large pattering raindrops. A raindrop two or three times larger than this could not exist, as it would separate into parts due to the rapidity of its fall through the air. The size of raindrops is usually determined by allowing them to fall into a layer of flour. By allowing drops of measured sizes to fall into the same flour and noting the size of the dough balls formed, the size of the raindrops can be determined. Raindrops are largest at the base of the cloud, and diminish in size, due to the evaporation, as the earth's surface is approached.

The size of raindrops.

Of the nine processes which may lead to cloud formation, only the three major processes are vigorous enough to cause sufficient condensation to produce precipitation. These are warm winds over cold surfaces, convection, and forced ascent by a barrier or storm. Thus whenever precipitation falls from a nimbus cloud, it has been formed by one or more of these three processes.

The three cloud-forming processes which may lead to precipitation.

242. Snow. — When condensation is sufficiently vigorous to cause precipitation while the temperature is below the freezing point, snowflakes instead of raindrops are formed. The structure of snowflakes may be carefully studied by catching them on a piece of black cloth and observing them through a magnifying glass. They have also been observed and photographed through a microscope. Many observers have sketched the varied forms of snow-

The structure of snowflakes.

Fig. 105. — Snowflakes.

flakes, but the most complete study of them has been made by Mr. Wilson A. Bentley, of Nashville, Vt., who has observed them for more than twenty years, and whose microphotographs have been reproduced in the *Monthly Weather Review* for 1901 and 1902. The six examples pictured in Fig. 105 are taken from this collection. If the temperatures are low, the snowflakes are always small, flat, and regular. They always have angles of 60° or 120°, which are characteristic of crystallized water. They bear every evidence of having been formed about a single center by continuous condensation or by the addition of small cloud particles. Attempts have been made to correlate their varied appearance with the temperature, the type of storm, the rapidity of condensation, etc., but the effect of each factor which enters in has not yet been determined. If the temperature is very low — at least below zero Fahrenheit — fine ice needles are formed instead of snowflakes. If the temperature is near the freezing point, particularly in the lower layers of the atmosphere, the snowflakes often mat together and form large clots. If the temperature is still higher, the snowflakes sometimes partially melt. Much of the rain which falls in the winter time probably left the cloud as snow and melted during the descent. *The effect of temperature on the form of the flake.*

243. Hail. — There are three kinds of hail, and each occurs at a different time of year and is formed in a different way. *Three kinds of hail.*

The hail which occurs during the winter consists of small, clear pellets of ice of about the size of large raindrops — in fact, they are frozen raindrops. Hail of this kind occurs when the temperature of the cloud where the raindrop is formed is above 32° F. while the lower layers of the air are still below the freezing point. As a result, the raindrop freezes during its fall, and reaches the ground as a hailstone. The quantity is usually small, although on rare occasions the ground may be covered to a depth of three or four inches by hail of this kind. *Winter hail.*

The so-called soft hail consists of small white pellets of what looks like compacted snow. It occurs usually in very small quantities during March and April, and occasionally during the autumn. It falls nearly always from an overgrown cumulus cloud. It is supposed to be formed from frozen cloud particles mixed with raindrops and compacted by a high wind. *Soft hail.*

In the summer time hail never occurs except during a thundershower. The hailstones are usually large, in some cases several inches in diameter, and they consist of concentric layers of compact *Summer hail.*

R

snow and ice. Hail nearly always falls at the beginning of a shower, and sometimes great damage is done. There are records when the hail-stones during a single shower have covered the ground to a depth of more than a foot. The full consideration of the formation of this kind of hail will be deferred until the mechanism of a thundershower has been studied, but the structure of the hailstones would seem to show that they had been formed in a cloud whirling about a horizontal axis. The nucleus is carried up and coated with snow; it then falls and is coated with water; it is then carried up again, the water freezes, and it is once more coated with snow. This process continues, adding coat after coat, until the hailstone becomes too heavy to be longer sustained, and it falls to the ground. As will be seen later, it is in the squall cloud at the front of a thundershower that these conditions are actually realized.

Section 119 of the instructions for preparing meteorological forms of the U. S. Weather Bureau says: " Care should be taken in deter-

Weather Bureau definition of sleet. mining the character of precipitation when in the form of sleet or hail. Only the precipitation that occurs in the form of frozen rain should be called sleet. Hail is formed by accretions consisting of concentric layers of ice, or of alternate layers of ice and snow. It frequently happens that snow falls in the form of small, round pellets, which are opaque, having the same appear-ance as snow when packed. This should never be recorded as sleet." The above is perfectly definite, and in use at all Weather Bureau stations. According to it, the winter hail or frozen raindrops should be called sleet. The soft hail should be called snow, and only the summer hail should be recognized as hail. Unfortunately, in most dictionaries, books on meteorology, and the popular mind, snow mixed with rain is considered sleet, and several kinds of hail are recognized.

244. Ice storms. — It sometimes happens that it rains very soon after a continued period of cold while the temperature of the ground

The charac-teristics of an ice storm. and the layer of air next it is still considerably below the freezing point. As a result, the rain freezes to everything that it touches, and trees, shrubs, vines, and the ground itself become covered with a layer of ice. This is known as an ice storm, and considerable damage is sometimes done by breaking down trees and vines on account of the weight of the ice which forms on them. These storms are particularly prevalent in New England, and Fig. 106 illustrates such a storm.[1]

[1] See *Monthly Weather Review*, December, 1900.

242

Fig. 106. — The Effects of an Ice Storm in New England.

(Reproduced from Harrington's *About the Weather*, copyright, 1899, D. Appleton and Co.)

245. Rain-making. — It has been a favorite popular belief that rain can be produced by cannonading and heavy artillery fire. The frequent occurrence of thundershowers on July 4 has been used as an argument in favor of this belief, and also the fact that rain has followed so many of the great battles. As regards July 4, statistics at many stations for many years show that there are no more thundershowers on the Fourth than on the third or fifth. In connection with rain following battles, it has been pointed out that the fact was mentioned long before gunpowder was invented. There are two possible reasons why the fact has been so often noted. In the first place, the discomfort and suffering brought about by the rain would surely cause it to be mentioned. And again, an army usually gets into position during good weather while the roads are good, so that by the time the battle begins a rain period would be due. *Rain on July 4 and after battles.*

Influenced, perhaps, by this popular belief, many attempts have been made by the so-called rain-makers to cause rain by artificial means, and in fact considerable money has been expended. The methods employed are either to cause violent explosions in the upper air or to liberate a large amount of gas in the upper air. The only way in which an explosion could cause rain would seem to be by furnishing through the smoke particles numerous nuclei of condensations, or by causing cloud particles to coalesce, due to the wave motion caused by the explosion. The liberation of gas might cause convection. But all these effects would seem to be on too diminutive a scale to influence the immense masses of air which must be affected in order to produce rain over a considerable area. As far as results are concerned, it has never been proven that any rain has fallen which would not have fallen if the experiments had not been tried. *Rain-making.* *The results.*

246. Cooling produced by precipitation. — As soon as it begins to rain, particularly in summer, the air is usually much cooler. One reason for this is the fact that the raindrops are from 3° to 15° F. colder than the air at the earth's surface. It is to be noted in this connection that a raindrop in falling from a cloud is not compressed and heated by the compression, as air would be, but retains its temperature during the descent. There are other causes, however, of the cooling caused by precipitation, such as the cutting off of insolation by the clouds, the coming down of cooler air from above, and the cooling due to evaporation from the wet ground. *Raindrops are colder than the air.* *Other causes of cooling.*

The Determination of Precipitation and the Results of Observation

247. The measurement of a rainfall. — The only measurement made in connection with a rainfall is the amount, that is, the thickness of the layer of water which the rainfall would produce on a level surface, provided none were lost. The instrument for determining this is called a rain gauge, and consists in its simplest form of a vessel for catching the rain and a measuring rod for determining its depth. Since the rain gauge is so simple in construction and rain is one of the most important of the meteorological elements, it is no wonder that rain gauges of one form or another have been in use for nearly three hundred years.

The amount is measured by a rain gauge.

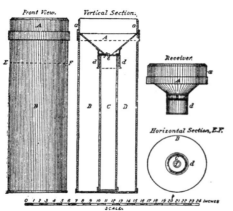

Front View. *Vertical Section.* *Receiver.* *Horizontal Section, E.F.*

SCALE.

Fig. 107. — The U. S. Weather Bureau Rain Gauge.

The U. S. Weather Bureau standard rain gauge, as illustrated in Fig. 107, consists of a galvanized iron cylindrical can eight inches in diameter and about twenty inches high. It is provided with a funnel-shaped cover or receiver, with a beveled rim sharp on the inside and accurately circular in order to catch the amount of rain which falls on a definite area. The shape is that of a funnel with a small opening, in order to prevent evaporation. The funnel opens into an inside brass cylinder which has just one tenth the area of the outer can. The depth of the water in this cylinder is determined by inserting a measuring rod and noting the height to which it is wetted. The measurements are made to the tenth of an inch, and thus the amount of the precipitation determined to the hundredth.

Description of the U. S. Weather Bureau rain gauge.

It is ordinarily stated that the rain gauge should be exposed in the open from three to six feet above the ground, and at a considerable distance from trees, buildings, or any obstruction. The disadvantage of this exposure is that there is no protection against

Exposure of a rain gauge.

the wind, and the eddy caused by the wind passing over and around the rain gauge itself lessens the amount of precipitation which is received. If the rain gauge is exposed at a greater height, wind velocities are larger, and the loss would be still greater.   A rain gauge should not be exposed on a roof, as it is impossible to know what effect the eddies caused by the building will have on the amount collected.   Ordinarily a roof exposure increases the amount collected somewhat.   The rain gauge of the U. S. Weather Bureau stations are usually located on the flat roofs of buildings in large cities, and thus, on account of their location, the indications of the gauges may differ 5 or even 10 per cent from the correct amount.

Recording rain gauges have also been devised, and these work either on the float or tipping bucket principle.   In the case of the float instruments, as the water rises the float is carried up and makes a Recording record on a revolving drum.   In the case of the tipping rain gauges. bucket instruments, the bucket becomes filled whenever a hundredth of an inch of precipitation has fallen.   It then tips over, brings another bucket into place, empties itself, and makes a mark on a revolving drum.

In addition to the amount, the time of beginning and ending is also noted in connection with a rainfall.   If the amount is too Other ob- small to measure, it is recorded as T (trace) in the records. servations.

248. The measurement of a snowfall. — Two measurements are made in connection with a snowfall: one is the depth of the snow which has fallen, and the other is the water equivalent of the snow- Two meas- fall.   The depth is determined simply by measuring it urements with a measuring rod.   The only difficulty is to find some made. place where the depth has not been changed by drifting.   For this reason it is customary to measure the depth at three or four different places which seem to be as free from drifting as possible, and take the average.   The depth is ordinarily recorded to the tenth of an inch.

In order to determine the water equivalent, the snow is sometimes caught in the outer can of the rain gauge when the funnel and inner cylinder have been removed.   It is then melted down by How the placing it in a warm room or by adding a known quantity water equiv- of warm water, which is deducted as soon as it has melted. alent is determined. The water is then poured into the inner cylinder and measured, and the water equivalent thus determined.   If it is thought that too much snow has been blown out of the rain gauge, another method may be used. A sample may be taken by inverting the can and pressing it down in the soft snow in some place free from drifting. By passing a piece of tin or a shingle underneath, a sample may be secured and melted down as before. It would, of course, be more accurate to repeat this several times.

It has been found that the number of inches of snow which corresponds to an inch of water is by no means constant. It requires all the way from 6 to 30 inches of snow to make an inch of water, depending on the lightness of the snow. The average value, however, is about ten.

Ratio of snow to water.

249. Observations of precipitation. — The observations which are made of precipitation at both the regular and coöperative stations of the U. S. Weather Bureau are the same. These are: the kind of precipitation; the time of beginning and ending; the amount of a rainfall; the depth and water equivalent of a snowfall; the amount of snow on the ground each day.

The observations made.

There are also many special stations which observe practically nothing else except rain and snow.

250. Normal values and precipitation data. — From these observations of precipitation, three sets of normals may be computed. These are: the normal hourly, daily, monthly, and annual amount of precipitation (rain and melted snow); the normal monthly and annual amount of snowfall; the normal monthly and annual number of days with precipitation. Since all of these quantities are mere numbers, these normals are computed in the regular way.

The various normals of precipitation.

The graph which represents the daily variation in the amount of precipitation is determined by plotting to scale the normal hourly precipitations. Its form is usually quite irregular, and is very different for different places and different times of the year.

The daily variation.

251. The accompanying table gives the amount of the precipitation (rain and melted snow) for the various months and for the year for several years, and also the normal values for Albany, N.Y.

THE AMOUNT OF PRECIPITATION FOR THE VARIOUS MONTHS AND FOR THE YEAR FOR SEVERAL YEARS, AND THE NORMAL VALUES, FOR ALBANY, N.Y.

	JAN.	FEB.	MAR.	APR.	MAY	JUNE	JULY	AUG.	SEPT.	OCT.	NOV.	DEC.	YEAR
1874	3.61	2.90	1.97	**4.97**	2.32	4.71	**6.78**	1.94	4.01	1.77	2.19	.76	37.93
1875	2.14	1.65	3.27	3.63	2.57	3.98	2.46	6.	.6	.97	.29	1.1	.25
1876	1.57	4.09	4.28	3.51	2.96	4.40	4.97	.	.1	.64	.65	2.4	.19
1877	1.95	.86	3.33	1.42	2.77	4.60	4.00	4.	.8	.86	.70	.7	.09
1878	4.45	4.12	2.18	3.99	3.65	4.54	5.52	3.	.2	.37	.43	6.1	.37
1879	2.54	2.80	3.79	3.17	.89	4.62	5.10	4.	.4	.24	.56	.2	.66
1880	2.96	2.67	2.17	2.75	3.35	2.21	3.78	2.	.8	.45	.49	.0	.54
1881	2.86	2.50	3.80	1.34	3.90	3.76	2.22	2.	.3	.19	.44	.8	.34
1882	2.64	3.31	1.79	1.27	4.15	3.98	3.97	1.	.7	.27	.97	.2	.76
1883	2.43	3.00	1.77	2.65	3.20	6.30	5.96	3.	.1	3.49	1.14	.5	.37
1884	2.98	3.85	4.00	2.09	2.79	1.80	5.04	5.	.8	8.64	.44	.2	.90
1885	3.09	1.38	.62	2.89	1.92	1.98	1.98	7.	.00	.54	.90	.5	.39
1886	3.66	1.40	2.73	3.67	3.40	3.19	2.56	.	.1	.43	.40	.1	.01
1887	3.02	2.86	2.90	2.49	2.27	2.99	4.61	4.		.22	.36	.4	.70
1888	3.04	2.07	**5.62**	1.95	2.98	3.18	2.52	4.		.10	.48	.3	.66
1889	2.83	1.81	1.76	1.25	3.32	6.43	4.19	3.		.48	.00	.1	.51
1890	2.28	2.52	3.72	1.64	5.19	2.72	2.37	5.		.76	.18	.9	.89
1891	**6.12**	4.14	3.12	2.27	1.69	2.65	6.11	5.		.13	.40	.2	.68
1892	4.08	2.13	1.64	.56	**5.30**	4.41	4.22	6.		.60	.29	.8	.83
1893	1.31	**4.63**	2.00	2.10	5.08	2.92	1.82	7.		1.67	.91	2.5	.39
1894	2.54	2.61	.85	2.02	4.64	3.29	2.96	2.		.62	1.96	3.1	.11
1895	1.65	1.63	1.31	3.09	1.72	1.72	4.02	3.		.35	.78	.5	.80
1896	.98	4.03	4.66	.98	1.55	2.49	3.57	2.		.53	.80	.7	.88
1897	1.62	2.05	1.85	3.12	4.69	4.45	6.67	4.		.01	.65	4.3	.79
1898	2.96	3.57	1.08	2.63	4.07	5.58	1.07	6.		.29	.22	3.0	.77
1899	2.50	2.92	3.97	1.03	2.23	1.61	2.69	1.		.85	.47	.5	.92
1900	2.33	2.84	4.62	1.31	1.36	3.54	3.41	2.		1.83	.95	.8	.56
1901	1.59	.56	4.14	4.66	4.79	3.14	4.26	4.	2	8.48	.02	.4	.53
1902	.67	3.04	3.24	2.33	1.91	3.91	5.37	3.	4.	.80	.95	.1	.48
1903	1.83	2.05	3.55	.79	.15	**6.44**	3.51	5.		.09	1.65	.5	.09
1904	2.51	1.17	1.94	2.87	2.16	5.48	2.96	2.		.09	.64	.7	.26
1905	2.66	.80	2.43	2.12	.96	3.58	2.00	3.		.38	1.49	.3	.98
1906	.97	2.09	2.54	2.20	3.90	5.80	3.91	2.		.62	6	.9	.51
1907	1.71	1.15	.87	2.33	3.21	3.29	4.14	.		.71	.5	.5	.63
1908	1.36	2.77	1.43	2.62	4.26	2.32	5.33	3.		.07	.0	.5	.41
Sums	87.43	87.47	94.94	83.71	105.30	132.01	136.05	134.18	112.78	105.54	94.81	90.96	1265.18
Normals	2.50	2.50	2.71	2.39	3.01	3.77	**3.89**	3.83	3.22	3.02	2.71	2.60	36.15

It will be seen that the various months may depart widely from normal, as the amount varies all the way from almost nothing to more than twice the normal amount. In the case of the annual amount a departure of 20 per cent from normal is not uncommon. A departure of three inches is common, of six inches is unusual, and of eleven inches is generally record breaking. The maximum occurs in July and the minimum in February. The reason for this is the large amount of rain which falls during the summer thunder showers. Similar data can be computed for every station.

The annual variation in precipitation at Albany.

252. The accompanying tables give the normal monthly and annual precipitation for various stations in the United States and the world.

NORMAL PRECIPITATION IN INCHES FOR THE VARIOUS MONTHS AND THE YEAR

STATION	LENGTH OF RECORD	JAN.	FEB.	MAR.	APR.	MAY	JUNE	JULY	AUG.	SEPT.	OCT.	NOV.	DEC.	ANNUAL
Albany, N.Y.	84	2.61	2.47	2.75	2.67	3.50	4.03	4.12	3.82	3.37	3.41	2.97	2.67	38.39
Atlanta, Ga.	44	4.78	5.05	5.5	4.0	3.3	3.9	4.	4.47	3.52	2.31	3.43	4.64	49.47
Baltimore, Md. . . .	39	3.23	3.54	3.9	3.1	3.5	3.7	4.	4.20	3.86	2.96	2.83	3.21	43.06
Bismarck, N. Dak. . .	34	0.53	0.53	1.0	1.7	2.4	3.3	2.	1.95	1.17	1.04	0.67	0.59	17.50
Boise, Idaho	32	1.86	1.51	1.54	1.1	1.4	0.9	0.	0.18	0.48	1.20	1.07	1.64	13.27
Boston, Mass.	91	3.74	3.51	4.14	3.8	3.6	3.1	3.	4.15	3.44	3.71	4.11	3.78	44.70
Buffalo, N.Y.	53	3.07	2.94	2.9	2.4	3.1	2.9	3.	2.98	3.34	3.48	3.25	3.38	37.19
Charleston, S.C. . . .	120	3.08	3.08	3.3	2.40	3.4	5.2	6.	6.65	5.22	3.90	2.70	3.34	48.59
Chattanooga, Tenn. .	31	5.24	5.06	5.9	4.36	3.7	4.0	3.	3.88	3.43	2.72	3.42	4.43	50.07
Cheyenne, Wyo. . . .	40	0.38	0.54	0.9	1.7	2.4	1.6	2.	1.47	1.00	0.71	0.41	0.39	13.80
Chicago, Ill.	39	2.08	2.30	2.5	2.7	3.6	3.5	3.	3.02	3.06	2.43	2.52	2.05	33.54
Cincinnati, Ohio . .	75	3.27	3.18	3.7	3.2	3.9	4.2	3.	3.61	2.79	2.59	3.19	3.38	40.91
Cleveland, Ohio . .	38	2.53	2.68	2.84	2.2	3.2	3.5	3.64	2.94	3.28	2.73	2.59	2.57	34.90
Columbus, Ohio . .	34	2.97	3.01	3.49	2.8	3.80	3.4	3.65	3.21	2.41	2.32	2.91	2.66	36.68
Davenport, Iowa . .	39	1.66	1.58	2.9	2.6	4.26	4.0	3.6	3.73	3.15	2.29	1.83	1.53	32.64
Denver, Col.	37	0.48	0.47	0.	2.0	2.5	1.4	1.64	1.35	0.90	0.97	0.56	0.60	14.05
El Paso, Tex.	55	0.42	0.47	0.	0.1	0.2	0.5	1.67	1.88	1.64	0.87	0.57	0.44	9.31
Erie, Pa.	35	2.99	2.90	2.	2.4	3.5	3.7	3.10	3.11	3.54	3.68	3.50	2.92	38.15
Flagstaff, Ariz. . .	21	2.91	2.62	2.	1.3	1.5	0.4	2.45	2.89	1.43	1.18	1.61	2.49	23.87
Galveston, Tex. . .	37	3.54	3.05	2.	3.1	3.2	4.3	3.96	4.79	5.84	4.35	4.03	3.75	46.78
Havre, Mont.	30	0.73	0.50	0.	0.8	2.0	2.9	1.89	1.21	1.06	0.62	0.68	0.54	13.63
Helena, Mont. . . .	29	1.00	0.66	0.	1.0	2.1	2.2	1.13	0.69	1.13	0.77	0.76	0.79	13.21
Indianapolis, Ind. . .	38	2.96	3.09	4.04	3.3	4.00	4.2	4.06	3.21	2.99	2.66	3.48	2.98	41.09
Jacksonville, Fla. .	56	2.82	3.25	3.39	2.7	3.93	5.6	6.37	6.60	8.16	4.60	2.21	2.86	52.53
Kansas City, Mo. . .	38	1.23	1.68	2.	3.1	4.5	4.9	4.47	4.31	3.71	3.06	1.96	1.36	36.94
Key West, Fla. . .	72	2.02	1.58	1.	1.20	3.0	4.6	3.56	4.89	6.49	5.11	2.13	1.94	38.26
Lincoln, Neb. . . .	34	0.66	0.95	1.	2.60	4.4	4.6	4.31	3.48	2.53	2.08	0.79	0.70	28.43
Los Angeles, Cal. . .	32	3.03	3.00	3.	1.	0.4	0.0	0.01	0.03	0.10	0.75	1.33	2.85	15.75
Madison, Wis. . . .	51	1.63	1.50	2.	2.	3.6	4.0	3.80	3.15	3.08	2.32	1.76	1.72	31.25
Memphis, Tenn. . .	49	4.99	4.63	5.	5.	4.3	4.3	3.39	3.33	2.94	2.55	4.44	4.30	50.22
Milwaukee, Wis. . .	39	1.63	1.50	2.	2.	3.6	4.0	3.80	3.15	3.08	2.32	1.76	1.72	31.25
Minneapolis, Minn. .	47	1.04	0.96	1.	2.	3.6	4.2	3.34	3.61	3.56	2.14	1.40	1.31	29.35
Mobile, Ala.	38	4.80	5.45	7.	4.	4.2	5.6	6.68	6.90	5.12	3.15	3.65	4.56	61.80
New Orleans, La. . .	64	4.54	4.28	4.	4.	4.0	5.3	6.53	5.65	4.49	3.25	3.81	4.54	55.63
New York, N.Y. . .	84	3.29	3.27	3.	3.	3.5	3.4	4.08	4.38	3.44	3.42	3.55	3.30	42.47
Omaha, Neb.	51	0.62	0.81	1.	2.	4.3	5.1	4.43	3.45	2.95	2.47	1.00	0.89	30.46
Philadelphia, Pa. . .	39	3.23	3.35	3.	2.	3.3	3.2	4.14	4.69	3.36	3.01	3.11	3.07	40.88
Phœnix, Ariz. . . .	32	0.85	0.87	0.	0.	0.1	0.0	0.89	0.94	0.69	0.37	0.67	0.86	7.27
Portland, Me. . . .	44	3.76	3.45	3.	3.	3.5	3.2	3.47	3.55	3.31	3.70	3.67	3.77	42.63
Portland, Ore. . . .	60	6.32	4.96	4.	2.	2.3	1.7	0.75	4.58	1.69	3.11	6.00	7.21	42.45
St. Louis, Mo. . . .	75	2.27	2.64	3.	4.	4.5	4.8	3.69	3.48	3.00	2.83	2.99	2.62	40.10
St. Paul, Minn. . . .	73	0.89	0.79	1.	2.	3.44	4.1	3.53	3.47	3.39	1.99	1.38	0.97	27.80
Salt Lake City, Utah .	36	1.33	1.48	2.	2.	2.16	0.7	0.50	0.85	0.91	1.43	1.39	1.40	16.33
San Francisco, Cal. .	61	4.82	3.63	3.	1.	0.74	0.0	0.02	0.02	0.31	1.03	2.62	4.64	22.96
Seattle, Wash. . . .	20	4.42	3.97	3.	2.	2.22	1.5	0.68	0.49	1.91	2.70	5.94	5.94	35.68
Tampa, Fla.	56	2.56	2.88	2.	1.	2.73	7.5	9.36	9.02	6.32	2.41	1.71	2.29	51.49
Topeka, Kan.	22	1.21	1.50	2.	2.	5.09	4.7	4.79	4.57	3.33	2.01	1.27	0.84	34.07
Washington, D.C. . .	73	3.13	3.09	3.	3.	3.71	3.7	4.34	4.08	3.25	3.12	2.59	3.01	40.80
Yuma, Ariz.	38	0.42	0.53	0.	0.	0.03	T	0.15	0.49	0.17	0.21	0.29	0.41	3.13

The year 1908 is the last one included in these normals. T signifies trace — less than one hundredth of an inch.

Station	Length of Record	Jan.	Feb.	Mar.	Apr.	My	June	July	Aug.	Se.	Oc.	No.	Dec.	Annual
St. Petersburg .	66	0.87	0.83	0.91	0.94	1.69	1.81	2.68	**2.72**	2.01	1.69	1.42	1.18	16.77
Stockholm . . .	35	0.79	0.	0.	0.	1.	1.	2.2	**2.44**	1.8	.	1.38	1.	17.21
London	40	2.01	1.	1.	1.	2	2.4	2.40	2.4	.	2.	2	3	25.47
Berlin	30	1.54	1.	1.	1.	2	**2.7**	2.24	1.6	.	1.85	1	3	22.84
Vienna	30	1.34	1.	1.	2.	2	2.64	2.68	1.6	.	1.81	1	9	25.08
Constantinople .	48	3.43	2.	1.	1.	1	1.06	1.65	2.0	.	4.	2 80	8.86
Athens	37	2.20	1.	0.	0.	0	0.32	0.3	0.5	.	2. 9	2 48	5.83
Jerusalem . . .	32	6.38	5.	1.	0.	0 00	0.00	0.00	0.04	.	2. 8	5 1	5.24
Paris	30	1.42	1.30	1.	1.	2 13	2.05	2.13	1.97	.40	1. 7	1 1	1.93
Rome	55	2.87	2.32	2.	2.	1	0.6	1.0	2.7	.09	4. 5	3 7	9.92
Capetown . . .	43	0.67	0.85	1.	3.	4	3.5	3. 1	2.1	.	1. 0	0 9	24.88
Adelaide . . .	50	0.75	0.	1.	2.	2	2.7	2. 8	1.9	.	1. 4	0 1	21.14
Peking	37	0.12	0.	0.	1.	3	9.4	6. 4	2.5	.	0. 8	0 8	24.96
Hong-kong . .	20	1.34	1.	5.	13.	16	13.3	14. 1	8.1	.	1. 9	1 2	84.72
Tokyo . . .		2.17	2.	5.04	5.	6	5.1	4. 9	7.9	.	4. 9	2 3	58.1
Manila	38	1.14	0.	1.10	4.	9	15.00	14. 1	14.7	.	5. 5	2 8	36.3
Bombay . . .	84	0.12	0.00	0.62	0.	20	24.57	14. 8	10.94	.	0. 7	0 04	3.9
Havana	30	2.72	2.28	2.	4.	7	5.04	6. 2	6.69	.	3. 7	2 17	1.6
Rio de Janeiro .	40	4.69	4.88	4.	3.	1	1.61	1. 5	2.2	.	4. 5	5 90	2.9
Buenos Aires . .	40	2.91	2.	2.	2.	2	2.1	2. 2	3.1	.	2. 7	3	6.7

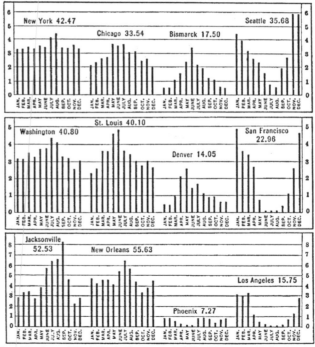

Fig. 108. — The Annual Variation in the Amount of Precipitation at 12 Stations in the U. S.

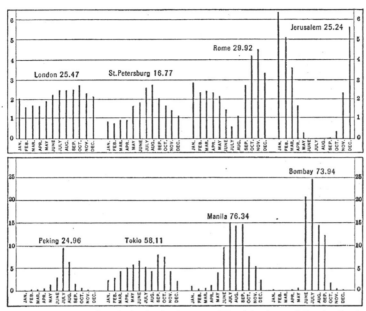

Fɪɢ. 109. — The Annual Variation in the Amount of Precipitation at 8 Foreign Stations.

In Figs. 108 and 109 some of these values are shown graphically. It
will be seen at once that the annual variation is very different
at different places. The chief factors which determine the
amount of precipitation and its distribution throughout the
year are : the general wind system ; the temperature changes
and also the contrast in temperature between the place in question and
these regions from which the prevailing winds come ; elevation and inclosure by mountains ; nearness to bodies of
water ; the characteristics of the storms which occur. At
San Francisco, for example, the precipitation occurs nearly
all during the winter, and almost no rain falls during the
summer. San Francisco is located in the region of prevailing westerlies, and furthermore, a continent is warmer than the
adjoining ocean during the summer and colder in winter.
Thus, during the winter we have the moisture-laden
prevailing westerlies blowing from the warmer ocean over the
colder land. Thus, condensation occurs, and the precipitation

Normal monthly and annual precipitation.

The factors which determine the distribution of precipitation throughout the year.

San Francisco.

is copious. During the summer, the prevailing westerlies are blowing from a colder ocean over a warmer land. As a result, there is practically no precipitation. At New York the precipitation is caused almost entirely by storms, extratropical cyclones, **New York.** and ·thundershowers, and as a result, the distribution throughout the year is quite uniform. The maximum occurs during the summer when thundershowers are most prevalent. Thus by considering these different factors, the amount of precipitation and its distribution throughout the year can be explained.

253. The accompanying table gives for Albany, N.Y., the amount of snowfall for the various months and for the year for several years, and also the normal values. **Snowfall at Albany.**

SNOWFALL FOR THE VARIOUS MONTHS AND FOR THE YEAR FOR SEVERAL
YEARS, AND THE NORMAL VALUES, FOR ALBANY, N.Y.

	JAN.	FEB.	MAR.	APR.	MAY	JUNE	JULY	AUG.	SEPT.	OCT.	NOV.	DEC.	YEAR
1886	19.7	2.4	8.0	1.1	0	0	0	0	0	0	22.3	11.1	64.6
1887	20.5	15.4	22.2	3.5	0	0	0	0	0	0	1.8	27.2	90.6
1888	20.9	9.2	50.9	T	0	0	0	0	0	T	9.5	3.0	93.5
1889	12.2	4.0	T	0.2	0	0	0	0	0	0	T	6.0	22.4
1890	5.0	5.2	11.0	0	0	0	0	0	0	0	0	25.9	47.1
1891	26.0	10.0	2.0	11.0	T	0	0	0	0	T	1.0	1.0	51.0
1892	10.0	9.0	5.2	T	0	0	0	0	0	0	T	1.0	25.2
1893	8.1	40.7	3.0	4.8	0	0	0	0	0	T	0.2	12.4	69.2
1894	14.9	24.7	1.0	2.5	0	0	0	0	0	T	8.6	24.4	76.1
1895	13.5	15.6	6.3	T	T	0	0	0	0	T	2.2	T	37.6
1896	7.1	15.3	23.8	3.2	0	0	0	0	0	0	0.1	5.0	54.5
1897	11.8	9.4	2.6	T	0	0	0	0	0	0	4.5	9.6	37.9
1898	19.8	23.5	2.0	1.2	0	0	0	0	0	0	10.3	7.3	64.1
1899	9.4	27.5	20.8	T	0	0	0	0	0	T	T	2.2	59.9
1900	7.9	9.6	22.8	T	T	0	0	0	0	0	14.4	1.5	56.2
1901	13.4	4.2	7.0	T	0	0	0	0	0	0	6.0	21.6	52.2
1902	2.1	16.8	9.1	T	T	0	0	0	0	T	1.8	33.5	63.3
1903	4.8	9.3	2.6	1.0	T	0	0	0	0	0.3	T	12.5	30.5
1904	20.8	8.2	11.1	5.0	0	0	0	0	0	T	6.2	12.5	63.8
1905	18.2	7.8	8.2	2.0	0	0	0	0	0	T	0.4	2.1	38.7
1906	2.5	16.6	15.8	6.7	0	0	0	0	0	T	10.4	6.7	58.7
1907	7.8	10.8	2.1	8.5	0.5	0	0	0	0	0.1	6.5	10.6	46.9
1908	5.1	20.6	5.6	0.4	0	0	0	0	0	T	1.7	8.7	42.1
Sums	281.5	315.8	243.1	51.1	0.5	0	0	0	0	0.4	107.9	245.8	1246.1
Normals	12.2	13.7	10.6	2.2	T	0	0	0	0	T	4.7	10.7	54.2

It will be seen that the monthly amounts may depart widely from normal, as the amount varies all the way from practically nothing **Normal** to more than twice the normal amount. In the case of the **snowfall.** annual amount, a departure of 40 per cent from normal is not unusual.

In the accompanying table are given the normal monthly and annual amount of snowfall for several stations in the United States.

THE NORMAL AMOUNT OF SNOWFALL IN INCHES FOR THE VARIOUS
MONTHS AND FOR THE YEAR

STATION	LENGTH OF RECORD	JAN.	FEB.	MAR.	APR.	MAY	JUNE	JULY	AUG.	SEPT.	OCT.	NOV.	DEC.	ANNUAL
* Albany, N.Y.	19	12.8	14.0	11.1	1.2	T	0.0	0.0	0.0	0.0	T	4.7	10.0	54.7
* Bismarck, N. Dak.	29	5.4	4.9	7.7	2.5	1.4	0.0	0.0	0.0	0.0	0.8	6.3	4.8	33.8
* Boise, Idaho	5	5.8	6.4	3.3	2.2	0.1	0.0	0.0	0.0	0.0	0.0	2.2	4.5	24.5
Boston, Mass.	36	11.9	11.9	7.6	2.7	0.0	0.0	0.0	0.0	0.0	0.0	2.5	8.8	45.4
* Buffalo, N.Y.	33	19.2	16.4	8.7	3.8	0.1	0.0	0.0	0.0	0.0	0.3	6.5	16.3	71.3
* Charleston, S.C.	19	T	T	T	0.0	0.0	0.0	0.0	0.0	0.0	0.0	0.0	T	T
* Chicago, Ill.	19	10.2	11.5	5.0	0.8	T	0.0	0.0	0.0	0.0	T	2.5	6.6	36.6
Columbus, Ohio	24	6.9	5.1	4.0	1.3	T	0.0	0.0	0.0	0.0	T	1.6	3.8	22.7
Denver, Col.	25	4.4	7.3	10.9	9.9	2.0	0.0	0.0	0.0	0.8	3.8	4.5	6.8	50.4
* El Paso, Tex.	20	0.9	0.5	0.1	0.0	0.0	0.0	0.0	0.0	0.0	0.0	0.3	0.6	2.4
* Galveston, Tex.	33	0.2	0.5	0.0	0.0	0.0	0.0	0.0	0.0	0.0	0.0	0.0	T	0.7
Havre, Mont.	10	6.0	5.0	5.0	2.0	5.0	T	0.0	0.0	T	3.0	7.0	4.0	37.0
Helena, Mont.	29	11.5	8.0	9.6	5.8	1.4	T	0.0	0.0	0.6	3.1	6.6	8.1	54.7
Indianapolis, Ind.	24	6.5	5.0	4.3	1.0	1.0	0.0	0.0	0.0	0.0	T	1.4	5.0	23.3
New Orleans, La.	39	0.2	0.3	0.0	0.0	0.0	0.0	0.0	0.0	0.0	0.0	0.0	T	0.5
* New York, N.Y.	19	8.7	11.5	8.2	0.9	T	0.0	0.0	0.0	0.0	T	1.8	5.9	37.0
* Omaha, Neb.	33	5.0	5.0	5.3	0.2	0.0	0.0	0.0	0.0	0.0	0.6	2.4	5.4	23.9
Philadelphia, Pa.		6.5	8.3	4.0	0.3	0.0	0.0	0.0	0.0	0.0	T	0.9	4.0	24.0
* Phœnix, Ariz.	8	0.0	0.0	0.0	0.0	0.0	0.0	0.0	0.0	0.0	0.0	0.0	0.6	0.6
Portland, Me.	35	18.8	20.3	14.4	4.0	T	0.0	0.0	0.0	0.0	T	4.8	12.0	74.3
Portland, Ore.	37	5.2	3.6	1.1	T	0.0	0.0	0.0	0.0	0.0	0.0	0.8	4.5	15.2
* St. Louis, Mo.	20	6.6	6.2	3.5	0.8	0.0	0.0	0.0	0.0	0.0	T	0.8	3.2	21.1
St. Paul, Minn.	24	7.7	6.2	8.8	3.6	0.2	0.0	0.0	0.0	T	0.2	4.7	5.7	37.1
* Salt Lake City, Utah	30	11.3	10.8	8.6	2.5	0.5	T	0.0	0.0	T	1.0	5.8	9.7	50.2
* San Francisco, Cal.	32	0.0	0.0	0.0	0.0	0.0	0.0	0.0	0.0	0.0	0.0	0.0	0.1	0.1
Topeka, Kan.	21	4.2	6.1	3.0	0.8	0.1	0.0	0.0	0.0	T	0.1	1.3	3.6	19.2
Washington, D.C.	26	6.0	7.9	4.6	0.4	T	0.0	0.0	0.0	0.0	T	1.0	3.5	23.4

T indicates a trace ; less than one tenth of an inch. Ordinarily the last year included in the normals is 1908. If the station has a * the last year is 1903.

254. In the accompanying table will be found the normal number of days with precipitation for the various months and for the year, for several stations in the United States. It will be seen that for the northeastern part of the country, precipitation occurs on nearly half of the days of a year.

Normal number of days with precipitation.

THE NORMAL NUMBER OF DAYS WITH PRECIPITATION FOR THE VARIOUS MONTHS AND FOR THE YEAR

STATION	LENGTH OF RECORD	JAN.	FEB.	MAR.	APR.	MAY	JUNE	JULY	AUG.	SEPT.	OCT.	NOV.	DEC.	ANNUAL
Albany, N.Y.	31	13	12	13	11	13	13	13	11	10	10	12	12	143
Bismarck, N. Dak.	34	7	8	8	8	11	12	10	8	6	6	7	7	98
Boise, Idaho	28	11	10	10	7	8	6	2	2	3	7	8	11	85
Boston, Mass.	36	12	11	13	11	11	10	11	10	9	10	11	11	130
Buffalo, N.Y.	33	19	17	16	12	13	11	11	10	11	13	19	18	170
Charleston, S.C.	38	10	10	10	8	9	11	12	13	10	8	8	9	118
Chicago, Ill.	38	11	11	12	11	12	11	10	9	9	9	10	11	126
Columbus, Ohio	30	15	13	14	12	13	12	11	10	9	9	11	14	142
Denver, Col.	37	4	5	7	9	11	7	9	9	5	5	4	5	80
El Paso, Tex.	25	3	3	2	1	2	4	8	8	6	4	3	3	47
Galveston, Tex.	33	11	10	9	7	6	7	9	10	7	8		11	105
Havre, Mont.	19	8	7	7	6	9	10	7	7	6	5	6	7	85
Helena, Mont.	28	9	7	8	8	12	12	8	5	7	6	7	8	97
Indianapolis, Ind	38	13	11	13	12	13	12	10	9	8	9	11	12	133
Key West, Fla.	38	4	7	5	4	8	12	13	13	16	13	8	7	110
Los Angeles Cal.														36
New Orleans, La.	39	11	10	9	7	8	13	15	14	11	6	8	10	122
New York, N.Y.	33	12	11	13	11	11	11	13	10	9	10	10	11	132
Omaha, Neb.	39	7	7	8	10	12	11	10	8	8	7	5	7	100
Philadelphia, Pa.		12	11	13	11	12	10	12	11	9	9	9	10	129
Phœnix, Ariz.	8	4	3	3	1	1	1	5	6	4	2	2	2	34
Portland, Me.	35	12	11	13	11	12	11	12	11	10	10	11	11	135
Portland, Ore.	35	20	17	18	15	14	11	4	4	8	13	17	20	164
St. Louis, Mo.	38	9	9	11	11	12	11	10	8	7	7	9	10	114
St. Paul, Minn.	38	9	8	10	10	12	12	10	10	9	9	8	9	114
Salt Lake City, Utah	36	10	10	10	9	8	5	4	6	4	7	7	10	89
San Francisco, Cal.		12	11	11	7	4	2	1	0	2	5	6	10	71
Seattle, Wash.		19	17	16	14	14	10	5	4	9	12	17	19	156
Topeka, Kan.	21	6	8	8	10	12	11	9	9	8	7	6	6	100
Washington, D.C.	38	12	11	12	11	12	11	12	11	8	9	10	10	129

The year 1908 is the last one included in the normals.

255. In addition to the various normals which have been fully discussed, the following precipitation data may be computed from the observations taken. *Precipitation data.*

(1) Unusual amounts of precipitation. It is customary at various stations to note the largest amount of precipitation which has fallen during a given time interval for each month, or for the year, or for the whole period during which observations have been taken. The various time intervals chosen for which to note the largest amount of precipitation are usually five minutes, ten minutes, one hour, one day, or one rainfall. A number of these record precipitations are mentioned in GREELY'S *American Weather*.

(2) Number of consecutive days without rain. Unless it is longer than three weeks, it is usually not noted.

(3) Number of consecutive days with rain every day.

(4) Duration of a rainfall — long continued precipitation.

256. At the regular stations of the U. S. Weather Bureau the following tables are kept constantly filled out and computed to date:

Data at U. S. Weather Bureau stations. (1) Precipitation (inches and hundredths) and departure from normal.

(2) Greatest in 24 hours; amount and date.

(3) One inch an hour and over; amount and date.

(4) Number of days with .01 inch and over; .04 inch and over.

(5) Number of days with .25 inch or over; 1.00 inch or more.

(6) Total snowfall (inches and tenths); number of days with snow.

(7) Greatest snowfall in 24 hours (inches and tenths); depth on ground at end of month.

(8) Greatest depth of snow on ground and date (inches and tenths).

(9) Daily precipitation (inches and hundredths).

(10) Daily snowfall (inches and tenths).

The first eight are kept for the various months and the year; the last two are daily.

THE DISTRIBUTION AND EFFECTS OF PRECIPITATION

257. **Geographical distribution of precipitation.** — The various normals of precipitation have been well determined at many stations both in this country and other parts of the world. There are still, however, large areas for which only scanty material is available, and, in the case of the oceans, definite observational material is almost entirely lacking except at island stations.

The normal annual precipitation for the world.

For this reason more than ordinary skill is necessary in preparing a chart which will exhibit correctly the normal annual precipitation for the world. Such a chart has been recently (1898) prepared by Alexander Supan and Chart XXII is based upon this.

258. The close correlation which exists between the general wind system and the normal annual precipitation is at once apparent. The equatorial belt of low pressure, the doldrums, is the belt of largest rainfall. The trade winds coming into this belt from either side bring large quantities of moisture-laden air at fairly high temperatures. Here the air loiters, and finally rises to begin its poleward journey on the outside of the atmosphere.

The precipitation in the doldrums.

The air is sultry, the sky is usually cloud-covered, and local convection causes frequent downpours of rain. A rainfall far above 100 inches is common in this belt.

In the region of the trade winds the precipitation is much less. If they blow over the ocean, they gain moisture rapidly as the doldrums are approached, but the air also rises in temperature, and Trade wind its increased capacity for moisture more than keeps pace with precipita- the increase in moisture. If the trade winds blow from the tion. ocean or are forced to rise by the continental elevations, copious precipitation is the result. On account of the direction of the trade winds (northeast in the northern hemisphere, southeast in the southern), it is the eastern coast of a country which receives the copious precipitations. Good examples of this are to be found in the relatively large precipitation on the coasts of Florida and Mexico in the northern hemisphere and on the east coast of southern Brazil and the east coast of northern Australia in the southern hemisphere.

The horse latitudes (30° to 35° N. and 30° S. latitude) are regions of high barometric pressure and descending air currents. The precipitation in these two belts is very scant. The arid portion of Central Scanty pre- Siberia, the desert of Sahara in northern Africa, Nevada and cipitation in Arizona in the United States, and the dry belt which stretches the horse across the lower part of South America and Australia, all latitudes. lie in or near the horse latitudes.

In the region of the prevailing westerlies, the precipitation is moderate. It is caused chiefly by the west winds blowing from the ocean over the land or by the many storms which invade the prevailing Precipitation westerlies. When these winds blow from the ocean over the in the pre- land and find the land colder or are forced to rise by conti- vailing west- neutal elevations, it is the west coast which receives the erlies. copious precipitation. Fine examples of this are the relatively large values of precipitation in North America from Oregon to Alaska, in the extreme southern part of South America, and in Scandinavia in Europe.

259. The chief factors which determine the amount of precipitation received at any station and its distribution throughout the The factors year are of sufficient importance to bear repetition. They which deter- are: the general wind system; the temperature changes, mine the amount of but chiefly the contrast in temperature between the place precipitation in question and those regions from which the prevailing and its dis- winds come; elevation and inclosure by mountains; near- tribution. ness to bodies of water; the characteristics of the storms which

occur.[1] As an illustration of the importance of these factors consider a strip across North America from the Pacific to the Atlantic along the northern boundary of the United States. The most copious precipita-

An illus- tion, about 100 inches, is found on the Pacific coast, where the

tration. moisture-laden prevailing westerlies come from the ocean over the land and are forced to rise by the Rocky Mountains. Nearly all of the precipitation occurs during the winter, because it is then that the land is colder than the ocean. After the Rocky Mountains are passed, the amount of precipitation is the least in this strip — only ten to twenty inches. The reasons are: increasing distance from the ocean, elevation, and mountain inclosure. As the Atlantic coast is approached, the amount steadily increases, and reaches about forty inches at the coast. This precipitation is nearly all caused by storms, extratropical cyclones, and thundershowers, and thus is fairly evenly distributed throughout the year. Near the coast there is a summer maximum and a winter minimum, since thundershowers are more prevalent in the summer time. In the same way the amount of precipitation at any station and its distribution throughout the year can be fully explained by considering these determining factors.

260. In Charts XXIII, XXIV, and XXV are given the normal annual precipitation and the normal precipitation for January,

Some pre- February, and March, and for July, August, and September,

cipitation for the United States. All of the varied characteristics

charts for of these charts can be readily explained on the basis of the

the United factors discussed above.

States.

Chart XXVI gives the normal annual snowfall for the United States.

261. Other precipitation charts. — Since precipitation is such an important climatic or meteorological element, nearly all of the normals

The more of precipitation and precipitation data mentioned in sections

important 255 and 256 have been determined for a sufficient number of

charts. stations to make possible the charting of them for various countries, and in some cases for the world. The chief ones for which charts for various countries and sometimes for the world are constructed are : the normal amount of precipitation for the various months and the year; the normal amount of snowfall for the various months and for the year; the normal number of days with precipitation for the various months and for the year; greatest number of consecutive days without precipitation; greatest number of consecutive days with rain every day.

[1] See *Monthly Weather Review*, April, 1902, p. 204.

262. Variation in the amount of precipitation with altitude. — If one rain gauge is placed in the open a few feet above the ground and another identical instrument placed at a height of, say, 200 feet, as much in the open as possible, it has been found that No change the gauge at the greater height will catch but a little more in the than one half the amount caught by the lower gauge. It was amount with formerly thought that precipitation decreased with altitude, small altitudes. but careful experimental investigation has shown that this discrepancy is due not to an actual difference in the amount of precipitation, but to the decided increase in wind velocity with elevation. It is the eddy formed by the wind in passing over and around the rain gauge which causes the deficit in the amount caught.

There is probably, however, a slight increase in the amount of precipitation with altitude. The raindrops are largest when they leave the base of the cloud, and decrease in size, due to evaporation, during their fall to the earth's surface. It would thus be The amount expected that the amount of precipitation would be largest increases at the average height of the rain-causing clouds, that is, slightly with about 4000 feet in winter and considerably higher in summer. The few altitude. different and uncertain observations which have been made would seem to confirm this expectation.

In a mountainous country, the regions around the mountains always have the largest precipitation. This is so uniformly true that it is often said that the rainfall map of a mountainous country is Mountain- almost identical with the topographic map. This increase ous regions is not due to any great extent to altitude, but to the fact that have copious the mountain itself by its presence increases the amount of precipitation. precipitation. When the air is forced to rise in going over the mountain, the windward side is usually deluged with rain, while the leeward side receives but little. If a mountain is situated in a country where the wind direction may be from almost any point of the compass during precipitation, then the whole region about the mountain will have a large amount of precipitation. The increase in precipitation in regions surrounding mountains is thus due not so much to elevation, but primarily to the presence of the mountain itself.

263. Relation of rainfall to agriculture. — In expressing the relation of rainfall to agriculture, it is often stated that more than 100 inches produces vegetation too luxuriant for agriculture; that from Limiting 100 to 18 inches is most favorable; that from 18 to 21 inches values. is suitable for grazing only; that below 12 inches the country is a desert.

s

Although these statements are in the main true, yet certain modifications are necessary. In the first place, high temperature as well as excessive precipitation is necessary in order to produce too luxuriant vegetation. In a cold region where a good part of the precipitation comes in winter as snow, 100 inches would not be unfavorable. The distribution of the precipitation throughout the year is also quite as important as the amount. Forty inches of precipitation which comes entirely as snow during three months in the winter would not be favorable for agriculture, even if the summer were sufficiently warm.

Distribution also important.

264. **Relation of rainfall and forests.** — There is a widespread popular belief that deforestation decreases rainfall and that wholesale tree-planting will increase it. The results of observation, however, are not in accord with this opinion. A comparison of the earliest records of rainfall in this country with the present observations shows that the amount of precipitation and its distribution throughout the year in the early colonial days, when most of the country was still covered with virgin forests, was practically the same as at present. Differences in the location and surroundings of the rain gauge probably cause far greater differences in the amount of precipitation observed than the presence or absence of forests. There is also no evidence that the wholesale tree planting in Egypt, now nearly a hundred years ago, has had any effect whatever on the rainfall of that country. More recent observations in Mauritius show that deforestation may reduce the number of rainy days slightly without changing appreciably the total amount of precipitation. The factors which do determine the amount of precipitation have been fully discussed; and as all of these are unchanging, there is no reason to believe that the destruction of forest ought to influence rainfall. To state that the presence of forest, particularly in a hilly and mountainous country, preserves the fertility of the valleys is an entirely different matter. The forests do cause a large part of the rainfall to be retained in the soil. As soon as the forests are cut, the water drains rapidly into the valleys after every storm, and the floods thus caused wash over the good soil and often destroy its fertility. Forests may also retain the snows of winter, and thus have an effect on irrigation or agriculture; but this is entirely apart from their influence on rainfall.

Deforestation does not reduce rainfall.

It does affect fertility.

265. **Effects of snowfall.** — A large part of the earth is covered during the winter by a layer of snow, varying from a few inches to perhaps

many feet in thickness. There are several results of this snow layer which deserve mention in passing. (1) It prevents deep freezing. Snow is a very poor conductor of heat, particularly if it is not compacted. As a result, a layer of snow of even small thickness prevents the ground from freezing to the depth to which it would, if it were bare. (2) Snow also lowers the temperature of the air. It is a good reflector, and reflects some 40 per cent of the insolation which falls upon it. As a result, temperatures are usually lower when the ground is snow-covered. (3) It retards the coming of spring, for the layer of snow must all be melted before the temperature can go much above 32° F. or the frost come out of the ground. (4) It causes floods, because the coming of a sudden thaw or a warm rain often delivers to the streams the precipitation which has fallen during several storms and has collected on the ground in the form of snow.

The four effects of a layer of snow.

In the arctic regions and in high mountains, the snow drifts or slides into the ravines and valleys, and is there compacted by the wind, surface melting, and an occasional rainstorm until a glacier is produced. Snow thus leads to the formation of glaciers.

Snow forms glaciers.

QUESTIONS

(1) Define evaporation and condensation. (2) State the sources of the water vapor of the atmosphere. (3) Which is the heavier, air or water vapor? (4) Define latent heat. (5) What are the two sources of latent heat? (6) State the different ways in which latent heat makes itself felt. (7) Upon what does the amount of evaporation depend? (8) How is the amount of evaporation determined? (9) Describe the three processes by which water vapor is distributed. (10) Upon what does the capacity of the air for water vapor depend? (11) When is air said to be saturated? (12) Distinguish between absolute and relative humidity. (13) Define the dew point. (14) How are the four quantities, temperature, absolute humidity, relative humidity, and dew point related? (15) Describe the chemical hygrometer. (16) Describe the construction and action of a hair hygrometer. (17) How accurate is the instrument, and how may it be standardized? (18) What other apparatus may be used to determine relative humidity? (19) Describe the dew point hygrometer. (20) Describe the construction and action of a psychrometer. (21) How is the effect of the wind obviated? (22) What moisture observations are made at the weather bureau stations? (23) Describe the typical daily variation in absolute humidity. (24) State and explain the characteristics of the annual variation in absolute humidity. (25) How are the general wind system and absolute humidity correlated? (26) Describe and explain the characteristics of the daily variation in relative humidity. (27) Describe and explain the characteristics of the annual variation in relative humidity. (28) Describe the geographical variation in relative humidity. (29) What is the effect of water vapor on the general circulation of the atmosphere? (30) What are the seven forms of condensation? (31) In what two

ways may condensation be brought about? (32) How may air be sufficiently cooled to cause condensation? (33) Describe and explain the formation of dew. (34) State the amount and sources of dew. (35) What are the conditions for the formation of dew? (36) Describe the appearance of frost. (37) Distinguish between light and killing frosts. (38) How are the destructive frosts of spring and autumn caused? (39) Describe in detail the methods of frost prediction. (40) What protection from frost may be used? (41) What observations of frost are made? (42) How is fog formed? (43) Distinguish between transportation and radiation fogs? (44) How do city fogs and country fogs differ? (45) What observations of fog are made? (46) State the early history of cloud classification. (47) Describe the introduction of the international system. (48) Treat fully the international system. (49) Describe in detail the thirteen cloud forms. (50) What is the usual sequence of cloud forms? (51) Describe the methods of determining the height of clouds. (52) How is the direction and velocity of motion of clouds determined? (53) Describe the construction and action of a nephoscope. (54) Define cloudiness; how is it estimated? (55) Define the five adjectives used to designate cloudiness. (56) Describe the three kinds of sunshine recorders. (57) What observations of the clouds, cloudiness, and sunshine are made? (58) What are the characteristics of the annual and daily variation in cloudiness? (59) Are nuclei of condensation necessary for the formation of cloud? (60) How large are cloud particles? (61) What holds cloud particles in suspension? (62) Describe the two kinds of haze. (63) Treat in detail the nine processes in cloud formation. (64) What conditions favor a clear sky? (65) How is a raindrop formed? (66) Why does not every cloud yield rain? (67) What cloud-forming processes may lead to precipitation? (68) Describe the structure of snowflakes. (69) Describe in detail the three kinds of hail. (70) What are the characteristics of an ice storm? (71) What rain-making experiments have been tried, and are they successful? (72) Why is it cooler after it has rained? (73) How is rainfall measured? (74) How is a snowfall measured? (75) What is the water equivalent of snow? (76) What observations of precipitation are made? (77) Describe the daily and annual variation in precipitation. (78) What are the factors which determine the distribution throughout the year? (79) What precipitation data are usually collected? (80) Describe in detail the geographical distribution of precipitation. (81) Trace the correlation between the general wind system and the distribution of precipitation. (82) How does the amount of precipitation vary with the altitude? (83) What is the relation of rainfall and agriculture? (84) What is the relation of rainfall and forests? (85) What are the effects of snowfall?

TOPICS FOR INVESTIGATION

(1) The methods of determining the amount of evaporation.
(2) Cobalt chlorid as a measurer of relative humidity.
(3) The theory of the psychrometer.
(4) The history of the instruments for measuring moisture.
(5) Frost prediction.
(6) The destructive frosts of spring and autumn.
(7) Protection from frosts.
(8) Nuclei of condensation for fog and cloud.
(9) The temperature required for the destruction by frost of different kinds of plants.
(10) History of cloud classification.

(11) The methods of photographing clouds.
(12) The methods of determining cloud heights.
(13) Sunshine recorders.
(14) Haze.
(15) Snow crystals.
(16) The structure of summer hail.
(17) Rain-making experiments.
(18) The relation of rainfall and forests.

PRACTICAL EXERCISES

(1) Compare the amount evaporated by a Piche evaporimeter and by a free water surface under different conditions.

(2) Determine the effect of the character of the surface on the amount evaporated. .

(3) Determine the moisture in the atmosphere by all the different methods, and compare the results under different conditions.

(4) Study critically the behavior of a hair hygrometer.

(5) Study the amount of dew deposited and the characteristics of the night.

(6) Whenever a fog occurs, determine exactly how it was caused.

(7) Determine by the shadow method the height of a few cumulus clouds.

(8) Observe for a number of days at a definite time the kind of cloud, and explain exactly how it was formed.

(9) Determine the size of raindrops.

(10) Determine the temperature of the raindrops as compared with the air temperature.

(11) Work out the effect of a building on the amount of rain caught by a rain gauge in different locations.

(12) Determine whether July 4 is more rainy than July 3 or 5.

In connection with absolute humidity, relative humidity, frost, fog, cloudiness, sunshine, and precipitation, the various normals and data mentioned in this chapter may be worked out for one or more stations. The graphs representing the diurnal and annual variations should be plotted and explained. If these determinations have been made for many stations in a state, charts may be prepared.

REFERENCES

For the description, illustration, construction, and use of apparatus to measure absolute humidity, relative humidity, dew point, the altitude of clouds, direction and velocity of motion of clouds, sunshine, and precipitation, see:

ABBE, CLEVELAND, *Meteorological Apparatus and Methods*, Washington, 1888.
MARVIN, C. F., Circular G (instructions for the care and management of sunshine recorders) and E (measurement of precipitation), Instrument Division, U. S. Weather Bureau.
MOORE, JOHN W., *Meteorology*, 2d ed., pp. 109–113, 175–189, 220–241.
Report of the Chief of the Weather Bureau, 1898–1899, Vol. II. (Report on the International Cloud Observations by F. H. BIGELOW.)
Sunshine Recorders and their Indications by R. H. CURTIS. Quarterly Journal of the Royal Met. Soc., XXIV, 1898.
Apparatus catalogues of various firms. (See p. 110.)
See also the various guides to observers mentioned in Appendix IX, in group (2) B.

For the observations of moisture, frost, fog, cloudiness, sunshine, and precipitation consult the publications mentioned on pages 110 and 111.

For moisture, cloud, and precipitation normals for various places, see:
HANN, *Handbuch der Klimatologie.*
VAN BEBBER, *Handbuch der Meteorologie.*
BLODGET, *Climatology of the United States.*
Climatology of the United States. Bulletin Q of the U. S. Weather Bureau by A. J. HENRY.
Summary of the Climatological Data for the United States by Sections (106 are to be issued).

For moisture only:
Report of the Chief of the Weather Bureau, 1896–1897, 1901–1902.
HENRY, ALFRED J., Report on the Relative Humidity of Southern New England and other Localities. Bulletin 19 of the U. S. Weather Bureau.
Temperature and Relative Humidity Data. Bulletin O of the U. S. Weather bureau by WILLIAM B. STOCKMAN.
Report on the Temperature and Vapor Tensions in the United States. Bulletin S of the U. S. Weather Bureau by FRANK H. BIGELOW.

For clouds only:
Report of the Chief of the Weather Bureau, 1896–1897.

For precipitation only:
FRITZSCHE, RICHARD, *Niederschlag, Abfluss, und Verdunstung auf. den Landflächen der Erde,* Halle, 1906.
HEBERTSON, ANDREW J., *The Distribution of Rainfall over the Land,* London, 1901.
SUPAN, ALEXANDER, *Die Verteilung des Niederschlags auf der festen Erdoberfläche,* Gotha, 1898.
Great Britain, Rainfall Tables for the British Islands, London, 1897. (Official; No. 114.)
HELLMANN, G., *Die Niederschläge in den norddeutschen Stromgebieten,* Berlin, 1906.
SANDERRA, MASÓ MIGUEL, *The Rainfall in the Philippines,* Manila, 1907.
VOSS, ERNST LUDWIG, *Die Niederschlägsverhältnisse von Südamerika,* Gotha, 1907.
WILD, H., *Die Regenverhältnisse des Russischen Reiches,* St. Petersburg, 1887.
Tables and Results of the Precipitation in Rain and Snow in the United States. (Smithsonian Contributions to Knowledge, 353.) 4°, 2d ed., xx + 249 pp., Washington, 1881.
Report of the Chief of the Weather Bureau, 1891–1892, 1896–1897.
HARRINGTON, MARK W., Rainfall and Snow of the United States, computed to the end of 1891. Bulletin C of the U. S. Weather Bureau.
HENRY, ALFRED J., Rainfall of the United States. Bulletin D of the U. S. Weather Bureau.
BIGELOW, FRANK H., The daily normal Temperature and the daily normal Precipitation in the United States. Bulletin R of the U. S. Weather Bureau.

For charts of moisture, cloud, sunshine, and precipitation, see:
BARTHOLOMEW, *Physical Atlas,* Vol. III, Atlas of Meteorology.
HANN, *Atlas der Meteorologie,* 1887.
ELIOT, SIR JOHN, *Climatological Atlas of India,* Edinburgh, 1906.

Russia, *Atlas climatologique de l'empire de Russie*, St. Pétersbourg, 1900.
BLODGET, LORIN, *Climatology of the United States*, Philadelphia, 1857.
LOOMIS, *Contributions to Meteorology*.
GREELY, GEN. A. W., *American Weather*, New York, 1888.
Climatic charts of the United States, Weather Bureau, Washington, D.C., 1904.
Climatology of the United States.   Bulletin Q, by A. J. HENRY (W. B. 361).
CLARK, KENNETH McR., "A new set of Cloudiness Charts for the United States," *Quart. Jour. Roy. Met. Soc.*, Vol. 37, No. 158, April, 1911.

For moisture only   :

Report of the Chief of the Weather Bureau, 1896–1897 and 1901–1902.

For precipitation only:

HERBERTSON, ANDREW J., *The Distribution of Rainfall over the Land*, London, 1901.
SUPAN, ALEXANDER, *Die Verteilung des Niederschlags auf der festen Erdoberfläche*, Gotha, 1898.
HELLMANN, G., *Die Niederschläge in den norddeutschen Stromgebieten*, Berlin, 1906.
VOSS, ERNST LUDWIG, *Die Niederschlägsverhältnisse von Südamerika*, Gotha, 1907.
DUNWOODY, H. H. C., Geographical Distribution of Rainfall in the United States.   Professional Papers of the Signal Service, No. 9.
HENRY, ALFRED J., Rainfall of the United States.   Bulletin D of the U.S. Weather Bureau, 1897.
Report of the chief of the Weather Bureau, 1896–1897.

For information about clouds and cloud classification, see:

Atlas de las Nubes, para el servico meteorologico Republica Mexicana, 1906.
BARBER, *The Cloud World, its Features and Significance*, London, 1903.
CLAYDEN, *Cloud Studies*, E. P. Dutton and Co., New York, 1905.
CLAYTON, Observations made at the Blue Hill Observatory.   (Published in the Annals of the Harvard College Observatory.)
Illustrative Cloud Forms.   (Issued by the Weather Bureau.)
International Met'l Committee, *International Cloud Atlas*, Paris, 1896.
LEY, *Cloudland*, London, 1894.
VINCENT, *Atlas des nuages*, Bruxelles, 1907.
See also other books mentioned in Appendix IX.

For a discussion of the destructive frosts of spring and autumn, see:

CLINE, Irregularities in Frost and Temperature in Neighboring Localities, Third Convention of Weather Bureau Officials, Proceedings, Washington, 1904, p. 250.
DAY, Frost Data of the United States and Length of the Crop-Growing Season, Bulletin V of the U. S. Weather Bureau.
GARRIOTT, Notes on Frost, Farmers Bulletin No. 104.
HAMMON, Frost, Weather Bureau Publication No. 186.
McADIE, Frost Fighting, Weather Bureau Publication No. 187.
Monthly Weather Review, August, 1908.

For recent articles on evaporation, see:

BIGELOW, Monthly Weather Review, 1907 on.
KIMBALL, Monthly Weather Review, December, 1904.

CHAPTER VI

THE SECONDARY CIRCULATION OF THE ATMOSPHERE

A. TROPICAL CYCLONES

A. TROPICAL CYCLONES

DEFINITION AND DESCRIPTION

266. **Definition and chief characteristics.** — A tropical cyclone is a storm which can be best defined by stating its chief characteristics.

Definition of a tropical cyclone. It is a vast atmospheric whirl, turning counterclockwise in the northern hemisphere, clockwise in the southern, with spirally inflowing winds which nearly always attain destructive velocities. The pressure is low in the center and it is attended by a large cloud area from which rain pours in torrents, sometimes with thunder and lightning. The whole formation is from 300 to 600 miles in diameter and is not stationary, but moves with a very moderate velocity over a fairly well-defined course.

Other names for the tropical cyclone. These storms, technically known as tropical cyclones, because they are whirling storms which originate within the tropics, have various names in different parts of the world. They are usually called hurricanes in the West Indies, typhoons in the China Sea, and baguois in the Philippine Islands.

The historical rise of our present information. The whirling nature of these storms was first recognized by Varenius in his *Geographica Generalis* in 1650. The famous sea captain, Dampier, who experienced one in 1687, correctly describes it and states that a typhoon is a kind of violent whirl which occurs on the coasts of Tonkin in July, August, and September. The first definite and exact information concerning the nature and characteristics of these storms comes, however, from Redfield in this country, Reid in England, and Piddington in Calcutta, in the first part of the last century. Other investigators who have added much to our knowledge of these storms are: Dové, Espy, Ferrel, Vines, Doberck, Meldrum, Blanford, Poey, Algué, and Eliot.

267. Description of the approach and passage of a tropical cyclone. — Tropical cyclones have been vividly described by many sea captains who have passed through them, and observers who have experienced them on land. The first signs of the approach of the storm are to be found in both sea and sky.①The sky is covered with a thin cirrus haze which causes lurid red sunsets and halos or rings about the sun by day and the moon by night. ②The air is still, moisture-laden, sultry, and oppressive.③ The barometer rises unduly high or remains stationary when the daily drop is expected.④The wind disappears and the long rolling swell of ominous import appears on the ocean.⑤ Soon the barometer begins to fall.⑥ A breeze springs up, but the air is still sultry and oppressive.⑦The cirrus haze becomes true cirrus, which usually stretches in bands across the sky and begins to thicken into cirro-stratus or sometimes cirro-cumulus. ⑧The barometer begins to fall more rapidly, the wind increases, and on the horizon the dark rain cloud, shieldlike, has appeared.⑨ The barometer now falls with startling rapidity; the blue-black rain cloud rushes overhead; rain falls in torrents, cooling the air; the wind has increased to full hurricane. strength, a hundred miles an hour or more; the sea is lashed into fury. This may continue many hours, when⑩ suddenly the wind ceases, the clouds break through, the temperature rises, the moisture grows less, and the barometer is at its lowest, for the calm central eye of the storm has been reached. The respite is but brief, perhaps twenty or thirty minutes when the wind changes to the opposite direction and increases to full hurricane strength as suddenly as it ceased.   Rain again falls in torrents, and everything is as before except that the barometer is rising.   After several hours the end of the storm is reached, the wind dies down, the rain ceases, the nimbus clouds break through and give place to cirrus, the temperature rises somewhat.   A while later and the nimbus clouds sink, shield like, below the horizon, the cirrus retreats after it, the wind is a gentle breeze, the barometer has reached its accustomed height; and but for the wreckage and the ominous heaving of the ocean one would not know that a storm had passed.

The coming of the cyclone.

The calm central eye.

The disappearing of the cyclone.

268. Distribution of the meteorological elements about a tropical cyclone. — The distribution of the meteorological elements (temperature, pressure, wind, moisture, cloud, and precipitation) about a tropical cyclone — in short the whole structure of the storm — can be well shown by means of two diagrams.   The first, Fig. 110, contains the pressure, wind, cloud,

The distribution of the elements is illustrated by two diagrams.

and precipitation, while the second, Fig. 111, shows the temperatur
and moisture.

269. The isobars are often nearly circular, although at times oval in
form.   The longer axis of the oval usually lies in the direction of motion

The distri- of the cyclone, although it may make any angle with this
bution of the direction.   The central pressure averages about 28.5 inches,
elements. although pressures as low as 27 inches have been observed.[1]
The belt of slightly higher pressure surrounding the cyclones and called

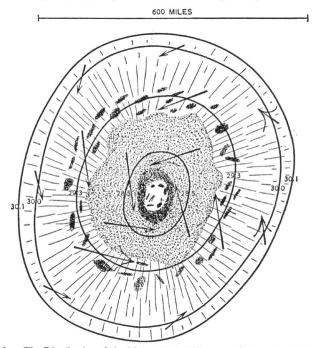

600 MILES

Fig. 110. — The Distribution of the Meteorological Elements about a Tropical Cyclone.

the pericyclonic ring usually has a pressure of about 30.1 inches.   It
will be remembered that the occurrence of this high pressure was one
of the first signs of the coming of the cyclone.

The wind blows spirally inward, turning counterclockwise in the
northern hemisphere.   Its direction makes an angle of about 30° with

[1] See *Meteorologische Zeitschrift*, 1902, p. 474, for a barograph trace during a typhoon.

the isobaric lines.   The value of the angle is nearly the same all the way round, although somewhat greater, 35° to 40°, in the northeast quadrant and somewhat less, 20° to 25°, in the southwest quadrant.   The wind velocity is small on the outside, nothing in the calm central eye, and attains its greatest velocity in the middle of the rain area.   The maximum velocity nearly always reaches 100 miles an hour and sometimes probably nearly 200.   The arrows in the diagram by their direction and length show the direction and velocity of the wind.

The cloud area is shaded in the diagram and is concentric with the isobars. . On the outside, the cirrus radiates in all directions, while the

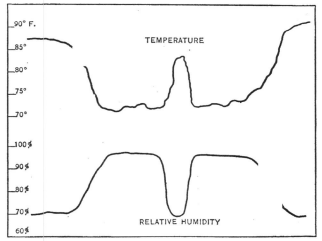

Fig. 111. — Temperature and Moisture Changes in a Tropical Cyclone.

nimbus clouds occupy the inside portion.   The transition clouds are usually cirro-stratus, perhaps cirro-cumulus.   The calm central eye is sometimes almost cloudless.   The precipitation is excessive, often several inches, and covers the same area as the nimbus clouds.   It is sometimes accompanied by thunder and lightning, although this is most common at the end of the storm.   It is said that the most violent cyclones are never accompanied by thunder and lightning.

The temperature and moisture are practically the same in all quadrants, and thus would be represented in the diagram by ovals concentric with the isobars.   As this would complicate the diagram too much, they are shown in section in Fig. 111.   The temperature is high, perhaps 80° F., before the coming of the cyclone. It drops to perhaps 70° when the nimbus cloud comes, due to the cooling caused by the precipitation. In the calm central eye it rises nearly as high as before the coming of the cyclone and drops again when the precipitation begins afresh. The relative humidity is high at the beginning, is perhaps 95 per cent during the precipitation, and drops in the calm central eye to perhaps 70 per cent, depending on the rise in temperature.

The whole formation is from 300 to 600 miles in diameter, while the central eye has a diameter from fifteen to twenty-five miles.

The distribution of the six meteorological elements is thus completely portrayed by means of these two diagrams. The sequence of the changes which might be expected in the meteorological elements due to the passage of a cyclone centrally or obliquely over a station can be determined by considering these diagrams as free and moving them centrally or obliquely over a given point.

270. Some special tropical cyclones. — Hundreds, yes, more than a thousand, of these storms have been more or less carefully observed in

The number of these which have been observed.
different parts of the world since 1400 A.D. It is thus impossible to write up the life history of all of them or even the most important of them. Various books on meteorology often give partial lists of the most destructive of them and perhaps a more detailed description of one or two. For all the known facts about any one, the reader must be referred to the periodical literature of the subject.[1] (See Appendix IX.)

271. The last very destructive hurricane in the West Indies is the one which caused such appalling losses at Galveston, September 8, 1900.

The Galveston hurricane on Sept. 8, 1900.
It is estimated that here above 6000 lives were lost and over \$30,000,000 worth of property was destroyed. This cyclone, unfortunately, is not very typical as regards its behavior or the path which it followed. It appeared first as a small storm on the morning of September 1 southeast of the island of Hayti. It passed over Cuba on the 5th and reached southwestern Florida on the 6th. Here it turned abruptly to the left and crossed the Gulf of Mexico, reaching Galveston on the evening of the 8th with its destructive energy at its maximum. It was then about 500 miles in diameter with a central pressure of less than 28.48 inches, and a wind velocity of more than a hundred miles an hour. The center passed south of Galveston but within less than 40 miles of the city.

[1] For pictures of the wreckage caused by a hurricane see *Monthly Weather Review*, Sept., 1906.

At Galveston, the lurid sunsets and the brick dust sky as heralds of an approaching hurricane were entirely lacking. The cirrus clouds made their appearance on the morning of the 7th, coming from the southeast, and the high tides and the heavy ocean swell appeared during the afternoon of the same day. The cirrus clouds became a mixture of cirrus, alto-stratus, and cumulus during the afternoon of the 7th, and later changed to strato-cumulus. On the 8th, in the early morning, the cloud forms were fracto-stratus and strato-cumulus with here and there a little blue sky, but this was soon followed by showery weather. The· dense rain cloud came at noon and continued until midnight. The wind went to the northeast on the 7th and blew with constantly increasing velocity. On the afternoon of the 8th it was still northeast and blowing a gale. By 5 in the afternoon, it had reached full hurricane strength. Between 5 and 6.15 it blew at the rate of 84 miles an hour for several five-minute periods. At 6.15 the anemometer and other meteorological instruments blew away after a wind velocity of 100 miles per hour had been recorded for two minutes. By estimation, the wind velocity reached 120 miles per hour between 6.15 and 8.30 P.M. After 8.30 the wind still continued of hurricane strength, but shifted first to the east, then the southeast, and finally reached the south by 11 P.M. From that time on, the direction continued south and the velocity steadily decreased. The pressure dropped rapidly all during the 8th and reached its lowest, 28.48 inches, at 8.30 P.M. As the storm passed a little south of Galveston, the central pressure must have been a little less than this. After 8.30, the barometer rose rapidly. That all this would be the natural sequence of events can be seen by moving the diagram given as Fig. 110 from right to left below a given point. The destruction of life and property at Galveston was caused by a storm wave as well as by the excessive wind velocity. The water rose steadily all during the 8th, flooding a large part of the city, and at 7.30 P.M. there was a sudden rise of four feet in a few minutes.

After passing Galveston, the storm turned to the right, passed inland and up the Mississippi Valley as far as Nebraska, where it then again turned, crossing Lake Michigan and Maine and passing out over Nova Scotia on the 12th. As soon as it went inland it lost its violence and many of the characteristics of a tropical cyclone and was gradually transformed into an extratropical cyclone. It also grew larger, and by the time it reached the St. Lawrence Valley it was more than a thousand miles in diameter. It retained, however, a low central pressure, rather high winds, and a compact form all through its life history. Figure

112 shows its path in detail, and Chart XXVII reproduces the 8 A.M. weather map for September 8, 1900, when it was near Galveston.

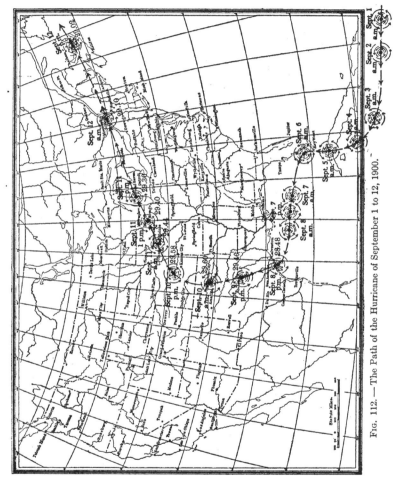

FIG. 112. — The Path of the Hurricane of September 1 to 12, 1900.

272. Rules for mariners. — The rules for mariners which are ordinarily given in connection with tropical cyclones are: first, to avoid running before the wind, particularly when the center of the cyclone is to the westward, as this would bring the vessel towards the center and across the track of the coming cyclone; secondly, to so direct the vessel as to avoid as far as possible the so-called " dan_ gerous half " of the cyclone.

To apply these rules necessitates the locating of the center of the cyclone, and this can be done from the wind direction. When cyclones were first studied, it was thought that the winds were truly **Locating the** circular, that is, followed the isobars; and if this were the **storm** case, the application of Buys Ballot's famous law: "Stand **center.** with your back to the wind and the low pressure will be on your left hand," would give at once the direction of the storm center. But it is now known that the wind direction makes an angle with the isobars, and for this reason the storm center can be more readily located by means of a diagram. Let A (Fig. 113) represent the wind direction,` then B must be the position of the isobaric line and the line C, at right angles to B, must give the direction toward the center of the cyclone. The direction of the center is thus known, and from observations of the sea and sky a close estimation of its distance can be made. The direction of the cirrus bands across the sky, the location of the nimbus cloud on the horizon, and the direction of motion of the ocean waves are also guides as to the direction in which the storm center is located.

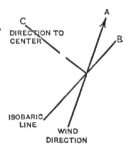

FIG. 113. — Locating the Center of a Tropical Cyclone.

The so-called " dangerous half " of a tropical cyclone is the north and northeast portion when the cyclone is moving northwest and the south and southeast portion when it is moving northeast. **The danger-** All this applies to the northern hemisphere. These portions **ous half of a** are considered the more dangerous because the wind ve- **cyclone.** locities are somewhat larger. The reason for this is because in these portions the velocity of the wind about the center is combined with (added to) the velocity of the permanent winds in the region through which the cyclone is moving. On the other side of the storm, these two velocities are opposed. (See figure 115.)

THE REGIONS AND TIME OF OCCURRENCE

273. Regions of occurrence. — There are five regions of the earth where tropical cyclones occur, and these are shown in Fig. 114.

They are: (1) The West Indies, the Gulf of Mexico, and the coast of
The five   Florida; (2) The China sea, Philippine Islands, and Japan;
regions of   (3) each side of India in the Bay of Bengal and in the
occurrence.   Arabian Sea; (4) east of Madagascar near the islands
of Mauritius and Réunion; (5) east of Australia, near Samoa. The

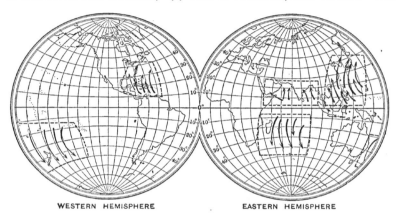

WESTERN HEMISPHERE   EASTERN HEMISPHERE
FIG. 114. — The Five Regions of Occurrence of Tropical Cyclones.

tropical cyclones always occur on the west side of an ocean. This is
true of the north Atlantic, the north Pacific, the south Pacific,
                and the south Indian oceans. Tropical cyclones never
Tropical cy-   originate on land, and if they run ashore, they weaken,
clones occur
on the west   lose their destructive violence, and are soon transformed
side of an   into extratropical cyclones. A range of mountains 3000
ocean, never
on land or   feet high is often sufficient to completely destroy a tropical
in the south   cyclone. It is also a very significant fact that tropical
Atlantic.   cyclones never occur in the south Atlantic Ocean.

274. **Tracks of tropical cyclones.** — The tropical cyclones originate
in the doldrums, not directly at the equator, but from 8 to 12° from it
The form of   on either side. Those in the northern hemisphere move
the track of   northwest through the trade wind belt with a velocity of
a tropical   from 6 to 12 miles an hour. They curve to the right in
cyclone.   about 30° north latitude, moving first due north and then
northeast through the region of the prevailing westerlies. They grow
somewhat larger, and the velocity of motion increases to 20, 30, or even
40 miles per hour. The path has somewhat the form of a parabola with

its vertex in 20° north latitude. In the southern hemisphere they move southwest at first, and then recurve in about 25° south latitude and move southeast through the region of the prevailing westerlies. Figure 115 shows the direction of rotation of these storms, the dangerous half (shaded), and the form of the path for both hemispheres.

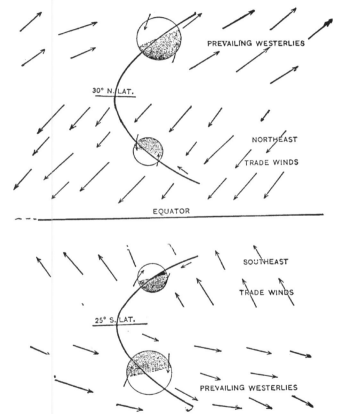

Fig. 115. — The Direction of Rotation, Dangerous Half, and Path of Tropical Cyclones.

The characteristics of the paths followed by the West Indies hurricanes are well shown in Fig. 116, which gives the paths followed by all recorded hurricanes during September from 1878 to 1900. The mean track is also indicated.

The tracks of the West Indies hurricanes.

In the case of the tropical cyclones which occur each side of

The tracks of the cyclones near India are irregular. India in the Bay of Bengal and in the Arabian Sea, this characteristic parabolic path is lacking. Here the course followed is more irregular, and the paths are much shorter and more nearly straight lines.

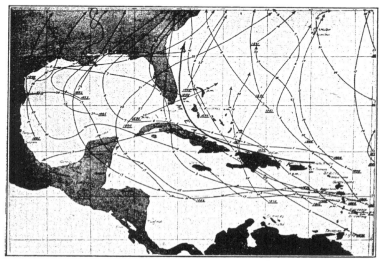

FIG. 116. — The Hurricanes of the West Indies during September, from 1878 to 1900.

(After GARRIOTT, U. S. Weather Bureau.)

275. Frequency at different times of year. — The frequency of occurrence at different times of year is exhibited in the accompanying table for the five regions of the earth where tropical cy-

The frequency at different times a year. clones occur. This table gives the percentage frequency for the different months, the name of the investigator who collected the statistics, and in some cases the period covered. The total number of these storms is also added, but these numbers are not comparable for the different regions, because the length of time is not the same and there was no common standard of intensity for determining how violent a storm must have been to be considered a tropical cyclone.

It will be seen that in connection with the cyclones in the West Indies, and in the China Sea and Philippines, the greatest number occurs during the months of July, August, September, and October; that is, in the late summer. **The maximum number occurs during the late summer.** In the case of the south Pacific and south Indian oceans, the largest number occurs during December, January, February, and March, again in the late summer of that hemisphere. In the case of the cyclones of the Bay of Bengal and the Arabian Sea there are two maximums; in **Near India they occur between the monsoons.** April, May, and June, and again in October and November. Here they occur during the light, baffling breezes which prevail between the periods of the steadily blowing monsoons.

THE ORIGIN OF TROPICAL CYCLONES

276. The convectional theory. —The distribution of the meteorological elements about a tropical cyclone has been fully considered, and the facts of observation concerning the regions of occurrence, the path followed, and the frequency at different times of year have been stated. It now remains to discuss the origin and development of these storms, and to account for all these facts of observation. The so-called convectional theory is the one usually **The convectional theory will be presented from the deductive standpoint.**

	Jan.	Feb.	Mar.	Apr.	May	June	July	Aug.	Sept.	Oct.	Nov.	Dec.	Total Number
(1) { West Indies Poey 1493-1855	1	2	3	2	1	3	12	27	22	20	5	2	355
{ West Indies 1878-1900	0	0	0	0	1	3	3	26	26	34	4	3	95
(2) { China Sea Doberck 1880-1901	0	0	0	2	4	10	19	18	23	13	9	2	34
{ Philippines Algué 1880-1901	2	0	1	2	5	9	16	17	19	14	11	5	468
(3) { Bay of Bengal Blanford and	2	0	2	8	18	9	3	3	5	27	16	8	139
{ Arabian Sea Dallas	6	0	4	13	19	21	1	0	2	11	21	2	53
(4) Mauritius { Piddington and Meldrum } 1851-1885	22	19	18	15	6	1	0	0	0	1	8	10	38
(5) South Pacific Knipping	30	18	28	6	1	0	0	0	1	1	3	12	125

given, and it will be presented here from the deductive standpoint. The fundamental principles in connection with convection, the effect of the earth's rotation on moving air, and the generalizations in connection with the general wind system of the world will be used as a basis; and from these, theoretical deductions will be drawn as to the varied characteristics which a storm thus produced should have. An exact agreement between the theoretical deductions and the observed facts will stamp both the deductions and the observational work as correct.

277. Convection is most vigorous in stagnant, overheated, moisture-laden air. Suppose that a large mass of quiet, warm, moist air exists somewhere. At some point, due probably to local causes, a convectional rise will take place. The warm, moist air as it rises will expand, grow cooler, reach its dew point, become cloudy, and perhaps yield precipitation. The latent heat liberated by the condensation of the water vapor into cloud and rain will lessen the rate of cooling of the rising air and cause still further rise. Due to the excess of temperature and moisture, the pressure would be slightly less than in surrounding regions. The cooler air, forcing its way in to replace the rising air, would be deflected to the right, in the northern hemisphere, by the earth's rotation, and would approach the center in a counterclockwise-turning spiral. This rotation of the air about the center would cause centrifugal force, which would hold the air away from the center, thus causing a still lower barometric pressure. This increased difference in pressure between the center and surrounding regions would drive in the air with greater vigor and cause a further rise of air in the center. This would cause more clouds, more precipitation, more latent heat, and a still more vigorous rise. This in turn would cause still more air to come towards the center, which would make the whirl still more vigorous, causing more centrifugal force and a still lower barometric pressure in the center. This again would drive the whole circulation with still greater vigor, and in this way the violence becomes greater and greater until a tropical cyclone is the result. The air rising in the center would continue to rise until it cooled, due to expansion, to the temperature of its surroundings, when it would spread out laterally, still carrying a slight excess of moisture with it. It would then settle down in surrounding regions, causing a slightly higher barometric pressure.

One would thus expect a convection-caused storm of this kind to consist of an area of low barometric pressure, with violent winds blowing spirally inward and turning counterclockwise in the northern hemi-

The origin and development of a tropical cyclone.

sphere.   One would furthermore expect to find dense nimbus clouds near the center with copious rain.   Due to the outflow aloft, one would expect to find thin clouds, cirrus or cirro-stratus, radiating from the center, and the whole formation surrounded by a ring of slightly higher pressure due to the congestion and settling down of this air discharged aloft.

278. The rapidity of rotation and the centrifugal force grow greater as the radius grows smaller.   Thus, if a tropical cyclone were particu_ larly violent, it might be expected that many miles from the center the centrifugal force would be sufficient to balance the pressure gradient, tending to drive the air in towards the center; and the result would be that the spirally inflowing air would not penetrate all the way to the center, but the cyclone would *How the central calm would be produced.* have a calm central eye.   The outflow of water from a circular wash-bowl through a central vent is exactly analogous.   The outflowing water begins to rotate and rotates faster and faster until a hollow core is often formed at the center.   In the case of the violent tropical cyclone, this calm central cylinder of air would be surrounded by a ring of rapidly whirling and slowly rising air.   This would entangle some of the air lying next to it and carry it up with it.   As a result, there would form a vacuum in the center of a tropical cyclone if the air were not replaced by a gentle descending air current from above.   This air coming from the upper atmosphere would be fairly moisture free and would build no clouds as it is descending, not ascending.   It would be heated both by compression and by absorbing the insolation of the sun, and thus should have a fairly high temperature and a correspondingly low relative humidity.

One would thus expect, in the center of a violent tropical cyclone, a calm area several miles in diameter with a gently descending air current, without clouds, and with comparatively high temperature and low relative humidity.

279. Where can a mass of calm, warm, moist air be found which might serve as the starting point of a tropical cyclone?   Surely not in the region of the prevailing westerlies or of the trade winds, because here the air is far from calm and the mixing is too thorough to allow excessive temperature or amounts of moisture; surely not in the horse latitude, because here the temperatures are too low and we have to do with descending *Where tropical cyclones would be expected to originate.* air currents which are always dry.   It is only in the perpetual summer sultriness of the doldrums, with their frequent calms, high temperature, excessive moisture, and violent local convection that tropical cyclones

could be expected to originate. One would furthermore expect tropical cyclones to form over the ocean rather than over the land. It is the ocean which can readily furnish the necessary moisture, and a land surface is too irregular and offers too much friction to permit the building of the regular violent wind circulation necessary for the building of such a storm. If the trade winds blow from the land over the ocean, they bring but little moisture with them to the doldrums. If, however, they have come a long distance over the ocean, they bring with them to the doldrums immense quantities of warm, moisture-laden air. Since the trade winds blow from the northeast in the northern hemisphere and the southeast in the southern, it is the west side of an ocean which would have the largest amount of moisture and thus be the most likely place of origin of tropical cyclones. The doldrums never invade the south Atlantic, and one would thus expect this ocean to be free from tropical cyclones.

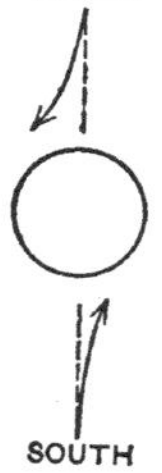

One would thus expect the tropical cyclones to originate in the doldrums, always over the ocean, on the west side of an ocean, and never in the south Atlantic.

280. What would be the course naturally followed by a tropical cyclone? According to Ferrel's law (see section 145), if air starts to move on the earth's surface, it will be deflected to the right in the northern hemisphere and the amount of the deviation will depend upon the velocity and latitude. Furthermore, it can be seen from the table there given (section 145) that for any given velocity the deviation is nothing at the equator and steadily increases with increasing latitude. If this is applied to the air coming into a tropical cyclone from the north and from the south, as is indicated in Fig. 117, and was first shown by Ferrel, it will be seen that the air coming from the south moves more directly towards the storm center than the air which comes from the north. As a result, the whole formation is pushed away from the equator. The other factor which would cause a motion of the tropical cyclone is the movement of the wind system in which it may find itself. One would thus expect a tropical cyclone to be pushed steadily away from the equator and at the same time to drift with the wind system in which it finds itself. One would thus expect a parabolic course convex towards the west, with the turning point or vertex in the horse latitudes, that is,

NORTH

SOUTH

The track which tropical cyclones would be expected to follow.

FIG. 117. — The Application of Ferrel's Law to the Air coming towards a Tropical Cyclone.

30° north and about 25° south latitude. One would not expect a tropical cyclone to form very near the equator, because the deviation of the air coming into it would be too slight to cause the vigorous whirl which generates the centrifugal force and the low barometric pressure upon which the life of the storm depends. At the equator, the air would come straight towards the center, filling up the depression, and the cyclone would never develop. Near India in the Bay of Bengal and in the Arabian Sea, the tropical cyclones never escape from the doldrums on account of the inclosure by land. As a result, one would expect the paths to be short and somewhat irregular with a northerly tendency.

One would thus expect, as regards path (except in the case of those near India), a parabolic form starting out, not at the equator, but some 8 to 12° from it and recurving in the horse latitudes.

281. At what time of year would tropical cyclones be most frequent? Since tropical cyclones cannot originate at the equator, one would expect the largest number when the doldrums in their migration are farthest from the equator. The doldrums are farthest north of the equator in the late summer, August, September, and October, and for the northern hemisphere the maximum number of cyclones should occur then. The doldrums are farthest south in February, March, and April, *The time of year when tropical cyclones would be expected to occur.* the late summer of the southern hemisphere, and in that hemisphere the maximum number of cyclones should occur then. Near India the monsoon is the all-controlling wind. Here one would expect tropical cyclones when the air is calm, warm, and moist; that is, during the frequent calms which exist in the intermissions between the monsoon periods. As there are two such transition periods each year, one would expect two periods of maximum frequency of tropical cyclones.

One would thus expect, as regards time of occurrence, that the maximum number (except in the case of India) would occur during the late summer, with almost none during the other half of the year.

282. Comparison with the observed facts. — Starting with the fundamental principles in connection with convection, the effect of the earth's rotation on moving air, and the generalizations in connection with the general wind system of the world, it has been determined in connection with a convection-caused storm what one would expect as regards the distribution of the elements about it, the regions of occurrence, the track *The comparison of deduction and observation.* followed, and the time of occurrence. If these theoretical deductions are compared with the facts of observation, an exact agreement will be

found. This stamps the whole process of reasoning, the deductions, and the observational work as correct.

One caution should perhaps have been given here. The convectional theory requires the temperature at any given level in a tropical cyclone to be higher than at the same level in surrounding regions. This has never been verified by observation, and some doubt it. If it should be found that it is not so, it would necessitate the remodeling of the present theory, and perhaps some other theory would have to be substituted for it.

THE COMPARISON OF A TROPICAL CYCLONE AND THE CIRCUMPOLAR WHIRL IN THE GENERAL WIND SYSTEM

283. A tropical cyclone and the circumpolar whirls in the general wind system of the world have many things in common, although in some respects they stand in sharp contrast.

In the first place, both are areas of low pressure with spirally inflowing winds, a calm center, and surrounded by a ring or belt of high pres-

A tropical cyclone and the circum-polar whirls have many things in common.

sure. In the case of the tropical cyclone, it is the whirl caused by the violent, spirally inflowing winds above the surface of the earth which develops the centrifugal force, which causes the low central pressure. The central calm may be from fifteen to thirty miles in diameter, and the pericyclonic ring is the belt of high pressure which surrounds the whole formation. In the case of the circumpolar whirls it is the centrifugal force due to the air currents moving spirally towards the poles on the outside of the atmosphere which causes the low polar pressures. The air is calm at the poles, and the horse latitudes are the surrounding belts of high pressure.

Land also affects both in the same way. If a tropical cyclone runs ashore, it weakens, grows less violent, and is perhaps entirely destroyed. The absence of land in the southern hemisphere is the reason for the lower barometric pressure and the more violent winds in the southern hemisphere.

In the matter of temperature, the two formations stand in sharp contrast. The tropical cyclone has a warm center, while the circum-

They are in sharp con-trast in others.

polar whirls have cold centers. The air in a tropical cyclone is accordingly discharged upward, and spreads out laterally aloft. In the case of the circumpolar whirls, the discharge is downward, and the air makes its way equatorward in the middle layer of the atmosphere.

B. EXTRATROPICAL CYCLONES AND ANTICYCLONES

DEFINITION AND DESCRIPTION OF AN EXTRATROPICAL CYCLONE

284. **Definition and chief characteristics.** — An extratropical cyclone, as the name implies, is a whirling storm which occurs outside the tropics; that is, in the temperate latitudes. These storms are depicted on the daily weather maps as areas of low pressure, and appear in great number and in almost infinite variety as regards position and form. The ceaseless changes in our weather are due almost entirely to the approach and passage of these areas of low pressure. For this reason they are sometimes spoken of as the lows of the weather map, or simply "lows." They have many things in common with the whirling storms which occur within the tropics, that is, the tropical cyclones, but, as will be seen later, in many respects they stand in sharp contrast.

The extratropical cyclones are the lows of the weather map.

An extratropical cyclone can be best defined and described by stating its chief characteristics. It is an area of low barometric pressure with spirally inflowing winds turning counterclockwise in the northern hemisphere and clockwise in the southern. The wind velocity is generally moderate; the accompanying cloud area is immense; precipitation usually falls; the changes in temperature and moisture are large and well marked. The whole formation is from a few hundred to several thousand miles in diameter and moves with moderate velocity from some westerly to some easterly quarter.

Definition and description in outline.

285. **Distribution of the meteorological elements about a typical extratropical cyclone.** — The exact structure and nature of an extratropical cyclone can best be learned by studying the distribution of the meteorological elements about one that is fully developed and typical, and this distribution is shown graphically in the accompanying diagram, Fig. 118.

The isobars are usually oval in form, the ratio of the two axes being 1.9 to 1, and the direction of the longer axis is northeast-southwest. The central pressure averages about 29.6 inches, although this varies all the way from a little less than 30 to even below 28 inches in some cases. The isobars are generally packed a little closer together in the southwest quadrant, and are farthest apart in the northeast quadrant. This means that the pressure gradient is greatest in the southwest quadrant.

The winds blow spirally inward towards the center, turning counterclockwise in the northern hemisphere and clockwise in the southern. The wind direction makes an angle with the isobaric lines which is greatest in the northeast quadrant, where it averages from 30 to 40°

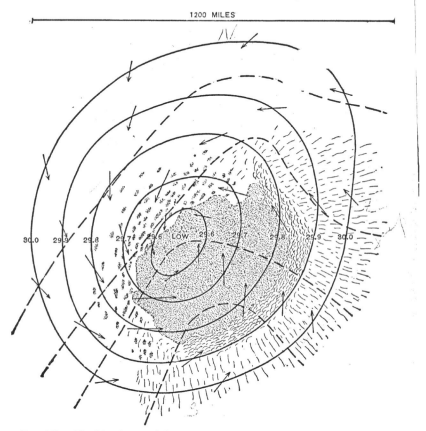

1200 MILES

30.0 29.9 29.8 29.7 29.6 LOW 29.6 29.7 29.8 29.9 30.0

Fig. 118. — The Distribution of the Meteorological Elements about an Extratropical Cyclone.

and is least in the southwest quadrant, where its value is from 15 to 25°. The wind velocity is usually moderate, and only in rare cases enough to be destructive. It is least on the outside and near the center, and greatest in between.

There is a marked rise of temperature on the south and east side of the storm, where the winds blow from some southerly quarter, and a decided drop in temperature on the west side, where the winds blow from some northerly quarter. The moisture changes are not shown in the diagram, but they follow the temperature changes in a general way. On the southern and eastern sides of the storm, the absolute humidity increases very rapidly with the rise of temperature, and it increases so fast that even the relative humidity sometimes increases' in spite of the increased capacity of the air for moisture, due to the higher temperature. In the central part of the storm, both the absolute and relative humidity are high. On the west side, the absolute humidity is very small, where the winds are from the north and the temperature is low, but the relative humidity remains high on account of the small capacity of the air for water vapor.

A detailed description of the distribution of the meteorological elements about an extratropical cyclone or low.

The cirrus clouds are almost entirely lacking on the west side, but extend far out to the east. The nimbus cloud area which also marks the region of precipitation, is not concentric with the isobars, but is located chiefly in the southeast quadrant. There are two series of transition clouds from the cirrus to the nimbus. The cirrus may become heavier, and then cirro-stratus, next alto-stratus, then stratus or fracto-stratus, and finally nimbus. The cirrus may also become cirro-cumulus, then alto-cumulus, next strato-cumulus, and finally nimbus. Both forms of transition may sometimes be seen in different parts of the sky at the same time. On the west side, the transition from nimbus to clear sky is usually this: the nimbus becomes fracto-nimbus, disclosing often an upper cloud area of cirrus or cirro-cumulus. The upper cloud area extends but a short distance from the center, and then ceases. The fracto-nimbus usually becomes strato-cumulus, then fracto-cumulus, and finally cumulus or a clear sky.

The diameter of the formation averages about 1200 miles and varies all the way from a few hundred miles to several thousand.

The weather map for 8 A.M., Dec. 30, 1907, which is given as Chart XXVIII, shows a very typical extratropical cyclone or low with its center over the Great Lakes. The weather maps for April 3, 1905; Jan. 3, 1906; Oct. 9, 1909; Jan. 18, 1910, also show very typical lows.

286. It must be held in mind that the distribution of the meteorological elements which was pictured in the diagram and has been so fully described is the typical one and applies to the eastern part of the United States. This typical distribution is slightly different in dif-

ferent parts of the world, is somewhat different in summer than in winter, and the actual distribution in the case of any individual storm may depart widely from the type form.[1]

The ratio of the longer to the shorter axis of the oval is about 1.9 to 1 in the United States, 1.7 to 1 in the Atlantic Ocean, and 1.8 to 1 in Europe. The central pressure is lower over the ocean than in either Europe or America. The direction of the longer axis is more towards the east in Europe than America. The wind direction makes a smaller angle with the isobars on the ocean. The chief differences, however, in the distribution of the elements are to be found in the location of the nimbus cloud area and the angle which the wind direction makes with the isobars in the various quadrants. These are very different in different places, and the explanation is to be found in the surface topography, *i.e.* in the nearness to the ocean, the direction towards the ocean, the presence and characteristics of mountain chains, etc. All of these differences are, however, comparatively small and unimportant. A low is essentially the same formation the world round.

The typical distribution is slightly different in different parts of the world.

In the winter the isobars are more nearly circular, and the central pressure is less than in summer. In winter, the wind velocity is usually larger, and the angle made with the isobars larger than in summer. The temperature difference between the west and east side of the storm is usually much larger in winter than in summer. The only important difference between winter and summer is in the nimbus cloud area. In winter, there is large, continuous nimbus cloud area mostly in the southeast quadrant, and precipitation falls steadily over this area. In summer, this nimbus cloud area is usually lacking. In its place are cirrus, cirro-stratus, and cirro-cumulus clouds, and, whenever local convection takes place, a thundershower is the result. Thus in winter over this area a continuous nimbus cloud with steadily falling precipitation is expected, while in summer cirriform clouds, sultry weather, and thundershowers are to be expected. The reason for this is the higher temperatures of summer, as the increased capacity of the air for water vapor permits it to be held as invisible water vapor, and it is not condensed into a nimbus cloud until convection takes place. The lows of spring and autumn are sometimes of one type and sometimes of

The typical distribution is somewhat different in summer than in winter.

---

[1] See *Meteorologische Zeitschrift,* Juli, 1903, pp. 307, for the distribution about "Lows," at St. Louis, U.S.A., and for references to other articles. See also *Annals of Harvard College Observatory,* Vol. XXX.

the other. The winter type may also occur in summer, but the oppo-site is exceedingly rare.

Each individual extratropical cyclone or low has its own peculiar characteristics and distribution of the meteorological elements, and it may differ much from the normal or type form. In order to gain anything like a full understanding of these storms, **The individual low may differ from the type form.** many individual lows as portrayed on the daily weather maps must be critically studied. It would probably be better to defer beginning this study until the first part of Chapter VIII has been reached. At least twenty-five separate storms well distributed throughout the year should be carefully considered. In each case the actual distribution of the elements should be exactly noted and compared with the type form. If there is any discrepancy, it should be explained. There are three factors which cause a departure from the type form: (1) The surface topog- **There are three factors which cause the discrepancies.** raphy of the country; such facts as the location of bodies of water, nearness to the ocean, the height, position, and direction of mountain chains, etc. (2) The meteorological condition of the country; such facts as to whether the ground is snow covered or not, whether the temperature and moisture are large or not, etc. (3) The neighboring meteorological formations; that is, the posi-tion of the various lows and their antitheses, the highs, with reference to the low in question. All departures from a normal or type form in connection with the distribution of the meteorological elements can be explained on the basis of these three factors.

287. There is one modification of the distribution of the meteoro-logical elements about an extratropical cyclone as just given, which is so marked, of so much practical importance, and of such **Many lows have a wind shift line.** general occurrence that it is worthy of special considera-tion. In the southern quadrants of a low, a so-called wind shift line often develops. This occurs in connection with about one low in seven in this country, and is much more common in Europe. The distribution of the meteorological elements about a low with a well-developed typical wind shift line is shown in Fig. 119.

The oval isobars have a projection in the southern quadrants which appears like a pocket or a V-shaped bulge pointing, ally towards the southwest. On the east side of this line, the wind continues from the south with small velocity; the precipitation area usually disappears, and the nimbus cloud is replaced by the cirriform transition clouds; the temperature and moisture continue high. On the west side of this line, there is a narrow belt of nimbus cloud with strong northwest winds. The temperature lines are also packed close together in this precipitation area. Apart from this modification in the southern quadrants, the distribution of the elements is the same as before. The weather map

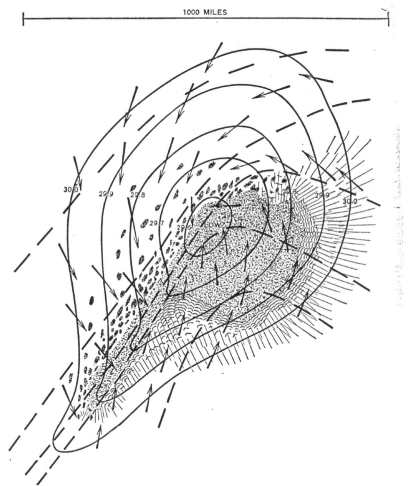

Fig. 119. — Distribution of the Meteorological Elements about a Low with a Typical Wind Shift Line.

for 8 A.M., March 3, 1904, which is reproduced as Chart XXIX, shows an extratropical cyclone with a pronounced wind shift line. The weather maps for Jan. 23, 1906; Jan. 4, 1907; Jan. 21, 1908; May 6, 1909; Nov. 17, 1909; Nov. 23, 1909; Aug. 30, 1910, also show lows with typical well-marked wind shift lines.

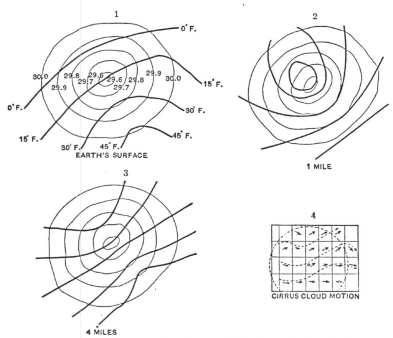

FIG. 120. — The Structure of an Extratropical Cyclone at Various Levels.

288. The distribution of the meteorological elements about an extra-tropical cyclone, its characteristics and structure at the earth's surface, have been fully considered. It now remains, in order to complete our knowledge of this formation, to consider its structure at various levels above the earth's surface. This information has been gained from observations made on mountains and by means of balloons and kites and by noting the direction and velocity of motion of clouds. The four small diagrams given as Fig. 120 will help to make clear the

The structure of an extratropical cyclone at various levels.

U

structure of this storm at various levels. The first diagram shows again the well-known distribution of pressure and temperature about an extratropical cyclone at the earth's surface. The central pressure has been assumed to be 29.6 inches, and the temperature difference between the northwest and southeast parts has been considered 45° F. This difference is rather large, but not at all unusual. The second diagram has been obtained from the first by computation and shows the isobaric lines at the height of one mile above the earth's surface. It has been assumed in making the computation that the vertical decrease in temperature was everywhere the same and equal to the average vertical temperature gradient; namely, 1° F. for 300 feet. Since the air in the southeast quadrant is warm, and thus light, the drop in pressure due to one mile of elevation will be less there than in the northeast quadrant, where the air is cold and heavy. As a result, the longer axis of the oval isobars now has a northwest-southeast direction, which is the direction of the greatest temperature contrast. The oval has also opened up so that the northwest portion is missing, and the center has also been displaced towards the northwest. This displacement of the center has been a matter of frequent observation. On the summit of Mt. Washington (elevation 6279 feet) the storm center passes on the average about three hours later than at the base of the mountain. This corresponds to a displacement of between one and two hundred miles in this small height. The third diagram was determined from the first in the same way as the second, and shows the isobaric lines at a height of about four miles. The oval form has now practically disappeared, and only a certain amount of distortion in the lines is now apparent. With increasing height this distortion grows less and less, and the isobaric lines become more nearly straight, running from west to east.

At the earth's surface the air motion is a whirl about the area of low pressure, blowing spirally inward, making an angle with the oval isobaric lines and turning counterclockwise in the northern hemisphere. With increasing elevation, the isobars become circular, even before the level of the clouds is reached, and the air moves nearly tangential to the isobars, rising slowly. By the time a height of one mile has been reached, as has been already shown, the isobars have again become oval, but with the axis now northwest-southeast, the center has been displaced, and the northwest portion of the storm is missing. An atmospheric whirl can now barely be said to exist, as the northwest portion is missing. The air motion approximates the form of two currents flowing side by side in opposite directions. On the right there is a warm current coming from the southeast, and on the left a cold current coming from the northwest. With increasing height the isobars become less and less oval, and finally become distorted lines which gradually grow straighter and straighter, as was shown in the diagrams. While this is taking place, the whirling nature of the air motion becomes less and less apparent, and the two currents flowing in opposite directions become more pronounced. There is also a motion outward from the center, as the rising air must be injected into the general circulation aloft. With increasing elevation these air currents grow weaker, and by the time the level of the cirrus clouds has been reached (five miles on) only a slight distortion of the west to east motion of the outer layer of the atmosphere is apparent. The fourth diagram gives the motion of the cirrus clouds about an area of low pressure as observed at the Blue Hill Observatory near Boston, under the direction of Mr. A. L. Rotch.[1] The distortion caused by the currents is only slightly apparent. An extratropical cyclone is thus a formation of the lower part of the atmosphere. It is often stated that their influence ceases at five miles, but, considering the fact that some cross the Rocky Mountains, a greater height must be inferred in some cases at least.

289. A comparison of the tropical and extratropical cyclones. — A tropical cyclone and an extratropical cyclone when considered superficially may seem very similar, but when the details in their structure are considered, they often stand in sharp contrast. Superficially considered, they are alike. Both are areas of low pressure, with spirally inflowing winds turning counterclockwise in the northern hemisphere. Both are attended by immense cloud areas and precipitation. They differ, however, in many ways: (1) In the case of an extratropical cyclone, the isobars are oval, and they lie closer together in the southwest quadrant, while in the case of a tropical cyclone, the isobars are more nearly circular and equidistant from each other. They differ in many details. The central pressure is also lower in the case of a tropical cyclone. (2) Wind velocities are much higher, and the angle made with the isobars less, in the case of a tropical cyclone. (3) In the case of a tropical cyclone, the temperature and moisture are the same in all quadrants; that is, they are symmetrical with respect to the center. In the case of an extratropical cyclone, the values of temperature and moisture are much higher in the eastern part than in the western. (4) In the case of a tropical cyclone, the cirrus cloud extends out in all directions from

[1] See *Annals of the Harvard College Observatory*, Vol. XXX., also *Monthly Weather Review*, March, 1907.

the center, and the transition clouds are the same in every quadrant. In the case of an extratropical cyclone, the cirrus cloud is found only on the east side, and the transition clouds are entirely different on the east and west sides. (5) The rain area is concentric with the low in the case of the tropical cyclone, while it lies mostly in the southeast quadrant in the case of an extratropical cyclone. All these contrasts are well brought out if the two diagrams (Figs. 110 and 118) which represent the distribution of the meteorological elements about a tropical and an extratropical cyclone are carefully compared.

The great difference in the two storms is noticed in connection with the central calm or eye. A tropical cyclone usually has a calm central area where the pressure is lowest, the clouds break through,

The great difference is the presence of a calm central eye. the temperature rises, and the relative humidity is much less. An extratropical cyclone probably never has such a central calm or eye. Some have thought that if the extratropical cyclone is very violent, an approximation to a calm central eye exists. This is probably not the case, however, and the suggestion of an eye was given by the fact that the low in question had a pronounced wind shift line. If such a low passes a little to the north of an observer, the precipitation may stop, the clouds change from nimbus to a cirriform variety, the temperature and moisture may remain high, and the wind may continue to blow gently from the south. Soon after it may be again raining, with the wind blowing sharply from the northwest and the temperature dropping rapidly. This would give the impression of an eye, but it is really only the passage of a wind shift line. How this sequence of changes is possible can be readily seen from the diagram (Fig. 119), which gives the distribution of the meteorological elements about a low with a typical wind shift line.

290. Description of the approach and passage of an extratropical cyclone. — The first signs of the approach of an extratropical cyclone or low are usually these: The wind begins to blow gently

The sequence of weather changes due to the approach and passage of a low. from the east, the pressure decreases slightly, cirrus clouds make their appearance, and the temperature and moisture begin to increase. Next the barometer drops a little more, the wind direction changes to the southeast and the velocity becomes a little greater, the cirrus clouds thicken to cirrostratus or cirro-cumulus, and the temperature and moisture continue to rise. In the winter time, as the popular phrase goes, the weather has begun to moderate. In the summer time, it is the beginning of a period of sultriness. The pressure now drops still more, the wind veers a little and blows harder, the cirriform clouds go through their regular transition into nimbus, and the temperature and moisture are high and increasing.   Next comes a period of rain or snow, with the barometer still dropping and finally reaching its lowest.   The wind, meantime, has slackened somewhat and veered a little and is perhaps now blowing from the south or southwest.   The temperature and moisture still continue high.   The wind now veers rather quickly into the southwest, then west, and finally northwest.   The barometer begins to rise, the precipitation grows less, and the temperature and moisture decrease.   Soon the nimbus clouds break up into fracto-nimbus, perhaps disclosing the upper cloud area.   The fracto-nimbus then changes into strato-cumulus, and finally cumulus or fracto-cumulus, with a clear sky at night.   In the meantime, the wind blows from the northwest with increasing velocity, the barometer is rising, and the temperature drops rapidly.   The air also becomes much dryer.   In the summer, the dry, cool, northwest wind has replaced the oppressive sultriness of a few days before.   In the winter, the thaw or warm spell has been replaced by a cold snap.   How often this sequence of weather changes has been noted by every one without realizing that it was but the approach and passage of an extratropical cyclone.   This series of weather changes requires from two to four days in winter for its completion and nearly double the time in summer.

The series of weather changes which has just been described applies to a nearly central passage of a low.   If it passes to the north or south of an observer, the sequence would be different and depend **Veering and** upon the distance of the center of the low.   Just what the **backing** sequence would be, can be determined at once by consider- **winds.** ing the diagram given as Fig. 118, movable and moving it from left to right centrally over or above or below a given point.   It is worthy of mention that if the low passes north of the observer, the wind veers, while if it passes south of the observer, the wind backs.   Since most lows pass north of the United States, veering winds are the rule and backing the unusual.   If the low has a wind shift line, the sequence of weather changes can be determined in the same way by using the diagram given as Fig. 119.

In the early part of the last century, when this sequence of weather changes was first noticed, it was explained, as was first done **The cause** by Davé, by assuming that there were two currents of air, **of weather** one warm and moisture laden from the south and the other **changes.** cold and dry from the north.   These were supposed to flow above each other in the tropics and side by side in the temperate latitudes. It was to the ceaseless struggle of these opposing air currents and to the temporary victory of first one, then the other, that the weather changes were ascribed. This gave rise to the popular expression that the wind makes the weather. It is now known that the mechanism back of the ceaseless changes in our weather is the approach and passage of extra-tropical cyclones or lows.

ANTICYCLONES

291. Just as two valleys are an impossibility without a hill or ridge of land between them, in the same way two areas of low pressure cannot exist without a region of high pressure between. These areas of high pressure stand in many ways in sharp contrast with the lows, and have many characteristics which are exactly the opposite. For this reason they are called anticyclones, or, sometimes, simply highs.

Anticyclones or highs are the opposite of lows.

An anticyclone or high can be best defined and described by stating its chief characteristics. It is an area of high barometric pressure with spirally outflowing winds turning clockwise in the northern hemisphere and the opposite in the southern; the wind velocity is usually very moderate, and calms are frequent; but few clouds are to be seen, and precipitation is usually lacking; the changes in temperature and moisture are large and well marked. The whole formation is from several hundred to several thousand miles in diameter, and moves with moderate velocity, sometimes loitering for a day or two, from some westerly to some easterly quarter.

Outline description.

292. The distribution of the meteorological elements about a typical anticyclone is shown in Fig. 121.

The isobars are often irregular in form, and in the center there may be several highs instead of a single peak of pressure. The more usual form, however, is the oval, and the longer axis generally extends northeast-southwest, north and south, or northwest-southeast. The central pressure averages about 30.6 inches, although this varies all the way from a little over 30 to more than 31 inches.

Description of the distribution of the meteorological elements about an anticyclone or high.

The winds blow spirally outward from the center, turning clockwise in the northern hemisphere. The wind direction makes an angle with the isobars which is least in the northeast portion, where it averages about 20 or 30° and is greatest in the southwest portion, where the average is from 60 to 70°. The wind velocity is very moderate and decreases towards the center, where calms are frequent, particularly at night.

There is a decided drop in temperature in the northeast portion, where the winds are from the north, and a decided rise in temperature

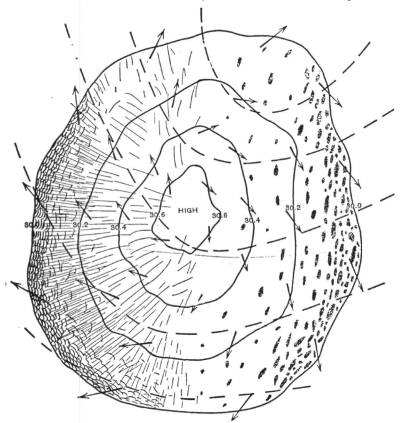

Fig. 121. — The Distribution of the Meteorological Elements about a Typical Anticyclone or High.

in the southwest and west portions, where the winds are from the south. The moisture changes are not shown in the diagram, but they follow the temperature. In the northeast portion, both the absolute and the relative humidity are low, while in the southwest portion both are high.

On the east side the strato-cumulus clouds of the departing low may be still visible. In the middle portion, convection-caused cumulus or fracto-cumulus clouds may be found. These usually disappear at night and are most numerous on the east side of the center. In the winter, a thin alto-stratus cloud due to radiation may be formed during the night and be visible in the early morning. On the west side, the cirrus and cirro-stratus of the coming low may have already made their appearance. Precipitation very seldom falls in connection with an area of high pressure. Occasionally a cumulus cloud may become sufficiently overgrown to yield a snow flurry in winter or a few raindrops in summer.

The diameter of the whole formation averages perhaps 2000 miles and varies from several hundred to several thousand miles.

The weather map for 8 A.M., April 23, 1906, which is given as Chart XXX, shows a very typical anticyclone or high with its center over Illinois. The weather maps for May 1, 1905; Oct. 28, 1907; Jan. 3, 1908; Jan. 30, 1909; March 22, 1909; April 10, 1909; June 18, 1909; Sept. 26, 1909; Oct. 29, 1909; March 14, 1910; Dec. 21, 1910, also show typical highs at various times of year.

293. It must be remembered that the distribution of the meteorological elements about an anticyclone or high which has just been so fully

The typical distribution is different in different countries and at different times of year. stated is the typical one, and applies to the eastern part of the United States. This typical distribution is slightly different in different parts of the world and at different times of year, and the actual distribution in the case of any individual high may depart widely from the type form.

In Europe, the direction of the longer axis of the oval is more northeast-southwest than in the United States, and the quadrant in which the wind direction makes the smallest angle with the isobars is also different.

In winter, the highs are larger and the central pressure is higher than in summer.

Each individual high has its own peculiar characteristics, and distribution of the meteorological elements, and it may depart much from the

Individual highs differ from the typical form. type form. The only way to gain detailed knowledge about this formation is to study carefully many individual cases as they occur on the daily weather maps. In each case, note carefully the actual distribution of the elements, compare this with the type form, and then explain all discrepancies. The three causes of departure from type are the same as for lows; namely,

(1) the surface topography of the country, (2) the meteorological condi_ tion of the country, (3) the neighboring meteorological formations.

294. The characteristics of a high at the earth's surface have now been fully stated, and it remains to consider its structure at various levels above the earth's surface. The change in form with elevation is very similar to that in the case of a low, and for that reason it will only be sketched in outline here. At the earth's surface the air motion is a whirl about an area of high pressure. The air moves spirally outward, making an **The structure of an anticyclone at various levels.** angle with the oval isobaric lines, and turning clockwise in the northern hemisphere. Since the northern portion is much colder than the southern, the pressure decrease with elevation will be greater in the northern portion than in the southern. By the time a height of one mile has been reached, the center has been displaced towards the cold area (that is, towards the north), the oval isobars have opened out, and the northern portion of the formation is lacking. An atmospheric whirl can now be hardly said to exist, and the air motion approximates the form of two currents flowing side by side in opposite directions. On the right there is a cold current coming from the north, and on the left a warm current coming from the south. With increasing elevation, the oval isobars

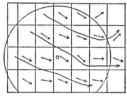

Fig. 122. — The Motion of the Cirrus Clouds about an Area of High Pressure.

open out more and more, and finally become distorted lines which tend to become straighter.

While this is taking place, the whirling nature of the air motion becomes less apparent, and the two currents, flowing in opposite directions, become more pronounced. There is also a component of motion in towards the center, for the air which moves spirally outward at the earth's surface is replaced by a gentle descending air current which must be supplied from above. With ever increasing elevation, the isobars become straighter and straighter and the two air currents become less and less pronounced, until, by the time the level of the cirrus clouds have been reached (five miles on), only a slight distortion of the west to east motion of the outer layer of the atmosphere is apparent. The motion of the cirrus clouds about an area of high pressure as observed at Blue Hill is given in Fig. 122. This distortion caused by the air currents beneath is only slightly apparent. Thus the high, like the low, is a formation of the lower atmosphere.

295. The sequence of weather changes brought about by the approach and passage of an anticyclone or high has been experienced many times by every one without knowing the cause of the changes. A low has probably just passed; its rain area has gone by; the cloud form has changed to strato-cumulus; the wind has gone to the northwest; the temperature has dropped markedly; the air has become much dryer; the barometer is rising. All this may be expressed meteorologically by saying that a low has passed and the weather control is being transferred to a coming high. It may be expressed popularly by saying that it has cleared off and several days of good weather are in store. The wind continues in the northwest and decreases in velocity. In fact the nights are almost perfectly calm and the wind blows only during the day. The barometric pressure continues to rise, usually holding about steady during the daytime and rising quite a little during the night and early morning. The strato-cumulus clouds become cumulus or fracto-cumulus and decrease in number and size. Since these are convection-formed, they disappear at night and are visible only during the daytime. The temperature drops fairly low at night and rises rapidly during the day so that the daily range is excessive. Soon the center of the high is reached. There is an almost perfect calm both day and night. The pressure is at its highest. The sky is cloudless or at most shows a few cumulus clouds. The air is dry and the daily range in temperature continues large. Then the barometer begins to fall. The wind goes to the northeast or east and, blowing at first very gently, soon increases in velocity. The temperature begins to rise, and soon the cirrus cloud puts in its appearance. The barometer continues to fall; the wind blows harder and perhaps shifts to the southeast; the temperature rises rapidly; the air becomes more moist; the cirrus clouds thicken into cirro-stratus or cirro-cumulus; and the weather control passes from the departing high to the coming low. This series of weather changes requires from two to four days in winter for its completion and nearly double the time in summer.

The series of weather changes which has just been described applies to the nearly central passage of a high. If it passes to the north or south of an observer, the sequence would be somewhat different. Just what the sequence would be can be determined by moving the diagram given as Fig. 121 from left to right centrally over or above or below a given point.

The sequence of weather changes due to the approach and passage of a high.

In summer, if a high loiters for several days, there is a series of days with rather high temperature during the day but with cool nights. If it loiters unduly long, a spell of dry weather is the result. In winter the low temperatures, sometimes many degrees below zero, occur during the first two days of a high. In autumn the morning fogs occur usually during anticyclonic weather, because it is then that the daily range of temperature is so large. In spring the destructive frosts usually occur just after the weather control has passed from a low to a strong coming high.

The Tracks and Velocity of Motion of Extratropical Cyclones

296. Tracks in the northern hemisphere. — If an observer far above the earth could look down upon the whole northern hemisphere, he would see a ceaseless procession of lows between the thirtieth and eightieth parallels of latitude, moving eastward and encircling the poles. Their big cloud areas would gleam white in the reflected sunlight. In Fig. 123 (after Loomis) the tracks of a number of lows are shown. Taking the northern hemisphere as a whole, the tracks followed by lows may be characterized by stating that they move eastward and slightly poleward. The reason for this eastward motion is because this is the prevailing direction of both the surface winds and the fast-moving upper air currents in these latitudes. The lows thus drift with the general wind system. *Lows move eastward and slightly poleward.*

Lows may originate anywhere. A good part of those which visit Europe originate over the Atlantic Ocean. In the United States, a favorite place of origin is the Mississippi Valley just east of the Rocky Mountains. The length of the path followed varies from a few hundred miles to, in a few rare cases, more than half of the circumference of the globe. *They may originate anywhere.*

The tropical cyclones come up from the tropics and join the extratropical ones in two places: over the West Indies and over the Philippines and Japan. As soon as a tropical cyclone enters the extratropical region it loses its violence; the central eye disappears; the rain area becomes excentric; its velocity of motion is larger, and it soon assumes all the characteristics of an extratropical cyclone. In fact, in a few cases, it has been known to merge with an extratropical cyclone. *Tropical cyclones sometimes join them.*

297. Tracks across Europe. — If a smaller portion of the world is considered, as for example Europe, it is found that not all areas are

covered by the same number of lows. The lows seem to prefer to
travel along certain rather definite paths; and if the actual
paths followed by the lows for a number of years are generalized or summarized, it is found that a more or less definite
system of storm tracks is the result. This has been done

The Van Bebber system of tracks for Europe.

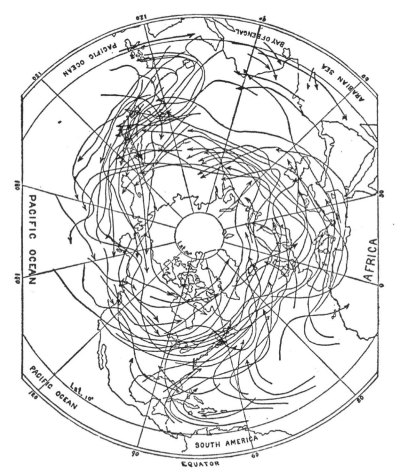

Fig. 123. — The Tracks of a Number of Lows in the Northern Hemisphere.
(After Loomis.)

with particular skill by Van Bebber, the head of the German meteorological service, and the resulting storm tracks for Europe are shown in
Fig. 124. The width
of the track is pro-
portional to the fre-
quence with which
it is followed by lows.
Track I is used more
than any other, and
chiefly in the autumn
and winter. It is
but little frequented
in spring. Tracks
II, III, and IV are
traversed at all times
of the year. Tracks
II and III are used
a little more during
the cold portion of
the year and IV in
summer and autumn.
Track Va is almost
never traversed in summer. Tracks Vb, Vc, and Vd are frequented at
all times of year.

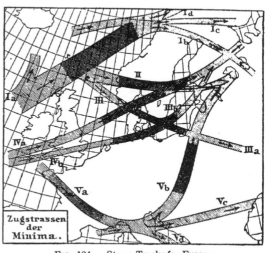

FIG. 124. — Storm Tracks for Europe.
(From BEBBER's *Lehrbuch der Meteorologie*.)

298. Tracks across the United States. — The tracks of lows across the
United States have been generalized by Bigelow, Russell, and Van Cleef
and the results of all three investigators are presented here.

In Fig. 125 the Bigelow system of tracks is shown. *The three
systems of tracks
across the
United
States.*
The main track follows the northern boundary of the United
States across the Great Lakes and out the St. Lawrence Valley.
This main track is joined by three others coming up from
the south. One comes up from Colorado and Utah and
joins it near Lake Superior. Another comes up from Texas and joins
it near Lake Huron. The third comes up the Atlantic coast and joins
it near Nova Scotia. There is a second main track across *The Bigelow system
of tracks.*
Texas and the Gulf States to the Atlantic coast, where it
either turns northward or goes out over the ocean. The
broken lines show the average daily movement. This system of tracks
is rather too highly generalized. It has been made simple by throwing
out too many tracks as erratic or exceptional. The percentage of lows
which will follow these tracks has thus been reduced.

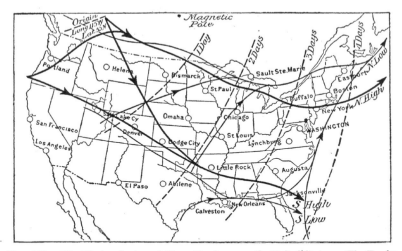

FIG. 125. — The Bigelow System of Storm Tracks across the United States. (The Tracks of Highs are also shown.) (U. S. Weather Bureau.)

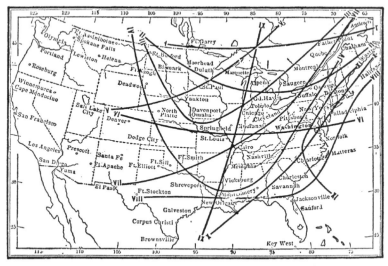

FIG. 126. — The Russell System of Tracks across the United States. (From RUSSELL'S *Meteorology*.)

299. In Fig. 126 the Russell system of tracks is shown. Since there are eleven tracks, this system is not as highly generalized as the first one, and a larger per cent of the lows will follow some one of these tracks.

The Russell system of tracks.

300. In Fig. 127 the Van Cleef system of tracks is shown. This is the least generalized of the three and thus accounts for the paths followed by the largest number of lows. In preparing this diagram, the tracks of the lows from 1896 to 1905, 1160 in number, were used. Of these 1160 only 57 were erratic and did not follow some one of the tracks. The frequency with which any track is traversed is indicated by its width.

The Van Cleef system of tracks.

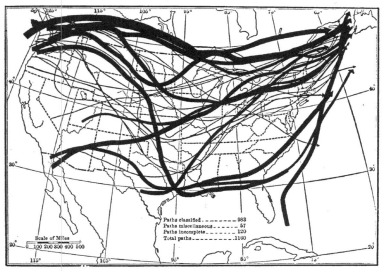

FIG. 127. — The Van Cleef System of Storm Tracks across the United States. (Twenty-seven Tracks are represented.) (U. S. Weather Bureau.)

301. If a low does not originate in the United States, it enters it from the northwest, west, or south. Those which enter north of the middle of the country have a tendency, in crossing the Mississippi Valley, to move towards the south and then recurve towards the northeast. In approaching the Atlantic coast, all lows move towards the northeast. Although lows may originate within the country or enter it from various directions, they nearly all

A summary of the three systems.

1882–1891

THE NUMBER AND VELOCITY OF MOTION OF LOWS ON THE ELEVEN DIFFERENT TRACKS ACROSS THE UNITED STATES (RUSSELL SYSTEM)

	I	II	III	IV	V	VI	VII	VIII	IX	X	XI	Total Number	Average Velocity
Jan.	6	6	23	3	7	3	2	11	2	6	1	63	40.0
Feb.	7	8	16	1	6	3	5	13	6	2	2	62	38.3
March	7	7	16		11	1	2	8	1	5		69	33.0
April	3	7	6		16	5	2	7		3	1	50	28.7
May	8	5	7	1	12	6		6	1	2	1	49	26.6
June	9	2	3	2	9	3	2	3		2	1	33	24.9
July	8	1	3	1	11	6		1		2		45	24.9
Aug.	11	9	8		8	3	1	3	3	1	1	47	28.1
Sept.	11	7	7		7	2		2	2	5		42	25.4
Oct.	11	11	7		11	3	1	5		4		54	29.2
Nov.	10	7	11		10	1		9	5	2		52	32.3
Dec.	25	5						13				71	36.6
Total	116	85	114	8	116	36	15	81	21	36	9	637	
Average Velocity	31.1	33.2	34.6	35.6	30.5	30.8	27.1	32.5	27.1	29.0	24.2		31.7

leave it by way of the St. Lawrence Valley and Newfoundland. The phrase " all roads lead to Rome " might be paraphrased in connection with lows into " all storm tracks lead to New England."

It must not be supposed that all lows follow one or the other of these many tracks without exception. A low may loiter in one place for a day or two, or turn sharply aside in one direction or another, or move erratically in almost any direction. These tracks simply represent normal behavior. In order to gain familiarity with the paths followed by lows, many individual cases, as indicated on the daily weather maps, must be studied.

302. **Velocity of motion.** — Several important facts in connection with the velocity of motion of lows are brought out by the accompanying table, which gives the number and velocity of motion

The number and velocity of motion of lows.

of lows for the ten years (1882 to 1891), classified according to the Russell system of tracks.

The average velocity for all tracks for all times of year is 31.7 miles per hour. The winter velocity is, however, nearly twice as large as the summer (January, 40; June and July, 24.9). The velocity of motion does not differ much in the different tracks and is greatest for the tracks which have the greatest curvature. Some tracks are much more frequented than others; compare, for example, I or V with IV or XI. Some tracks are also more frequented at certain times of year; I, for example, with 3 in April and 25 in December, or VIII with 1 in July and 13 during both December and February. It is also a very important fact that the number of lows in winter is more than double the number in summer; compare, for example, December, 71 with June, 33.

Generalizations as to the velocity of motion of lows.

Lows move faster across the United States than any other country. One authority, which gives the average velocity for the United States as 26 miles per hour, finds for Japan a velocity of 24; Russia, 21; the north Atlantic, 18; and Europe, 16. That lows move faster in winter than in summer is found to be true for all parts of the world.

Velocity in various countries.

It must not be supposed that the velocity of motion of individual lows always conforms to the normal. A low may loiter for a day or two, drifting, perhaps, but one or two hundred miles in that time. Another low may rush across the country, covering a distance of sixteen or seventeen hundred miles in a single day. As has been so often stated, the only way to learn the characteristics of lows and highs, their tracks and velocity of motion, is to study them as they are portrayed from day to day on the weather maps. In making this study, the normal behavior should be held in mind, comparisons made, and discrepancies explained.

Individual lows are often erratic.

THE TRACKS AND VELOCITY OF MOTION OF ANTICYCLONES

303. Anticyclones or highs are more erratic than lows, both as regards the track followed and the velocity of motion.

In Fig. 125, the Bigelow system of tracks for highs was shown. Highs usually enter the country from the extreme northwest or over California. To those entering from the extreme northwest, two courses are open. They may move eastward and slightly southward along the northern boundary of the

The Bigelow system of tracks for highs.

x

United States until the Atlantic coast is reached, where they turn towards the northeast and proceed in the direction of Iceland. The highs entering from the northwest may also move southeast over Kansas and the Gulf states to the Atlantic coast near Florida. The highs coming in over California usually move southeast and join this track just south of Kansas. After leaving the Atlantic coast, near Florida, the highs usually continue to move towards the southeast in the direction of Bermuda.

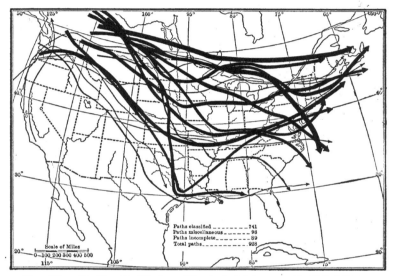

Fig. 128. — The Van Cleef System of Tracks for Highs across the United States.
(U. S. Weather Bureau.)

The Van Cleef system of tracks for highs is shown in figure 128. This system of tracks is based upon the highs from 1896 to 1905, 928 in number. Only 98 of the 928 were eratic and did not follow one of the indicated tracks. The width of the track is proportional to the frequency with which it is traversed.

The Van Cleef system of tracks for highs.

304. The velocity of motion of highs averages somewhat less than that of lows, but this is because a large high often remains almost stationary for a day or two and perhaps a week, and this brings down the average velocity of motion. Small highs, particularly those wedged in between lows, usually move with about the same velocity as the lows themselves.

The velocity of motion of highs.

The following table, compiled by C. F. von Herrmann, is taken from the Bulletin of the Mount Weather Observatory, Vol. II, part 4, p. 196. It contains the average, highest, and least velocity of motion of both highs and lows for the various months and for the year. The data summarized cover the twenty-seven years from 1878 to 1904. For the purpose of comparison, the values for lows determined by Loomis for the period 1872–1884 are also added. The truth of the generalizations which have been stated for the velocity of motion of both highs and lows is at once apparent from this table.

305. The four weather maps for 8 A.M., Jan. 30, Jan. 31, Feb. 1, and Feb. 2, 1908, which are given as Charts XXXI– XXXIV, will show the actual distribution of the meteorological elements about highs and lows in individual cases, and will also illustrate the statements. which have been made concerning the tracks followed and the velocity of motion on the part of both highs and lows.[1] The weather maps for Dec. 26 to 29, 1904; Jan. 2 to 5, 1906; Jan 14 to 17, 1906; Dec. 8 to 12, 1907; Dec. 21 to 24, 1907; Dec. 28 to 31, 1907; Jan. 10 to 13, 1908; Feb. 3 to 7, 1908; Feb. 13 to

Four illustrative weather maps.

[1] For other sets of weather maps to illustrate typical lows, see: Monthly Weather Review 1904, end of volume; Climatic charts of the United States; Climatology of the United States (Bulletin Q of the U. S. Weather Bureau).

VELOCITY OF LOWS AND HIGHS IN MILES PER HOUR

			JAN.	FEB.	MAR.	APR.	MAY	JUNE	JULY	AUG.	SEPT.	OCT.	NOV.	DEC.	YEAR
Loomis for lows 1872–1884			33.8	34.2	31.5	27.5	25.5	24.4	24.6	22.6	24.7	27.6	29.9	33.4	28.4
Lows	Average		34.8	34.8	31.6	26.9	24.3	24.0	24.4	24.6	24.8	27.4	30.7	34.9	28.6
	Highest		67.9	81.0	57.4	68.8	54.7	52.0	50.0	55.0	50.0	65.0	55.0	79.0	
	Lowest		4.2	8.3	6.5	8.0	7.0	7.0	8.0	4.2	4.0	4.0	4.3	6.3	
Highs	Average		29.5	28.2	26.7	25.2	25.4	23.7	22.2	22.1	24.7	24.7	27.1	27.4	25.6
	Highest		66.7	50.0	50.0	50.8	58.0	57.7	39.0	41.0	52.1	52.1	52.4	66.7	
	Lowest		2.0	5.0	11.1	7.0	8.3	10.0	9.4	8.3	8.0	7.0	10.0	6.0	

C. F. von Herrmann 1878–1904

16, 1908 ; Feb. 16 to 20, 1908 ; Jan. 28 to 31, 1909 ; Feb. 17 to 20, 1909 ; Feb. 22 to 25, 1909 ; April 5 to 8, 1909 ; April 28 to May 1, 1909 ; Oct. 31 to Nov. 3, 1909, show more or less typical highs and lows following well-recognized paths.

Statistics on Extratropical Cyclones and Anticyclones

306. It is probably safe to affirm that no subject in meteorology has been more thoroughly studied by the statistical method, that is, by generalizing statistics, than have the highs and lows. Since the files of daily weather maps, particularly for the United States and Europe, cover a period of from thirty to forty years, the material for such a study is ample. All the statements which have been made concerning the distribution of the meteorological elements about the highs and lows, their tracks and velocity of motion, have been gained by the inductive method of reasoning and are based upon the general conclusions which have been drawn from tables of statistics about highs and lows. Many other tables of statistics have been prepared and general conclusion drawn. Some of these have been incorporated in various books on meteorology, but for most of them the reader must be referred to the periodical literature of the subject.

Many generalizations have been made.

The Origin and Growth of an Extratropical Cyclone and Anticyclone

307. Theories as to the origin of extratropical cyclones and anticyclones. — When the tropical cyclone was being considered, the convectional theory of its origin and growth was given, and it was found that the requirements of the theory and the observed facts were in absolute agreement. Since tropical cyclones and extratropical cyclones, when superficially considered, are so similar and have so many things in common, it is but natural to ascribe to both the same cause, namely, convection. In fact, for a long time, even during the last part of the last century, all cyclones and lows were referred to this same cause.

The convectional theory of the origin of lows.

If convection is the cause of extratropical cyclones, they must originate in pockets of warm, moist, quiet air. Now, as a matter of fact, lows may originate anywhere. They form over the frozen snow-covered areas of the extreme northwest when the temperatures are far below zero and the air is very dry. They also

The three objections.

originate in the fast-moving winds over the Atlantic Ocean. In short, they originate in large numbers when pockets of warm, moist, quiet air are certainly lacking, in fact, when just the opposite conditions are present. Furthermore, if they are of convectional origin, there ought to be more in summer than in winter, for the quieter air, the higher temperature, and the larger amounts of moisture during the summer time should be more conducive to their formation than the opposite conditions which exist during the winter. Statistics, however, show that there are more than twice as many lows during the winter as during the summer. Again, if lows are of convectional origin, the temperature at any given level above the earth's surface in the center of a low ought to be higher than at the same level in surrounding regions. It will be remembered that it is the liberation of latent heat when the moisture condenses to form cloud and precipitation which heats the air, maintains it at a higher temperature than its surroundings, causes still further rise, and thus supplies the energy for growth or continuance. Now, observations of the temperature in the center of a low and in surrounding regions at the same level have been made at mountain observatories and by means of balloons and kites, and it is found that the temperature in the center of a low is often not as high as in surrounding regions at the same level. Thus, on account of those three objections, convection as the origin of all extratropical cyclones or lows must be abandoned. Lows have been found with centers warmer than their surroundings. It may be that most of the lows of summer and some of those of winter are convection-caused, but convection as the universal cause of all lows must certainly be given up.

An attempt has been made to obviate some of these difficulties by ascribing the origin of lows to convection, not at the earth's surface, but high up in the atmosphere. It is known from balloon ascensions that there are many layers in the atmosphere of very different temperature and moisture. If a warm, moist layer should find itself underneath a colder and less moist layer, there would be unstable equilibrium and convection would surely take place. The convectional motion high up in the atmosphere would make itself felt at the earth's surface as an area of low pressure. Since in winter the layers of the atmosphere have been found to differ most in temperature and moisture, it is then that the most convection high up in the atmosphere would be expected.

The convection might occur high up in the atmosphere.

308. The origin of lows is sometimes ascribed to eddies formed in the general wind system. If a stream of water flows into a quiet pond or if

two streams flow past or over each other in different directions or with different velocities, eddies are often formed. Now the general
The driven circulation of the atmosphere in extratropical regions takes
eddy theory place in three layers. In the northern hemisphere, the
of the origin upper layer moves from the west, the middle layer from
of lows. the northwest, and the surface layer from a little south of west. These layers also have different velocities of motion, so that it is easy to think of eddies as forming where these layers meet.

If a stone is dropped into a fast-flowing stream, an eddy often results. Now the air which moves from the equator poleward in the outer layer of the atmosphere must drop down in extratropical regions to commence its journey back to the equator. This dropping down of air masses might readily cause eddies.

If a stream flows over a rough, irregular bed, eddies are usually formed. The surface of the continents is very rough and irregular and the air moving over it might readily have eddies formed in it.

There are thus at least three ways in which the formation of eddies in the atmosphere is easily thinkable. Each eddy would develop a small amount of centrifugal force, and this would cause a slightly lower barometric pressure and a low has thus originated. Since the circulation of the atmosphere is more vigorous in winter than in summer, eddies would form in greater numbers in winter than in summer.

This objection at once suggests itself: the direction of rotation of an eddy would be entirely a matter of chance, so that as many would rotate
An objec- one way as the other. Thus, when fully grown, the whirl
tion and its about lows should be clockwise as well as counterclockwise.
answer. Observations, however, show that the whirl about a low is always counterclockwise in the northern hemisphere. A partial answer can be given to this objection. Due to the earth's rotation, the deviation is always to the right, and the whirl must be counterclockwise. If an eddy turned in the other direction, it would be immediately stopped and put out of existence, so that only those eddies turning in the right direction could continue to exist and grow.

309. Another way of accounting for lows is to consider them simply the antitheses or results of highs. An area of high pressure is accom-
Lows may panied by descending air currents and winds blowing spirally
be caused outward at the earth's surface. Unless the air is to congest
by highs. at the earth's surface, there must be some upward escape. At some point the air rises to relieve this congestion, and thus a low is formed. If this explanation of the origin of lows is accepted, the

question has simply changed from how lows originate to how highs originate.

310. During the last few years another very important theory as to the origin of lows has been developed. This is the work of Bigelow of the U. S. Weather Bureau and is sometimes called the counter current theory. According to this theory, the origin of a low is to be found in two great air currents of opposite direction and very different temperature. Bigelow's counter current theory of the origin of lows.

It was shown, when the general wind system of the world was being considered, that if the surface of the earth were level and the same everywhere, the belts of high and low pressure would be uniform and would extend all around the earth and the exchange of air between the equator and pole in the extratropical regions would take place in three regular layers, one above the other. The surface of the earth is, however, not level, and it is a diversified surface, being part land and part water. As a result, these belts of pressure break up into peaks and depressions, that is, permanent highs and lows, which change their intensity and position between summer and winter. The prevailing westerly air currents in the outer layer of the atmosphere seem to preserve their identity and regularity in spite of this breaking up of the belts of pressure, but the two lower layers are very much confused and mixed, so that here the exchange of air between the equatorial and polar regions takes the form of great jets or air currents which move now equatorward, now poleward. Those from the equator are, of course, warm and moisture-laden, while those from the polar regions are relatively cold and dry. The cause of the counter currents.

. The region of strongest air currents, and thus most vigorous exchange of air between equator and pole, seems to be at an elevation of about 1.5 miles above the earth's surface. Suppose that there are two great air currents at this level, flowing side by side, but of opposite direction and of very different temperature, the one on the right being warm and flowing poleward and the one on the left being cold and flowing equatorward. Due to this difference of temperature, at the earth's surface there will exist an area of low pressure with spirally inflowing winds turning counterclockwise in the northern hemisphere. The truth of these statements becomes evident when the structure of a low above the earth's surface as given in section 288 is considered. It was there shown that the atmospheric whirl which exists at the earth's surface, degenerates at a height of a little more than a mile into two counter currents, flowing in different How the counter currents produce a low.

directions with very different temperatures. The converse is equally true. Two currents in the upper atmosphere necessitate a low with inflowing winds at the earth's surface. A low, then, with its spirally inflowing winds may be considered simply as the surface effect of two powerful counter currents of very different temperature in the upper atmosphere. Since the exchange of air between equator and pole is more vigorous in winter than in summer, the number and intensity of lows should be greater in winter than in summer. This theory, then, seems to meet all the observed facts, to be free from objections, and to be plausible, but not so easy to picture as some of the other theories.

311. Highs and lows are very closely related, so that the origin of anticyclones or highs must be accounted for as well as the origin of extratropical cyclones or lows. One way of explaining the origin of highs is to consider them the antitheses or results of lows. The air rises in areas of low pressure. This air, ejected from lows, must collect somewhere and increase the pressure or form a high. A high, then, may originate in the congestion of the air ejected from lows. If this explanation of the high is accepted, the real question has become the question as to the origin of the lows.

Highs may be caused by lows.

312. Another way of accounting for highs is to ascribe their origin to the congestion of air currents in the outer layer of the atmosphere due to the convergence of the meridians. In the outer layer of the atmosphere the air moves spirally from the equator towards the pole. All meridians (that is, north and south lines) converge towards the pole so that the area steadily decreases with increasing latitude. As a result the air must congest and finally drop or be forced down to start equatorward in the lower layer. This congesting of the air currents may give rise to areas of high pressure.

Highs may be caused by the congestion of the poleward moving air currents.

313. A high may also be caused by radiation. If there is an area particularly free from clouds and moisture, the radiation of heat from the ground and the lower air will go on with greater rapidity, particularly at night, than in surrounding regions. As a result, the air here will become particularly cold, dense, and heavy. It will contract somewhat and air will come in aloft to take its place. As a result, an area of high pressure has formed.

Highs may be caused by radiation.

314. Bigelow's counter current theory may also be used to account for highs. If at the height of a mile or so there are two counter currents, one on the right, cold and dry, flowing equatorward and one on the left, warm and moist, flowing

Bigelow's counter current theory of the origin of highs.

poleward, there will exist of necessity at the earth's surface an area of high pressure with spirally outflowing winds turning clockwise in the northern hemisphere.   When the structure of a high above the earth's surface was being considered, it was found that it too degenerated from an atmospheric whirl at the earth's surface to two counter currents at an elevation of a little more than a mile.   A high, then, can be considered simply as the surface effect of two powerful counter currents in the upper air.

Four theories as to the origin of a low have been given and four for the origin of a high.   These theories can be better contrasted and compared after the methods of growth, development, and maintenance of both highs and lows have been considered.

315. **Growth of an extratropical cyclone.** — The crucial question in connection with the growth or continued existence of an extratropical cyclone is the source of the energy.   A full-grown extra- How is the tropical cyclone has an activity of millions of horse power energy and does work at this rate for days or perhaps weeks.   This gained? immense amount of energy must be gained from some source, and the method by which this energy is gained will tell the story of the growth and development of an extratropical cyclone.   There are three possible sources of this energy : the latent heat liberated by condensation of The three moisture to form cloud and precipitation ; the energy of sources of motion of the general wind system of the world ; the relative energy. displacement of masses of cold and warm air.

If the energy is gained from the condensation of moisture, then whatever may have been the origin of a low, it will build itself up in accordance with the convectional theory and will be essentially a How the convectional formation.   The method of growth has been energy from treated in full in connection with the tropical cyclone.   A the latent heat may small area of low pressure has originated with rising air build up a currents.   The rising air cools, reaches the dew point, low. forms cloud and precipitation, and liberates an immense amount of latent heat.   This heats the rising air, maintains it at a higher temperature than its surroundings, causes further rise, and thus acts like forced draft in a chimney.   A more violent indraft at the earth's surface now takes place.   It becomes a whirl due to the earth's rotation. Centrifugal force is developed and the central pressure becomes lower. The formation has grown, due to the latent heat, and acquired more energy.   It builds up and continues until the supply of energy lessens, when it gradually dies out.   There have been violent lows with very little cloud and no precipitation. It would thus seem that they must have some other source of energy. Furthermore lows are most violent in winter, when the temperatures are lowest and the air is dryest and thus the supply of available energy is least. It would thus seem that the latent heat due to condensation cannot be the only source of energy for lows.

When the eddy theory for the origin of lows was being worked out, it was thought that these eddies might gain their energy from the general

The energy might come from the general wind system. winds of the world, just as water eddies gain their energy from the currents in which they are formed. If this were the case, the energy acquired by a low would be taken from the general wind system and the circulation would thereby be hindered and slowed down. It seems incredible, however, that a low five miles thick and 3000 miles wide, thus resembling a piece of paper, and of a size comparable with the great air currents themselves, could receive all its energy from these currents, just as water eddies receive their energy from the currents which cause them.

If the Bigelow counter current theory of lows is accepted, then the

The energy might come from the displacement of warm and cold air. energy comes from the relative displacement of masses of cold and warm air. These two counter currents, due to their temperature difference, produce the area of low pressure and the atmospheric whirl at the earth's surface. The more vigorous these currents, the greater the temperature differences and thus the more vigorous the effect, namely, the low, which they produce at the earth's surface.

There is nothing to prevent all three of these sources of energy from

All three sources of energy might contribute at the same time. contributing to the growth of a low at the same time and such is probably the case. Careful observations and many of them, particularly in connection with the beginning of lows and at various levels in the atmosphere, would permit the relative importance of these three sources of energy to be determined. The last one is sufficient in itself to account for lows, the first two are not.

316. **Growth of an anticyclone.** — In connection with the growth of

There are three sources of energy for the growth of highs. an anticyclone or high, the source of the energy is again the chief thing to be considered. If a high is due to the accumulation of air from lows or to radiation, it is gravity which supplies the energy; and either of these processes which originated the high may continue indefinitely and become more vigorous and thus account for the growth or continuance of the high.

If highs are due to the congestion of poleward moving air currents, then the energy comes from the general wind system, and this process can continue an indefinite time.

If highs are due to counter currents in the upper air, then the source of energy is the relative displacement of masses of warm and cold air. The high at the earth's surface is simply the effect of these currents, and the more vigorous the currents the more vigorous the high.

There is nothing to prevent all four of these processes from working simultaneously to build up and develop a high. Sometimes it would seem that one or the other of the processes stands out conspicuously as the origin of a particular high, but it is impossible to say that the others are not present. Any one of the four is sufficient by itself to produce a high.

317. **The characteristics which lows and highs should have and the comparison with the observed facts.** — The various theories as to the origin of lows and highs and the sources of the energy which causes their growth and development have been fully considered. The general features of both the lows and highs have been accounted for. There now remain many details in connection with the distribution of the meteorological elements about typical lows and highs, their paths, and velocity of motion in connection with which the demands of theory on the one hand and observed facts on the other must be noted and a comparison made.

The details in connection with lows and highs must now be explained.

On the east side of an area of low pressure the wind is from some southerly quarter, and this transports warm, moisture-laden air, raising the temperature and humidity. On the west side, the wind direction is from some northerly quarter and here the air is cold and dry. The great difference in temperature and moisture between the two sides of a low is thus due to the wind direction. The oval form of the isobars and the direction of the oval is due to the temperature difference between the two sides. With increasing altitude this northeast-southwest oval first becomes circular in form, then again oval, but with the longer axis northwest-southeast. All this change in form is due simply to the temperature differences on the two sides. The cloud and rain area is located chiefly in the southeast quadrant because here the temperature and moisture have their largest values. The cirrus cloud is formed only on the east side because the air which rises in a low is injected into the rapidly moving westerly air currents of the outer layer of the atmosphere. This air, still containing

The characteristics of a low are explained.

considerable moisture, is carried rapidly eastward, but makes very little headway towards the west against the air current.

In connection with an area of high pressure, the cloud forms on the east side are the last reminders of a departing low, while the cloud forms on the west side are the first heralds of a coming low. Only convection-caused cumulus clouds or radiation clouds in the early morning are found in the central part of a typical high. The northeast portion of a high is the coldest and the southwest the warmest, due to the transportation of air by the wind.

The characteristics of a high are explained.

Lows and highs move in general eastward. This is because both the surface winds and the upper air currents move from a little south of west towards the east. In other words lows and highs drift in the general wind system. If a small area is considered, as for example the United States or Europe, the path followed by any individual low or high is determined by four factors: (1) the topography of the country, (2) the meteorological condition of the country, (3) the surrounding highs and lows, (4) the distribution of the meteorological elements about the individual low or high itself. These four factors and the motion of individual lows and highs will be considered more fully in the chapter on weather prediction. Since these four factors are for the most part constant or subject to periodic variations, it is to be expected that lows and highs will follow more or less exactly a rather definite system of tracks.

The direction and velocity of motion of lows and highs are explained.

Lows and highs move faster in winter than in summer. This is because the general winds of the world, which carry the lows and highs, move faster in winter than in summer.

THE EFFECT OF PROGRESSION ON THE DIRECTION AND VELOCITY OF CYCLONIC WINDS

318. The conception which is often held of a low (or a high) is that the air which constitutes the formation moves bodily forward over the earth's surface. A low moves usually eastward and slightly poleward. In the southern quadrants of a low the air motion with respect to the center is from the southwest to the northeast. Here, then, the motion of progression and the motion of revolution about the center agree and the velocities would be added. In the northern quadrants of a low the air motion with respect to the center is from the northeast to the southwest. Here the motion of progression and the

If the air accompanied a low, the wind velocity on the north and south side would be very different.

motion of revolution are opposed and the velocities would be subtracted. To make the picture more definite, suppose that a low is moving towards the northeast with a velocity of thirty miles an hour. Suppose also that the velocity of revolution is thirty miles an hour. If the air moved with the formation, there would be then a wind velocity of sixty miles an hour in the southern quadrants and a dead calm in the northern quadrants. Now, no such difference in wind velocity as this exists between the northern and southern quadrants.

The truth of the matter is that the low, the formation, goes forward, but the air which constitutes it at any moment does not. A low takes in air at the front, moves it a short distance, and abandons it. The low as a formation moves forward, but the air which constitutes it at different stages in its journey is entirely different. *The formation moves; the air does not.*

This is exactly analogous to the progress of an ocean wave in deep water. Each mass of water describes a little oval path and comes back nearly to where it was before, but the wave has gone on. The wave as a formation has gone on, but the water which constitutes it is entirely different. *A wave analogy.*

Over the ocean, in connection with tropical cyclones, and at considerable altitudes, the velocity of the wind is found to be greater in the southern quadrants of lows than in the northern. This means that in these cases there is a slight carrying forward of the air which constitutes the low. Over the land, however, and near the earth's surface the friction is too great to permit the actual transportation of air by the low.

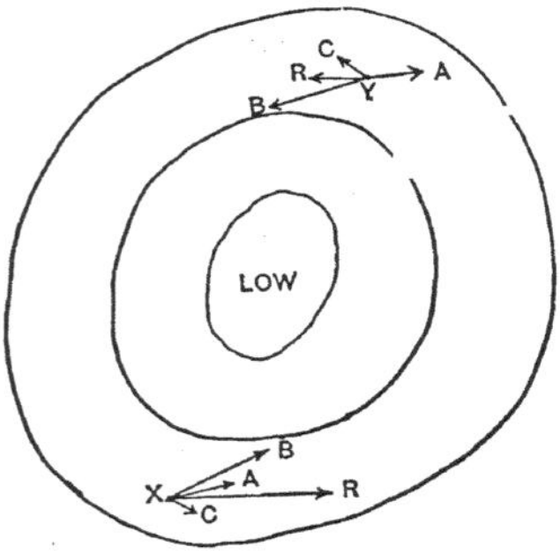

319. If a low progresses as a formation, it now remains to consider what actually determines the direction of motion of the air and its velocity at any point when a low passes. Let us consider, for the sake of definiteness, the two points X and Y in Fig. 129, at the earth's surface *The factors which determine the direction of the wind and its velocity at any point.*

FIG. 129. — Diagram Illustrating the Determination of the Wind Direction and Velocity at any point near a Passing Low.

on opposite sides of a passing low in the United States. These points are located in the region of the prevailing westerly winds and thus at each point there would be the component A, which, in direction and length, is to represent these prevailing westerly winds in which the low drifts. Of course, this component marked A is really the combination of two things: the exchange of air between equator and pole and the deviation due to the earth's rotation. If the earth did not rotate, the surface wind direction in extratropical regions in the northern hemisphere would be south. This is deviated to the right by the earth's rotation and becomes the prevailing westerly winds with a direction from a little south of west. Since a low is near, there is a pressure gradient towards the center, and the air tends to move along the gradient at right angles to the isobaric lines, but it is deviated to the right by the earth's rotation. This cyclonic component is indicated in the figure by B in each case, and is the combination of two things: the pressure gradient towards the center, and the deviation due to the earth's rotation. On account of friction, the air has its greatest velocity of motion about a low, not at the earth's surface, but some distance above, where the isobars are nearly circular and the direction of the air motion is nearly tangential to the isobars. Due to fluid friction this more rapid motion of the upper air would drag around the surface air to a slight extent with it. If, for example, an iron ring were suspended in a tank of water and made to revolve rapidly, the water would have a certain amount of motion imparted to it through friction. In the same way the rapidly revolving ring of air above the earth's surface will impart a certain amount of motion to the more slowly moving air above and below. This component is indicated by C in each case. If these three components are summed up, the resultant R in each case is obtained. It will be seen that the velocity is larger at X than at Y and that the direction is more nearly tangential to the isobars at X than at Y. If the distribution of the meteorological elements about a low is considered, it will be seen that all this is in exact accord with the results of observation.

If other points at different levels in connection with lows and highs are considered, two more components must at times be added. In the upper part of a low the rising air is forced out and injected into the surrounding air. There is here a component directed directly from the center. In the case of the upper part of a high, air is drawn in to take the place of the descending air current. Here there is a component directed directly in towards the center. The other component to be considered is due to the forward motion of the low or high. If no air were carried along with the formation, this component would not exist, but as a small amount particularly at higher levels and over the ocean is carried along with the formation, there is this last component which has the direction of the low or high.

If it is desired to determine the wind direction and velocity at any level and at any point in connection with either a low or a high, it is only necessary to determine these five components in direction and relative magnitude and find their resultant.

THE CORRELATION OF THE METEOROLOGICAL ELEMENTS

320. Suppose, for example, the wind direction and the fact as to whether precipitation was falling or not have been recorded for the same hour each day for many years. A table of statistics can be prepared which would show for each wind direction the number of times precipitation had occurred with that direction. It would be found for the northeastern part of the United States that precipitation occurred with the greatest frequency when the wind was southeast. East, south, and southwest would also stand high. It would also be found that precipitation occurred the least number of times when the wind direction was northwest. West, north, and northeast would also stand low. There is thus a decided connection or correlation between the wind direction and the frequency of precipitation. The nature of the correlation is determined by summarizing statistics.

The meteorological elements are all correlated.

The correlation of any observation of any element with any other can be worked out. Many of these correlations are of no practical value or importance. A few of those which are of some interest are the following: maximum temperature and wind direction; minimum temperature and wind direction; pressure and temperature; temperature and precipitation.

The reason why a correlation of the meteorological elements should exist must now be considered. To do this, the question as to what causes our weather must be raised and answered. Our weather consists of two things: the typical weather which would exist if there were no disturbances, and the influence of the passing lows and highs. Now, the typical weather which would exist if there were no disturbances is indicated by the normal values of all the meteorological elements together with their diurnal and annual variations. Lows and highs are definite, well-known formations moving along well-recognized tracks. A correlation of the meteorological

The reason for the correlation.

elements therefore exists because the weather is not haphazard or a matter of chance, but is the resultant of two factors, both of which are definite, well-known, and nearly always the same.

C. THUNDERSHOWERS

DEFINITION AND DESCRIPTION

321. Definition and chief characteristics. — A shower is usually defined as a copious rainfall of short duration. The intermittent down-

Shower and thunder- shower de- fined.

pours which occur during certain types of spring weather are, for example, spoken of as April showers. If a shower is accompanied by thunder and lightning, it is considered a thundershower or thunderstorm. Some would reserve the word thunderstorm for a more than usually violent thundershower, but the distinction is seldom made. The presence of thunder and light-

Thunder and light- ning accom- pany many occurrences.

ning while it is raining does not necessarily mean that a thundershower is in progress, for all violent rains which fall from thick clouds are accompanied by thunder and light- ning. Thunder and lightning accompany tropical cyclonesl tornadoes, desert whirlwinds, and volcanic eruptions as well as thundershowers. It was formerly thought that thunder and light- ning were the essential things in connection with a thundershower;

Thunder and light- ning a result, not a cause.

in short, they were considered the cause of the storm. It is now known, however, that they are extremely secondary and play a very unimportant part. In fact, they are a result and not a cause. They are the inevitable consequence and accompaniment of copious condensation in a thick cloud, and play no part whatever in the mechanism of a thundershower.

A thundershower is a too well-known phenomenon to need a careful definition and a full description. It can be best defined and described

Outline de- scription of a thunder- shower.

by stating its chief characteristics. A thundershower is an immense cumulo-nimbus cloud accompanied by copious precipitation, a marked drop in temperature, and a peculiar, often violent, outrushing, squall wind which just precedes the rainfall. Thunder and lightning are always present and hail some- times falls. It is a violent, local storm, covering a comparatively small area and lasting but a short time. Damage is often caused by wind, hail, and lightning.

322. Description of the approach and passage of a thundershower. — Thundershowers occur everywhere from the equator to the pole and have thus been more generally observed than any other storm. It is the typical thundershower on a hot summer afternoon whose approach and passage is here described in outline.

It has been a hot, sultry, oppressive day in summer. The air has been very quiet, perhaps alarmingly quiet, interrupted now and then by a gentle breeze from the south. The pressure has been gradually growing less. The sky is hazy; cirrus clouds are visible; here and there they thicken to cirro-stratus or cirro-cumulus. The temperature has risen very high and the absolute humidity is very large, but owing to the high temperature, the relative humidity has decreased some- what. The combination of high moisture and temperature and but little wind has made the day intensely sultry and oppressive. In the early hours of the afternoon, amid the horizon haze and cirro-stratus clouds in the west, the big cumulus clouds, the thunderheads, appear. Soon distant thunder is heard, the lightning flashes are visible, and the dark rain cloud beneath comes into view. As the thundershower approaches, the wind dies down or becomes a gentle breeze blowing directly towards the storm. The temperature perhaps drops a little as the sun is obscured by the clouds, but the sultriness and oppressive- ness remain as before. The thundershower comes nearer, and the big cumulus clouds with sharp outlines rise like domes and turrets one above the other. Perhaps the loftiest summits are capped with a fleecy, cirrus- like veil which extends out beyond them. If seen from the side, the familiar anvil form of the cloud mass is noticed. Just beneath the thunderheads is the narrow, turbulent, blue-drab squall cloud. The patches of cloud are now falling, now rising, now moving hither and thither as if in the greatest commotion. Beyond the squall cloud is the dark rain cloud, half hidden from view by the curtain of rain. The thunderheads and squall clouds are now just passing overhead. The lightning flashes, the thunder rolls, big, pattering raindrops begin to fall or perhaps, instead of these, damage-causing hailstones. The gentle breeze has changed to the violent outrushing squall wind, blowing directly from the storm, and the temperature is dropping as if by magic. Soon the rain descends in torrents, shutting out everything from view. After a time, the wind dies down but continues from the west or northwest; the rain decreases in intensity; the lightning flashes follow each other at longer intervals. An hour or two has passed; it is growing lighter in the west; the wind has died down; the rain has almost stopped. Soon the rain ceases entirely; the clouds break through and become

Description of the approach and passage of a typical thunder shower.

fracto-stratus or cirriform; the temperature rises somewhat, but it is still cool and pleasant; the wind has become very light and has shifted back to the southwest or south. Now the domes and turrets of the retreating shower are visible in the east; perhaps a rainbow spans the sky; the roll of the thunder becomes more distant; the storm has passed, and all nature is refreshed.

323. **Distribution of the meteorological elements about a thundershower.** — The distribution of the meteorological elements about a thundershower is best shown by means of the accompanying diagram, Fig. 130, which depicts the changes in the elements during a hot summer day with a typical thundershower between three and five in the afternoon.

The changes in the meteorological elements which occur on a hot summer day with a typical thundershower in the afternoon.

The temperature rises unduly high during the day, and reaches a maximum just before the coming of the storm. It drops slightly when the sun is obscured by the coming clouds, but the large drop in temperature, which may amount to from 6° to 20° F., occurs during the first twenty minutes of the thundershower. The temperature then continues to drop slowly until the end of the thundershower is reached. After the storm passes, the temperature usually rises again, but does not begin to attain the height reached just before its coming.

The pressure usually sags somewhat during the day. When the squall wind begins to blow, there is a sudden increase in pressure of six or seven hundredths of an inch and the pressure oscillates up and down slightly all through the storm. At its end, the pressure is generally a little higher than just before the beginning. The pressure sometimes begins to fall after the shower has passed, but it generally rises, particularly if the shower has been a heavy one and no more are following in quick succession. These oscillations of pressure are easily traceable in the indications of a barograph and from an interesting record of a thundershower.[1]

The wind is light and from the south during the day. As the thundershower approaches, it shifts to the east and blows gently directly towards the approaching storm. This is suddenly replaced by the violent, sometimes damage-causing, squall wind which blows from the west or northwest directly from the thundershower. During the thundershower the wind holds its direction but steadily lessens in velocity. After the storm passes, it is light and often shifts back to the southwest or south.

[1] For thundershower barograph curves, see *Monthly Weather Review*, 1898, Vol. XXVI, p. 592.

The relative humidity is high, but drops during the day, due to the great rise in the temperature. When the thundershower commences and the temperature drops, it rises rapidly and attains a value of 85 or 90 per cent. After the storm passes it drops again slowly, due to the

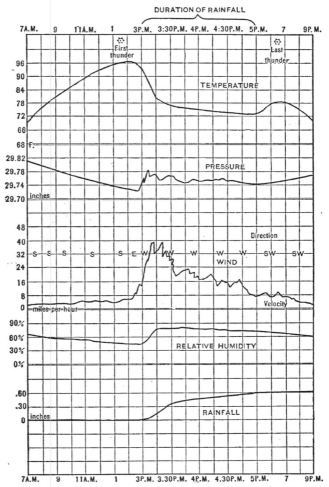

Fig. 130. — The Change in the Meteorological Elements during a Hot Summer Day with a Typical Thundershower in the Afternoon.

rise in temperature.   Haze and cirriform clouds are prevalent during the day.   As the storm passes, the thunderheads, squall cloud, and nimbus clouds follow in succession, the whole being called a cumulo-nimbus cloud.   After the thundershower passes, stratiform clouds exist for a while, but the cirriform clouds usually again make their appearance.

The rain starts with a few large pattering drops, then falls in torrents, and then gradually lessens in intensity until the end of the thundershower is reached.

The average duration of a thundershower is a little less than two hours.   The first thunder is ordinarily heard about an hour, or a little more, before the coming of the rain, and the last thunder is heard about the same time after the cessation of the rain.

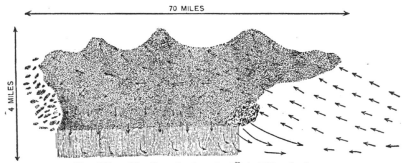

FIG. 131. — The Cross Section of a Typical Thundershower.

It must always be held in mind that the changes in the meteorological elements which have been so fully described and illustrated in the diagram apply to the typical thundershower.   Thundershowers are slightly different in different parts of the world, and the thundershowers which occur in winter and at night are somewhat different from those which occur on a hot summer day.   Each individual thundershower may also differ widely from the type form.   There is no more interesting or profitable study than to watch the progress of a thundershower with the type form always in mind.

324. **Cross section of a thundershower.** — In the cross section of a typical thundershower, Fig. 131, the air circulation and the distribution of the cloud masses are shown.   The rising air which builds the storm, the descending and forward moving air currents underneath the thundershower, and the whirl in the squall cloud with the peculiar outrushing squall wind are

The cross section of a typical thundershower.

all shown. The thunderheads with their cirruslike caps, the immense nimbus cloud, the stratiform clouds at the back, and the turbulent squall cloud are also depicted. The vertical and horizontal scales are not the same because a thundershower has an average length of perhaps seventy miles and an average height of only about four miles. The anvil-like form of the immense cumulo-nimbus cloud is here clearly seen.

Two cumulo-nimbus clouds are represented among the illustrations of the cloud forms in connection with section 217. In one the thunderhead is especially prominent while in the other, which is a side view of a thundershower, the anvil-like form of the cloud is particularly noticeable.

325. The observations of thundershowers. — Many special stations for a certain period of years have been maintained by several states in the United States and by several countries in Europe for the detailed study of thundershowers. Most of the information which we have has been gained from the observations made at such stations. Since a thundershower is such a small *Special stations have been maintained.* local formation, the regular and coöperative stations of the U. S. Weather Bureau are too far apart to make a detailed study of them possible.

If instruments are not used, the observations ordinarily made are: the time of beginning and ending; the direction of motion; the violence of the thunder and lightning; the presence of hail; the intensity of the precipitation; the amount of the drop in temperature; the violence of the wind. If instruments are at hand, the observations include a record of the changes in *The observations which are made.* all the meteorological elements, together with the observation of the time of beginning and ending, direction of motion, thunder and lightning, and hail. In determining the time of beginning and ending in the case of a thundershower, it is the time of occurrence of the first and last thunder that is usually taken. This is better than to use the time of occurrence of the first and last lightning, as the lightning can be seen so much farther at night than during the daytime. In fact, the so-called " heat lightning " usually indicates the presence of a thundershower, for it is probably the reflection from clouds or the hazy sky of lightning which is accompanying a shower which is below the horizon of the observer.

THE REGIONS AND TIME OF OCCURRENCE

326. Geographical distribution. — Thundershowers occur in nearly every part of the world, but the number decreases rapidly from the

Thunder-
showers
occur every-
where.

equator towards the pole. Within the tropics there are many places where there are nearly 200 days in the course of a year with thundershowers. The number of days with thundershowers decreases rapidly with latitude, until in the polar regions but one or two thundershowers in the course of several years may be recorded. Fewer thundershowers occur over the ocean than over the land, and mountainous regions have far more than level country.

In the United States the largest number occurs in the Gulf States, where there are on the average about sixty days in the course of a year with thundershowers. The number decreases both north and west. In New England the average is not much over fifteen. The accompanying table gives the normal number of days with thundershowers for several stations in the United States.[1] These normals are based on the ten years, 1901 to 1910, inclusive.

Statistics as to the number of days with thundershowers or the number of thundershowers are very unreliable and cannot be compared. The

Statistics
are unre-
liable.

reason is because there is no uniformity among observers as to what constitutes the presence of a thundershower. Some count only the presence of rain, others the audibility of thunder or the visibility of lightning or the occurrence even of the so-called heat lightning.

327. Relation to extratropical cyclones and V-shaped depressions. — Nearly all thundershowers which occur in extratropical regions are to

Thunder-
showers
occur in the
southern
quadrants
of lows.

be found in the southern quadrants of a low. The thundershowers which are due to purely local causes, such as a mountain or a peculiarity of the general wind system, are, of course, to be expected. It will be remembered that the great difference between a winter and summer low was the fact that the nimbus cloud area is unusually absent in summer and its place is taken by cirriform clouds, with sultry weather and thundershowers. Chart XXXV, which gives the daily weather map

[1] For the number of thundershower days in the United States during 1903 see *Monthly Weather Review*, 1903, end of volume. For a chart of the normal number of thundershower days in the United States, see: Climatology of the United States, by A. J. HENRY (Bulletin Q of the U. S. Weather Bureau). For the number of thundershowers at Albany, N.Y., each year from 1884 to 1910, see section 28.

NORMAL NUMBERS OF DAYS WITH THUNDERSHOWERS FOR THE VARIOUS MONTHS AND FOR THE YEAR

Station	Jan.	Feb.	Mar.	Apr.	May	June	July	Aug.	Sept.	Oct.	Nov.	Dec.	Annual	Largest Annual	Smallest Annual
Albany, N.Y.	0.0	0.1	0.7	1.3	3.3	6.3	7.5	4.5	2.4	1.2	0.0	0.0	27.3	35	21
Bismarck, N. Dak.	0.0	0.0	0.0	0.9	3.2	7.1	7.2	6.0	2.4	1.1	0.0	0.0	27.9	37	20
Boise, Idaho	0.1	0.1	1.1	1.2	2.7	3.4	2.7	1.9	2.2	0.8	0.1	0.3	16.6	26	7
Boston, Mass.	0.3	0.0	0.5	1.0	1.9	3.0	4.3	3.1	1.9	0.4	0.0	0.0	16.4	23	13
Buffalo, N.Y.	0.4	0.3	1.3	0.8	4.5	4.8	7.7	4.2	2.9	2.1	0.5	0.0	30.7	40	23
Charleston, S.C.	0.8	1.9	2.6	4.6	7.6	10.7	14.9	14.2	6.3	1.0	1.0	0.8	66.4	92	48
Chicago, Ill.	0.6	0.5	2.6	3.2	6.0	7.4	7.5	5.9	4.2	1.6	0.8	0.0	40.3	51	31
Columbus, Ohio	0.3	0.6	2.6	2.7	6.1	9.2	8.8	5.6	3.7	1.1	1.0	0.0	41.8	49	35
Denver, Colo.	0.0	0.2	0.1	0.7	7.4	9.7	11.8	10.9	4.7	0.6	0.0	0.0	46.6	58	39
El Paso, Tex.	0.3	0.6	2.2	0.9	1.0	3.2	7.0	10.1	2.9	1.6	0.7	0.0	28.4	37	20
Galveston, Tex.	0.7	2.4	0.0	4.0	4.7	6.8	8.3	7.3	5.2	2.2	1.8	1.6	44.6	55	35
Havre, Mont.	0.0	0.0	0.3	0.6	2.0	7.5	6.0	5.9	1.9	0.0	0.0	0.0	23.3	34	15
Helena, Mont.	0.2	0.3	2.7	0.9	5.0	8.2	10.6	6.7	2.8	0.4	0.3	0.0	34.7	42	24
Indianapolis, Ind.	0.3	1.4	1.1	3.0	6.1	8.0	7.1	6.1	3.5	1.5	1.2	0.4	40.4	52	32
Key West, Fla.	0.6	0.4	0.7	1.8	4.1		8.5		8.1	2.4	0.7	1.3	48.8	68	30
Los Angeles, Cal.	0.3			0.6	0.2	0.3	0.1	0.5	0.6	0.4	0.1	0.0	4.2	8	1
New Orleans, La.	2.0	2.9	3.6	3.9	6.4	9.7	15.0	13.2	8.8	1.6	1.0	2.3	70.4	77	66
New York, N.Y.	0.1	0.1	0.4	1.8	3.5	5.5	7.8	5.1	2.2	0.7	0.0	0.0	27.2	37	15
Omaha, Neb.	0.1	0.3	1.2	3.8	7.4	11.0	10.1	8.3	5.7	3.1	0.9	0.1	52.0	64	33
Philadelphia, Pa.	0.1	0.6	1.7	2.0	5.0	5.6	8.4	5.8	2.1	1.0	0.2	0.3	32.6	51	18
Phoenix, Ariz.	0.1	0.7	1.5	0.9	1.6	1.1	8.2	11.3	4.0	0.7	0.9	0.3	31.3	48	17
Portland, Me.	0.1	0.1	0.1	0.4	1.4	2.7	4.2	2.3	1.4	0.7	0.0	0.0	13.7	22	7
Portland, Ore.	0.3	0.3	0.1	0.1	1.0	0.6	0.7	0.8	0.4	0.3	0.0	0.0	4.4	13	1
St. Louis, Mo.	0.0	0.9	3.5	4.7	7.6	8.3	8.5	8.0	4.3	2.0	1.0	0.3	49.4	62	26
St. Paul, Minn.		0.0	0.5	1.3	4.8	6.8	5.8	5.9	4.2	1.3	0.1	0.0	30.7	37	15
Salt Lake City, Utah	0.8	0.4	1.8	2.3	3.7	4.7	6.8	9.4	3.6	1.2	0.2	0.1	35.0	50	20
San Francisco, Cal.	0.0	0.3	0.1	0.1	0.0	0.0	0.0	0.2	0.3	0.2	0.1	0.1	1.4	5	0
Seattle, Wash.	0.0	0.2	0.4	0.3	0.6	1.1	0.0	0.4	0.3	0.1	0.1	0.0	4.5	8	0
Topeka, Kan.	0.4	0.4	2.7	3.4	7.9	8.5	9.4	8.4	5.4	2.2	0.9	0.3	49.9	60	39
Washington, D.C.	0.2	0.3	1.2	2.8	5.1	7.0	10.9	6.8	3.7	0.6	0.4	0.1	39.1	48	31

for 8 A.M., Sept. 2, 1904, shows a low with several thundershowers in its southern quadrants. The presence of a thundershower is indicated by ⌐⌐. The weather maps for June 30, 1904; July 14, 1904; March 1, 1907; July 17, 1908; April 30, 1909; May 6, 1909; May 15, 1909, are also particularly good illustrations of this.

Thundershowers also occur in great numbers in connection with V-shaped depressions. Whenever an isobaric line, instead of being **V-shaped** straight or uniformly curved, is bent so as to have the form **depressions.** of a pocket or trough, it is spoken of as a V-shaped depression. Whenever a low has a pronounced well-developed wind shift line, the isobars in the southern quadrants nearly always have this **Thunder-** peculiar V-shaped bulge. This has been fully described and **showers are** illustrated in section 287. If a low with a wind shift line **numerous** **with all V-** and this peculiar V-shaped bulge crosses the country during **shaped de-** the summer, it is nearly always attended by thundershowers **pressions.** particularly along the wind shift line. A V-shaped depression sometimes indicates that a secondary low is forming. These secondary lows are very common in Europe, but not so frequent in this country. They sometimes develop all the characteristics of a regular low and are even more violent. They usually form in the southern quadrants of a low, and their motion is eastward and northward usually with a greater velocity than the parent low. As a result, they seem to circle about it in a counterclockwise direction. Whenever these secondaries occur in summer, they are nearly always attended by thundershowers. Sometimes thundershowers will occur in connection with a V-shaped depression when it is neither a wind shift line or the beginning of a secondary low. Thus whenever a weather map is being studied, all V-shaped depressions should be carefully noted and the characteristics of the meteorological elements about them critically examined.

328. Path across a country. — When an overgrown cumulus cloud first develops into a cumulo-nimbus cloud

FIG. 132. — The **Form of a** and becomes a thundershower, it usually Typical Form **thunder-** covers but a small area. It is perhaps a of a Thunder- **shower.** shower. few miles long and a mile or two wide.

As it moves across the country, it becomes constantly larger, so that at the end of six or seven hours, which is about the average life of a thundershower, it has a front some 150 or 200 miles long and is perhaps 40 miles wide. Its form is that shown in Fig. 132.

The typical path of a shower across the country is thus pear-shaped and is shown in Fig. 133. Much larger individual thundershowers have been noted. In some cases **Form of** they have lasted more than twelve **path.** hours and have covered a path more than 500 miles long and nearly as wide in its widest part. A large thundershower, which traversed Germany August 9, 1881, has been carefully charted by Köppen and is shown in Fig. 134.

FIG. 133. — The Typical Path of a Thundershower.

329. Direction and velocity of motion. — Thundershowers may move in any direction, but in extratropical regions the great majority of them move from west to east. **Direction of** This agrees with the direction of motion of the upper air **motion.** currents and also with the direction of motion of the surface winds in the southern quadrants of a low. Those due to purely local causes, as a mountain, sometimes remain stationary.

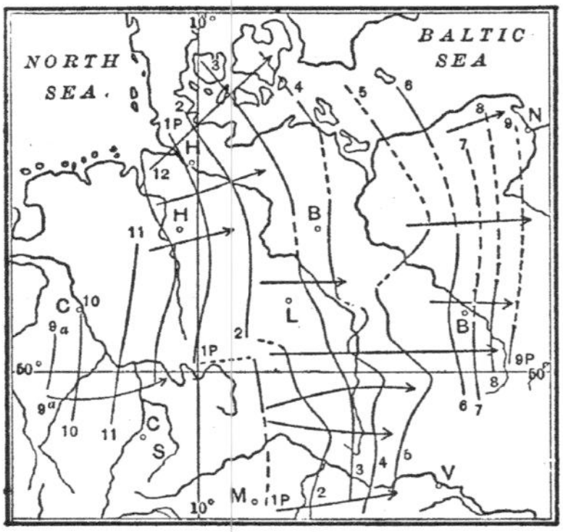

FIG. 134. — The Path of a Large Thundershower across Germany.

The average velocity for the United States is between 30 and 40 miles per hour; in **Velocity of** Europe, it is between **motion.** 20 and 30 miles. The velocity of motion is greater over the ocean than on land and is greater in winter and at night than in summer and during the daytime.

Individual showers often depart widely from the type as regards direction and velocity of motion.

330. Time of day and season of occurrence. — Thundershowers may occur any hour of the day or night, and any month in the year. The great majority, however, occur during the warmest **Time of day** months in the year, June, July, August, and during the **and season** hottest part of the day, 3 to 5 P.M. **of occurrence.**

Most stations in the United States have a pronounced maximum between 2 and 6 in the afternoon, and a small secondary maximum between 3 and 5 in the early morning. The forenoon and the hours near midnight show the smallest number.

Over the ocean more occur during the night than during the day, and the same is true of Iceland and some coast stations.

331. Periodicity of thundershowers. — In addition to the prominent well-marked diurnal and annual periodicity in the occurrence of thundershowers, it has been thought that other faintly marked periods

Four possible insignificant periods in the occurrence of thundershowers.

are also present. Some think that the moon influences the occurrence of thundershowers; the number being slightly larger during new moon and first quarter than during full moon and last quarter. Some think that there is a tidal influence, the number being greater during the high tide than during low tide. Others have thought that there was a 26-day period corresponding to the time of revolution of the sun and an 11-year period corresponding to the sun spot cycle. If there were uniform observations at many stations covering, say 100 years, it would be possible to answer the questions as to the existence of these periods at once. As it is, the observations are far from uniform and the records are often short and fragmentary. As a result, it seems impossible at present to be sure whether these periods exist or not. They are certainly not large or well marked.

THE ORIGIN AND GROWTH OF A THUNDERSHOWER

332. Three classes of thundershowers. — All thundershowers cannot be ascribed to the same cause, and for this reason, in studying the

The three classes of thundershowers.

origin of thundershowers, it is more convenient to divide them into three classes; namely, heat or convection thundershowers, cyclonic thundershowers, thundershowers due to local conditions. Each of these three classes will be considered separately. All thundershowers have this in common:

All due to the rising of air.

they are caused by moist rising air. This moist rising air cools due to expansion, reaches its dew point, and builds the immense cumulo-nimbus cloud with its copious precipitation. As a result of this copious condensation in a thick cloud, lightning and thunder occur.

333. Heat- or convection-caused thundershowers. — Heat- or convec-

The origin and development of a convection-caused thundershower.

tion-caused thundershowers have their origin in masses of warm, moist air. This air rises, due to convection, cools by reason of expansion, reaches its dew point, and builds first a cumulus cloud. If the mass of rising air is not too large, or if it rises in different places in smaller amounts, only cumulus clouds are the result. If the mass of rising air is large, the cumulus cloud becomes overgrówn and the condensation of moisture is sufficiently vigorous to cause precipitation. The cumulus cloud has now become a cumulo-nimbus cloud and underneath it a descending air current forms. There are three causes of this descending air current. In the first place, the falling raindrops carry down air with them. In the second place, the air underneath the cloud becomes cooler and heavier because of the cold raindrops which pass through it and also because it is shielded from insolation by the cloud. In the third place, there is a certain amount of reaction against the rising air, for a large amount of air is injected into the upper atmosphere and this must cause an outward movement in all directions. Thus, underneath the cumulo-nimbus cloud, there is a descending air current which is brushed forward when it reaches the earth's surface, both as a result of the forward movement of the shower and to take the place of the air which is rising due to convection. In between this rising moist warm air in front of the thundershower and this descending air current underneath the cloud mass, a vigorous eddy or whirl about a horizontal axis forms. This is seen in the turbulent squall cloud and the peculiar violent outrushing squall wind is a part of it. *The explanation of the squall.*

It might seem that the squall is very vigorous compared with the gentle air currents which build it, but it must be remembered that the squall cloud and wind are very small compared with the immense amount of rising and descending air. Furthermore, the squall cloud is near the axis of the whirl. All these air motions are shown well in Fig. 134. The air motion in a thundershower is very similar to what would be produced by placing a card edgewise on a table, inclining it backward so as to make an angle of about 60° with the table, and then moving it slowly forward. The air would rise over the card, descend back of it to fill the vacant space as it was moved forward, and an eddy about the upper horizontal edge of the card would also form.

The origin and growth of a heat- or convection-caused thundershower have now been fully stated and the reason for the structure of the storm and the changes in the meteorological elements during its approach and passage should be apparent. The barometric pressure rises as soon as the squall wind arrives. The reason for this is the downward movement of the air masses underneath a thundershower, and also the fact that the air beneath the cloud mass is cooler and thus denser and heavier. *The explanation of the changes in the meteorological elements.*

There are three reasons why the temperature drops rapidly when the squall wind comes. Underneath the cloud mass there are cold descending air currents. Furthermore the air is cooled by the colder raindrops which fall through it, and it is shielded from insolation by the cloud mass. The reasons for the change in the wind, moisture, cloud, and precipitation are too evident to need explanation.

This class of thundershowers ought to occur in the greatest number and with the most vigor when large masses of warm moist air are most

Time and place of occurrence. numerous. This would be during the hottest time of year and during the hottest time of day. Furthermore, it would be when a place was located in the southern quadrants of a low, for it is then that the southerly winds are transporting the largest amount of warm, moist air. It will be seen that all these requirements of theory are in good accord with the observed facts.

There are two or three observed facts in connection with thundershowers which have been observed so often by professional meteorologists and others as well that an explanation of them should

Rivers hinder thundershowers. be given. One is the fact that a thundershower often seems to be unable to cross a large river. It advances to one bank, remains stationary, and perhaps weakens and disappears or builds sidewise along the river. If it crosses the river, it is practically a new shower which builds on the other bank. This has been observed too often to question the fact. It must be that the river is colder than the surrounding country and is thus the seat of a gently descending air current, which, however, has sufficient vigor to prevent the convectional rise of air which is the condition of life and advance on the part of the thundershower. In winter, when the river is warmer than the land, this hindrance to the advance of thundershowers should not exist. It also should be much less at night.

Another fact, which has been often observed, is that it rains harder after each lightning flash. If this is true, it may be that the small rain-

It rains harder after a lightning flash. drops in the cloud are all charged with the same kind of electricity and kept from uniting by electrical repulsion. As soon as they are discharged by the lightning flash, they coalesce much more readily and build the larger drops which soon fall to the earth's surface. The interval of time between the lightning flash and the increase in the rainfall ought to give the time required for the raindrop to fall. It may also be that, for some reason, the small droplets suddenly unite to form large ones, and thus, as a result, the lightning flash occurs. In one case, the lightning flash is the cause; in the other, the result. Interesting observations in this connection could be made.

334. Convection can be caused by cooling the top of a layer of air as well as by heating the bottom. Thus thundershowers might be caused by cooling the upper layers of the atmosphere as well as by heating unduly the surface layer. This probably often occurs. Excessive radiation during the night from the upper layers of the atmosphere, particularly if there is a thin cloud layer, may cool the upper air sufficiently to cause unstable equilibrium and thus convection and a thundershower. These thundershowers would occur chiefly in the early morning when the air was coldest. The secondary maximum in the frequency of thundershowers in the early morning which has been observed at most stations in the United States is probably due to thundershowers formed in this way.

Another method of beginning of convection-caused thundershowers.

335. **Hail.** — Hail falls wherever thundershowers occur. It is estimated that from one half to one tenth of all thundershowers are accompanied by hail. Hail practically never occurs except during the hottest part of the year and the hottest time of day. In this respect, it shows a much more marked annual and daily periodicity than do thundershowers. Hail never falls except during the beginning of a thundershower. The area covered by a fall of hail is very much smaller than the area covered by the thundershower which accompanies it. It is usually not more than 6 or 7 miles wide and perhaps 40 or 50 miles long. Thundershowers which are accompanied by hail usually have a very well-developed squall cloud and are more than usually violent. The structure and characteristics of this particular kind of hail have already been fully stated in section 243.

The characteristics of a fall of hail.

The hailstones are formed in the whirling squall cloud of a thundershower. The nucleus is carried up and coated with snow; it then falls or is carried down, and is coated with water; it is then carried up again; the process continues, adding coat after coat, until the hailstone becomes too heavy to be longer sustained and it falls to the ground. It will be seen at once that all of the facts concerning the structure of a hailstone and concerning the time and place of occurrence of hail are in accord with this explanation.

The formation of the hailstones.

The possibility of the existence of temperatures sufficiently low to cause hail to be formed might be questioned. Suppose a thundershower is 4 miles thick, and that the cloud commences one mile above the earth's surface, and suppose, furthermore, that the temperature is that of the rising air. This air cools 1.6° F. for every 300 feet until the cloud level is reached. Then the rate of cooling

The cause of the low temperature.

will be roughly one half of that until the freezing point is reached and then still less. A rough calculation will show that the top of the cloud ought to be from 75° to 100° F. colder than the temperatures at the surface. Thus, even if surface temperatures are well up towards 90° F., the top of a thundershower must be well below the freezing point and even in the neighborhood of zero and thus composed of snowflakes or ice crystals. There is no difficulty then in finding sufficiently low temperatures for the formation of hail.

336. **Cyclonic thundershowers.** — Cyclonic thundershowers are due to the passing of a low and the coming of a high; they are thus conditioned on the change in the weather control from an area of low pressure to an area of high pressure. If a low goes by north of a station so that its southern quadrants pass over the place in question, warm moist air is brought to the place by the southerly winds which accompany the low. A coming high is heralded by rather brisk, cold, dry jets of air from the northwest.

The transition from one to the other is usually slow and gradual. There are times, however, when it is abrupt, and this is particularly the case when the low has developed a prominent wind shift line in its southern quadrants. The jets of cold dry air from the northwest may overrun or underrun the warm moist air of the low. If the cold dry air overruns the warm moist air, then there is unstable equilibrium and all the conditions for convection are fulfilled, for there is warm air below and cold, dense air above. Convection will take place and, in summer, thundershowers will often form. These are really, in their growth and development, convection thundershowers, but their origin is cyclonic because it is conditioned upon the interaction of a high and a low. If the cold dry air of the coming high underruns the warm moist air of the departing low, this warm moist air will be raised bodily and forced to rise. This forced rise of warm moist air in summer, and even in winter, often causes thundershowers. Thundershowers formed in this way by the interaction of a high and low often have a very long front. It may extend even the whole length of the wind shift line, and in England these showers are called "line showers." Cyclonic thundershowers may occur any hour of the day or night or at any time of year.

337. **Thundershowers due to local conditions.** — Thundershowers due to local conditions are nearly always caused by the presence of a mountain and a sea breeze, or a mountain breeze, or a wind caused by a passing high or

low.   If a strong sea breeze blows against a mountain side near the shore and the air is forced to rise, a permanent cumulus cloud is formed over the mountains.   If the sea breeze is particularly strong or moisture-laden, the cumulus cloud may become overgrown and develop into a thundershower.   Such a thundershower would remain stationary near the mountain and disappear in the late afternoon with the dying down of the sea breeze.   A mountain breeze blowing up a mountain side during the daytime could produce a stationary thundershower in the same way.   Many of these stationary thundershowers over mountain peaks are reported in the Alps and in the regions near the Rocky Mountains.   Air set in motion by a passing high or low and forced to rise by a mountain could also produce a thundershower.   It will be seen in every case that the cause of the thundershower was air forced to rise by a barrier.   A cumulus cloud is first formed, and this may become overgrown and develop into a thundershower.

THUNDER AND LIGHTNING

338.   Since lightning and thunder are only attendant phenomena caused by the copious condensation in a thick cloud and have no part in the mechanism of a thundershower, they will be considered later in Chapter XI in connection with atmospheric electricity.

These will be considered later.

D. TORNADOES

DEFINITION AND DESCRIPTION

339.   **Definition and chief characteristics.** — The tornado is the most diminutive and yet the most violent and destructive of all storms.   It is peculiar to the United States, although in a slightly modified form, it at times occurs in other parts of the world.   The name is derived from the Spanish and refers to the twisting or rotating nature of the storm.   It is always associated with a violent thundershower, which is usually accompanied by hail, a pronounced squall wind, and violent thunder and lightning.   It occurs almost exclusively during the warmer months of the year and during the hottest part of the day.   The term cyclone is still often used popularly, and even in the newspapers, in referring to these storms, but they should be called tornadoes, and the term cyclone should be reserved for the tropical cyclone or the extratropical cyclone.

The chief characteristics of a tornado.

The most distinctive thing about a tornado is the peculiar black funnel-
shaped cloud which extends downward from the heavy cloud masses
above, usually reaches the earth's surface, and causes complete devasta-
tion wherever it touches.   In the United States, nearly a hundred lives
and several million dollars of property are lost annually by tornadoes,
while a single violent one may cause four or five times this amount of
loss.

340. **Description of the approach and passage of a tornado.** — Since
a tornado is always associated with a heavy thundershower, the charac-
The funnel   teristics of the day and the weather changes which precede
cloud.   the coming of a tornado are the same as those which herald
the coming of a violent thundershower on a hot, sultry summer afternoon.   Nearly all observers agree that just before the formation of the
funnel cloud, the clouds have an ominous greenish black appearance,
and seem to rush together and start whirling with great violence.   Then
the black funnel cloud appears, which drops lower and lower until the
surface of the ground is reached.   Here it enlarges slightly, so that some
have described the tornado cloud as having the form of an hourglass.
It usually sways slightly from side to side and often writhes and twists.
Sometimes the funnel cloud jumps a certain strip, only to touch the
ground again farther on.   The whole formation is usually less than
1000 feet in diameter and passes a given point in less than half a minute,
but this is sufficient for complete destruction.

The destruction wrought by a tornado seems to be caused both by
excessive wind velocities and also by an explosive action.   The explo-
Two causes   sive action is probably due to the sudden decrease in the
of the de-   barometric pressure.   The barometric pressure is normally
struction.   between fourteen and fifteen pounds per square inch.   If the
pressure were suddenly reduced one half, it would cause a pressure
of over seven pounds per square inch on the inside surface of all objects
containing air which could not quickly escape.   The destruction wrought
by a tornado is often weird and unusual as well as terrible.

Due to tornadic action, large trees are stripped of their branches,
broken off near the ground, or torn up by the roots; heavy brick and
The charac-   stone buildings are crushed and destroyed as if they were
teristics of   card houses; tin roofs are torn from buildings and carried
the destruc-   many miles through the air; loaded cars and even locomo-
tion.   tives have been blown from the track; heavy iron
girders have been carried over the tops of buildings; iron bridges
have been moved from their foundations.   Straws have been driven

336

Fig. 135. — Two Views of the same Tornado at Goddard, Kansas, May 26, 1903, about 4 P.M. (No barograph disturbance was noticed at Wichita about 18 miles distant.)

FIG. 136. — A Tornado at Oklahoma
City, May 12, 1896.

FIG. 137. — Damage caused by a Tornado at
Rochester, Minn., August 21, 1883.

FIG. 138. — Wreckage of Anchor Hall, Jefferson and Park Avenues,
St. Louis, May 27, 1896.

through boards, laths through trees, and small sticks of timber through iron plate. When the funnel cloud strikes a building, it often seems to explode. The roof is carried up and the side walls fly apart. Chests explode; corks are drawn from empty bottles; chickens are stripped of their feathers. Soot is often seen to rise in quantity from the chimneys of near-by houses and window panes often fly outward.

The noise which accompanies a tornado is tremendous. It has been likened to the noise of a thousand express trains rush- The accom-
ing through tunnels or to the sound produced by thousands panying
of wagons loaded with iron and moving rapidly over an noise.
uneven pavement. The uproar is so great that the crash of individual buildings is seldom heard.

The funnel cloud of a tornado is usually associated with the front of the violent thundershower. But little rain usually falls Relation to
before its coming, and hail usually follows it. The lightning the thunder-
at times is almost incessant so that the funnel cloud has shower.
a reddish, lurid look. After a tornado passes, all the characteristics of a violent thundershower usually appear.

In Figs. 135 and 136 photographs of distant tornadoes are repro-duced, and in Figs. 137 and 138 some of the damage caused by tor-nadoes is pictured.[1]

341. **Distribution of the meteorological elements about a tornado.**— The changes in the meteorological elements brought about by a thundershower have already been considered. Only the The pres-
changes brought about by the funnel cloud, the tornado sure and
proper, will be here stated. The barometric pressure in wind in a
the center of the funnel cloud of a tornado has never been tornado.
determined, as the instruments have always been smashed or exploded. Almost instantaneous drops in pressure of nearly an inch have been observed within a few hundred feet of a tornado, but the drop in the center of the funnel cloud is, without doubt, much larger than this. From its explosive effect, it has been estimated that the pressure per-haps drops to one half its value in the center. The wind velocity has never been measured. It certainly goes well over a hundred miles per hour, and may even reach 500 miles per hour, in violent tornadoes. The direction of the whirl about the center is always counterclockwise, and would thus seem to be determined by the rotation of the earth, or

[1] For additional pictures of tornadoes and the damage caused by them see *Monthly Weather Review*, July, 1899; September, 1905, p. 400; June, 1906, p. 276; June, 1907, p. 258.

z

perhaps the rotation of air about the low in the southern quadrants of which the thundershower and tornado usually form. The only distinctive cloud form is the blue-black funnel cloud which extends down from the cloud masses above to the earth's surface. No noticeable changes in temperature, moisture, or precipitation occur.

342. Observation of tornadoes. — Deliberate, careful observations are not often made during the passage of a tornado, not even if the path The obser- of destruction is several hundred feet from the observer. vations to be Such observations are, however, much needed. They should made. cover three things: (1) the appearance of the clouds, their color and motions, the direction and velocity of motion of the tornado, the characteristics of the thundershower which it accompanies; (2) the changes in the meteorological elements during the whole day, but more particularly during the passage of the thundershower and tornado; (3) the kind and peculiarities of the destruction wrought.

In making full notes of any meteorological occurrence, it is always better to record them in full on the spot and not trust at all to one's memory later.

THE REGIONS AND TIME OF OCCURRENCE

343. Geographical distribution. — Tornadoes occur almost exclusively in the United States, although in a slightly modified form they The place sometimes occur in the other countries where violent thunder- of occur- showers are common. They do not occur with the same rence of frequency in all parts of the United States, but visit chiefly tornadoes. the Mississippi Valley and certain of the southern states. They are practically unknown in the Rocky Mountain region and near the Appalachian system. They are most frequent over level country which is not heavily wooded and are practically unknown in mountainous or forest covered regions. In Fig. 139, the distribution of all recorded tornadoes from 1794 to 1881, as determined by Lieutenant J. F. Finley, is given.[1]

The number of observed tornadoes is increasing each year, but this does not necessarily mean that the number of tornadoes is increasing, The number as the country is becoming more thickly settled and the of tornadoes. occurrence and details of tornadoes are more fully published. During the year 1877 to 1887, on the average, 146 tornadoes occurred annually in the United States.

[1] See C. ABBE, "Tornado Frequency per Unit Area," *Monthly Weather Review*, June, 1897, p. 250.

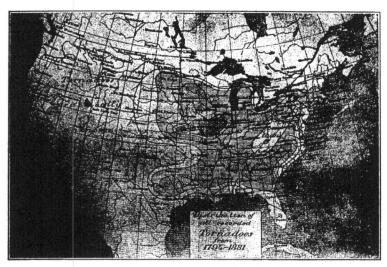

Fig. 139. — The Distribution of all Recorded Tornadoes from 1794 to 1881.
(From GREELEY'S *American Weather*.)

344. Relation to extratropical cyclones. — Tornadoes nearly always occur in the southern or southeastern portion of an extratropical cyclone and from 200 to 800 miles from its center. It will be remembered that it is here that violent thundershowers form in the largest numbers. It is sometimes said that the low which is attended by tornadoes has such definite and peculiar *The type of low which gives rise to tornadoes.* characteristics that the probability of a tornado can almost be predicted from the characteristics of the low. These peculiar characteristics are the following: The isobars are distinctly oval, and extend exactly north and south. The wind shift from south to northwest or west is very sharp, and often a V-shaped bulge in the isobars in the southern quadrants has developed. The temperature lines are packed very closely together along this wind shift line, and the bend in them is sometimes so sharp that they bend backward on themselves. The temperature difference between the east and west portion of the low is large and well-marked.

Charts **XXXVI** and **XXXVII** which give the daily weather map for 3 P.M. March 11, 1884 and 8 A.M. April 25, 1906, represent typical tornado lows. In the first case, there were *Tornado prediction.*

several distinct tornadoes in the southeast portion, and in the second case a tornado occurred in Texas. Unfortunately these characteristics are not always present when a low is attended by tornadoes. In fact, a tornado can occur when none of them are present. A careful study of, say, a hundred lows which have been attended by tornadoes will always leave the impression that it is impossible to do any prediction of the occurrence of a tornado from the characteristics of the low.[1]

345. Path across a country. — The path of a tornado is from a few feet to perhaps 2000 feet wide, and from a mile to sometimes 200 or 300 miles long. The direction of motion is easterly or south-easterly and the velocity of motion from 20 to 50 miles per hour. As a result a tornado passes a given point in less than half a minute.

The path.

346. Time of day and season of occurrence. — Tornadoes are most frequent from 3 to 5 in the afternoon, and least frequent from 7 to 9 in the morning. The diurnal variation is very large and well-marked. Of those occurring in the evening, most have originated during the afternoon and have continued their existence into the evening.

The time and season of occurrence.

Tornadoes occur chiefly during May, June, July, and August, but, particularly in the Southern states, they are common at all times of the year.

THE ORIGIN AND GROWTH OF A TORNADO

347. The origin of tornadoes. — The method of formation of the various kinds of thundershowers has already been fully treated. In considering the origin of a tornado it is only necessary to explain why the tornado funnel cloud forms underneath certain showers. The formation of a tornado is usually ascribed to one of two causes: violent local convection at a given point or the existence of an energetic eddy.

If a thundershower is viewed from the side, it is often noticed that one or two thunderheads rise much higher than the rest. This means a more violent convectional updraft at these points than elsewhere. If, at some point, the convectional updraft were especially energetic, there would develop a well-marked indraft at the earth's surface to replace the rising air. This moving air would be deviated to the right, due to the earth's rotation, and a vigorous atmospheric whirl would build itself up. The growth and

Convectional origin of a tornado.

[1] For a long series of weather maps when tornadoes occurred see, FINLEY, "Tornado Studies for 1884," Prof. Papers of the Signal Service, No. XVI.

development would be almost exactly the same as in the case of the tropical cyclone and would soon lead to a fully developed tornado. The direction of revolution would be here determined by the rotation of the earth.

At the level of the lower clouds there are numerous air currents with slightly different directions and very different velocities. These could easily form an energetic eddy which would make itself felt **Tornadoes** at the earth's surface. Since a tornado is so small and the **may be due** path covered so short as compared with the thundershower **to eddies.** which it accompanies, it is not unreasonable to think of it simply as an eddy. It is a well-known principle that when one eddy forms inside another, the small eddy takes the direction of rotation of the larger. Now tornadoes form in the air which is moving counterclockwise about an area of low pressure and thus would rotate in the same direction.

Many more exact observations of tornadoes are necessary before every step in the origin and development of a tornado can be fully stated.

348. Explanation of the facts of observation. — The theories as to the origin of a tornado account for the existence of the whirl, its direction of rotation, and the high velocity. There are several other **Explanation** facts of observation which need explanation. The rapid **of various** whirl causes centrifugal force, and this is the cause of the **facts of observation,** very low barometric pressure at the center, which in turn is **particularly** the cause of the tornado funnel cloud. The air has been **the funnel** relieved of perhaps one half of the pressure upon it; it has **cloud.** expanded quickly, cooled below its dew point, and produced a cloud extending the whole length of the whirl. The funnel cloud has often been observed to touch the earth's surface for a certain distance, then rise above the earth, leaving a strip without destruction, and then descend again to the earth's surface and continue its work of destruction. This simply means that the whirl weakened for a certain distance. The weakening of the whirl accounts at once for the absence of destruction. It also would mean less centrifugal force, a higher barometric pressure at the center, and thus less cooling due to expansion, and perhaps absence of the cloud as the dew point would not be reached.

PROTECTION FROM TORNADOES

349. The subject of protection from tornadoes can be treated from two entirely different points of view; first, the precautions to be taken in order to secure safety if one is overtaken by a tornado and, secondly, tornado insurance again loss caused by tornadoes.

If it is a frame building, the southwest corner of the cellar is probably the safest place. This is on the assumption that the building will be **The places** carried away by the first blast of wind. If the building is of **of greatest** stone or brick, or if one is caught out of doors, it is better to **safety.** lie down in the open. Do not seek safety under a tree. In places often visited by tornadoes, so-called " cyclone cellars " are sometimes constructed, and these are even provided by the more cautious with tools and certain supplies, in case everything is demolished, or the opening is covered with débris.

Insurance against loss by tornadoes is now provided by several tornado insurance companies. Until recently they were not on a very **Tornado** satisfactory basis, but at present the distribution of tornadoes **insurance.** and the amount of damage caused by them is sufficiently well known to permit the probable risk to be computed.[1]

E. WATERSPOUTS AND WHIRLWINDS

WATERSPOUTS

350. Waterspouts are simply tornadoes which occur over bodies of water. This is abundantly proved by the fact that whenever a tornado **Waterspouts** crosses a river, pond, or body of water, it immediately becomes **are simply** a waterspout; and waterspouts which run inland develop at **tornadoes.** once all the characteristics of tornadoes. Waterspouts are most common in the warmer and calmer seas, although they may occur wherever violent thundershowers are found. When the funnel cloud touches the surface of the water, it is greatly agitated, and the water has been observed to rise even eight or ten feet. This gives an idea of the diminution in the barometric pressure in the center of a waterspout. If it were a vacuum, the water would rise a little more than thirty feet. The spray is, of course, often carried up much higher, but the stories of large quantities of water being carried up from the sea into the clouds is pure myth. When a waterspout crosses a vessel, it has been found that the water in it is fresh. It is thus a condensation product, and has not been carried up from the sea.

Waterspouts have usually been observed from considerable distances

[1] See: Tornado Insurance, by HOWARD E. SIMPSON, Colby College, Waterville, Me., Insurance Monitor, February to September, 1883.

FIG. 140. — The Cottage City Waterspout, August 19, 1896, 1:02 P.M.

and with no great care. The waterspout off Cottage City, Mass., August 19, 1896, occurred under circumstances remarkably advantageous for making observations and photographs, and has been A notable for this reason more carefully studied than any other. It water-has been critically studied and treated by Frank H. Bigelow spout. in four articles in Vol. XXXIV (1906) of the Monthly Weather Review. One of the photographs there given is reproduced as Fig. 140. This waterspout occurred in connection with a thundershower which was due to the overrunning of the surface air by a layer of colder air, as the weather control passed from an area of low to an area of high pressure.

WHIRLWINDS

351. The violent sandstorms and the dusty whirlwinds of desert regions are probably only tornadoes of a slightly modified form and small intensity which are occurring under rather unusual Desert conditions. Over a dry desert region, the amount of mois- whirlwinds. ture in the air is so small that violent convection causes no precipitation or even cloud. The surface layer of air is heated to very high temperatures, as the sky is always cloudless, and energetic convection must often take place. If the sandstorm is a straight blow with rather long front, then it has all the characteristics of a thundershower, except that the cloud and precipitation are lacking. Whenever a dusty whirlwind with a small slender column appears, it is probably a tornado of small intensity, but without a funnel cloud, as there is not moisture enough to form it.

In fact, even the little whirlwinds which spring up along a dusty roadway on a hot summer afternoon and move a short distance and rise to a height of twenty or thirty feet, remind one of a diminutive tornado. Even when carefully considered, as to origin and characteristics, they still have some things in common with tornadoes.

F. CYCLONIC AND LOCAL WINDS

INTRODUCTION

352. In the last part of Chapter IV the general winds of the world were classified and discussed. It remains in connection with this chapter to treat those winds which are due to the passing of Cyclonic cyclonic and anticyclonic disturbances. There are three of winds. these which are well known in the United States. They are: first, the

warm moist south wind often called by its Italian name, the sirocco;
second, the cold dry northwest wind which accompanies the coming
There are of a cold wave, and may under certain circumstances be
three in the considered a blizzard; third, the chinook, which is perhaps
United
States. better known by its Swiss name, the foehn.

THE CYCLONIC AND LOCAL WINDS OF THE UNITED STATES

353. Warm wave (sirocco). — If a low is passing north of a station, warm moisture-laden air from the south is transported to the
Cause of a place in question. If the low moves slowly or if the in-
warm wave. draft is particularly energetic, a decided rise in temperature may take place, and this is spoken of as a warm wave. The passing of a cyclonic disturbance is particularly apt to cause a warm wave
in the states east of the Mississippi River, as the air transported from
the Gulf States is especially warm and moist.

In summer, the warm wave takes the form of several hot, sultry, oppressive days. The temperature and moisture are high, the wind blows
from the south, and the sky is hazy and partly cloud-covered,
The charac- usually with cirriform clouds. Relief comes when the center
teristics of a
warm wave of the low approaches sufficiently near to cause rain, or when
in summer the low passes and the wind shifts to the northwest. In
and winter. winter it takes the form of a thaw, and the sleighing may be
spoiled or the ground freed from snow. Ice storms often result from
the coming of this south wind, since the wind velocity is much greater
at the height of a mile or so above the earth's surface. This transports
warm moist air much more rapidly at this level, so that the upper air may
be much warmer than the air at the earth's surface.

This warm moist wind from the south which causes a warm wave is
often called by its Italian name, the sirocco, but this name is ordinarily
The sirocco. employed to designate a particularly warm moist wind, and
not to designate the south wind in front of any coming low.

354. Cold wave (blizzard). — The meaning of the term cold wave
has been made definite by the U. S. Weather Bureau. It is defined as
Definition of a drop of a certain number of degrees in twenty-four hours
a cold wave. with a minimum below a certain temperature. The amount
of the drop and the minimum are different for different times of year,
Cause of a and for different groups of states. A cold wave is brought
cold wave. about by the cold, dry, northwest wind which marks the
passing of a low and the coming of a high. In the popular mind, the

drop in temperature and the wind are associated together, although technically the cold wave has to do with the drop in temperature only. If it still continues to snow after the wind has gone to the northwest, and the temperature has dropped, or if the high northwest wind drifts the light snow, a blizzard is said to be in progress. The blizzard. There is no exact definition of a blizzard, but its characteristics are supposed to be high northwest wind, driving snow, and low and falling temperature.

The origin of the cold air which constitutes a cold wave has been ascribed to several causes. Some have claimed that the cold air was transported from the far north, and that cold waves thus The origin have their place of beginning in the extreme northwest. of cold Others have claimed that the cold air descends from the waves. upper atmosphere, while still others look for the cause in the continued radiation of heat to the clear sky night after night. That a cold wave is made on the spot, so to speak, seems more plausible than that the air is transported long distances.

355. Chinook (foehn). — The chinook occurs chiefly on the eastern side of the Rocky Mountains, and is particularly common in the states of Wyoming and Montana. It is a hot dry wind coming from the Description. west across the mountains. It usually makes its appearance suddenly, and the temperature may rise even 40° in fifteen minutes. The snow disappears as if by magic, for it evaporates, due to the dryness of the air, as well as melts. It often removes as much snow in a day as ordinary spring thawing would remove in two weeks. If it is common, it raises the normal annual temperature of a place by several degrees.

This wind was first noted and studied in northern Switzerland, where it was called the foehn. It occurs chiefly on the north side of the Alps, although it does appear on the south side as well. In some The foehn valleys, it blows from 30 to 50 days during the months from and its char- November to March and raises the normal annual tempera- acteristics. ture considerably. If the foehn continues for several days, or if it is particularly vigorous, not only does all the snow melt, but everything becomes so dry that special precautions are taken to prevent a general conflagration in case of fire. Sometimes all fires are extinguished while the foehn blows. This wind is also found in Greenland and New Zealand. In fact, it is found wherever a mountain chain and passing lows are associated together. This wind is usually called the chinook in the Western states, although it is better known as the foehn in other parts of the world, and this last name is fast becoming universal.

The cause of the foehn can best be stated in connection with a diagram (Fig. 141). The air is passing over a mountain towards a **The cause** cyclonic center and is forced to rise by the mountain. It **of the foehn.** expands and cools at the rate of 1.6° F. per 300 feet, soon reaches its dew point, becomes cloudy, and yields copious precipitation. The latent heat liberated by the condensation of the water vapor warms the rising air and prevents its cooling at so rapid a rate. In fact the rate is cut down from 1.6° F. per 300 feet to a little less than half this amount. After the top of the mountain is reached, the air begins the descent on the other side. It is now being compressed and grows warm at the rate of 1.6° F. per 300 feet. The precipitation ceases, the clouds disappear, and the air continues to grow warm at the same rate during the whole of the descent. As a result, it may reach the same level on the other side of the mountain 20° or even 40° warmer than before it started the ascent. It has also become very dry, as so much moisture was removed by precipitation. This explanation of the foehn was first given by Hann of Vienna. It was formerly thought that the Swiss foehn was due to hot dry air which came in some way from the desert of Sahara.

FIG. 141. — Diagram Illustrating the Formation of the Foehn Wind.

A chinook should occur on the eastern side of the Rocky Mountains whenever a well-developed cyclonic storm passes across the northern **The relation** part of the United States. It ought to occur on the western **of the** side of the mountains when the storm center passes across the **mountain** southern part of the country. As most cyclonic storms pass **and the low.** along the northern boundary of the United States, the chinook has been observed on the east side much more frequently than on the west. A cyclonic storm, central over Germany, ought to cause a well-marked foehn on the northern side of the Alps, and a foehn on the south side ought to be caused by a low, central over Italy or the Mediterranean Sea. As a matter of fact, the foehn is much better developed and of much more frequent occurrence on the north side of the Alps than on the south side. The reason is because the air coming north from Italy is already warm and dry, and thus lends itself much more readily to the formation of a foehn wind.

The Cyclonic and Local Winds of Other Countries

356. The various cyclonic and local winds which occur in different parts of the world may be divided in four groups according to cause, and each of these groups will be treated in order. *There are four groups*

(I) The first group includes those which correspond to *of these winds.* the warm, moist, south wind which is the cause of the warm waves in the United States, and is the well-known sirocco of Italy.

The solano is a very hot, dusty, southeast wind which *The sirocco* occurs in the Mediterranean, especially on the eastern *and similar* coast of Spain. *winds.*

The leveche is a hot, dry, southwest wind, which also occurs in Spain.

The leste is a very hot parching south wind which occurs in Madeira and northern Africa.

All these winds have the same origin. They are due to the indraft of warm air from the south towards the center of an advancing low. They will be dry or moist, according as the region over which they advance is dry or moist. In the southern hemisphere, the corresponding winds would come from the north.

The brickfielder of southern Australia is a hot, dry, north wind.

The zonda of the Argentine Republic is a hot northerly sirocco.

(II) The second group includes those winds which correspond to the dry, cold, northwest wind which ushers in the cold waves in the United States, and is often a blizzard. *The bliz-*

The buran, or purga, is a very cold northeast wind, which *zard and* occurs in Russia and central Asia. The snow is blown by *similar winds.* the wind, and it is often blizzard-like.

The pampero is a dry, cold, southwest wind which is common in southern Brazil, Argentina, and Uruguay.

The tormentos of Argentina is the same kind of a wind.

All of these winds are due to the cold dry northwest wind (northern hemisphere) which marks the passing of a low and the coming of a high.

(III) The winds of the third group are due to the bringing down of cold dry air from a plateau into a valley by a passing anticyclonic disturbance. Masses of air might be brought down from a *The mistral* plateau into a valley by a passing low as well as by a high, *and similar* but the air would not be cold and dry, as the presence of a *winds.* high is necessary to give the air over the plateau these characteristics. The air is warmed somewhat by compression during the descent, but if the descent is not too rapid, it still reaches the valley colder and dryer than the valley air.

The mistral (Italian: *magistrále* = masterly) is a dry, cold, northwest wind found in the Rhone Valley, and due to masses of air coming into the valley from the plateau in the southeastern part of France.

The bora (Greek: βόρεας = north wind) is a furious northerly wind, coming into the Adriatic Sea from the plateau of Carinthia.

The tramontana is a searching northerly blast found on the Italian side of the Adriatic.

The gregale of Malta is a cold, dry, unhealthy wind, and is due to the same cause.

The williwaus of Terra del Fuego, in Patagonia, is perhaps of the same origin. This is a hurricane-like wind coming down from the mountains and blowing with great violence only 8 or 10 seconds.

(IV) The fourth group includes those winds which get their peculiar characteristics from the topography and condition of the regions from **The khamsin and similar winds.** which the winds come. The wind is, of course, caused to blow from these particular directions, due to the passing of a high or low.

The harmattan is a hot, dusty, east wind, which is found in the west side of the desert of Sahara. It is most common during December, January, and February, and brings much sand with it.

The khamsin (Arabic: khamsûn = fifty) is found in Egypt and is a hot wind from the desert. Its direction is usually from the south or southeast, and it blows from 20 to 50 days in the course of a year.

It will be seen that all of these are cyclonic or local winds, in the sense that they are due to the passing of highs or lows, and in some cases to the characteristics of the surrounding region.

QUESTIONS

(1) State the chief characteristics of the tropical cyclone. (2) What names are applied to them in different parts of the world? (3) Trace the historical rise of our present information about tropical cyclones. (4) Describe the approach and passage of a tropical cyclone. (5) Describe in detail the distribution of the meteorological elements about a tropical cyclone. (6) Illustrate the distribution by means of two diagrams. (7) Give the life history of some tropical cyclone. (8) What are the rules for mariners in connection with tropical cyclones? (9) How is the storm center located? (10) Which is the dangerous half of the tropical cyclone and why? (11) Where are the regions of occurrence of tropical cyclones? (12) Describe the form of the track followed. (13) At what times of year do they occur? (14) State in full the convectional theory of the origin of tropical cyclones. (15) Explain the formation of the calm central eye. (16) Explain the path followed and the time of occurrence of

the tropical cyclone. (17) Why do tropical cyclones occur where they do? (18) Compare the tropical cyclone and the circumpolar whirl in the general wind system. (19) State the chief characteristics of the extratropical cyclone. (20) Describe in detail the distribution of the meteorological elements about a low. (21) How does the distribution differ at different times of a year and in different parts of the world? (22) What are the factors which cause a departure from the type form? (23) Describe the distribution of the meteorological elements about a low with a wind shift line. (24) Illustrate the distribution by means of a diagram. (25) Describe in detail the structure of an extratropical cyclone at various levels above the earth's surface. (26) Contrast the tropical and extratropical cyclone. (27) Describe the approach and passage of a low. (28) Define a high and state the distribution of the meteorological elements about it. (29) Describe the structure of a high at various levels above the earth's surface. (30) Describe the sequence of weather changes as a high approaches and passes. (31) Describe the tracks followed by lows in the northern hemisphere. (32) Where do they originate? (33) Describe the system of tracks across Europe. (34) Describe the three systems of tracks across the United States. (35) Treat the velocity of motion of lows across the United States. (36) Describe the tracks followed and the velocity of motion of highs. (37) Treat in full the theories as to the origin of extratropical cyclones and anticyclones. (38) Describe the methods of growth of an extratropical cyclone. (39) Describe the growth of an anticyclone. (40) In connection with both high and lows explain the peculiarities in the distribution of the elements and the direction and velocity of motion. (41) Does the air accompany a high or low? (42) What factors determine the direction of the wind and its velocity at any point near a low or high? (43) What is meant by the correlation of the meteorological elements and why does one exist? (44) Define a thundershower and state its chief characteristics. (45) Describe the approach and passage of a thundershower. (46) Describe and illustrate by means of a diagram the changes in the meteorological elements as a thundershower passes. (47) Draw a cross section of a thundershower. (48) What observations are made of thundershowers? (49) Where do thundershowers occur? (50) What is the relation of thundershowers to extratropical cyclones? (51) How are they related to V-shaped depressions? (52) Describe the form of a thundershower and the form of its path. (53) What is their direction and velocity of motion? (54) At what time of day and season of the year do they occur? (55) Are thundershowers periodic in their occurrence? (56) State the three classes of thundershowers. (57) Describe in detail the development of a heat- or convection-caused thundershower. (58) Describe the structure and formation of hail. (59) How are cyclonic thundershowers caused? (60) How are thundershowers due to local conditions formed? (61) Define a tornado and state its chief characteristics. (62) Describe the approach and passage of a tornado. (63) State the changes in the meteorological elements as the tornado passes. (64) When and where do tornadoes occur? (65) How are they related to extratropical cyclones? (66) Explain the origin and growth of a tornado. (67) Describe a waterspout and state its cause. (68) State the cause of the violent sand storms and the dusty whirlwinds of desert regions. (69) Name the three cyclonic winds peculiar to the United States. (70) Describe the wind which causes the warm wave and is often called the sirocco. (71) Describe the wind which causes the cold wave and may sometimes be considered a blizzard. (72) What is the chinook wind? (73) Where does the foehn occur and what are its characteristics? (74) State the cause of the foehn wind. (75) Name the four classes of cyclonic or local winds and describe some examples in each class.

TOPICS FOR INVESTIGATION

(1) The complete life history and characteristics of some tropical cyclone.

(2) The calm central eye of a tropical cyclone and the evidence of its existence.

(3) The extent to which a tropical cyclone retains peculiar characteristics after entering extratropical regions.

(4) The difference in the distribution of the meteorological elements about a typical low in Europe and in America.

(5) General laws derived from statistics on highs and lows.

(6) The theories as to the origin of highs and lows.

(7) The complete life history of some thundershower.

(8) The periodicity of thundershowers.

(9) The theories as to the formation of hail.

(10) The size and structure of hailstones.

(11) The complete life history and characteristics of some tornado.

(12) The characteristics of the low which is accompanied by tornadoes.

(13) Tornado insurance.

(14) The complete life history and characteristics of some waterspout.

(15) The characteristics of the blizzard.

PRACTICAL EXERCISES

(1) Note the distribution of the meteorological elements about several highs and lows and in each case explain in full any departure from the type form.

(2) Note the path followed by several highs and lows and in each case explain any departure from the normal path.

(3) From a 10-year file of the daily weather maps work up statistics on some point in connection with highs and lows.

(4) Work out the correlation of the meteorological elements for one or more stations.

(5) Determine from a file of weather maps the location of thundershowers with reference to the lows which they accompany.

(6) Determine for one or more stations the normal number of days with thundershowers for the various months and for the year.

(7) Work up statistics as to the time of day and season of occurrence of thundershowers.

(8) If a thundershower with hail occurs, study critically the hail stones.

REFERENCES

TROPICAL CYCLONES

ALEXANDER, WILLIAM H., Hurricanes. Bulletin 32, U. S. Weather Bureau, 1902.

ALGUÉ, JOSÉ, *The Cyclones of the Far East*, 2d ed., 4°, 283 pp., Manila, 1904.

BERGHOLZ, PAUL, *Orkane des fernen Ostens*, xii + 260 pp., Bremen, 1900.

DAVIS, WILLIAM M., *Whirlwinds, Cyclones, and Tornadoes*, 24°, 90 pp., Boston, 1884.

DOBERCK, W., *The Law of Storms in the Eastern Seas*, 4th ed., 8°, 44 pp., Hongkong, 1904.

ELIOT, SIR JOHN, *Handbook of Cyclonic Storms in the Bay of Bengal for the Use of Sailors*, 2d ed., 8°, 2v., Calcutta, 1900–1901.

FISHER, ALFRED, *Die Hurricanes oder Drehstürme Westindiens*, 4°, 70 pp., Gotha, 1908.
GARRIOTT, E. B., West Indian Hurricanes, 4°, 69 pp., 1900. Bulletin H, U. S. Weather Bureau.
PIDDINGTON, HENRY, *The sailors Horn-book for the Law of Storms*, 6th ed., 8°, 408 pp., London, 1876.

For the life history of various tropical cyclones and other storms see the periodical literature.

EXTRATROPICAL CYCLONES AND ANTICYCLONES

For the theories as to the origin and nature of extratropical cyclones and anti-cyclones, see:

ABBE, CLEVELAND, *Mechanics of the Earth's Atmosphere* (2 vols. of collected papers).
BIGELOW, FRANK H., Monthly Weather Review (various articles from 1902 on).
HILDEBRANDSSON, H. H., ET TEISSERENC DE BORT, *Les bases de la météorologie dynamique*, Paris, 1900–1907.
Report of the chief of the Weather Bureau, 1898–1899. (Report on the international cloud observations.)
STREIT, A., *Das Wesen der Cyclonen*, Wien, 1906.

For the distribution of the meteorological elements about highs and lows, see:
Annals of Harvard College Observatory, Vol. XXX.
HANN, JULIUS, *Lehrbuch der Meteorologie*.
HANZLIK, STANISLAV, *Die räumliche Verteilung der meteorologischen Elemente in den Antizyklonen*, 94 pp., Wien, 1898.
LOCKYER, WILLIAM J. S., *Southern Hemisphere Surface Air Circulation*, 4°, 109 pp., 1910.
Meteorologische Zeitschrift, p. 307, Juli, 1903.
Monthly Weather Review, March, 1907.

For the tracks followed by highs and lows, see:
BIGELOW, FRANK H., Storms, Storm Tracks, and Weather Forecasting, 8°, 87 pp., Washington, 1897. Bulletin 20, U. S. Weather Bureau.
DUNWOODY, H. C., Summary of International meteorological Observations, 1878–1887. Bulletin A, U. S. Weather Bureau.
VAN CLEEF, Monthly Weather Review, Vol. 36, 1908.

THUNDERSHOWERS

CONGER, N. B., Report on the Forecasting of Thunderstorms during the Summer of 1892. Bulletin 9, U. S. Weather Bureau, 1893.
GOCKEL, ALBERT, *Das Gewitter*, 246 pp., Köln, 1905.
PLUMADON, J. R., *Les orages et la grêle*, 8°, 192 pp., Paris, 1901.
TOMLINSON, CHARLES, *The Thunderstorm*, London, 1877.

TORNADOES ·

American Meteorological Journal, August, 1890. (Four prize essays on tornadoes.)
FERREL, WILLIAM, Cyclones, Tornadoes, and Waterspouts. Professional Papers, U. S. Signal Service, No. XII, Washington, ·1882.

FINLEY, J. P., Report of the Tornadoes of May 29 and 30, 1879, in Kansas, Nebraska, Missouri, and Iowa. Professional Papers, U. S. Signal Service, No. IV, 1881.

FINLEY, J. P., Characteristics of six hundred tornadoes. Professional papers, U. S. Signal Service, No. VII, 1884.

FINLEY, J. P., *Tornadoes*, New York, 12°, 196 pp., 1887.

FINLEY, J. P., Tornado studies for 1884. Professional papers, U. S. Signal Service, No. XVI, 1885.

HAZEN, H. A., *The Tornado*, New York, 12°, 143 pp., 1890.

Report of the Chief of the Weather Bureau, 1895–1896. (A Study of the Tornadoes, 1889–1896.)

WATERSPOUTS

BIGELOW, F. H., four articles in the Monthly Weather Review, 1896, Vol. 34.

CHAPTER VII

WEATHER BUREAUS AND THEIR WORK

A Brief History of the U. S. Weather Bureau

357. In 1747 Benjamin Franklin made a very important discovery in connection with storms. He had arranged with his brother in Boston to make some observations of an eclipse of the moon, while he took simultaneous observations in Philadelphia. Shortly before the occurrence of the eclipse, a strong northeast wind set in in Philadelphia, bringing with it clouds and rain, so that the observations could not be secured. Since the wind came from the northeast, he supposed, of course, that the observations had not been made in Boston. Great was his surprise when, several days later, he received word that the observations had been secured in Boston, and that a heavy storm had commenced the following morning. From further observations collected in connection with this storm, and from other observations as well, he came to the conclusion that a storm was a moving formation and that, although the wind usually commenced to blow from the east or northeast, its motion was from some westerly

Benjamin Franklin observes that a storm is a moving formation.

to some easterly quarter. As soon as the fact was fully recognized that a storm was a moving formation, the desire at once arose to keep track of it and herald its coming, but the means of communication were too slow and uncertain to make this practicable. After the invention of the electric telegraph, in 1837, the receiving of simultaneous observations from different parts of the country, and the heralding of storm, was put into partial operation, but this was brought to an end by the breaking out of the Civil War.

The storms on the Great Lakes had attracted attention on account of their severity and the losses caused by them. In 1869 Professor Cleveland Abbe, who was then Director of the Observatory at Cincinnati, was asked by the Board of Trade of that city, to undertake the forecasting of these storms, and the Western Union Telegraph Co. transmitted the messages free of charge.

The founding of the weather service by Abbe.

The work was so successful, and the results so satisfactory that the attention of the whole country was attracted to it. In 1870 a bill was introduced into Congress by Hon. H. E. Paine, from Wisconsin, perhaps at the suggestion of Professor I. A. Lapham of Milwaukee, to appropriate $20,000 to make the weather service national. The bill was passed, and the weather service of the United States was thus inaugurated. It was made part of the Signal Service and placed under the War Department, and thus General Albert J. Myer became its head. The first warning was issued November 1, 1870, and the first daily weather map was issued January 1, 1871. This made the United States the fourth country to issue daily weather maps, as the Netherlands, England, and France had already done so.

The first weather map.

General Myer continued to be the head of the weather service until his death, August 24, 1880. He was a strong man of great executive ability and under him the weather service developed rapidly. He also enjoyed popular approval and confidence. He was succeeded by General William B. Hazen who continued in office until his death, January 16, 1887. His administration was characterized by a fostering of the scientific side rather than by a development of the executive side. He was succeeded by General, then Captain, A. W. Greely, who continued in office until the weather service and signal service were separated. This separation took place July, 1891, and the weather service was then made a separate bureau and placed under the Department of Agriculture. Mark W. Harrington, who was then Professor of Astronomy and Director of the Observa-

The different heads of the weather service.

tory at the University of Michigan, became the first chief of the U. S. Weather Bureau. He was succeeded July 4, 1895, by Willis L. Moore, who is at present the able head of the service.

The Weather Bureau has always been characterized by steady advance, development, and improvement, along both scientific and executive lines. In 1870 there were only 24 stations. In 1893 there were nearly 136 stations, and the annual cost was about $900,000. At present there are nearly 200 stations, and the annual cost is about $1,600,000. It is often stated, even by foreign scientists and weather service officials, that the U. S. Weather Bureau is, in many respects, a model weather service.

The Weather Bureau is characterized by steady advance.

THE PRESENT ORGANIZATION

358. The central station of the U. S. Weather Bureau is, of course, located at Washington, and the rest of the country is divided into districts in two entirely different ways, for two different purposes. In the first place, there are twelve climatological districts, conforming to the twelve principal drainage areas of the United States. This scheme affords the best system of territorial units for the compilation and discussion of climatological data. These twelve districts are: (1) North Atlantic States, (2) South Atlantic and East Gulf States, (3) Ohio Valley, (4) Lake Region, (5) Upper Mississippi Valley, (6) Missouri Valley, (7) Lower Mississippi Valley, (8) Texas and Rio Grande Valley, (9) Colorado Valley, (10) Great Basin, (11) California, (12) Columbia Valley.

The twelve climatological districts.

The country is also divided into six (formerly eight) forecast districts for the purpose of forecasting the weather and predicting storms.

The six forecast districts.

Prior to 1909 there were twenty-one climatic subdivisions for the purpose of preparing climatological data and statistics. The observations were summarized for each of these districts as a whole. These were

New England	Lower Lake	Southern Slope
Middle Atlantic States	Upper Lake	Southern Plateau
South Atlantic States	North Dakota	Middle Plateau
Florida Peninsula	Upper Mississippi Valley	Northern Plateau
East Gulf States	Missouri Valley	North Pacific Coast Region
West Gulf States	Northern Slope	Middle Pacific Coast Region
Ohio Valley and Tennessee	Middle Slope	South Pacific Coast Region

The country was also divided into forty-five sections for the collection and dissemination of weather observations. Each state was usually

The former method of subdivision. a section, but Maine, New Hampshire, Vermont, Massa-chusetts, Connecticut, and Rhode Island were grouped together as New England, and Maryland and Delaware, and Oklahoma and Indian Territory, were considered a single section in each case. Alaska, Hawaii, and Porto Rico were included in the forty-five. Up to 1909, each section, with the exception of Alaska, published a monthly climatological report, and a weather bulletin weekly during the summer. At present the section practically does not exist, although in the supervision of substations and in the collection of observations the section directors still perform their old duties within their respective stations.

Scattered throughout the country are Weather Bureau stations of three different kinds. There are, first, the regular stations of the

The three kinds of weather bureau stations. Weather Bureau, of which there are nearly 200. There are, in addition, nearly 3000 coöperative stations, and about 500 special stations. Observations are also received from many stations in Canada, Mexico, and the West Indies, and from some stations in Europe, Asia, and various islands.

The accompanying map (Fig. 142) shows the 12 climatological districts and the 6 forecast sections into which the country is divided. The centers of the 12 districts are indicated by a • while the centers of the 6 forecast sections are underlined. Only about half of the regular stations are shown. For a full list of the regular stations, see Appendix VII.

359. The chief of the U. S. Weather Bureau has his headquarters at Washington, and closely associated with him in executive and scientific

The organization at Washington. work are the assistant chief, the chief clerk, the professors, the heads of the divisions into which the work at the central station is divided, and the inspectors. The assistant chief acts in place of the chief when he so desires or is absent. The chief clerk has charge of the correspondence and all questions of personnel. The professors, of whom there are eight at present, are particularly well-trained men who are connected with the scientific, rather than the executive, work. The ten divisions into which the work at the central station at Washington is divided are: forecast, river and flood, instrument, publications, supplies, telegraph, accounts, marine, library, and climatological, and each is presided over by an efficient head.[1] It

---

[1] The number of divisions, however, is a matter which is subject to frequent change. Very recently (1911) the forecast, marine, and river and flood divisions have been practically consolidated into one, the division of observations and reports.

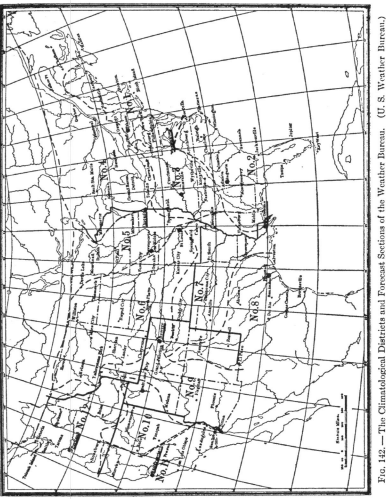

FIG. 142. — The Climatological Districts and Forecast Sections of the Weather Bureau.   (U. S. Weather Bureau.)

The 12 districts are numbered and drawn with broken lines.   The centers are indicated by a ●.   The forecast sections are drawn with solid lines and the centers are underlined.

is the duty of the inspectors to visit the regular stations from time to time, both to make investigations in case anything has gone wrong, and also to make suggestions as to how the efficiency and usefulness of the station may be increased.

At the regular Weather Bureau stations outside of Washington from one to fourteen men are employed. The head of each station receives the title of official in charge, and the one next in authority to him is usually called the first assistant.

At the central station at Washington about 200 are employed, and at all the regular stations outside of Washington about 500, so that there **Number of** are, all told, about 700 commissioned employees in the service **employees.** of the Weather Bureau. In addition, nearly 100 display men and observers receive compensation for their services. The salary of the chief is $6000, and that of the assistant chief and chief clerk, $3000 each. At the regular weather bureau stations outside of Washington the salaries run from $3500 down. The observers at coöperative stations receive no compensation for their services, except the publications of the bureau.

THE STATION EQUIPMENT AND THE OBSERVATIONS TAKEN

360. The central station of the U. S. Weather Bureau is pictured in Fig. 143, which is used as the frontispiece of this book, and consists of **The central** a main building and several adjoining buildings located at **station at** 24th and M streets in Washington. On the ground floor of **Washington.** the main building are located the rooms for the weather forecasting and the library. The library now contains nearly 30,000 volumes. It is the best meteorological library in this country, and one of the best in the world. On this same floor are located the accounts, telegraph, and river and flood divisions. On the second floor are located the rooms for the marine work, and the offices of the chief, assistant chief, and chief clerk. The instrument division, the printing outfit, the climatological division, and the division of supplies are located in the adjoining buildings. The observations taken and the instruments used are essentially the same as those at all the regular stations of the Weather Bureau. There are, however, several special instruments in use at Washington.

The regular stations of the U. S. Weather Bureau are usually located in the larger cities, as the weather maps and weather forecasts can be more quickly distributed and reach more people. From three to eight or ten rooms are usually occupied by a Weather Bureau station. In some cases the Weather Bureau owns the building, but more often the rooms are simply rented.   They are generally located on the top floor of some high office building, and a flat roof must also be available for exposing the instruments. In New York the rooms were on the 20th floor of 100 Broadway,[1] and the instruments were on the 23d floor.

The equip-ment of a regular Wea-ther Bureau station.

In Boston, the Weather Bureau is located in the main tower of the post office building.   On the roof will be found a thermometer shelter, containing a maximum and a minimum thermometer, wet and dry bulb thermometers which can be whirled, a Richard Frères thermograph, and sometimes a recording hygrometer. A tipping bucket rain gauge, a Robinson cup anemometer, an electric contact sunshine recorder, and a contact-making wind vane will also be found suitably exposed.   If the station has but three rooms, one would probably contain the desks of the official in charge, or local forecaster official, and his first assistant, the library, and perhaps files and supplies. A second room would contain the apparatus in actual use and additional apparatus for demonstration purposes and to replace anything which might become broken. Here would, at least, be found the " triple register " for recording the wind velocity, wind direction, sunshine, and precipitation ; a Richard Frères barograph; and two good mercury barometers.   The third room would be a work room for preparing the weather map and repairing apparatus. Here would be found the printing presses and the addressograph for addressing the daily weather maps and other publications. At the section centers and at stations located in large cities more room is necessary, as there is a larger number of employees, but the general arrangement is the same.

At a coöperative station, only a maximum and minimum thermometer in a thermometer shelter and a rain gauge are necessary so that the equipment is very simple. The thermometer shelter is usually located in the open over sod, and some five or six feet above it. It may be fastened to the north side of some unheated building, leaving ample space for ventilation between it and the building.   The rain gauge is usually located in the open, a few feet above the ground.

The equip-ment of a coöperative station.

361. At a regular station of the U. S. Weather Bureau the following observations are taken :

[1] May 1, 1911, the station was moved to the Whitehall Building, 17 Battery Place. The office rooms are now on the 29th floor of a 31-story building.

Continuous

Pressure with barograph.*

Wind direction ⎫
Wind velocity ⎬ (Recorded on the same revolving drum; instruments in the open.)
Sunshine ⎪
Precipitation ⎭

Moisture with hygrometer.† Temperature with thermograph.†

8 A.M. and 8 P.M.

Maximum and minimum temperature.† Pressure.*
Temperature.† Clouds (upper and lower).
Moisture (wet bulb thermometer).† Precipitation.

Miscellaneous

Fog, frost, hail, thunder, halo, aurora, sunset and sunrise colors, smoke, haze.

* Instrument in the office. † Instrument in the thermometer shelter.

Form No. 1083 — Met'l.

U. S. DEPARTMENT OF AGRICULTURE, WEATHER BUREAU.

(Station) ---

(Date) --, *19*
(75th Meridian Time.)

	8 A.M.	8 P.M.
Dry thermometer	....	....
Wet thermometer	....	....
Maximum thermometer	....	....
Minimum thermometer	....	....
Direction of wind	....	....
Precipitation...............................	....	....
Clouds { Upper	....	....
{ Lower	....	....
Barometer —		
Attached thermometer	....	....
Observed reading	....	....
Total correction...........................	....	....
Station....................................	....	....
Reduced	....	....

Compared with Form No. 1001 — Met'l.

....

NOTES:

The observations at 8 A.M. and 8 P.M. are taken down on form 1083, which is here reproduced. The data on this form and values obtained from the sheets taken from the recording instruments are copied into forms 1001 and 1014. Form 1001 and 1083 together with the original triple register, thermograph, and barograph sheets are sent monthly to Washington. The necessary data are also copied into the " Climatological Record," a book retained by the local station. A monthly summary is also printed by each regular station. A pamphlet entitled " Instructions for Preparing Meteorological Forms " and the " Station Regulations of the Weather Bureau " give detailed information as to the taking and recording of the observations. The Weather Bureau also publishes a large number of blank forms, and form 4024 is a check list of the various forms. Samples of these various forms may be secured from the regular Weather Bureau stations and are very valuable illustrative material of the work of a station.

The observations taken at a regular station and where they are recorded.

At a coöperative station, the following observations are required:

Maximum temperature.	Snow on ground.
Minimum temperature.	Duration of precipitation.
Set maximum.	Prevailing wind.
Precipitation.	Cloudiness.
Snowfall.	Miscellaneous phenomena.

Only a monthly report is required, and this is made out in triplicate. One copy is retained by the observer and two are sent to the section center. One of these is eventually sent to Washington for filing.

The observations taken at a coöperative station.

At a special station, those observations are taken for which the station was founded.

The Development of the Daily Weather Map

362. In 1780 the Meteorological Society of the Palatinate, with its center at Manheim, began the collection and publication of detailed meteorological observations for Europe. A few reports from America were also included and the work continued for about thirteen years. Based upon these observations, the first synoptic charts or weather maps were made by H. W. Brandes in 1820. He constructed them for the whole of the year 1783, and described the results, but did not publish the maps. From 1821 until his death in 1857, William Redfield of New York

The first weather map was constructed by Brandes.

collected meteorological observations and studied storms by means of weather maps. He seems to be the first one in this country, and Redfield's the first one after Brandes, to have charted simultaneous work. observations in order to study storms and atmospheric movements and changes. In this country, E. Loomis and J. Espy also made many weather maps during the first half of the last century. In England, Mr. James Glaisher began constructing weather maps in 1848, and during the London World's Fair in 1851 the first weather maps based upon observations received by telegraph were produced. After 1851, however, the work was discontinued. In 1863 Leverrier of the The first Paris Observatory began the publication of the series of regular daily weather maps for Europe, based upon telegraphic series of daily observations, which has continued unbroken until the present weather time. The series of daily weather maps for the United States maps. was begun in 1871, which makes the United States the fourth country to issue regular daily maps, as the Netherlands, England, and France had already done so. At first the maps were issued in this country three times a day, at 7 A.M., 2 P.M., and 9 P.M. The change was made from three maps to two maps, issued 8 A.M. and 8 P.M., January 1, 1889. Since September 30, 1895, only one daily map, the 8 A.M. map, has been issued.

In 1868 Leverrier began the publication of a daily weather map for the whole globe, but the work was discontinued after several years. Weather From 1875 to 1895, the U. S. Weather Bureau made a maps for weather map for the whole northern hemisphere, and the the northern maps were published for the first ten years and form a very hemisphere. valuable collection. This map is still made in manuscript at Washington, and one of them is reproduced as Chart L. At present, the Deutsche Seewarte, at Hamburg, and the Danish Meteorological Office, at Copenhagen, publish jointly a daily map of the North Atlantic. This together ,with the maps for Europe and the United States cover a good part of the northern hemisphere.

THE DAILY WEATHER SERVICE OF THE U. S. WEATHER BUREAU

363. The taking and sending of the observations. — At all the regular The obser- stations of the U. S. Weather Bureau, and at the Canadian vations are and Mexican stations as well, the observations are taken at taken at eight, East- 8 A.M. and 8 P.M. Eastern Standard Time, that is, five hours ern Stand- from Greenwich. This means, for example, that they are ard Time. taken in the morning at 7 A.M. in Chicago, at 6ʽ A.M. in

Denver, and at 5 A.M. in San Francisco. In fact, the observations are commenced a little before eight, so that the work may be finished and the telegram, which is to convey them to other stations, prepared by 8 o'clock.

The observations are not telegraphed as figures, but they are reduced by means of an elaborate, yet very ingenious, complete, and A code is satisfactory code, to a series of words. According to the used for their transcode book, the data for transmission by telegraph may con- mission. sist of fifteen words and they are to be enciphered in the following order:

1. Name of station (or telegraphic designation therefor, as furnished from the central office).

2. Pressure and temperature (corrected readings).

3. Precipitation.

4. Direction of wind, state of weather, and maximum temperature.

5. Current wind velocity and minimum temperature, or current wind velocity and maximum temperature.

6. Minimum or maximum temperature (in twenty-four hours) which occurred more than twelve hours previous to observation.

7. Marked rise or fall of pressure (from stations specially designated).

8. Report on the river observation (from stations specially designated and in certain authorized reports).

9. Frost (light, heavy, or killing).

10. Thunderstorms.

11. Fog, haze, or smoke.

12. Upper clouds.

13. Lower clouds (when sent).

14. Maximum wind velocity and direction (in accordance with special instructions).

15. Special monthly or weekly reports.

The code is of such a character that one familiar with it is able to translate it without using a code book or memorizing words. The figures are conveyed, in general, by the first vowel and the consonant which precedes it. The following illustrations of the second word in the telegram will make this clear:

PRESSURE	TEMPERATURE	WORD
30.24	50	De *mu* lsion
30.24	58	De *mo* crat
30.24	88	De *so* late
30.54	50	Me l *my*
30.54	58	Me *mo* ry
30.54	88	Me *so* type
30.56	88	Mi *so* gamy

Not all of the fifteen words are usually sent.   Numbers 1, 2, 3, 4, 5, 14, are nearly always sent; number 6, 9, 10, are generally sent; and the others only occasionally.

The purpose of using a code instead of figures is primarily to secure brevity and thus save expense.   As it is, the telegraphic tolls of the **The purpose** Weather Bureau amount to more than $100,000 annually **of the code.** and constitute one of the largest bills of the bureau.   The use of a code also prevents the observations from being tampered with or read in transit, and at the same time greater accuracy is secured, as words are always transmitted with fewer mistakes than figures.

The weather telegram containing from six to fifteen words is thus filed at the local telegraph office by each regular station at eight o'clock. **How the** These messages have precedence over all others, so that **telegrams** usually before 9.30 o'clock Washington, New York, and **are sent.** other large stations are in possession of the observations made at all regular stations, and all stations which produce weather maps have received a sufficient number of reports to construct an accurate weather map.   This exchange of telegrams is accomplished by a circuit system.   The telegrams are not sent individually from each station to every other station, but are collected at some center and then sent as a unit through a series of stations connected up as a continuous circuit.   It is not customary to wait until all the telegrams have been collected and then form the circuit.   As soon as a sufficient number has been collected, they are sent around the circuit.   At map-producing stations, the observations thus ordinarily came in in three or four installments.   The last ones usually get in before 10 o'clock.   The system of circuits is changed from time to time, and is worked out with the greatest care, so as to make the amount of telegraphing, and thus the expense, a minimum.   At both 8 in the morning and 8 at night, the observations are sent out as just described.

364. **The charting of the observations.** — The central station at Washington and many of the regular stations of the U. S. Weather Bureau issue daily weather maps which give, both in the form of a chart and in tables, the observations taken at 8 in the morning.   A weather map, based upon the 8 P.M. observations, is prepared in manuscript at Washington and a few other stations, but it is never published except by the newspapers.

The weather map (form DD) was used, prior to 1910, at nearly all of the map-producing stations, and is still used at a few of them.   It is 16 × 11 inches and contains an outline map of the United States ·

10 × 6½ inches. This outline map is printed in brown and contains the names of the regular stations. On this map are charted the pressure, temperature, wind direction, and state of the weather. The pressure is indicated by a series of solid lines called isobars, drawn through places having the same pressure. The lines are drawn for every tenth of an inch difference in pressure and the pressure is indicated at each end of the line. These **How the observations are charted in the daily weather map.**
lines inclose areas of high and low pressure, and the areas of highest and lowest pressure are marked with the words " High " and " Low." These are the extratropical cyclones and anticyclones which have been studied in a previous chapter. Dotted lines, called isotherms, connect places having the same temperature. These lines are drawn for every ten degrees, and the temperature of each line is indicated at both ends of it. The wind direction is indicated by an arrow which flies with the wind. The state of the weather is indicated by a set of symbols. O indicates a clear sky, ◑ partly cloudy, ● cloudy, ⊛ indicates rain, ⑤ snow, Ⓜ that the report is missing. /⌇ indicates that a thundershower occurred at the station during the preceding twelve hours. These explanations are stated briefly in the " Explanatory Notes " which are printed in one corner of the map and are here reproduced.

EXPLANATORY NOTES

Observations taken at 8 A.M., seventy-fifth meridian time. Air pressure reduced to sea level.

ISOBARS, or continuous lines, pass through points of equal air pressure.

ISOTHERMS, or dotted lines, pass through points of equal temperature.

SYMBOLS indicate state of weather: O clear; ⊖ partly cloudy; ⊕ cloudy; R rain; S snow; M report missing; /⌇ thunderstorm. ARROWS fly with the wind.

Shaded areas when used show regions of precipitation during past 24 hours.

"T" in table, indicates amount too small to measure.

The various weather maps which have been reproduced in the chapter on storms will illustrate the method of charting the observations. Further information is also conveyed by means of statistics in tabular form in the lower right-hand part of the map. Here are given temperature at 8 A.M., 75th meridian time, temperature change in 24 hours, maximum temperature in last 24 hours, minimum temperature in last 12 hours, wind velocity 8 A.M., precipitation in last 24 hours. On the lower left-hand side of the map are given the weather predictions for the particular region where the station is located and a general discussion of the weather conditions.

Since 1910 most of the map-producing stations issue the so-called " commercial weather map," which is just the same as the maps which are published in the newspapers. The size is the same as The commercial weather map. before, only the color is blue instead of brown. The isobars, the highs and lows, the method of indicating the direction of the wind, and the method of indicating the state of the weather are the same as before. Only four isotherms are drawn; for zero, freezing, 90° F., and 100° F. In addition, the temperature at 8 A.M., the amount of the precipitation, and the wind velocity are printed on the face of the map near each station. This, of course, necessitates a change in the statistical material given in tabular form. The following legend is placed on these maps and makes the slight differences clear:

Observations taken at 8 A.M., seventy-fifth meridian time.

ISOBARS, or continuous lines, pass through points of equal air pressure.

ISOTHERMS, or dotted lines, pass through point of equal temperature; they will be drawn only for zero, freezing, 90°, and 100°.

SYMBOLS indicate state of weather: ○ clear; ◐ partly cloudy; ● cloudy; ⓡ rain; ⓢ snow; ⓜ report missing.

Arrows fly with the wind. First figure, temperature; second, 24-hour rainfall, if it equals .01 inch; third, wind velocity of 10 miles per hour or more.

Many newspapers now publish a map of this kind, together with statistical material in tabular form and the predictions. In the large cities it is nearly always possible to get an evening paper which contains the morning map and often a morning paper which contains the evening map.

365. The weather map (form C) published at Washington, is somewhat larger (24 × 16 inches) and contains more information. The The Washington map. outline map is here printed in blue and the isotherms are continuous lines and are printed in red. On these maps three additional things are charted: the area where precipitation has fallen, areas of marked rise or fall in temperature, and the path of the " lows." The area where precipitation has fallen is indicated by shading. Areas where the temperature has risen or fallen 20° are inclosed by red dots. The path of a prominent low is indicated by a series of arrows. In the tables, two additional columns containing the pressure reduced to sea level and standard gravity, and the change in pressure

in twelve hours, are given. The predictions issued at Washington and at the various forecast centers, viz., Chicago, New Orleans, Denver, Portland, Ore., and San Francisco — cover the whole country.

At Washington, about 1500 copies of the weather map are issued daily. At other map-producing stations, from a few hundred to 3000 maps are distributed daily. The newspapers, of course, make the weather map widely accessible.

366. The construction of the weather map. — The construction of the daily weather map at Washington is an interesting operation, and usually occupies about two hours, from 8.30 to 10.30 in the morning. As soon as the weather telegrams are received they are sent at once to the rooms of the forecast division. Here some one familiar with the code translates the messages into figures and reads them aloud. As the telegrams are being read, one clerk on an outline map of the United States and adjoining countries constructs a chart showing the change which has taken place in the temperature during the past twenty-four hours. A second clerk constructs a chart showing the change in the barometric pressure during the past twenty-four hours. A third clerk constructs two charts, one showing the humidity and the other showing the amount, kind, and direction of motion of the clouds. The forecast official himself, or some one under his direct supervision, constructs a fourth chart, which may be called the general weather chart, and is later published as part of the weather map. This shows the pressure, temperature, direction and velocity of the wind, the rain or snow which has fallen, and the state of the weather for each station. As soon as the weather telegrams have all been received and read, and the observations have been recorded, the isobars and isotherms are drawn in, the areas of high and low pressure are so marked, the direction of motion of the lows is indicated, the areas where precipitation has fallen are shaded, and the areas where the temperature has risen or fallen twenty degrees are inclosed. The map is then hurried to the lithographers to be put upon the stone. In the meantime, a force of typesetters have set up, from the dictation, the material for the tabular form. As soon as the map is ready to be lithographed, the forecast official immediately begins to dictate his forecasts for the various parts of the country. These are immediately set up in type, and shortly after he has finished the proof of the chart, tables, and forecasts is ready. The completed weather map is then printed as rapidly as possible, and the whole operation does not occupy much more than two hours.

How the weather map is constructed at Washington.

At the regular stations of the U. S. Weather Bureau, where daily weather maps are produced, very much the same thing takes place,

The chalk plate process. usually on a smaller scale, with the exception that the chart part of the map is not lithographed, but is produced by the chalk plate process. A plate of steel the size of the chart is covered with a layer of chalk about one eighth of an inch thick. In this is cut, by means of suitable instruments, the various symbols and lines for indicating the various things which are depicted on the chart. This chalk plate is then locked in an iron frame and stereotyped in the usual way by pouring in molten type metal. The plate containing the impression is then trimmed, any imperfections are remedied, and necessary corrections are made. The tables and forecasts are set up in type as before, and the map is printed in the regular way. At these stations, only one chart, besides the one which is reproduced, is usually constructed as the observations come in.

At a few regular stations of the U. S. Weather Bureau, the weather map (chart, tables, and forecasts) is still mimeographed. In this case,

The mimeograph is sometimes used. the symbols for the chart are usually cut into the waxed sheet by a specially constructed typewriter. The isobars and isotherms are drawn in free hand. These maps are just as useful, but are not as neat in appearance, as the lithographed maps or those prepared by the chalk plate process.

367. **The distribution of the map.** — In certain cities, a few weather maps may be carried by special messengers to particular places, but

The distribution of the weather maps. these may be left out of consideration, and it may be said that they are practically all distributed by mail. The weather maps are folded and placed in wrappers which have been addressed the day before, by means of an addressograph. The outside of the wrapper carries this inscription:

IMMEDIATE — U. S. Weather Report. U. S. Department of
By authority of the Post Office Department, Agriculture,
June 18, 1881, this Report will be
treated in all respects as letter mail. Weather Bureau.

OFFICIAL BUSINESS

Penalty for Private Use $300.00

This insures that, anywhere within a hundred miles of a map-producing station, the daily weather map will be received in the middle or later part of the afternoon at the latest.

368. Other methods of distributing the maps, forecasts, data, and warnings. — In addition to issuing daily weather maps, there are several other methods used by the U. S. Weather Bureau in distributing its meteorological information, forecasts, and warnings. These are by means of the newspapers, flags, cards, the telephone, and special messages.

Four additional ways of distributing forecasts.

The agents of the various press associations, and the reporters for the various newspapers visit the Weather Bureau stations as soon as the observations have been received and charted. They are furnished with the weather maps, forecasts, and all the data which they care to use. Nearly all the daily papers print the weather forecast at the top of the first page, and then devote more space on some inside page to weather data in tabular form, a general discussion of the weather conditions, and a more detailed description of the impending weather changes. The probable coming of any damage-causing storm or weather change usually receives a special write-up. Since 1910 many newspapers are publishing the weather map complete, *i.e.* the chart, the statistics in tabular form, a discussion of the weather conditions, and the forecast. These newspapers are to be particularly commended in this respect, for they make accessible to a large number of readers the full meteorological material in complete and definite scientific form. Whenever a newspaper undertakes this work, the nearest Weather Bureau station usually furnishes the stereotyped plate from which the chart is printed.

The work of the newspapers.

Flags are also used, particularly along the seacoast and in the lake regions, to make known the weather forecasts, and to announce the coming of storms. The various flags used at the U. S. Weather Bureau display stations and their explanation are here given.

The various flags and their meaning.

The weather forecasts are also printed on small cards and these receive a wide distribution, particularly through the mails. About 25,000,000 of these cards are distributed annually. They are placed in small metal frames and exposed in public places, particularly in post offices. If there are several rural free delivery lines going out from any post office, the forecasts will be telegraphed from the nearest map-producing station of the Weather Bureau to the post office at about 10.30 in the morning. The forecast is then stamped by means of rubber stamps on these cards, which are exposed in public places and also distributed by the rural mail carriers.

The card service.

2 B

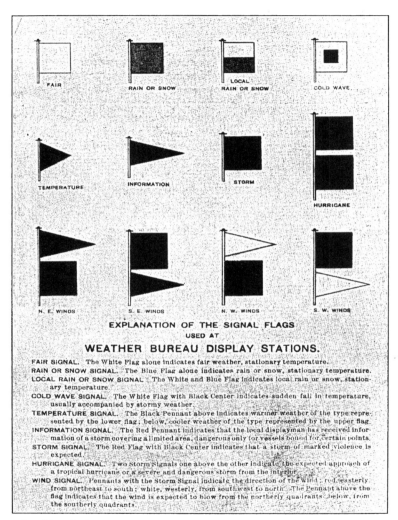

EXPLANATION OF THE SIGNAL FLAGS
USED AT
WEATHER BUREAU DISPLAY STATIONS.

FAIR SIGNAL. The White Flag alone indicates fair weather, stationary temperature.

RAIN OR SNOW SIGNAL. The Blue Flag alone indicates rain or snow, stationary temperature.

LOCAL RAIN OR SNOW SIGNAL. The White and Blue Flag indicates local rain or snow, stationary temperature.

COLD WAVE SIGNAL. The White Flag with Black Center indicates sudden fall in temperature, usually accompanied by stormy weather.

TEMPERATURE SIGNAL. The Black Pennant above indicates warmer weather of the type represented by the lower flag; below, cooler weather of the type represented by the upper flag.

INFORMATION SIGNAL. The Red Pennant indicates that the local displayman has received information of a storm covering a limited area, dangerous only for vessels bound for certain points.

STORM SIGNAL. The Red Flag with Black Center indicates that a storm of marked violence is expected.

HURRICANE SIGNAL. Two Storm Signals one above the other indicate the expected approach of a tropical hurricane or a severe and dangerous storm from the interior.

WIND SIGNAL. Pennants with the Storm Signal indicate the direction of the wind; red, easterly, from northeast to south; white, westerly, from southwest to north. The Pennant above the flag indicates that the wind is expected to blow from the northerly quadrants; below, from the southerly quadrants.

By night a Red Light indicates Easterly Winds, and a White Light below a Red Light Westerly Winds. No night Hurricane Warnings are displayed.

If a particularly violent or damage-causing storm is expected, special warnings are often sent out by messengers to the captains of all vessels in a harbor, and to those who will be particularly inter- Special ested in, and affected by, the storm. messages.

369. The weather service in the evening and on Sundays and holidays. — The observations taken at 8 in the evening are telegraphed to Washington and other selected stations of the Weather The evening Bureau in the regular way. At these stations, a manuscript service. weather map is made and the forecasts, data, and warnings are sent out, but no weather map is published, except by some newspaper.

On Sundays and holidays, the observations are taken and sent by telegraph in the usual way. At Washington, the manuscript weather maps are made, but the one for Sunday morning is not The service published until Monday afternoon, and the morning map on Sundays for a holiday is published a day or two later. At the other and holidays. map-producing stations of the Weather Bureau a manuscript may is sometimes constructed for the Sundays and holidays, but they are never published.

OTHER WORK AND PUBLICATIONS OF THE U. S. WEATHER BUREAU

370. The routine work at a regular station of the U. S. Weather Bureau consumes most of the time of the officials. The apparatus must be kept in good working condition. The observations The routine must be taken at the appointed times, and the weather work of the telegrams prepared for transmission. The charts on the Weather revolving drums of the automatically recording instruments Bureau. must be changed at stated intervals. The observations must be entered in appropriate forms for transmission to Washington and in the records which are retained at the station. If it is a map-producing station, all the work incident to the construction and distribution of weather maps must be performed. The forecasts, data, and warnings must be sent out. The monthly report must be prepared. All this leaves but little time for other work.

Considerable instructional work is also done by the officials at Weather Bureau stations. If the station is located in a city where there is a college or university, it is often the local forecast official who Educational gives the course or courses in meteorology in the institution. and research Many public lectures are also given, and frequently classes work. from various schools visit the station in order that the equipment and work may be explained to them.

Research work is also encouraged on the part of the Weather Bureau officials. Very little, however, is done, and the chief reason is lack of time.

371. The publications issued by the U. S. Weather Bureau, in addition to the daily weather maps and forecast cards, may be divided into two groups, the periodical publications and those which appear at irregular intervals. The periodical publications are:

The publications of the Weather Bureau.

Monthly Weather Review, 4°, 1872 to date.
Bulletin of the Mount Weather Observatory, 8°, 1908 to date.
National Weather Bulletin (weekly during the summer, monthly during the winter).
Snow and Ice Bulletin (weekly during the winter).
Meteorological Charts of the Great Lakes.
Annual Report of the Chief.

The publications which appear from time to time are:

Numbered Bulletins (nearly 40 have now appeared).
Lettered Bulletins (Bulletin V has recently appeared).
W. B. Publications (these include many of the others).
Miscellaneous Publications.
Summary of the Climatological Data for the United States by Sections (106 are to be issued).

Prior to 1909, forty-four section centers published a weekly weather bulletin during the summer and a monthly climatological report, but these were then discontinued. Each regular Weather Bureau station publishes a monthly summary of a single page. For further details in connection with all these publications, see Appendix IX.

The publications of the U. S. Weather Bureau are distributed with a free hand. The daily weather maps are sent free of charge to schools, post offices, factories, large office buildings, in fact, wherever they will be exposed so that the information which they contain may reach quickly a large number of people. The other publications of the Weather Bureau are sent free of charge to libraries and institutions of learning. They may all be obtained by any one on payment of prices which are not excessive.

The distribution of the publications.

The Weather Bureaus of Other Countries

372. Nearly all the large civilized nations of the world now maintain a national weather service and publish daily weather maps and forecasts. This is true of all the larger countries of Europe, Japan, China, Canada, India, Australia, New Zealand, Argentine Republic, and Algeria. *The countries which maintain a national weather service.*

The construction and appearance of the daily weather map is much the same in all countries. In the details, however, there is considerable difference. In all cases, the observations are taken at the same time at many stations. These are transmitted by telegraph, using a code, to one or more central stations where the daily weather maps are produced. The weather map always contains an outline map on which is charted some of *The weather maps of other countries.* the observational material. More of it is presented in tabular forms and forecasts and a general discussion of the weather conditions is always included. The observations are taken at different hours in different countries, but always in the early morning. The code used generally consists of a series of numbers, and does not make use of words. Wind velocity is often charted on the map as well as wind direction. This is usually done by placing a different number of barbs on the arrows. Sometimes temperature and pressure are charted on separate maps. Although the details may be different, one familiar with weather maps would find no difficulty in interpreting the weather map of any other country. The weather maps of Great Britain, France, Germany, Japan, and India are particularly interesting. Samples of the weather maps issued by the various countries may occasionally be secured from the U. S. Weather Bureau at Washington, by any one who will make good use of the illustrative material.

The English weather service is managed by the Meteorological Office at South Kensington, London, S. W. There are upwards of 200 stations scattered over the British Isles and divided into *The English weather service.* five classes. In addition, the Royal Meteorological Society and the Scottish Meteorological Society have covered the country with a network of climatological and phenological stations. The weather map is issued daily, and is based on the observations received by telegraph from twenty-nine home and forty-four foreign stations. The home observations are taken at 7 A.M. and 6 P.M. West European time. The foreign observations are all taken in the early morning, but not at the same time. The telegram which conveys the observations consists of six numbers of five figures each. The daily weather report consists of four quarto pages. Pages one and four contain data in tabular form, while pages two and three contain charts of the various meteorological elements. The Meteorological Office also issues a weekly and a monthly weather report.

The central office of the weather service for the Dominion of Canada is located at Toronto. Here are received twice daily, by telegraph, the observations from forty-three Canadian stations and sixty-four stations in the United States. From these, the daily weather forecasts and bulletins are prepared and distributed from various centers.

The Canadian weather service.

The preparation of the daily weather map, particularly in Europe, requires international coöperation. This has lead to International Meteorological Congresses and to uniformity in nomenclature, time of observation, and the like. The adoption of the international system of cloud classification in 1891 is an example of this. There is also a set of international meteorological symbols for the various things observed.[1] It has also been proposed to have a set of international storm warnings. The system shown in Fig. 144 will probably be universally adopted.

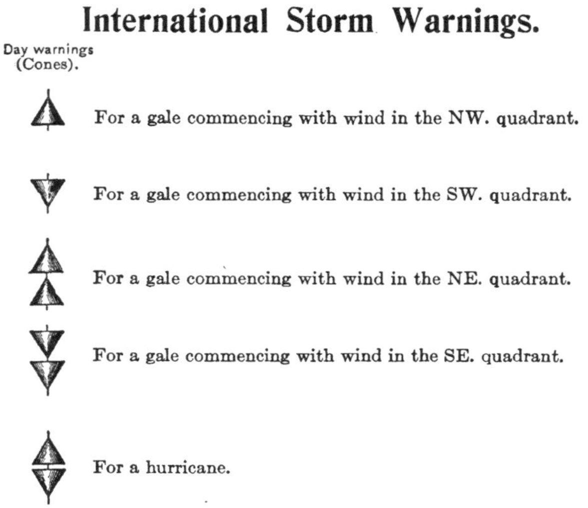

International co-operation.

International Storm Warnings.

Day warnings
(Cones).

 For a gale commencing with wind in the NW. quadrant.

 For a gale commencing with wind in the SW. quadrant.

 For a gale commencing with wind in the NE. quadrant.

 For a gale commencing with wind in the SE. quadrant.

 For a hurricane.

FIG. 144. — International Storm Warnings.

[1] See *Monthly Weather Review*, December, 1905, p. 524.

Some Special Meteorological Observatories; their Equipment and Work

373. The two most important meteorological observatories in this country which are not directly connected with the routine work of a Weather Bureau station are the Blue Hill Meteorological **Blue Hill** Observatory near Boston, and the Mount Weather Research **and Mount** Observatory at Mount Weather, in Virginia. The Blue **Weather.** Hill Observatory is a private observatory, and was founded by Mr. A. L. Rotch in 1885. A few years later it became associated with the Astronomical Observatory of Harvard College and its records and results are printed in the Annals of that observatory. Particularly noteworthy has been the work of this observatory in the use of kites to explore the upper air and in cloud observations. The Mount Weather Research Observatory was founded by the U. S. Weather Bureau, and most of the time is to be devoted to balloon and kite work, a study of the solar heat, and the electrical condition of the atmosphere.

Other observatories, some of them connected with institutions of learning, are making valuable series of observations and publishing results, but they are undertaking no elaborate pieces of research work. Among these may be mentioned the Experiment Station of the Massachusetts Agricultural College at Amherst, Mass. The two most famous mountain observatories in this country were those on Pike's Peak in Colorado, and on Mt. Washington in New Hampshire. Both have now been discontinued, but valuable observations were secured.

In Europe, the most famous meteorological observatories are, perhaps, those at Pottsdam and Lindenburg near Berlin; Pavlovsk near St. Petersburg; Trappes, Mont Souris, and Parc St. **The famous** Maur in France; and Kew and Greenwich in England. **meteorologi-** The most famous mountain observatories are probably **cal obser-** Sonnblick and Hoch Obir in Austria, Santis in Switzerland, **vatories** Pic du Midi and Puy de Dôme in France, Wendelstein and Brocken **of Europe.** in Germany, Etna in Italy, and Ben Nevis in Great Britain.

There are now so many observatories in addition to the regular stations of the weather service of the various countries, that only the more important ones can be named. To attempt to describe their equipment and work in small space is out of the question.

QUESTIONS

(1) What important discovery did Benjamin Franklin make in connection with storms? (2) State in outline the history of the U. S. Weather Bureau. (3) Name the chiefs of the Weather Bureau and the characteristics of the administration of each. (4) What has been the cost of the Weather Bureau? (5) Describe the three different ways in which the United States is divided into districts by the Weather Bureau. (6) How many kinds of Weather Bureau stations are there? (7) Describe the organization of the Weather Bureau at Washington. (8) How is the work at the central station at Washington subdivided? (9) How many are employed by the U. S. Weather Bureau and with what salaries? (10) Describe the Weather Bureau station at Washington. (11) Describe a typical Weather Bureau station. (12) Describe the instrument equipment of a Weather Bureau station. (13) What observations are taken at a regular Weather Bureau station? (14) What observations are required at a coöperative station? (15) Describe the development of the daily weather map. (16) Describe the development of the weather maps for the northern hemisphere. (17) Describe the taking and sending of the observations for the daily weather map. (18) Describe the code for sending them. (19) How are the observations charted on a weather map? (20) How does the Washington weather map differ from those issued at regular stations? (21) Describe the construction of the weather map at Washington. (22) Describe the chalk plate process of producing the weather map. (23) Describe the different ways in which weather forecasts, data, and warnings are distributed. (24) Describe the weather service in the evenings and on Sundays and holidays. (25) Describe the educational and research work of the Weather Bureau officials. (26) What are the periodical publications of the Weather Bureau? (27) What publications appear from time to time? (28) What differences are found in the weather maps of other countries? (29) Name and describe some special meteorological observatories in this country and in Europe.

TOPICS FOR INVESTIGATION

(1) The history of the weather map to 1850.
(2) The various codes used to transmit weather observations.
(3) The processes used in constructing a weather map.
(4) The history and present organization of the weather service of some country.

PRACTICAL EXERCISES

(1) Construct a weather map for the United States. (The necessary observations can probably be secured from any map-producing station of the U. S. Weather Bureau. Material for six maps can be found in WARD'S *Practical Exercises in Elementary Meteorology*.)

REFERENCES

For the history of the U. S. Weather Bureau and its present organization, consult: MOORE, JOHN W., *Meteorology*, 2d ed., London, 1910. (Chapter V, pp. 44–68.)

MOORE, WILLIS L., "Forecasting the Weather and Storms," *The National Geographic Magazine*, June, 1905.

POLIS, DR. P., *Der Wetterdienst und die Meteorologie in den Vereinigten Staaten von Amerika und in Canada*, Berlin, 1908.

Two pamphlets published at Washington and each entitled ∷ The Weather Bureau."

The annual " Reports of the Chief of the Weather Bureau."

For a description of the station equipment and the methods of taking and recording the observations, see:

For the United States:

Instructions for Coöperative Observers. Station Regulations. Instructions for Preparing Meteorological Forms. (All published by the U. S. Weather Bureau at Washington, D.C.)

For Great Britain:

SCOTT, R. H., *The Observer's Handbook*.

For France:

ANGOT, ALFRED, *Instructions Météorologiques*.

See also Appendix IX.

For the equipment and work of special meteorological observatories, see:

WALDO, FRANK, *Modern Meteorology*, London, 1893 (pp. 160–203).

For a description of the weather services of other countries, see:

CAMPBELL, BAYARD, "Government Meteorological Organizations in Various Parts of the World," *Quart. Jour. R. Met. Soc.*, April, 1899.

BEBBER, W. J. VAN, *Lehrbuch der Meteorologie*, Stuttgart, 1890 (pp. 359–384).

BÖRNSTEIN, R., *Leitfaden der Wetterkunde*, Braunschweig, 1906 (pp. 178–200).

MOORE, JOHN W., *Meteorology*, 2d ed., London, 1910 (pp. 26–43, 69–74).

PERNTER, J. M., *Wetterprognose in Österreich*, Wien, 1907.

Monthly Weather Review, August, 1907, p. 364.

CHAPTER VIII

WEATHER PREDICTIONS

Introduction

374. The general method of weather forecasting. — Weather has been defined as the condition of the atmosphere at any particular time and place. The condition of the atmosphere is determined by the six so-called meteorological or weather elements; namely, temperature, pressure, wind, moisture, cloud, and precipitation. The best way, then, to describe the weather or to depict exactly the condition of the atmosphere is to state the numerical values for the meteorological elements. To forecast or predict the weather is thus to foretell the values which the meteorological elements are expected to have.

Weather and weather prediction defined.

There are three factors which determine the condition of the atmosphere. The weather may thus be looked upon as the composite or resultant of three things: (1) the typical or normal condition of the atmosphere which would exist if there were no disturbances or local influences; (2) the disturbances caused by such passing meteorological formations as extratropical cyclones, anticyclones, thundershowers, tornadoes, and the like; (3) local influences such as land and sea breeze, the presence of large bodies of warm water, mountains, etc.

The weather is a composite of three things.

The typical or normal weather which would exist if there were no disturbances due to local causes or the passing of meteorological formations would be different for different parts of the country. For the northeastern part of the United States, it would be something like this. The average daily temperature would be highest the last part of July, and then decrease steadily, and gradually, day by day, until the last of January, when it would be least. It would then begin to steadily and gradually increase. Each day there would be the regular daily oscillation of temperature with its maximum in the early afternoon and its minimum at sunrise. The pressure would be very constant, slightly higher in winter than in summer. There would also be the small daily oscillation with its chief maximum at 10 A.M. and its chief minimum at 4 P.M. The wind would be always moderate in velocity, blowing harder in the winter than in summer, and by day than by night. The wind direction would shift from northwest in winter to southwest in summer and back again. The absolute and relative humidity would also show a regular daily and annual variation. There would be very few clouds; perhaps now and then a cumulus due to convection. Precipitation would be entirely lacking.

Normal or typical weather.

In short, the regular daily and annual variations in the meteorological elements would be present, but there would be no irregular fluctuations. It would be difficult to give numerical values to this theoretical state of things. They would be different from the normals found from observation for any station, as cloud and precipitation are lacking. However, in practical forecasting, one is not concerned with this theoretical state of things, but with the normals as derived from observation, since cloud and precipitation are actually present. The various normals, then, in connection with the meteorological elements, may be considered to represent normal or typical weather.

The passing meteorological formations which exert the chief influence in the United States are the lows and highs (the extratropical cyclones **The passing** and anticyclones). V-shaped depressions and other second- **meteorologi-** ary forms of isobars should perhaps also be mentioned. **cal forma-** Occasionally, a tropical cyclone visits the Gulf States. **tions.** Thundershowers and tornadoes should also be included, but thundershowers always accompany a low or V-shaped depression, and tornadoes are always associated with thundershowers. Thus, extratropical cyclones and anticyclones are by far the most important formations to consider. In fact, New England is crossed by such a ceaseless procession of these two formations, that the weather is nearly always dominated by one or the other, and can hardly ever be said to be typical. All of these formations have already been critically studied in the chapter on storms.

Local influences are not usually very numerous or of great importance. If a place is located on the seacoast, then land and sea breeze must be **Local in-** taken into account. In forecasting the weather for New **fluences.** York State, the presence of the Great Lakes play an important part, particularly in the late autumn and early winter, when they are still much warmer than the land and are putting large quantities of moisture into the atmosphere. A large river also sometimes influences weather conditions.

375. The various normal values, when once determined for the different meteorological elements, hold for all time. Thus normal **The lows** weather is known for weeks or years ahead. The local **and highs** influences are usually unimportant, and can be estimated **are the im-** with fair precision. The chief difficulty, and practically the **portant** **things in** only difficulty, is thus to determine the influence of the **forecasting.** passing meteorological formation. If the lows and highs were even always typical, weather forecasting would be an easy matter.

That is, if they all followed typical courses with known velocities and were typical as regards the distribution of the meteorological elements about them, their influences could be readily estimated. As it is, weather forecasting is by no means an easy matter, as the lows and highs are seldom typical as regards path, velocity, or characteristics. Weather forecasting, then, really turns on this one thing, the estimation of the influence of the passing meteorological formations. The general method to be followed in predicting the weather is thus apparent. The first thing to do is to estimate the influence which the passing meteorological formation is going to exert one or two days in the future. The local influences must next be estimated. When these two have been combined with normal weather, the resultant is the expected or predicted weather. The details in this process will be considered a little later.

376. The work of the U. S. Weather Bureau. — For the purpose of weather forecasting, the United States is divided by the U. S. Weather Bureau into six forecast districts with centers at Washington, Chicago, Denver, San Francisco, Portland, and New Orleans. These districts are shown on the map in figure 142. Based on the 8 A.M. observations, the officials in charge of each forecast district issue weather forecasts and warnings for the respective districts. As soon as made, these are forwarded by telegraph to the Central Office at Washington, and forecasts for all districts appear on the Washington weather map. Based on the 8 P.M. observations, forecasts and warnings are issued from Portland and San Francisco only for the respective districts. The forecasts and warnings are issued from Washington for all the other districts. A local forecast official at a map-producing station issues forecasts for the immediate vicinity of the station only. On the weather map, this forecast appears, and also the forecasts for the state, and perhaps adjoining states, which have come from the respective forecast centers. The prediction based on the 8 A.M. observations are for 36 hours, while those based on the 8 P.M. observations are for the following 48 hours.

Where the forecasts of the U. S. Weather Bureau are made.

The forecasts consist of predictions of temperature, wind, and state of the weather. This last includes the amount of cloud and the kind and quantity of precipitation. Pressure and moisture are never forecasted. Cold waves, frosts, and high wind velocities (storms) are also predicted and the appropriate warnings sent out. A local forecast official at a map-producing station forecasts temperature, wind, and state of the weather only. All fore-

Of what the forecasts consist.

casts and warnings in connection with high wind velocities (storms), cold waves, and frosts come from the forecast centers.

These forecasts and warnings of the Weather Bureau are not only printed on the daily weather map, but they receive a wide distribution by means of newspapers, cards, flags, and special messengers. These various methods of distribution have already been discussed in section 368.

377. Long training is required on the part of the Weather Bureau officials before they are allowed to issue forecasts. An official in charge of a station where forecasts are not issued and first assistants at large stations are permitted to make practice forecasts. After this has been continued with fair success for a year, the authority to issue forecasts may be given. Each official who is authorized to make local forecasts is still required to make practice forecasts for the state in which the station is located. He must also make practice storm, cold wave, and frost forecasts for the local station where he is. All of these practice forecasts are immediately mailed to Washington for verification.

The training of a forecast official.

Ability to forecast well depends upon characteristics of mind as well as careful training. Greely in his *American Weather* says: "The skill of a weather predictor arises largely from his alert comprehensiveness of mind, accurate and retentive memory, phlegmatic but confident temperament, and long experience." A weather forecaster must take in many details at a glance; he must recollect past occurrences; he must not lose confidence in himself or his ability, even if an occasional prediction goes wrong.

The characteristics of a successful forecaster.

378. The general method of weather prediction in other countries is the same as in this; although the details may be quite different. Most countries issue but one set of forecasts and warnings each day. This is often in the early afternoon, instead of the morning. Some countries get supplementary telegrams from a few stations just before the forecasts are issued. The forecasts are usually issued from a central station for the whole country, although this is not true for the larger countries. The signals used and the methods of distributing the forecasts are often quite different.

The methods of preparing forecasts in other countries.

WEATHER PREDICTION CONSIDERING THE LOW AS THE DOMINATING FORMATION

379. Locating the storm center twenty-four hours ahead. — Since the estimation of the disturbances caused by passing meteorological formations is the thing of chief importance in forecasting the coming weather, weather prediction may be considered under three heads: (1) when a passing low is the dominating formation; (2) when a V-shaped depression or some secondary isobaric form is exerting the chief influence; (3) when a passing high is exerting the dominating influence. *Three passing formations must be considered.*

If it is a low which is going to dominate the weather for the immediate future, the first thing to do is to determine where the center is expected to be twenty-four and forty-eight hours ahead. A fairly exact estimation of where the center will be twenty-four hours ahead must be made; the position forty-eight hours ahead need be only roughly estimated. The general rule is to determine the track which the low seems to be following, and *The first step in locating a low twenty-four hours ahead.* then put it ahead on this track the distance normally covered in twenty-four and forty-eight hours. If a low has just formed, or has just come into the field of observation, the past is no guide as to the track. The safest rule in these cases is to assume that the low probably will follow the most frequented track which passes through the locality where it appeared. For this the Van Cleef system of tracks (see section 300) will probably be the most satisfactory. If the forecasting is being done for the northeastern part of the United States, the lows have usually been observed and charted on the weather map a day or two before they become the dominating influence. If this is the case, note which track the low has been following and put it ahead on this track. Either the Bigelow, Russell, or Van Cleef system of tracks may be used. It probably would be better to use that system which contains the track which the low seems to be following most exactly. The normal velocity of motion is about 900 miles a day in winter and about 600 in summer and intermediate values in other seasons. (See section 302.)

After it has been determined where the center of the low would be twenty-four and forty-eight hours ahead, if it followed its track with normal velocity, it is customary for the skilled forecaster to modify this estimation by taking account of other considerations. This simply means that as a result of his long experience, he has formed certain empirical rules for his own guidance. This shows the necessity of long experience before reliable forecasts can be made and also the difficulty in explaining exactly how a weather forecast is made. Some of these **Some empir-** rules are the following: (1) Lows near each other tend to **ical rules** coalesce. This is particularly true if one low is in the **for modify-** northern Mississippi Valley and the other in the southern, **ing the** **normal posi-** or if one low is coming up the Atlantic Coast and the other **tion.** is coming eastward across the Great Lakes. They usually coalesce at an intermediate point and become more intense than either component. Often the appearance of coalescing may be given by the fading away of one low, while the other becomes more intense and dominant. (2) Lows and highs repel each other. If a high is directly on the track of a coming low, the low is likely to be retarded or to be deflected from its normal track. (3) Lows tend to follow the record of previous days. This means that if a low is a slow-moving one, it tends to retain that characteristic all through its life history. (4) Lows tend to move toward the areas of greatest rainfall during the preceding twenty-four hours. (5) Lows tend to move toward the areas of highest dewpoint. (6) Lows tend to move toward areas of least wind velocity. (7) Lows tend to grow more intense as they approach bodies of water as the Great Lakes or Atlantic coast. (8) Lows tend to move faster as the pressure grows less. (9) Lows with above normal winds weaken, while lows with below normal winds develop lower pressure. (10) Lows on a curved track tend to move faster after they begin to move northeastward. This set of rules makes no pretense at being complete. Only the well-recognized and more important ones have been given. Each forecaster, as he acquires experience, will prefer to formulate his own experience. They will be found very useful, however, by a beginner.

The method, then, of determining where the center of a low is expected to be twenty-four and forty-eight hours ahead is to put the low forward **The method** on the track which it seems to be following the normal dis- **briefly** tance, and then to modify this estimation by taking account **stated.** of certain general principles which have been learned from experience and perhaps formulated as rules. As a result of this process, the forecaster will come to a very definite conclusion as to where the center of the low is expected to be twenty-four hours ahead, and he will also have a general idea of where he expects it to be forty-eight hours ahead.

380. In this connection, the question can be well raised as to what really determines the track which a low is to follow and its velocity

of motion. The path and velocity are probably completely determined by the general drift of the atmosphere, the characteristics of the distribution of the meteorological elements about the formation itself, the location and characteristics of the surrounding meteorological formations, the surface topography of the country, and the meteorological condition of the country. It will be seen at once that the factors which enter in are large in number and very complex. If the value and relative importance of all of those factors could be determined from a study of many weather maps, it then would be possible in any particular case to determine the amount of motion, and the direction of motion, which each factor would cause, and the resultant for all the factors would give the direction of motion and the velocity of motion of the low. This can, however, probably never be done with precision. These factors are, however, of very different importance. Some are of prime importance, and many have such insignificant influence that they can almost be neglected. The general drift of the atmosphere and the distribution of pressure about the low itself are probably the most important factors in determining the path and velocity of motion of a low. The distribution of the temperature over a country is probably the next most important factor, while the distribution of the winds about the low as regards direction and velocity, and whether the low is accompanied by unusually high or low wind velocities, are also very important considerations. The location of the rain area during the preceding twenty-four hours and the direction and intensity of the surrounding highs and lows perhaps stand next in importance.

The things which determine the track and velocity of motion of a low.

A complete solution of the problem is impossible.

As an approximation to the solution of the general problem, the path and velocity of motion of a low might be worked out, using the two most important factors only. This has been done by Mr. Edward H. Bowie, who was then local forecast official at St. Louis, Mo., and has been recently transferred to Washington to be forecaster there. His investigation will be found in the Monthly Weather Review for February, 1906 (Vol. XXXIV, p. 61), and the reader must be referred to this article for a full discussion of his method. He first determines, for the various months in the year, the twenty-four hour drift of the atmosphere for all parts of the United States. Then, in the case of the low in question, the pressure gradient toward the center from the north, northeast, east, southeast, etc., is found from the weather map. The resultant of these eight pressure

The Bowie method of locating a low.

2 c

gradients is then found, and this represents the direction and magnitude of the unbalanced pressure which is forcing the low to move. The combination of this unbalanced pressure with the general drift of the atmosphere gives the direction of motion and the velocity of motion of the low. Only two of the general factors are taken account of in this method, but the high degree of accuracy which Mr. Bowie has attained in forecasting, by the use of his methods, attests the prime importance of these two factors and the value of his method. Even when the Bowie system is not used in full, the underlying principle is of great value to the forecaster. One can notice very easily in which quadrant of a low the isobars are most closely packed together. Here the pressure gradient will be largest, and the low will, in general, move away from this area.

Trabert[1] has recently called attention to the fact that the distribution of temperature over a country may determine the direction of motion of a low. Usually the temperature increases from the north toward the south. In this case, the winds on the eastern side of a low are bringing warmer, moisture-laden air from the south, while the winds on the western side are bringing cold, dry air from the north. The pressure will thus be falling on the eastern side and rising on the western side, and the low will thus move towards the east. If the temperature distribution over a country were different, a different direction of motion might result. The motion, in general, would be at right angles to the temperature gradient.

In 1905 the Belgian Astronomical Society instituted an international competition in forecasting at Liège, and the first prize was awarded to **The Guilbert rules.** Gabriel Guilbert of Caen. Later (1909) Guilbert published his method of forecasting in book form under the title "Nouvelle méthode de prévision du temps." His system is really based upon three rules, all of which relate to the winds which surround the low. These rules may be summarized and stated as follows:[2]

(1) Every depression that gives birth to a wind stronger than the normal will fill up more or less rapidly. On the other hand, every depression that forms without giving rise to winds of corresponding force will deepen, and often depressions that are apparently feeble will be transformed into true storms.

(2) When a depression is surrounded by winds having varying degrees of excess or deficiency, as compared with the normal wind, it moves

---

[1] See: "Die Zugrichtung der Depressionen," by WILH. TRABERT, in *Das Wetter*, February, 1911.

[2] See *Monthly Weather Review*, May, 1907, p. 210.

towards the region of least resistance. These favorable areas are made up of regions in which the winds are relatively light, and especially of such as have divergent winds with respect to the center of the depression.

(3) The rise of pressure takes place along a direction normal to the wind that is relatively too high, and it proceeds from right to left; an excessive wind causes a rise of pressure on its left.

Here, again, we have a whole system of forecasting built upon one of the more important of the many factors which determine the path and velocity of motion of a low. The wind direction and its velocity and the pressure gradient are very closely related, so that, in a way, the Bowie system and Guilbert's rules rest on the same foundation. These rules are of great value to the forecaster, as they can be easily applied, and help to guide one's estimation as to the probable direction and velocity of motion of a low.

It will also be noticed that all of the empirical rules which were stated in section 379 for locating the center of a low twenty-four hours ahead depend on the influence of one or several of the various factors which determine the path of a low.

381. Determining the distribution of the meteorological elements.— The method of determining the probable location of the center of the dominating low twenty-four and forty-eight hours ahead *First study* has just been fully discussed. The next step is to determine *critically the* the probable distribution of the meteorological elements *distribution* about the low in its predicted location. In order to do this, *ments about* first study critically the distribution of the elements about *the low.* the low as indicated on the weather map which is serving as the basis for the forecast. Notice in detail just what the values of temperature, pressure, wind, moisture, cloud, and precipitation are in different parts of the area covered by the low. If there is any departure from the normal distribution about a low, try to explain each departure from normal. It might be well in this connection to emphasize the necessity of noting carefully whether the low has a wind shift line, or is accompanied by the winter or summer type of cloud area. Next decide whether the predicted location of the center of the low twenty- *Next esti-* four hours ahead is going to change the previous distribution *mate the* in any way. Nearness to a body of water or a range of *caused by* mountains and the meteorological condition of the country *the new* probably exert the greatest modifying influence. Finally *location.* assume that the distribution of the elements about the low, twenty-four hours ahead, will be the same as on the previous weather map with the exception of any changes which the new location of the low may be expected to cause. The rule, then, briefly expressed, is to assume

The rule briefly expressed. that the distribution of the elements will be what it was, with the exception of the changes caused by the new surroundings.

382. The prediction. — The various steps which must be taken before a weather forecast can be made have been discussed individually

The three factors must be summed up. and in detail. It has been seen that the weather is the composite or resultant of three factors: normal weather, local influence, and the disturbances caused by passing meteorological formations. Normal weather is known for days and months in advance. The local influences are usually insignificant. No rules can be laid down for these. They must be learned from experience for each individual place. In the case of a low, the method of locating its center and determining the distribution of the elements about it, and thus its influence twenty-four or forty-eight hours ahead, has been explained. It now remains simply to sum up these three factors in order to make a weather prediction.

It is the best of practice for a beginner, or even for one who has acquired considerable skill in forecasting, to force oneself to form an exact

The making of exact forecasts gives valuable training. picture of what the weather is expected to be. This exactness can well be carried, in some cases, to the point of estimating the numerical value which each of the elements is expected to have. It must not be expected, however, that such a forecast will verify in every detail. Something will almost always be wrong, but fortunately the forecaster is not left to helplessly wonder what went wrong. As soon as the weather has occurred, and the next weather map has been received, it is always possible to see exactly what was overestimated or underestimated, and these very mistakes become stepping stones for the acquirement of more skill, accuracy, and confidence in forecasting. It will bear repetition that these exact forecasts have great value as a training. If such an exact forecast is to be made, the best order in which to predict the values of the meteorological elements is perhaps this: (1) pressure; (2) wind direction and velocity; (3) clouds — amount and kind; (4) moisture; (5) temperature; (6) precipitation — kind and amount.

The forecasts of temperature, wind, and state of the weather issued by the U. S. Weather Bureau are definite and explicit, but do not usually contain estimates of the numerical values that the elements are expected to have. The various phrases which may be used, and their exact meaning, are laid down in the " Station Regulations " of the U. S. Weather Bureau. Some of these regulations are here quoted. " Forecasts based on the 8 A.M. reports will be for the night The fore- of the current day, 8 P.M. to 8 A.M., and for the following casts as day, 8 A.M. to 8 P.M. They may cover the afternoon of the made by the U. S. current day only when marked changes are expected. Am- Weather biguous expressions will be avoided in the preparation of Bureau. forecasts. When conditions are so uncertain as to make the accuracy of a forecast doubtful, the word ' probably ' or ' possibly ' may be used.

" Forecasts of temperature changes will be made when the 24-hour changes at 8 A.M. and 8 P.M. of the following day are expected to equal or exceed 6° in the months, June, July, August, and September; 8° in April, May, October and November; and 10° in December, January, February, and March. The terms ' warmer, colder, decidedly warmer, decidedly colder ' will be used in describing the corresponding changes. Forecasts of fair, partly cloudy, or cloudy, will be made when pre- cipitation to the amount of 0.01 inch or more is not expected. When precipitation of 0.01 inch or more is expected, its character will be indicated by the use of such terms as ' rain, snow, local rain, showers, local snow, snow flurries, thundershowers, thunder- storms,' etc."

383. The following example will serve to illustrate the method of making an exact prediction and also a more general one, such as would be issued by a local forecast official. The weather maps for The mate- 8 A.M., Sunday, Dec. 8, 1907, and for 8 A.M., Monday, rial upon Dec. 9, are reproduced as Charts XXXVIII and XXXIX. which the forecast The first table contains various observations made at many is based. regular weather bureau stations at 8 A.M. on Monday, while the second table contains the detailed observations made at Albany, N.Y., at 8 A.M. and 8 P.M. on Sunday and at 8 A.M. on Monday. The problem is, on the basis of this material, to form a definite picture of the weather changes expected at Albany during the next thirty-six hours, and then to formulate such a forecast as would be issued by a local forecast official. The forecast must be made separately for Monday night (8 P.M. Monday to 8 A.M. Tuesday) and for Tuesday (8 A.M. to 8 P.M. of that day), thus covering a period of thirty-six hours from the time of the last observations which can be utilized. Similar observational material and the weather maps are always available whenever a fore- cast is made.

OBSERVATIONS 8 A.M. MONDAY, DEC. 9, 1907, AT VARIOUS WEATHER BUREAU STATIONS

STATIONS	TEMPERATURE Min. in Last 12 Hours	Max. in Last 24 Hours	WIND, MILES PER HOUR AT 8 A.M.	MAX. WIND REPORTED IN LAST 12 HOURS	PRECIPITATION LAST 24 HOURS. INCHES.
Abilene	46	70	20		
ALBANY	26	44	4		
Alpena	34	38	8		.12
Amarillo	30	60	24		
Asheville	50	60	12		.36
Binghamton	30	42	6		
Bismarck	16	40	16	32	.04
Boston	40	48	4		
Buffalo	44	32	8		
Cairo	54	58	10		.22
Canton	36	44	4		T
Charleston	56	64	14		.02
Chattanooga	52	60	4		.02
Chicago	44	52	12		.02
Cincinnati	48	62	6		.30
Cleveland	42	54	24		.04
Columbus	42	58	14		.24
Davenport	46	52	4		.02
Denver	22	52	6	36	
Detroit	40	52	18		.22
Dodge	30	50	30	36	.01
Duluth	32	38	12		.04
El Paso	36	62	4		
Erie	46	48	14		T
Escanaba	36	40	10		.18
Fort Smith	48	70	8		
Galveston	62	68	10		T
Grand Rapids	44	52	6		.38
Green Bay	34	44	12		.28
Havre	-2	24	0		.06
Helena	14	28	4		T
Houghton	32	38	18		.02
Huron	18	42	22		T
Indianapolis	48	50	12		.40
Jacksonville	64	72	10		
Kansas City	46	58	28	28	.02
Key West	72	80	12		
Knoxville	46	54	4		
La Crosse	40	46	4		.20
Lander	6	36	4		.22
Los Angeles	50	66	6		
Louisville	52	62	8		.16
Macon	58	66	6		.06

STATIONS	TEMPERATURE Min. in Last 12 Hours	Max. in Last 24 Hours	WIND, MILES PER HOUR AT 8 A.M.	MAX. WIND REPORTED IN LAST 12 HOURS	PRECIPITATION LAST 24 HOURS. INCHES.
Marquette	36	42	10		T
Memphis	56	60	12		.42
Miles City	—	—			
Milwaukee	42	44	4		.22
Modena	24	50	4		
Montgomery	58	66	4		.16
Montreal	36	38	6		T
Moorhead	12	32	14		
New Orleans	58	66	6		.64
New York	40	52	4		
Norfolk	42	64	6		
North Platte	24	52	24	36	.02
Oklahoma	38	68	30	38	.06
Omaha	36	48	38	40	T
Oswego	36	40	14		
Philadelphia	38	54	4		
Phœnix	40	66	4		
Pierre	22	44	24		
Pittsburg	42	60	6		T
Portland, Me.	30	42	4		
Portland, Ore.	40	46	4		.26
Rapid City	—	—	—		
Rochester	42	50	4		T
St. Louis	52	56	10		T
St. Paul	38	46	18		
Salt Lake	26	42	4		
San Diego	52	64	4		
San Francisco	58	58	4		
S. S. Marie	36	38	4		.14
Scranton	32	52	4		
Spokane	30	36	4		
Springfield, Ill.	50	56	4		.01
Springfield, Mo.	48	60	6		
Syracuse	36	48	8		T
Tampa	62	76	8		.02
Toledo	42	58	14		.42
Vicksburg	58	64	6		.26
Washington	32	62	4		
White River	—	—	—		
Williston	—	—	—		
Winnemucca	32	48	4		
Winnipeg	-6	20	12		
Yellowstone	4	30	6		.04

OBSERVATIONS AT ALBANY

		Dec. 8, 1907	Dec. 9, 1907
Temp. (F.)	8 A.M.	25.4	28.0
	Max. previous 12 hours	32.4	35.4
	Min. previous 12 hours	23.6	26.7
	8 P.M.	35.0	
	Max. previous 12 hours	44.0	
	Min. previous 12 hours	25.4	
Pressure reduced to sea level	8 A.M.	30.25	30.19
	Max. previous 12 hours	30.25	30.19
	Min. previous 12 hours	30.20	30.16
	8 P.M.	30.19	
	Max. previous 12 hours	30.25	
	Min. previous 12 hours	30.15	
Wind	Direction 8 A.M.	S.W.	N.
	Direction 8 P.M.	N.	
	Velocity 8 A.M.	2	2
	Velocity 8 P.M.	2	
Moisture	Rel. humid. 8 A.M.	94	100
	Rel. humid. 8 P.M.	91	
Clouds	Kind and direction of motion 8 A.M.	Alto-stratus from W.	Dense fog
	Cloudiness 8 A.M.	Pt. cloudy (6) light fog	Foggy
	Kind and direction of motion 8 P.M.	Light fog	
	Cloudiness 8 P.M.	Clear	
Precipitation in inches	Amount at 8 A.M. during previous 12 hours	0	0
	Amount at 8 P.M. during previous 12 hours	0	
	Kind		
Snow on ground 8 A.M.		0.5	Trace
Snow on ground 8 P.M.		0.3	

As soon as the two weather maps are consulted, it will be seen at once that it is an area of low pressure which is going to dominate the weather of the Middle Atlantic States during Monday night and Tuesday. On the Sunday map it will be seen that an area of high pressure of only moderate intensity is controlling the weather of the Atlantic seaboard, that an area of low pressure of marked intensity and regular form with its center just north of Texas is dominating the weather of the whole Mississippi Valley, and that an area of high pressure is just pushing its way in from the extreme northwest. By 8 A.M. on Mon-

The motion of the highs and lows during the preceding twenty-four hours.

day, as indicated on the map of that day, the area of high pressure has moved farther out over the Atlantic Ocean and has almost released its control of the weather of the Atlantic seaboard. The area of low pressure has moved southeast, recurved, and is now pursuing its course up the Mississippi Valley. It has become a strong, well-marked formation, and will surely dominate the weather of the Middle Atlantic states during the next two days. The northwestern area of high pressure has pushed its way in and become a well-marked, nearly typical high.

The first step is to form one's opinion as to the path which the low will probably follow, and the location of its center twenty-four hours ahead, that is, at 8 A.M. on Tuesday. In this case either the Bigelow, Russell, or Van Cleef system of tracks may be used. Each contains a well-marked track which corresponds with the motion of the low. One would expect it to move up the Mississippi Valley, across the Great Lakes, and down the St. Lawrence Valley. The rule for locating the center twenty-four hours ahead is to put it along on the track which it seems to be following the normal amount, and then to modify the estimation by other considerations. For December the normal amount is something less than 900 miles in twenty-four hours.[1] One would thus expect the center to be just north of lake Erie at 8 A.M., Tuesday. Now, are there any modifying circumstances? Near-by lows tend to coalesce and lows and highs repel each other. There is no near-by low with which to coalesce, and the Atlantic high is too weak and too far away to retard it. The western high is pushing in vigorously, and might tend to push the low eastward a little. Lows tend to follow the record of previous days. The low has hardly covered the 900 miles during the previous twenty-four hours, so that it seems to be one moving with a little less than normal speed. Lows tend to move faster after they recurve, and they also tend to move faster and grow a little deeper as they approach the Atlantic Ocean. It will thus be seen that the slow motion of the previous twenty-four hours has been explained, and one would expect greater speed during the next twenty-four hours. If the Bowie system of forecasting is used, it will be seen at once that the pressure lines are crowded together in the northwestern and western portion. This means that the formation would be pushed eastward. It will also be noticed that the region of high winds is between the low and the following high. If the Guilbert rules are held in mind, one would expect again that the formation would move more

The location of the center of the low twenty-four hours ahead.

[1] For the numerical data about highs and lows, see Chapter VI, part B.

than the normal amount eastward. As a result of all these modifying considerations, one would put the center a little farther along its course and somewhat eastward. As a final conclusion, then, Lake Ontario might be chosen as the probable location of the center of the low, 8 A.M. on Tuesday.

The Atlantic high will have moved far out over the ocean, and the western high will follow eastward behind the low; but neither of these formations will exert any influence on the weather at Albany during the time interval in question. *The probable motion of the other formations.*

The next question is this. What will be the probable distribution of the meteorological elements about the low on this Tuesday morning, when its center is over Lake Ontario. To answer this one must first study critically the present distribution and then estimate the probable changes caused by the new location. The central pressure is 29.6 inches. This is just about normal for a fully developed typical low, and the form of the isobars and the direction of the longer axis of the oval are also very *The distribution of the meteorological elements about the low.* typical. The wind directions form a very perfect counterclockwise spiral about the area of low pressure. At only a very few stations there are unusual or unexpected wind directions. In the northern, eastern, and southern quadrants the wind velocity is light or moderate. In the belt, at the west, between the low and the high, the wind velocities are large, reaching the velocity of a storm wind, forty or more miles per hour, at a few stations. The temperature lines are much distorted. The low is characterized by a very marked rise of temperature in front of it, and a sharp drop on the western side of the center. The cloud area extends far out to the east, reaching to the coast. On the western side, the nimbus cloud area quickly gives place to a clear sky. The precipitation has taken the form of rain, and has been general in the northern, eastern, and southern portions of the storm. The quantity has not been large, as it varies from a mere trace to not more than half an inch at any station. What changes may be expected in this distribution of the elements due to the new location of the center? As a low crosses the Great Lakes and nears the Atlantic coast, it comes into regions of greater moisture and usually becomes a little deeper. One would thus expect, Tuesday morning, to find the central pressure a little less, and the amount of precipitation somewhat larger, but in other respects without change.

During Tuesday, one would expect the low to pass down the St. Lawrence Valley and past the Albany station.

The definite picture of the expected changes in each element may now be formed.

Pressure. — At 8 A.M., Monday, at Albány, the pressure was 30.19, the wind direction was north, and its velocity was two miles per hour.

The definite picture of the expected weather changes at Albany. All this means that the departing high was just on the point of passing over the weather control to the coming low. One would thus expect the pressure to drop steadily during Monday and Tuesday. If the estimated central pressure of the low at 8 A.M. on Tuesday is placed at 29.4, then one would expect the pressure at Albany to be about 29.5 or 29.6 at this time. It should continue to fall during Tuesday, reaching its lowest, perhaps 29.4 or even a little lower, Tuesday afternoon or night. After the center of the low had passed nearest to Albany, the pressure would, of course, begin to rise.

Wind. — At 8 A.M., Monday, the wind was north with a velocity of only two miles per hour. It will be remembered that the wind velocities on the eastern side of the low were light or moderate. One would thus expect only light or moderate winds until the center of the low had passed. The direction should change to the east or southeast by Monday night. It would then shift more to the south, and should continue from some southerly quarter during most of Tuesday. Later on Tuesday the wind should shift to the southwest and west, and eventually northwest. The Hudson River Valley, however, runs north and south. Thus, due to a local influence, one would expect the easterly and westerly components to be lessened and the winds to be mostly southerly or northerly.

Cloud. — At 8 A.M., Monday, it was foggy almost to the point of rain, but no rain had fallen. One would expect it to remain totally cloudy during the whole of Monday and Tuesday. The nimbus cloud area ought to be reached before Monday night, and it ought to continue during the whole of Tuesday. If it should rain intermittently, the cloud would, of course, be called some form of stratus, instead of nimbus when there was no precipitation.

Moisture. — At 8 A.M., Monday, it was foggy with a relative humidity of 100 per cent. With the rising temperature, the relative humidity would grow somewhat less, but it should remain high all during the storm.

Temperature. — At 8 A.M., Monday, the temperature was 28.0°. The maximum during the previous day had been 44°. The coming storm was characterized by a marked rise in temperature. One would thus expect a maximum on Monday well above 44°, say 50°; but little drop during the night; a maximum on Tuesday even higher than on Monday, say 55° to 60°. By Tuesday night, the temperature should begin its rather sharp drop.

Precipitation. — Up to 8 A.M., Monday, there had been no precipitation. With the rapidly rising temperature, the kind of precipitation would, without doubt, be rain. The amount at various stations had been moderate, a trace to a half inch. The quantity should grow larger. One would thus expect some during Monday night and a half inch or more during Tuesday.

The definite picture of the expected weather changes during Monday night and Tuesday has now been given. It remains to formulate the forecast such as a local forecast official would issue. There **The official** are no predictions of a cold wave, frosts, or destructive winds **forecast.** to be made. The forecast will thus be concerned with temperature and state of the weather. A forecast in accord with the definite picture which has just been formed would be: For Albany and vicinity, rain and warmer to-night; Tuesday, rain with continued high temperatures.

The weather map for 8 A.M., Tuesday, Dec. 10, 1907, is reproduced as Chart XL, and the accompanying table contains the detailed observations at Albany at 8 P.M. Monday, and at 8 A.M. **The verifica-** and 8 P.M. Tuesday. By studying this map and the table **tion.** it becomes evident at once to what extent the predictions verify.

The illustration which has just been given will serve to elucidate the method of forecastings. There is no royal road to becoming a skillful forecaster. Practice is the essential thing. At the begin- **Experience** ning many mistakes will be made. Influences will be over- **is essential.** estimated or underestimated, or overlooked. When the next map is issued and the weather occurs, the forecaster is not left to wonder helplessly what went wrong. It is nearly always evident what was incorrectly estimated, and these very mistakes are the most valuable things in the acquirement of skill and experience.

OBSERVATIONS AT ALBANY

		Dec. 9, 1907	Dec. 10, 1907
Temp. (F.)	8 a.m.		53.0
	Max. previous 12 hours		53.0
	Min. previous 12 hours		45.5
	8 p.m.	45.5	48.2
	Max. previous 12 hours	51.0	55.0
	Min. previous 12 hours	27.0	48.2
Pressure reduced to sea level	8 a.m.		29.52
	Max. previous 12 hours		29.95
	Min. previous 12 hours		29.52
	8 p.m.	29.95	29.40
	Max. previous 12 hours	30.19	29.52
	Min. previous 12 hours	29.95	29.40
Wind	Direction 8 a.m.		S.
	Direction 8 p.m.	S.	N.
	Velocity 8 a.m.		13
	Velocity 8 p.m.	15	3
Moisture	Rel. humid. 8 a.m.		90
	Rel. humid. 8 p.m.	70	.92
Clouds	Kind and direction of motion 8 a.m.		Nimbus from S.
	Cloudiness 8 a.m.		Cloudy (10)
	Kind and direction of motion 8 p.m.	Stratus from S.	Nimbus from N.
	Cloudiness 8 p.m.	Cloudy (10)	Cloudy (10) light fog
Precipitation in inches	Amount at 8 a.m. during previous 12 hours		.10
	Amount at 8 p.m. during previous 12 hours	Trace	.58
	Kind	Rain	Rain
Snow on ground 8 a.m.			Trace
Snow on ground 8 p.m.		Trace	Trace

WEATHER PREDICTION WHEN V-SHAPED DEPRESSIONS AND OTHER SECONDARY ISOBARIC FORMS ARE PRESENT

384. Whenever a weather map is to serve as the basis of a weather forecast, the form of the isobaric lines must be critically studied. If these

All secondary isobaric forms must be noted.

lines have any other form than that of ovals surrounding areas of low or high pressure, they must be given particular attention, and this is especially the case if the general eastward motion of all meteorological formations will bring them near enough to the station in question to exert an influence during the following thirty-six or forty-eight hours. Two of these isobaric forms, other than the oval lows and highs, have received special names and merit consideration. These are *V-shaped depressions*, and *secondary lows*. Other isobaric forms have been named, and have, perhaps, individual characteristics, but their influence on weather is so slight that they do not merit consideration.

A V-shaped depression is a pocket-like bulge or projection in an isobaric line, and may have several different meanings. If two areas of low pressure are quite near together, the isobaric lines in crossing the ridge of higher pressure between them, must have this V-shaped bend. The same is true of the saddle which separates two areas of high pressure which are near together. In these two cases, the V-shaped bulge has no particular significance or importance. If the isobaric lines in the southern quadrant of a low all have this V-shaped bulge, it will probably be found that the low has a well-marked wind shift line. Lows with this peculiarity have already been fully discussed, and the significance of this V-shaped bulge, as far as coming weather is concerned, is evident. If an isobaric line surrounding an extensive high shows such a bulge, it often means that an area of low pressure is about to form, that is, it marks the origin of an extratropical cyclone. This is particularly liable to be the case if, in addition to this bulge, the wind is blowing from different directions at near-by stations and a small cloud area has formed. Sometimes a V-shaped depression which does not become a definite low will, however, cause thundershowers in summer or a snow flurry in winter. A V-shaped bulge in connection with an area of low pressure may mean that a secondary low is about to form.

The various kinds of V-shaped depressions and what they signify.

The general rule for weather forecasting, in connection with V-shaped depressions is, then, to note carefully if they exist and, if so, to remember that they may signify simply the dividing line between near-by lows or highs, the presence of a wind-shift line in connection with a low, the formation of a new area of low pressure, or the formation of a secondary low.

A secondary low is an area of low pressure of much smaller extent but sometimes of greater violence, which may exist in the southern quadrants of a much larger low. They are not very common in this country, but quite common in Europe. They usually move eastward and northward, faster than the big low, so that they appear to circle around it in a counterclockwise direction. Sometimes the big low fades away and the secondary assumes chief importance.

Secondary lows.

In weather prediction, it is best to consider the secondary low simply as another low, and forecast accordingly. The explanation of the popular expression that " the weather, instead of clearing off as it gave promise, has had a relapse " is often to be found in the existence of these secondary lows. Charts XLI and XLII, which represent the daily weather maps for Dec. 27 and 28, 1904, illustrate well the formation of a secondary low from a V-shaped bulge. On the 27th only a slight V-shaped bulge is in evidence, while on the 28th the secondary low is already fully formed. The weather maps for Jan. 27 and 28, 1908; May 20 and 21, 1908; April 30 and May 1, 1909; March 6 and 7, 1910; Oct. 27 and 28, 1910; Nov. 27 and 28, 1910, also illustrate well the formation of secondaries.

WEATHER PREDICTION CONSIDERING THE HIGH AS THE DOMINATING FORMATION

385. If it is a passing high instead of a low which is going to dominate the weather for the coming twenty-four or forty-eight hours, the general method of procedure, in making a forecast, is just the same as if it were a low. Determine first where the center of the high will be twenty-four and forty-eight hours ahead. This is done by noting the path which it seems to be following, and then putting it ahead on this path the normal distance covered in the given time. This estimated position is then modified by taking into account other considerations, which may be simply the general result of experience or may have taken the form of definite rules. Next assume that the distribution of the meteorological elements about it will be what it was, with the exception of any changes which its new location might cause. It is important in this connection to hold in mind the two chief ways in which an area of high pressure builds and becomes more intense. These are through the discharge of air from lows and through growing colder due chiefly to radiation to a clear sky. After the probable location of the high and the distribution of the elements have been determined, the method of arriving at the forecast is just the same as that for the low, which has been treated in detail.

An outline of the method of making the predictions.

WEATHER PREDICTION BY SIMILARITY WITH PREVIOUS MAPS

386. Prediction by similarity with previous maps is an entirely different method of weather forecasting from that which has just been described in detail. It might well be called the mechanical or automatic method of forecasting, because individual judgment plays no part. The first step is to find, in the series of weather This is the mechanical method. maps, that one which is the most similar to the map which is to be the basis of the forecast. Weather maps have been issued regularly in this country since 1871, so that The method stated. the file covers nearly forty years. In order to use this method of forecasting, it presupposes that the maps have been classified and indexed so that the similar map can be found. In this long series, there probably would always be at least one map which would be fairly similar to the map under consideration. The method of making the forecast is simply to assume that the weather changes which took place in the previous instance will again occur in the present case.

The six daily weather maps which are reproduced as Charts XLIII–XLVIII illustrate well this method of forecasting. The three maps for Thursday, Jan. 9, 1908, for Tuesday, Jan. 14, 1908, and for An illustration of the method. Monday, Jan. 27, 1908, are extremely similar. In each case, there is a high with a very moderate central pressure (30.1–30.3) with its center over the Gulf States, and this high, in each case, has a lip or projection extending over the Great Lakes. In each case, there is a low of considerable intensity which is departing by way of the lower St. Lawrence Valley, and another low of considerable intensity which is pushing in from the extreme northwest. The high is accompanied in each case by a decided drop in temperature over the Great Lakes. Further points of similarity can be detected by studying critically the distribution of the elements in the three cases. The three maps which follow are extremely similar. In each case, the high drifted eastward the same amount and the extension over the Great Lakes became more pronounced and was accompanied by a still further drop in temperature. In each case the coming low moved eastward about the same amount and developed a double center, becoming really a trough of low pressure extending from the Great Lakes to the Gulf. Weather prediction for the Middle Atlantic and New England states based upon the principle of similarity would have received very complete verification.

THE PREDICTION OF PECULIAR AND DANGEROUS OCCURRENCES

387. The prediction of frost. — The method of predicting the damage-causing frosts of the late spring and early autumn was described in

Chapter V, B, (sections 206–210). A frost is predicted by the officials of the U. S. Weather Bureau in exactly the same way as any other

The method of frost prediction. temperature. On the basis of the weather map the forecaster must estimate the probable minimum temperature (real air temperature in the thermometer shelter) on the following morning. In making this estimation he will be guided largely by the probable clearness of the sky during the night and the probable wind velocity. It will be remembered that a clear sky and absence of wind are essential for a large drop in temperature and a consequent frost.

If, after the probable minimum temperature in the thermometer shelter has been estimated, it is desired to determine the probable temperature of low-growing vegetation in the open at various

The causes of the difference between the temperature in a thermometer shelter and of vegetation under different conditions. points in a limited area surrounding the station in question, three things must be taken into account : (1) Plant temperatures go below the real air temperature because they are not sheltered and are free to radiate their heat. (2) Vegetation is located near the ground and not at the height of the thermometer shelter. (3) The variation in temperature over a limited area is often considerable. Thus the temperature of vegetation in the open, near the ground, and in the coldest part of a limited area may be expected to be from 5° to even, in extreme cases, 20° lower than the estimated minimum in the thermometer shelter. These facts are of vital importance and must not be overlooked.

Frost predictions and warnings are not issued by local forecast officials, but are issued only by the district forecasters from the respective centers.

What constitutes a frost in the Weather Bureau sense. Warnings of light and heavy frost will be verified by the occurrence of light and heavy frost respectively; and also by a reported minimum temperature of 40° and 32° respectively, accompanied by clear or partly cloudy weather and light winds or calm during the period for which frost is forecast. Thus what constitutes a frost has, in a certain sense, been defined by the Weather Bureau.

388. The prediction of cold waves. — The meaning of the term

What constitutes a cold wave. " cold wave " has been made definite by the U. S. Weather Bureau. Nos. 299–304 of the " Station Regulations," which are reproduced here, contain the definition.

" Cold wave warnings will be ordered when it is expected that a 24-hour fall in temperature, equalling or exceeding that specified for the district, will occur within the 36 hours following the observation upon which the order is based, accompanied by a minimum temperature of the required degree, or lower.   The districts and the respective temperature falls and minimum temperatures required to verify cold wave warnings during the different seasons are as follows :

" In northern Maine, northern New Hampshire, northern Vermont, northeastern New York, western Wisconsin, western Iowa, Minnesota, North Dakota, South Dakota, Nebraska, Montana, Wyoming, Idaho, eastern Washington, and eastern Oregon : a 24-hour fall of 20°, with a minimum of zero in December, January, and February, and a minimum of 16° from March to November inclusive.

"In southern Maine, southern New Hampshire, southern Vermont, Massachusetts, Rhode Island, Connecticut, New York, (except northeastern part), northern New Jersey, Pennsylvania, Ohio, Indiana, Michigan, Illinois (except Cairo), western Maryland, West Virginia northern Kentucky, Missouri, eastern Iowa, eastern Wisconsin, Kansas, Colorado, the Texas panhandle, northern New Mexico, northern Arizona, Utah, and Nevada; a 24-hour fall of 20° with a minimum of 10° in December, January, and February, and a minimum of 24° from March to November, inclusive.

" In southern New Jersey, Delaware, eastern Maryland, the District of Columbia, Virginia, western North Carolina, the northwestern quarter of South Carolina, northern Georgia, northern Alabama, northern Mississippi, Tenessee, Cairo, southern Kentucky, Arkansas, Oklahoma, Indian Territory, northern Texas (except the panhandle), southern New Mexico, western Washington, and western Oregon : a 24-hour fall of 20°, with a minimum of 20° in December, January, and February, and a minimum of 28° from March to November, inclusive.

" In eastern North Carolina, central South Carolina, central Georgia, central Alabama, central Mississippi, northern Louisiana, and central Texas : a 24-hour fall of 18° with a minimum of 25° in December, January, and February, and a minimum of 32° from March to November, inclusive.

"In the coast region of South Carolina and Georgia, extreme southern Georgia, Florida, extreme southern Mississippi, southern Louisiana, the Texas coast, California, and southern Arizona : a 24-hour fall of 15°, with a minimum of 32° in December, January, and February, and a minimum of 36° from March to November, inclusive."

The best precept to follow in making a cold wave forecast is perhaps the following, which is a slight modification of the one advocated by

2 D

the U. S. Weather Bureau. Through the coming area of high pressure (a cold wave is always caused by a coming high) draw two axes at

The area for which a cold wave is to be predicted. right angles to each other extending north and south, and east and west. The area for which to predict a cold wave will be an oval area located entirely in the southeast quadrant. A cold wave is particularly sure to occur if a passing low is located to the east of the high, and the high has an oval form with the larger axis extending northeast-southwest. The ac-

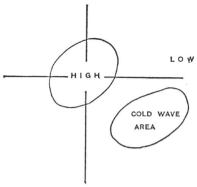

companying figure makes clear the area for which the cold wave is to be predicted. The brief rule is then to determine first if the probable drop in temperature will be sufficient to cause a cold wave, and then to predict it for an area located as stated above. Great caution should be used in predicting it for any other area, as many failures have resulted from so doing.

Cold wave forecasts and warnings are issued only by district forecasters, and not by local forecast officials. On the Washing-

Fig. 145. — Diagram Illustrating the Area for which to Predict a Cold Wave.

ton weather map, these forecasts are indicated by the letters *CW* printed near each station for which a cold wave is predicted.

How the prediction is indicated. Chart XLIX, which reproduces the daily weather map for Jan. 6, 1909, illustrates well the area for which a cold wave is predicted, and the method of indicating the prediction on the weather map.

Bulletin P (W. B. publication 355) of the U. S. Weather Bureau, by Edward B. Garriot, entitled "Cold Waves and Frosts in the United

Illustrations of cold waves. States," contains 328 charts which reproduce the daily weather maps and exhibit the meteorological conditions which attended the principal cold waves from 1888 to 1902 inclusive. This invaluable publication also contains a brief account of the origin and cause of cold waves, and presents a chronological account of the historic cold periods in the United States prior to 1888. No one wishing to make cold wave forecasts could do better than study critically this exhaustive treatise.

389. **The prediction of tornadoes.** — Since a tornado usually covers an area only a few hundred feet wide and a few miles long, it is decidedly undesirable to alarm a whole state or several states with the forecast that a tornado could occur. For this reason, tornadoes are never predicted by the U. S. Weather Bureau. No. 244 in the "Station Regulations" covers the point.

Predictions of tornadoes are not made.

" Forecasts of tornadoes are prohibited. When conditions are favorable for the occurrence of destructive local storms, the term severe thunderstorms' or 'severe local storms' may be used by district forecasters. The phrase 'conditions are favorable for the destructive local storms' will be used only by the Chief of Bureau, or, in his absence, from the Central Office, by the chief of the Forecast Division."

In a previous chapter (sections 343 to 346) there was given a full discussion of the portions of the country most frequented by tornadoes, the time of day and season of occurrence, the type of low which most frequently gives rise to the violent thundershowers which are accompanied by tornadoes, and the location of the tornado with reference to the low. In forming one's own opinion from a weather map as to whether a tornado might occur, there are three things to note : First, is the type of low one which is suggestive of tornadoes, secondly, is it the season of the year when they are likely to occur, thirdly, is the place where they could be expected to occur one much frequented by tornadoes.

When tornadoes are possible.

390. **The prediction of destructive wind velocities (storms).** — Storm warnings are displayed when the wind is expected to attain a velocity for a period of five minutes, equaling or exceeding the verifying velocity within the 24 hours following the time that the warning is ordered hoisted. The above may serve as a definition of what is meant by a destructive or storm wind. There are at present nearly 60 Weather Bureau stations on the Atlantic coast and Gulf coast, on the Great Lakes, and on the Pacific coast in addition to the many adjacent storm warning display stations, where these warnings are displayed. The verifying velocity is different for each station, and varies from roughly 20 to 60 miles per hour. For example, at New York it is 44 miles per hour, at Boston 32, at Portland, Me., 28, at Norfolk 26, at Buffalo 26, at Chicago 46, at Duluth 40, at Seattle 35, and at San Francisco 36.

What constitutes a storm wind.

The verifying velocities.

The probable occurrence of a verifying velocity is predicted in the same way as any other wind velocity. A storm warning should contain the location of the storm center,

The method of prediction.

and the probable direction in which it will move, with a forecast of the force, direction, and shifts of the wind. These warnings are issued only by a district forecaster, and the flags used to announce them are described in section 348.

Bulletin K (W. B. publication 288) of the U. S. Weather Bureau, by E. B. Garriott, entitled "Storms of the Great Lakes," contains 952 Illustrations. charts, which reproduce the daily weather maps and exhibit the meteorological conditions which attended the storms which were accompanied by a wind velocity of the verifying amount from 1876 to 1900 inclusive. This exhaustive treatise on the storms of the Great Lakes must be studied critically by any one who would predict wind velocities for this region.

391. The prediction of floods. — The method of predicting floods will be given in Chapter X, which treats of floods and river stages. The regula- The regulations concerning river and flood forecasts are tions con- the following: cerning river and flood "Flood warnings and forecasts of river stages will be issued forecasts. from forecast centers that are river centers, and from specially designated river centers. Copies of all river forecasts and warnings will be promptly transmitted to the Central Office; regular forecasts on the Daily River Forecast card (Form No. 1086 — Met'l), and special forecasts on the Special River Forecast card (form No. 1087 — Met'l). Flood warnings will be immediately transmitted by telegraph to the Central Office."

THE ACCURACY AND VERIFICATION OF PREDICTIONS

392. The terms used in official predictions. — The terms which may be used by the officials of the U. S. Weather Bureau in making The terms predictions are all prescribed and each has a definite mean- used in offi- ing. This, to a certain extent, hampers individuality cial predic- and a free expression of opinion as to the coming tions are all prescribed weather, but it is necessary in order to prevent am- by the biguity and hedging, and to make possible a systematic Weather Bureau. verification of the predictions. The terms which may be used, and the meaning of these terms in connection with forecasts of temperature and state of the weather, have been given in section 382. What constitutes a dangerous wind velocity (storm), a cold wave, and a frost has been stated in sections 390, 388, 387, respectively.

393. The system of verification. — All forecasts and warnings are sent to the Central Office of the Weather Bureau for verification, and the only official verification takes place there. In the case of the forecast districts, the forecasts and warnings are sent each day to Washington by the forecasters for the district. All local forecasts of temperature and state of the weather made at regular stations, and all practice forecasts in connection with temperature, state of the weather, storm winds, cold waves, and frosts are entered on the appropriate form (No. 1069 — Met'l) and forwarded to Washington. At the central office of the U. S. Weather Bureau at Washington, these forecasts and warnings are verified in accordance with a definite set of rules, and the percentage of accuracy of each forecaster can be determined. The forecasts and warnings are not verified as a whole, but on five different counts — that is, the forecasts of temperature, state of the weather, storm winds, cold waves, and frosts are all verified separately. These rules for verifying can best be given by quoting from the "Station Regulations."

All forecasts and warnings are sent to Washington for verification.

They are verified on five different counts.

"Forecasts and warnings and practice forecasts and warnings will be verified by the following rules:

"Rain or snow forecasts and all modifications thereof, containing the terms 'possibly,' 'probably,' etc., will be verified by the occurrence of 0.01 inch or more of precipitation.

The rules for verifying predictions.

"Precipitation to the amount of 0.01 inch or more which is not forecast, or which is forecast and does not occur, will be charged as a failure.

"A forecast of precipitation will be considered to be neither a success nor a failure when a trace of precipitation is recorded within the period specified.

"In determining the percentage of verification, the number of forecasts verified will constitute the dividend; the divisor will be the sum of the number of forecasts verified, the number of forecasts not verified, and the number of occurrences of 0.01 inch of precipitation or more which were not forecast.

"Forecasts of temperature change, and all modifications thereof containing the terms 'possibly,' 'probably,' etc., will be verified by the occurrence of a temperature change, of the kind forecast, equaling or exceeding the stationary limit for the season.

"Temperature changes equaling or exceeding the stationary limit which are not forecast, or which are forecast and do not occur, will be charged as failures.

"In determining the percentage of verification, the number of forecasts verified will constitute the dividend; the divisor will be the sum of the number of forecasts verified, the number of forecasts not verified, and the number of occurrences of temperature changes equaling or exceeding the stationary limit not forecast.

"A storm warning will be verified by the occurrence of a verifying velocity for a period of 5 minutes at the station where the warning is displayed, or by the occurrence of a verifying velocity within 150 miles of the station, within 24 hours after the warning was ordered hoisted, and without regard to changes in the original order that may be made during the period.

"The continuance of a verifying velocity for a period of 20 minutes without a warning will be considered a storm without a warning.

"Not more than one verification, or one storm without warning, will be counted for any one station during any 24-hour period.

"In determining the percentage of verification, the number of warnings verified will constitute the dividend; and the sum of the number of warnings verified, the number of warnings not verified, and the number of storms without warnings will constitute the divisor.

"Cold wave warnings will be verified by the occurrence of the required 24-hour fall in temperature, accompanied by the required minimum within the 36 hours following the regular observation on which the warning is based.

"The occurrence within a 24-hour period of the temperature change and minimum temperature required to verify a cold wave warning, for which a warning was not issued, will be counted a cold wave without warning.

"Not more than one verification, or one cold wave without warning, will be counted at any one station during any 24-hour period.

"In determining the percentage of verification, the number of cold wave warnings verified will constitute the dividend; and the sum of the number of warnings verified, the number of warnings not verified, and the number of cold waves without warnings will constitute the divisor."

"In determining the percentage of verification, the number of frost warnings verified will constitute the dividend; and the sum of the number of frost warnings verified, the number of warnings not verified, and the number of frosts and verifying temperatures without frost for which warnings were not issued will constitute the divisor."

394. The accuracy attained. — The statement is usually made that the accuracy attained by the official forecasters of the U. S. Weather Bureau is between 80 and 85 per cent. This is, however, a general average for all forecasters, all sections of the country, and all five of the lines along which forecasts and warnings are verified. The accuracy which can be attained in different parts of the country is very different. It is probably easier to forecast for California than any other part of the country, and the hardest part is perhaps New England. The forecasts of cold waves probably verify least often, as the accuracy is only about 70 per cent. Some forecasters are more skillful than others, but the accuracy of different forecasters predicting for the same locality would probably not differ as much as 10 per cent. An accuracy of 85 per cent means that on the average, one day in seven will see a complete failure of the weather predictions. It must be held in mind in this connection, however, that a mere guess ought to yield the accuracy of but 50 per cent.

The average accuracy is above 80 per cent.

The accuracy depends on the locality, the forecaster, and the thing predicted.

The accuracy attained in Europe is about the same as in this country, although the terms used in forecasting, and the method of verification, are quite different in different countries.

Weather forecasting, as regards accuracy, may be considered on about the same level as the practice of medicine. The forecaster can diagnose the present condition of the atmosphere with as much precision as a physician can determine the bodily condition of a patient. He can predict the coming weather certainly with as much certainty as a physician can predict the exact course and outcome of any well-known disease. Weather forecasting can probably never hope to attain the accuracy of astronomy in predicting celestial occurrences.

Weather forecasting on the same level as the practice of medicine.

The increase in accuracy during the last fifteen years has been small, and this would seem to indicate that present methods can yield no greater accuracy. This certainly does not mean that no elaborate investigations have been undertaken to determine if a new system could be devised or the present methods improved. The weather map for the whole northern hemisphere has been constructed daily, and the permanent highs and lows have been critically studied in order to determine if changes in them exert an influence on the passing highs and lows and thus determine their path, velocity, and characteristics. The weather maps for differ-

The accuracy has not increased with time.

ent levels above the earth's surface have been constructed in order to determine if the passing meteorological formations, as portrayed at these levels, are less erratic in their behavior than at the earth's surface. The areas of greatest and least pressure and temperature change have been studied critically to determine if these are any more regular in their behavior than highs or lows. The variations in the energy received from the sun have been determined and studied in order to ascertain if these influence the highs and lows. Of these, the study of the permanent highs and lows has yielded the greatest results, but none of these investigations has revolutionized the methods of forecasting or added much to the accuracy.

Many investigations have been undertaken to improve present methods.

395. The popular idea of the accuracy of weather forecasts. — The newspapers and periodicals are essentially just, fair, and considerate in their attitude towards the Weather Bureau, and the failure of some predictions to verify. It is very seldom that a hostile editorial is seen or a vindictive, sarcastic attitude or comment is met with. There are, of course, many jokes at the expense of the Weather Bureau and its failures. There are also many jokes about the repairs necessary to an automobile, but that does not prove that the automobile is not both a useful and pleasure-giving vehicle. There are a very few who sneer at the Weather Bureau and declare that it should be done away with. Those who have this attitude are usually those who really know very little about meteorology, the work of the Weather Bureau, or the value of its predictions. The Weather Bureau certainly saves to this country annually at least three and probably ten times its cost. It thus gives very handsome returns for what is expended, and any one who sneers at its work is purely a useless, destructive critic, unless he can propose some plan which would lead to greater returns.

The newspapers and periodicals are fair.

The Weather Bureau is a very profitable investment.

It has often been said that it would be very desirable if the weather map could receive such wide distribution that it would come to the notice of every one, and at the same time the public could be educated to interpret the map. Every person would then make his own predictions, and the attitude of a person towards what he does himself, even if imperfect, is always very different from what it is towards the same thing done, perhaps somewhat better, by some one else.

Long-range Predictions

396. Prediction from station normals. — The definite predictions made by the officials of the U. S. Weather Bureau on the basis of the daily weather map are for only thirty-six or forty-eight hours ahead, and it is impossible to extend the period with any certainty to more than three days, even in the case of the most typical maps. Recently the attempt has been made to issue a general prediction for a whole week, and the basis of this will be discussed later. The popular desire is, however, not for predictions for a few hours ahead or at most a week ahead, but for predictions for a month, or several months, or a whole year in advance. These are called long-range predictions.

What is meant by long-range prediction.

The attempt has sometimes been made to use the station normals for the various meteorological elements as the basis of long-range predictions. For example, for the northeastern part of New York State, it could be stated with certainty on the basis of these various normals that the temperature would fall below zero at least once during the winter, that there would be at least one thundershower during the summer, that at least two feet of snow would fall during the winter, etc. These statements are, however, not really long-range predictions, but merely a description of the climate of the locality. Furthermore, they give no indication as to what the weather is going to be on any definite date, or even whether the season is going to be too hot, or too cold, too wet, or too dry.

No definite predictions can be made from station normals.

The attempt has also been made to use departures from normal as the basis of long-range predictions. There is a strong popular belief that, for example, if a certain month is too hot or too dry, the next month will be too cold or too wet, in order to compensate. Now these departures from normal, particularly in the case of temperature and precipitation, have been studied critically for many stations and for many years,[1] and the conclusion drawn from such research work has always been that, from the departure from normal on the part of any particular month, no conclusion whatever can be drawn as to whether the following month will be too hot or too cold, too wet or too dry.

Departure from normal cannot be used as a basis for long-range weather predictions.

Much research work has also been done on the probable characteristics of one season as determined by the departure from normal of a

[1] See Bulletin U of the U. S. Weather Bureau.

previous season. Köppen has found, for the middle of Europe, that

Seasonal   the probability of a departure from normal in the other di-
sequence.   rection in the case of temperature, from one season to another, is as follows:

winter to spring	0.49		spring to summer	0.45
winter to summer	0.44		spring to autumn	0.40
summer to autumn	0.38		autumn to winter	0.45
summer to winter	0.50		autumn to spring	0.52

This means, for example, that the probability of a cold autumn following a warm summer is 38 out of 100. It will be seen at once that the
preponderance of probability is too slight to make this of much value in
forecasting. At any rate, no definite forecasts can be made. Hellmann
has found, for Berlin, that there is the greatest probability of the following successions, but the preponderance of probability was always small:

moderately mild winter . . . . . . cool summer
very mild winter . . . . warm summer
moderately cold winter . . . . . . cool summer
very cold winter . . . . very cool summer
moderately warm summer . . . . . moderately mild winter
very warm summer . . . . cold winter

Similar probable sequences could be worked out for any station on the
basis of the observations, and the preponderance of probability could
be determined for each sequence. It is of too little value in forecasting,
however, to make it worth while. If it has been done, one might say,
for example, after a cold winter had passed, that the probability of a
warm summer was perhaps 46 out of the 100. This would be very far
from a satisfying long-range prediction.

397. Weather cycles. — The daily and annual change in the values of
the meteorological elements is, of course, very marked. There is always

Several   a pronounced diurnal and annual variation, and the charac-
cycles have teristics and magnitude of these variations in the case of each
been de-   element have been fully discussed in previous chapters. Inves-
tected.   tigators have also thought they have found several minor
periods or cycles which underlie the changes in the values of the elements.
Cycles of 3 days, 5.5 days, 11 years, and 35 years without doubt exist.[1]

Cycles of 26.7 days, corresponding to the sun's rotation, 7 years,

A criterion 19 years, corresponding to the nutation period, and many
of value.   others have been announced by different investigators. In
fact, the number of cycles has become so large that it is a serious ques-

[1] See *Monthly Weather Review*, April, 1899, p. 156.

tion if some criterion should not be used to determine if a cycle is worthy of consideration. It could well be said that a cycle is not worth considering, unless (1) it is very well marked, if it occurs at only a few stations; (2) it occurs at many stations, if it is of small magnitude; (3) it corresponds to something else which has the same period and could be its cause.

The three-day cycle is probably the average interval between the passing of a low and the coming of the following high. The 5.5-day cycle is, in a certain sense, a double three-day cycle. It is **The 5.5-day** probably the average time interval between the passage of **cycle.** well-marked lows near any given station. It is evident to any one who follows the daily weather maps, day after day, that these two cycles may be said to exist, but they are very far from definite and could not be projected into the future more than two weeks at the very most, and only then with great uncertainty.

The eleven-year cycle corresponds to the sunspot period, and is very pronounced in terrestrial magnetism and atmospheric electricity. It is easily found in certain of the meteorological elements, **The eleven-** but the magnitude of the variation is extremely small, and **year cycle.** the observations must be carefully averaged at many stations for a long time, to detect it with certainty. Its presence has been detected chiefly in connection with the number of tropical cyclones, the amount of precipitation, and the tracks followed by lows.[1]

The cycle of 35 years is due to Brückner, and is particularly noticeable in connection with temperature and precipitation. Not only have direct observations been used in the discussion, but such **The thirty-** indirect data as the dates of harvests, the opening and clos- **five-year** ing of navigation, the height of inclosed seas, the severity of **cycle.** winters, etc. In fact, even the size of the circle of annual growth in the case of very old trees has been used.

European observations show that during the last two centuries —
1746–1755, 1791–1805, 1821–1835; 1851–1870 were relatively warm.
1731–1745, 1756–1790, 1806–1820, 1836–1850, 1871–1885, were relatively cold.
1756–1770, 1781–1805, 1826–1840, 1856–1870 were relatively dry.
1736–1755, 1771–1780, 1806–1825, 1841–1855, 1871–1885 were relatively wet.

The amount of the oscillation in the case of the precipitation is very different in different countries, and is usually greater in the interior than near

[1] See W. J. HUMPHREYS, *Astrophysical Journal*, September, 1910.

the coast. On the average, it amounts to about 20 per cent of the total amount. The period is not strictly 35, but varies from 34 to 36 years.

It will be seen at once that all of these cycles are too indefinite and

The cycles are of no value in forecasting.

of too little importance to play any part in weather forecasting or serve as a basis for long-range predictions.

398. **Tendency of a weather type to continue.** — If the ceaseless changes in the values of the meteorological elements are carefully studied, it will be apparent that the number of changes in the weather from one day to the next is smaller than the number of

There is a decided tendency for the existing weather to continue.

continuations of the same kind of weather. That is, after a cold day, there is a greater probability of another cold day than of a warm day, and after a rainy day there is a greater probability of another rainy day than of a fair day. There is thus a decided tendency for the existing weather to

continue. At Brussels, for example, the probability of a change after a certain number of days with a certain character as regards temperature and rain is given in the following table:

After a continuance of	1	2	3	4	5	6	7	10	15 days
Temperature	.25	.24	.22	.21	.17	.17	.15	.15	.13
Rain	.37	.32	.30	.26	.27	.24	.25	.23	.23

This means that if there have been four rainy days in succession, the probability of the next day being fair is only 26 out of 100. A similar table could be prepared for any station, and the same decided tendency of existing weather to continue would be found. This table also shows that if a weather forecaster in Brussels should each day systematically predict that the next day would be the same as regards temperature and rainfall, he would attain an accuracy in forecasting of about 75 per cent.

If longer periods of time are considered, it will be found that, as soon as a definite type of weather has become established, there is a decided

A type of weather also tends to continue.

tendency for this type to continue several days or perhaps even several weeks. A warm rainy period is apt to continue for a week or two, and cold dry weather usually lasts equally long. During the types, the highs and lows tend to follow

the same tracks and have the same characteristics. This is a matter

This is due to the permanent areas of high and low pressure.

of great practical importance in forecasting, and more and more attention is being paid to it. This tendency of a weather type to continue is probably the result of what Teisserenc de Bort calls the " centers of activity " of the atmosphere. If the chart which exhibits the isobaric lines for

January is studied, it will be seen that in the northern hemisphere there are four permanent highs, an immense one over Asia, a smaller one central over North America, and two still smaller ones over the Atlantic and Pacific oceans. There are also two permanent areas of low pressure with their centers over the north Atlantic near Iceland and over the north Pacific south of Alaska. These areas are the so-called centers of activity and remain practically fixed. They are the permanent landmarks of pressure, and stand in sharp contrast to the moving areas of low and high pressure. If any change occurs in the location or intensity of these permanent areas, the moving areas of high and low pressure have different characteristics and move over different tracks, and thus give rise to a different type of weather. Since these permanent areas of pressure change their characteristics slowly and reciprocally, there is a tendency for a given weather type to continue for several days, or even several weeks. The general forecasts which have lately been attempted at Washington for a week ahead, are based upon a study of these permanent areas of pressure and this tendency of a weather type to continue. These centers of action change with the time of year. A series of isobaric charts, exhibiting the normal pressure for every month in the year, would be necessary for the full discussion of the subject. The exact type of weather which exists for given characteristics on the part of these areas has not yet been fully determined. In fact, not much more has been done than to recognize the importance of these areas and the fact that the type of weather probably does depend, to a large extent on their characteristics.[1] The pressure distribution over the whole northern hemisphere for January 28, 1910, is given as chart L. This is an illustration of the daily pressure map as made at Washington for the whole northern hemisphere (see section 362). The Siberian high has an unusual intensity. The low over Europe is the strong persistent one which caused the excessive precipitation over France, which resulted in the great floods at Paris in January, 1910.

399. Popular superstitions and credulity. — In the foregoing paragraphs (sections 396, 397, 398), the science of meteorology has presented all that it has to offer in connection with long-range weather predictions. On the basis of the daily weather map, exact predictions for the coming 36 or 48 hours can be made, and these verify in about 80 to 85 per cent of the cases. Based on the tendency of a weather type to continue, and on a study of the centers of activity in the *A frank statement of what can be accomplished along the lines of weather prediction.*

[1] For a short bibliography of articles on this subject see *Bulletin of the Mount Weather Observatory*, vol. III, part 4, p. 237.

northern hemisphere, it is possible to make a general forecast for perhaps a week ahead. It was seen that long-range weather predictions could not be made with any desirable definiteness or certainty based on station normals, departures from normal, or weather cycles. Since this is the state of the case, the U. S. Weather Bureau and scientific meteorologists are frank and honest enough to admit that, at present, long-range weather predictions are an impossibility. Long-range weather predictions are very desirable, and every means of making them is being thoroughly investigated; but simple honesty demands that the admission be made that, at present, there is no way known of making them.

The popular desire, however, is for long-range weather predictions, and, this being the case, it is not remarkable that various attempts are **The various things which serve as a basis for long-range forecasts.** made to satisfy this desire in one way or another. Some try to make long-range predictions for themselves and friends, using as a basis the action of some animal, or the weather on some definite date, or something connected with the moon. Others, bolder and usually with a desire for financial gain, publish these forecasts in almanacs and the like. Some newspapers even publish these predictions, and worse than this, they actually pay for the privilege of publishing these worse than useless predictions. Now if these long-range predictions are not mere guesses, the various things which serve as a basis may be grouped under three heads: (1) the influence of the moon, the sun, or the planets — in short, astronomical control; (2) the actions of animals, birds, and plants; (3) the weather during certain days, months, or season. Now, in the case of the sun and moon, as was stated in connection with weather cycles, there may be a very slight influence on some phases of the weather, but the influence is so slight that it is almost impossible to find it with certainty. It is certainly so small that it can play absolutely no part in forecasting. As for the rest, there is, in the first place, absolutely no scientific reason why they should have any connection with the coming weather; and, in the second place, the observations at many stations have been averaged for many years to determine if they do have any influence, and the result has always been to find that they do not. Thus all long-range forecasts built upon such things are mere superstitions and have no foundation whatever. They are no better than mere guesses, and it should be remembered that a mere guess should be correct half the time, so that there should be no surprise at some chance verifications. It might perhaps be agreed that those who make these long-range forecasts

simply for themselves and friends, should be allowed the possible pleasure of this probably harmless self-delusion.   It is a very different
matter, however, in the case of those who publish their forecasts and
particularly those who publish them for financial gain.   They are not
only making gain at the expense of the superstition and credulity of the
public, but they are making the public believe that something can be
done which cannot be done.

These published forecasts are of all degrees of definiteness.   Sometimes, in some almanacs, the statement " Snow may be expected at this
time " will be found on the margin, and extending over a A typical
third of a winter month.   Since this would be verified by long-range
snow occurring anywhere in the United States at any time forecast.
during the period, it is needless to add that it is certain of verification.
Sometimes the more elaborate forecasts run something like this : February 21 to the end of the month, constitutes a storm period of marked
energy.   There will be snow in the northwest, gales over the Great
Lakes, and freezing weather on the Atlantic coast.   A storm of marked
energy will move from west to east across the country.   Its coming will
cause warmer weather with rain or snow.   It will be followed by northwest winds and colder with snow flurries in New England.   There will
be thundershowers in the Southern states and the month which closes
with this period will average too warm.   The above is copied from no
source, has no basis whatever, and can be applied to any year.   It is
simply a mixture of meteorological information, meteorological statistics, glittering generalities, and plain guesses ; and any forecaster of the
U.S. Weather Bureau, and any scientific meteorologist could write predictions like the above by the page and volume if he were willing to cheapen
and degrade his science to that extent.   There is never an in- A critical
terval of seven days at the end of February without a low of , analysis of
fair intensity crossing the United States somewhere.   Lows such a
always move from west to east, and are preceded by south forecast.
winds and warmer weather and are followed by northwest winds and
colder weather.   All this is simply meteorological information.   Now,
as regards the gales on the Great Lakes, snow flurries in New England,
and thundershowers in the Southern states, if a series of weather maps
covering these seven days for the last twenty years were carefully studied,
it would be seen that these occurrences took place about 18 out of the
20 years.   The chance is thus ten to one in favor of these things happen-
ing.   It is therefore an extremely safe prediction considering statistics.
Freezing weather on the Atlantic coast is a glittering generality, as the

32° F. line always intersects the Atlantic coast somewhere. That the month would average too warm is a plain guess, and thus stands an even chance of being right for any station. Furthermore, there are certain to be many localities where it will verify.

The widespread belief in the existence of an equinoctial storm and Indian summer comes, to a certain extent, under the head of popular superstitions. If the equinoctial storm is defined as a rain-storm, lasting at least three days, and occurring within two or three days of the 21st of September, then there is seldom a year when one occurs. If, however, the equinoctial storm is defined as any rainstorm lasting, say a day or longer, and occurring within two weeks of the 21st of September, then there is very seldom a year when several equinoctials do not occur. The reason for the belief in an equinoctial storm is probably the fact that about this time of year the first storms of the winter type, with steadily falling precipitation, make their appearance. They stand in sharp contrast to the summer type with the sultry weather and thundershowers. Storms of the winter type can occur, however, during any month of the summer. The amount of the precipitation near the 21st has been shown by averaging the observations at many stations to be no greater than before or after that date.[1]

The equinoctial storm and Indian summer.

The case is similar with Indian summer. If Indian summer is defined as a spell of peculiar weather in the autumn, characterized by great warmth, smokiness, and haziness, and lasting for several weeks, then Indian summer seldom occurs. If, however, Indian summer is defined as a few days of slightly greater warmth and haziness, which only serve to emphasize our otherwise delightful autumn weather, then Indian summer nearly always occurs.[2]

FORECASTS FROM LOCAL OBSERVATIONS AND APPEARANCES OF SKY

400. **Prediction from the readings of instruments and the appearance of the sky.** — The question is often raised, if it is possible to predict the weather from the sky appearance and the indications of meteorological instruments without using the daily weather maps. It can readily be shown that a good general idea of the coming weather can probably be formed in this way, but that exact and definite prediction requires the use of the

A general idea of the coming weather can be gained from sky appearance

[1] See *Monthly Weather Review*, November, 1901, p. 508.
[2] See *Monthly Weather Review*, January, 1902, p. 19.

daily weather map.   Suppose, for example, that it is January, and local and the wind has just changed to the southeast.   Suppose, observa- furthermore, that the temperature is rapidly rising, that tions. the moisture is increasing, that the barometer is falling, that the sky is hazy, and the cirrus clouds are visible, perhaps thickening to cirro-stratus or cirro-cumulus.   It is evident that an area of low pressure is about to dominate the weather.   The normal sequence of weather changes during the next two days would be precipitation probably in the form of snow, winds shifting to the northwest, and then clearing and colder.   A good general idea of the coming weather has thus been formed. If such questions as the probable amount of snowfall, the probable dura- tion of the snowfall, the probable rise in temperature, the possibility of the snow turning to rain, the probable drop in temperature after the storm passes, are to be answered, the daily weather maps are indispens- able.   As a second illustration, suppose it is again January, and the wind has just gone to the northwest.   Suppose, furthermore, that the barometer is rising, the moisture is decreasing, the temperature is falling rapidly, the clouds are breaking up into strato-cumulus and cumulus, and the air is becoming clear.   It is evident that the weather control is passing to a coming high.   The normal weather sequence would be two or more days with northwest winds, low temperature at night, and plenty of sunshine.

The sky appearance and the indications of local instruments indicate whether a coming or departing low, a coming or departing high, is domi- nating the weather.   As soon as this is determined, the probable sequence of weather changes for the next day or two is at once ap- parent.   To make a definite prediction, however, weather maps must be used.

The readings of local instruments are of such value in determining the location and direction of motion of a low, that the following is printed on the face of all weather maps issued by the U. S. Weather Bureau:

"When the wind sets in from points between south and southeast and the barometer falls steadily, a storm is approaching from the west or northwest, and its center will pass near or to the north of the observer within twelve to twenty-four hours, with winds shifting to northwest by way of southwest and west.   When the wind sets in from points between east and northeast, and the barometer falls steadily, a storm is approach- ing from the south or southwest, and its center will pass near or to the south or east of the observer within twelve to twenty-four hours, with winds shifting to northwest by way of north.   The rapidity of the

2 E

storm's approach and its intensity will be indicated by the rate and amount of the fall in the barometer."

401. **Weather proverbs and weather rules.** — Weather proverbs are as old as written language. The generalization of weather experience into proverbs and weather rules seems to have been one of the first acts of civilized man. Weather proverbs dating back to at least 4000 B.C. have been found on the clay tablets of Babylonia; there are many of them scattered through the oldest manuscripts; their number has increased through centuries; and at present there are hundreds of them. The following collection, in versified form, usually ascribed to Dr. Jenner, the discoverer of vaccination, is probably the most famous and interesting:

There are many weather proverbs.

Dr. Jenner's collection.

"The hollow winds begin to blow,
The clouds look black, the glass is low,
The soot falls down, the spaniels sleep,
And spiders from their cobwebs creep;
Last night the Sun went pale to bed,
The Moon in halos hid her head,
The boding shepherd heaves a sigh,
For see! a rainbow spans the sky;
The walls are damp, the ditches smell,
Closed is the pink-eyed pimpernel;
Hark how the chairs and tables crack!
Old Betty's joints are on the rack;
Her corns with shooting pains torment her
And to her bed untimely sent her;
Loud quack the ducks, the peacocks cry,
The distant hills are looking nigh;
How restless are the snorting swine l
The busy flies disturb the kine,
Low o'er the grass the swallow wings;
The cricket, too, how sharp he sings!
Puss on the hearth, with velvet paws,
Sits wiping o'er her whiskered jaws;
The smoke from chimneys right ascends,
Then spreading back to earth it bends;
The wind unsteady veers around,
Or setting in the South is found;
Through the clear stream the fishes rise,
And nimbly catch th' incautious flies;
The glowworms, num'rous, clear, and bright,

Illumed the dewy dell last night;
At dusk the squalid toad was seen
Hopping and crawling o'er the green;
The whirling dust the wind obeys,
And in the rapid eddy plays;
The frog has changed his yellow vest,
And in a russet coat is dressed;
The sky is green, the air is still,
The merry blackbird's voice is shrill,
The dog, so altered is his taste,
Quits mutton bones on grass to feast;
And see yon rooks, how odd their flight!
They imitate the gliding kite,
And seem precipitate to fall,
As if they felt the piercing ball.
The tender colts on back do lie,
Nor heed the traveler passing by.
In fiery red the Sun doth rise,
Then wades through clouds to mount the skies.
'Twill surely rain, — I see with sorrow
Our jaunt must be put off to-morrow."

Several fairly complete compilations of weather proverbs have been made, and in the references to the literature given at the end of this chapter some of them are mentioned. Various countries have different weather proverbs and sometimes those well known in one country will be unknown or will not apply at all in another. Weather proverbs, particularly those referring to the sky appearance and the meteorological elements, are often called weather rules, or prognostics.

402. By far the most important question in connection with weather proverbs is whether they have any foundation or not; that is, whether they are mere superstitions or whether they have a basis in fact.[1] In this regard, weather proverbs may be divided into five classes. The first two are fairly well founded, while the last three are mere superstitions. **The five classes of weather proverbs.**

The first class includes those which infer an impending weather change from the sky appearance and something connected with the meteorological elements — for example, " Rainbow in the morning, sailor take warning; rainbow at night, sailor's delight." Now a rainbow in the morning, means that the sun is shining **Illustrations of the classes.**

[1] See " Some Weather Proverbs and their Justification," by W. J. HUMPHREYS, *Popular Science Monthly*, 1911.

in the east, which is clear, and that it is raining in the west. Since thundershowers and storms move, in general, from west to east, this means that rainy weather is impending. Similarly a rainbow at night means that the sun is setting clear in the west while it is raining to eastward. This indicates a departing storm and that a period of good weather is at hand. A second example of this same class of proverbs is the following: "Mackerel sky and mares' tails make lofty ships carry low sails." Now mackerel sky and mares' tails, as popularly expressed, mean technically that cirro-cumulus clouds are present in the sky. This is a transition cloud form from cirrus to the coming nimbus. The storm cloud is usually accompanied over the ocean by high winds, and this will cause a vessel to carry but little canvass. It is both interesting and instructive in connection with weather proverbs to try to trace out the scientific basis for them.

The second class of weather proverbs are those which infer the coming weather from the behavior of animals, plants, and inanimate things. The coming of a low with its rain area and high shifting winds is usually heralded by an increase of temperature and moisture and a decrease of pressure. Drains are said to smell before rain. This may simply mean that the lower pressure causes some of the air to escape, thus causing the odor to become noticeable. The increase in temperature and moisture often causes a change in the behavior of animals. Their cries and actions are different, but this does not mean in any sense that the animals are endowed with prophetic vision. They are simply reacting to a changed present condition which is the forerunner of the rain period.

The three classes of weather proverbs which have no scientific basis whatever, are (1) those which infer the future weather at some distant date from the actions of animals or plants; (2) those which infer the future weather from the weather at some previous time; (3) those which infer the weather from some astronomical body. No credence whatever is to be placed in these sayings, because there is no reason why they should be true; and statistics show that they are not true. The following will serve as examples of these: Squirrels gather more nuts before a hard winter; If it rains St. Swithin's day, it will rain forty days; The moon and the weather change together.

A running commentary on Dr. Jenner's doggerel verses, quoted above, will illustrate the scientific basis for many of these prognos-

A commentary on Dr. Jenner's collection. tics. These are all rain prognostics, and belong to the first two classes of weather proverbs. Since rain is expected, it can be inferred at once that a low is approaching. This means that the temperature and moisture are increasing, the pressure is lessening, the wind is in the southeast, and haze and cirriform clouds are prevalent. The glass is low refers, of course, to the falling barometer. The red color of the sun at sunrise, the presence of clouds, the halo around the moon, the pale appearance of the sun, all indicate the hazy condition of the atmosphere and the presence of cirriform clouds. The shifting, rising wind indicates that the storm is coming steadily nearer. The falling of soot in chimneys and the dampness of walls simply indicate that the moisture is larger. To the same cause, joined with higher temperatures, may be attributed the closing of sensitive flowers, the rheumatic pains, the shooting of corns, the low flight of insects and of birds in search of them, the restlessness of animals, and the other changes in their cries and movements.

QUESTIONS

(1) Define weather and weather prediction. (2) Of what three things is weather the composite? (3) Describe the normal or typical weather for the northeastern part of the United States. (4) Name the passing meteorological formations which exert the chief influence on the weather. (5) State what is meant by local influences. (6) State in outline the general method of weather forecasting. (7) How is the United States divided for the purpose of forecasting? (8) What forecasts are issued by a local forecast official? (9) Of what does a forecast consist? (10) What training do the weather bureau forecasters receive? (11) What are the characteristics of a successful forecaster? (12) How do the methods of making weather forecasts differ in other countries? (13) Describe in full the method of locating the center of an area of low pressure 24 hours ahead. (14) What are some of the rules for modifying the estimated position? (15) What really determines the track of a low? (16) Which are the more important factors in determining the track of a low? (17) Describe the Bowie method of locating a low. (18) Describe in detail the method of determining the distribution of the meteorological elements about a low 24 hours ahead. (19) What are the advantages of making an exact forecast? (20) What terms may be used in the forecasts made by the U. S. Weather Bureau? (21) What is meant by secondary isobaric forms? (22) Describe in detail the various kinds of V-shaped depressions and their significance. (23) What are secondary lows? (24) Describe in outline the method of making a weather prediction when a high is the dominating formation. (25) Describe the method of weather prediction by similarity with previous maps. (26) How are frost predictions made? (27) How is the temperature of low-growing vegetation itself determined? (28) What constitutes a frost in the weather bureau sense? (29) What is meant by a cold wave? (30) How is the area for which a cold wave is predicted to be determined? (31) How is the cold wave prediction indicated on the weather map? (32) Why are tornadoes not predicted? (33) When is a tornado likely to occur? (34) What is meant by a destructive wind velocity or storm? (35) What are the regulations concerning river and flood forecasts? (36) What terms may be used in the official weather bureau predictions? (37) How are weather predictions verified? (38) On what counts are weather predictions verified? (39) What accuracy is attained by official forecasters? (40) What

investigations have been undertaken to improve the present methods of forecasting? (41) What is the attitude of newspapers and periodicals towards weather forecasts? (42) What is meant by a long-range prediction? (43) To what extent can the station normals be used in the making of long-range predictions? (44) To what extent can departure from normal be used as a basis for long-range predictions? (45) What are weather cycles? (46) To what extent can weather cycles be used in forecasting? (47) Describe in detail the tendency of a weather type to continue. (48) To what is this tendency due? (49) What can be accomplished along the line of scientific weather prediction? (50) What are the various things which serve as a basis for long-range weather forecasts? (51) Describe the typical long-range weather forecast. (52) To what extent can predictions from the readings of instruments and the appearance of the sky be made? (53) Describe weather proverbs. (54) Into what five classes may the weather proverbs be divided? (55) What basis have weather proverbs?

TOPICS FOR INVESTIGATION

(1) The Bowie method of locating the center of a low.
(2) The rules for locating a low 24 hours ahead.
(3) V-shaped depressions.
(4) Secondary lows and their influence on the weather.
(5) Local influences in forecasting.
(6) The area for which to predict a cold wave.
(7) The accuracy attained in weather forecasting.
(8) The terms used in expressing weather predictions and the system of verification in other countries.
(9) The methods used by the weather bureaus of other countries in making and distributing predictions.
(10) Centers of action.
(11) Long-range predictions.
(12) Weather cycles.
(13) Weather proverbs.

PRACTICAL EXERCISES

(1) Make several exact weather predictions and also in the form issued by the officials of the U. S. weather Bureau, when a low is dominant, when a high is dominant, and when secondary isobaric forms are present. In each case verify the forecast according to the regular rules. (For this work a file of weather maps will be very useful, although predicting for the coming day is always much more interesting.)

(2) In several cases locate the center of a low 24 hours ahead, using the Bowie method.

(3) From a file of weather maps select those which illustrate well the rules for modifying the estimated location of the center of a low.

(4) From a file of weather maps select those which illustrate well forecasting when the secondary isobaric forms are of importance.

(5) From a file of weather maps select those which illustrate well the prediction of dangerous or damage-causing occurrences.

(6) From a long series of observations (those at a regular Weather Bureau station would be necessary) compile statistics and determine if certain weather proverbs or weather rules or things used as the basis of long-range predictions have any foundation or value. Determine in each case the preponderence of probability.

REFERENCES

The following are the books and pamphlets which, as a whole or in part, deal with weather and weather forecasting in general:

ABBE, CLEVELAND, Preparatory Studies for Deductive Methods in Storm and Weather Predictions, 165 pp., Washington, 1890. (Annual Report of the Chief Signal Officer for 1889; Appendix 15.)

ABBE, CLEVELAND, The Aims and Methods of Meteorological Work, Baltimore, 1899. (Special publication of the Maryland Weather Service.)

ABERCROMBY, RALPH, Principles of Forecasting by Means of Weather Charts, 122 pp., London, 1885.

ABERCROMBY, RALPH, Weather, London, 1887.

BEBBER, W. J. VAN, Handbuch der ausübenden Witterungskunde, 8°, 2 parts, Stuttgart, 1885-6.

BEBBER, W. J. VAN, Beurteilung des Wetters auf mehrere Tage voraus, Stuttgart, 1896.

BEBBER, W. J. VAN, Die Wettervorhersage, 2d. ed., 219 pp., Stuttgart, 1898.

BIGELOW, FRANK H., Storms, Storm Tracks, and Weather Forecasting. U. S. Weather Bureau, Bulletin No. 20 (W. B. No. 114).

CHAMBERS, GEORGE F., The Story of the Weather, London, 1897.

DALLET, G., La prevision du temps.

FREYBE, OTTO, Praktische Wetterkunde, 173 pp., Berlin, 1906.

GRANGER, FRANCIS S., Weather Forecasting, 121 pp., Nottingham, 1909.

GUILBERT, GABRIEL, Nouvelle méthode de prévision du temps, 343 pp., Paris, 1909.

KLEIN, H., Wettervorhersage für jedermann, Stuttgart.

KUHLENBÄUMER, TH., Unser Wetter und seine Vorherbestimmung, 164 pp., Münster, 1909.

MOORE, WILLIS L., Weather Forecasting. U. S. Weather Bureau, Bulletin No. 25 (W. B. No. 191).

MOORE, WILLIS L., "Forecasting the Weather and Storms," The National Geographic Magazine, June, 1905, Vol. XVI, No. 6.

PERNTER, J. M., Wetterprognose in Osterreich, 61 pp., Wien, 1907.

PERNTER, J. M., "Methods of Forecasting the Weather," Monthly Weather Review, December, 1903.

SCOTT, ROBERT H., Weather Charts and Storm Warnings, 229 pp., London, 1887.

SCOTT, ROBERT H., Notes on Meteorology and Weather Forecasting, 40 pp., London, 1909.

SHAW, W. N., Forecasting Weather, 8°, xxvii + 380 pp., London, 1911.

Station Regulations of the U. S. Weather Bureau, Washington, 1905.

WARD, ROBERT DE C., Practical Exercises in Elementary Meteorology, Ginn & Co., 1899.

For the prediction of particular and damage-causing occurrences such as storm winds, cold waves, frost, floods, etc., see:

GARRIOTT, EDWARD B., Cold Waves and Frosts in the United States. U. S. Weather Bureau, Bulletin P (W. B. No. 355).

GARRIOTT, EDWARD B., Storms of the Great Lakes. U. S. Weather Bureau, Bulletin K (W. B. No. 288).

In connection with long-range weather predictions, consult:

GARRIOTT, EDWARD B., Long-range Weather Forecasts. U. S. Weather Bureau, Bulletin No. 35 (W. B. No. 322).

WALZ, F. J., " Fake Weather Forecasts," *Popular Science Monthly*, pp. 503–513, Vol. 67, 1905.

For collections of weather proverbs, weather rules, weather folklore, and weather prognostics, see :

CHAMBERS, GEORGE F., *The Story of the Weather*, London, 1897.

DUNWOODY, H. H. C., Weather Proverbs. Signal Service Notes, No. IX, Washington, 1883.

GARRIOTT, EDWARD B., Weather Folklore. U. S. Weather Bureau, Bulletin No. 33 (W. B. No. 294).

HORNER, D. W., *Observing and Forecasting the Weather*, 46 pp., London, 1907.

INWARDS, RICHARD, *Weather Lore*, 233 pp., London, 1898.

MICHELSON, W. A., *Wetterregeln*, 17 pp., Braunschweig, 1906.

SWAINSON, CHARLES, *A Handbook of Weather Folklore*, 275 pp, 1873.

PART II

PART I consists of eight chapters which treat in succession the atmosphere, the heating and cooling of the atmosphere, temperature, pressure and wind, moisture in all its forms, the various storms, weather bureaus and their work, and weather predictions. These chapters are presented in full, and the various topics are treated at length. This material is treated in full in practically all textbooks on meteorology, and makes a complete treatise in itself.

Part II consists of five chapters which are treated in full in some books and passed over with a few words or pages in others. If meteorology is taken in the largest sense, as the science of all atmospheric phenomena, and should include all the work of weather bureaus, then these chapters should be included in a treatise on the subject. These chapters treat climate, floods and river stages, atmospheric electricity, atmospheric optics, and atmospheric acoustics. It was at first intended to consider these five chapters as an appendix and present simply the syllabus of each chapter, and the references to the literature. If this book is used as a textbook, the student has had sufficient material presented to him in a predigested form, and it would be best for him to work up these various topics for himself. The general reader, however, would probably prefer to know what would be covered in these chapters without working up the subject from the literature. For that reason, most of the topics are sketched in outline, but no attempt is made at the fullness or completeness of Part I.

425

CHAPTER IX

CLIMATE

DISTINCTION BETWEEN WEATHER AND CLIMATE

403. Weather is the condition of the atmosphere at any particular time and place, and is best described by stating the numerical values of **Weather and climate defined.** the various meteorological elements. Climate is generalized weather, and has to do with a larger area and longer time. Climate is also variously defined as the totality of the weather or the customary course of the weather. Weather changes from moment to moment, but climate remains the same. Weather has to do with the particular values of the meteorological elements, while climate is concerned more with the normal values.

426

Climatology is the science of climate and includes, not only a description of the climate and a statement of the causes of its Climatology characteristics, but also its effect on animal and vegetable defined. life and its relation to the occupations and activities of man. .

CLIMATIC DATA AND CHARTS

404. The material necessary for a description of the climate of a locality may be presented as tables of statistics or by means of graphs and charts. It should include the normal hourly, daily, monthly, and yearly values of all the meteorological elements The tables of data and all the tables of data which could be worked up in connec- which tion with the various elements. It should also include tables should be of data concerning the composition of the atmosphere, the given. amount of evaporation, the temperature of the ground, solar radiation, thunderstorms, fogs, the electrical condition of the atmosphere, the haziness of the sky, and the like. It might also include tables of data in connection with the dates of harvest, the freezing over of rivers, the arrival and departure of birds, and the like.

Hann, in his *Lehrbuch der Klimatologie*, mentions thirty-six summaries of data which should always be included in a climatological study, and Abbe, in his *Aims and Methods in Meteorological Work* adds six more to the list. They are the following:

(1) The monthly and annual mean temperature of the air.

(2) The extent of the mean diurnal range of temperature for each month.

(3) The mean temperature at two specific hours, namely, the early morning and midafternoon.

(4) The extreme limits, or total secular range, of the mean temperatures of the individual months.

(5) The mean of the monthly and annual extreme temperatures, and the resulting non-periodic range.

(6) The absolute highest and lowest temperatures that occur within a long interval of time.

(7) The mean variability of the temperature as expressed by the differences of consecutive daily means.

(8) Mean limit, or date, of frosts in spring and fall, and the number of consecutive days free from frosts.

(9) The elements of solar radiation as measured by optical, chemical, and thermal effects.

(10) The elements of terrestrial radiation as measured by radiation thermometers.

(11) The temperature of the ground at the surface, and to a depth of one or two yards.

(12) The monthly means of the absolute quantity of moisture in the atmosphere.

(13) The monthly means of the relative humidity of the air.

(14) The total precipitation, as rain, snow, hail, dew, and frost, by monthly and annual sums.

(15) The maximum precipitation per day and per hour.

(16) The number of days having 0.01 inch or more precipitation, including dew or frost.

(17) The percentage of rainy days in each month or the probability of a rainy day.

(18) The number of days of snow, with the depth and duration of the snow covering.

(19) The dates of first and last snowfall.

(20) Similar data for the dates of hail.

(21) Similar data for the dates of thunderstorms.

(22) The amount of cloudy sky, expressed in decimals of the whole celestial hemisphere.

(23) The percentage of cloudiness by monthly means, for three or more specific hours of observation.

(24) The thickness of the cloud layer, or the amount of strong sunshine as shown by Campbell's sunshine recorder.

(25) The number of foggy days, or the total number of hours of fog.

(26) The number of nights with dew; also the quantity of dew.

(27) The monthly means or total of wind velocity, or estimated wind force.

(28) The frequency of winds from the eight principal points of the compass, and the frequency of calms.

(29) The frequency of winds for each hour of observation and the diurnal changes in the winds.

(30) The meteorological peculiarities of each wind direction, or the respective wind roses for temperature, moisture, cloudiness, and rainfall.

(31) The mean annual barometric pressure.

(32) The total evaporation, daily, and monthly, or some equivalent factor, such as the depression of the dewpoint combined with the velocity of the wind.

(33) Variations in the gases contained in the atmosphere, provided they are suspected to be of importance.

(34) Impurities in the atmosphere, such as the number of dust particles, and especially the number of spores or germs of organic life.

(35) The porportions of ozone, the peroxide of hydrogen, and nitric acid.

(36) The electrical condition of the atmosphere, if there is any method of obtaining it.

To these Abbe adds the following:

(37) The sensation experienced by the observer, such as mild, balmy, invigorating, depressing, and other terms used to express the effect of the weather upon mankind.

(38) The number of storm centers that pass over a given locality, or the storm frequency, monthly and annual.

(39) Frequency of severe local storms.

(40) The duration of twilight.

(41) The blueness or haziness of the sky.

(42) The number and extent of the sudden change from warm to cold, or moist to dry weather, and *vice versa*.

Even the above list is by no means complete, as many more summaries could be added. Most of this material can be presented by means of graphs and charts, as well as by tables of statistics.

The Factors which Determine Climate

405. The climate of different parts of the world is very different, and the chief factor which determines the climate is the latitude of the locality, for with the latitude varies the amount of insolation received from the sun. In fact, the word climate comes from the Greek and means inclination. Primarily, then, it was the varying inclination of the sun's rays at different latitudes which was considered. If latitude were the only factor in determining climate, all places with the same latitude would have the same climate. The climate which would exist if it depended on latitude alone is often called the solar climate. There are, however, several other climatic factors and the climate is not the same at all places having the same latitude. The solar climate as modified by these other factors is often called the physical climate. These modifying factors are: the relative distribution of land and water, the altitude above sea level, mountain ranges, and the topography of the locality. Under this last is included the nature of the soil, and whether the country is vegetation-covered or forested. *{Latitude is the chief climatic factor.}* *{There are four other climatic factors.}*

It is sometimes stated that the type of storm, the amount of rainfall, the direction of the prevailing winds, etc., are the factors which determine climate. This would seem to be a mistake, as it is these things which constitute climate. The climate is different in two localities, and this means that these things are different in the two localities. The climatic factors are the factors which determine why these differences exist, and the one major factor and the four minor factors have just been stated. *{Things which should not be considered climatic factors.}*

The Climatic Subdivisions of the World

406. Subdivision on the basis of latitude. — The division of the world into geographical or climatic zones on the basis of latitude goes back to

Parmenides originated the present subdivision into zones.

the time of the Greek philosophers. The division of the world into the five zones used at present is generally ascribed to Parmenides, who flourished about 450 B.C. These five zones are the torrid zone, the two temperate zones, and the two frigid zones. The torrid zone is divided into two equal parts by the equator, and is bounded on the north and south by the tropics of Cancer and Capricorn. The width of the zone is thus 47°,

The torrid, temperate, and frigid zones.

and the sun reaches the zenith on at least one day during the year at every place within this zone. The two frigid zones lie wholly within the Arctic and Antarctic circles and surround the two poles. The sun never rises at least one day in the year at all places within these zones. The two temperate zones lie between the frigid zones and the torrid zone, and each is 43° wide. The torrid zone covers 40 per cent, the two temperate zones 52 per cent, and the two frigid zones 8 per cent of the earth's surface. It will thus be seen that they are of very unequal size. The names of the zones are unfortunate, as they would seem to suggest a temperature basis for the subdivision, while in reality the subdivision is on the basis of latitude only. The torrid zone is more appropriately called the tropical zone, and the frigid zones the polar zones. This system of Parmenides which has come down to us was used by Aristotle (about 384 B.C.).

Various other methods of dividing the world into zones were proposed by different Greek philosophers. Eudoxus of Cnidus, who lived about 366 B.C., divided the quadrant of the earth into 15 parts, of

Other systems of subdivision on the basis of latitude.

which 4 were assigned to the torrid zone, 5 to the temperate, and 6 to the frigid. The tropics and polar circles were thus fixed at 24° and 54° of latitude respectively. Claudius Ptolemy, the great astronomer and geographer, who flourished at Alexandria in about 150 A.D., proposed a different system. Near the equator, the width of a zone was determined by a difference of 15 minutes in the length of the longest day. In higher latitudes, differences of half an hour, an hour, and finally a month, were used. Until within the last century or two all systems of subdivision were based on latitude only. This means that latitude was practically the only climatic factor which was recognized, although a few mentioned the importance of other things.

407. Subdivision on the basis of temperature. — Temperature is the most important of the meteorological elements in its influence on plant and animal life and the occupations and habits of life of man. Isothermal lines do not follow parallels of latitude, and it has thus been proposed to use them instead of parallels of latitude for bounding the various zones. *Isothermal lines do not follow parallels of latitude.*

According to Supan, the equatorial or torrid zone should be limited on either side by the normal annual isotherm of 68° F. This would cross the United States from the southern part of California to the northern part of Florida. The intermediate or temper- *Supan's* ate zone has for its poleward limit the isotherms of 50° F. *subdivision.* for the warmest month. This would cross North America from Alaska to a little north of Newfoundland. In this system of subdivision there would thus be five zones as before, only they would be bounded by isothermal lines instead of parallels of latitude.

Köppen has proposed a system of subdivision, based upon temperature, into nine belts or zones, a central zone and four on each side of it. Two of these are further subdivided into three parts. These *Köppen's* belts are as follows : *system of subdivision.*

(1) Tropical belt : All months hot, that is, with a normal temperature of over 68° F. during all months. It would extend roughly from 20° N. to 16° S. latitude.

(2) Subtropical belts : 4 to 11 months hot, that is, over 68° F. ; 1 to 8 months temperate (50° F. to 68° F.).

(3) Temperate belts : 4 to 12 months between 50° F. and 68° F.

(4) Cold belts : 1 to 4 months temperate ; the rest cold, that is with a normal monthly temperature below 50° F.

(5) Polar belts : all months below 50° F. The two temperate belts are subdivided into three parts with these characteristics : (*a*) constant temperature during the year, (*b*) hot summers, (*c*) moderate summers and cold winters.

408. Subdivision on the basis of the general wind system. — The importance of the general wind system in determining the geographical distribution of the various meteorological elements has been brought out in previous chapters. It was shown that it was *The nine-* the chief factor in determining the distribution of precipita- *division* tion, and the absolute and relative humidity were both *based on* closely correlated with it. It would thus seem that the *wind* world might be subdivided into climatic zones on the basis *system.* of the permanent wind system. When subdivided on this basis, nine zones are usually recognized, a central zone and four on each side of it. The central zone is called the subequatorial zone, and here would be experienced the calms of the doldrums and trade winds blowing in opposite directions at different times of year. Next to this central zone, on either side, would be the two trade wind zones. Next to these would come the two subtropical belts or zones. Here would be found the calms of the horse latitudes and winds blowing from opposite directions during different parts of the year. Beyond these zones would come the zones of the prevailing westerlies, but as they cover such a very large part of the earth's surface, it has seemed best to divide each of these zones into two by the polar circles. There would thus be nine zones, corresponding to the terrestrial wind system.

409. Subdivision on the basis of surface topography. — The climate of different localities in the same zone is by no means the same. This

A zone may be divided into six parts, using surface topography as a basis.

is true no matter if latitude, temperature, or the general wind system has been made the basis of subdivision into zones. It is thus necessary to subdivide the zones on the basis of the characteristics of the surface. Six subdivisions are usually made : ocean, west coast, east coast, plain, plateau, and mountain. The climate on the east coast of an ocean is usually somewhat different from that on the west coast. For this reason, two coast or littoral climates must be recognized. It is also not sufficient to say a climate is continental as distinguished from marine or littoral. The low-lying areas, the plateaus, and the mountain regions must be treated separately.

Thus, if the world is first divided into zones on the basis of latitude, or temperature, or the general wind system, and then if these zones are subdivided on the basis of nature of the surface, the resulting subdivision will be small enough that a general description of the climate of each one can be given, and what is said about the climate of one locality will hold for all other places in the same subdivision.

410. Other climatic subdivisions of the world. — There are four other climatic subdivisions of the world, each with a different basis, which deserve a brief consideration. These are the subdivisions of Supan, Köppen, Ravenstein, and Herbertson.

Supan arbitrarily divides the world into 35 so-called climatic provinces. The attempt is here made to group together those near-by places which

Supan's thirty-five climatic provinces.

have the same surface topography and where the meteorological elements have the same characteristics. The number of these provinces, namely 35, and their boundaries are purely arbitrary. Provinces No. 25, 26, 27, and 28 include the United States, and their characteristics are as follows:

No. 25, Californian province. Here it is relatively cool, especially in summer, and there is a marked subtropical rainy season.

No. 26, North American mountain and plateau province. Here there are great daily and yearly ranges in the values of the meteorological elements, and it is also dry.

No. 27, Atlantic province. Here the chief characteristics are contrast in temperature between north and south in winter, extreme climate even on the coast, a plentiful, evenly distributed rainfall, rapid changes.

No. 28, West Indian province. Here there is an equable temperature and rain at all seasons, but with a well-marked summer maximum.

Köppen's climatic subdivision of the world has a botanical basis. Five kinds of plants are recognized: (megotherms) those which need a continuously high temperature and abundant moisture; (xerophytes) those which like high temperature and dryness; (mesotherms) those which require moderate temperatures and a moderate amount of moisture; (mikrotherms) those which need lower temperatures; (hekistotherms) those which will live in low temperatures. The five main divisions are further subdivided until the whole number reaches twenty-four. *Köppen's botanical system.*

According to Ravenstein, the world is subdivided into sixteen climatic types, and the basis of the classification is temperature and relative humidity. *Ravenstein's system.*

According to Herbertson, the world is subdivided into six natural geographical regions, and these are subdivided until the whole number of recognized climates reaches fifteen. His basis of classification is a combination of temperature, rainfall, topography, and vegetation. *Herbertson's system.*

411. It will thus be seen that there are many methods of subdividing the world into smaller areas for the purpose of discussing the climate, and many different bases have been used for the various classifications. The purpose in each case has been to form a sufficient number of climatic provinces so that the climate at different localities in the same province would be essentially the same. In general, the larger the number of provinces, the more nearly alike will the climate be at different places in the same province. *The purpose of subdivision.*

2 F

The Climate of the United States

412. Introduction. — The United States is a country of such vast extent and with such a diversified surface that practically no statements

The climate in different parts of the country is very different.

can be made about the climate of the country as a whole. The statements which would be true for one part of the country would be entirely incorrect for another. In one part of the country, New England for example, the weather is dominated by an almost unbroken procession of passing highs and lows. As a result, great irregular changes in the values of the meteorological elements follow each other in quick succession. The precipitation is all storm-caused and quite evenly distributed throughout the year. The climate is thus extreme, very variable, and with copious, evenly distributed precipitation. In another part of the country, California for example, there is a well-marked rainy season. The precipitation here is caused mostly by the general wind system of the world. The irregular changes in the meteorological elements are few, so that during any one season, one day is much like another. The climate is thus very uniform with a well-marked wet and dry season.

Subdivision essential.

The first essential, then, in discussing the climate of the United States is to subdivide the country into smaller areas so that the climate of all places in the same district will be practically the same.

413. Subdivision. — The United States might be subdivided into smaller areas for the purpose of discussing the climate, using any one of

The three systems of subdivision used by the Weather Bureau.

the systems of subdivision and classification which have just been treated. Since, however, the U. S. Weather Bureau has collected practically all of the climatological data, it would be better to adopt the systems of subdivision which have been followed there. Formerly the country was divided into twenty-one climatic divisions (see section 357), and the climatological data and statistics were summarized for each of these districts separately. At present, these districts are still used to a slight extent in summarizing data. When the form of the Monthy Weather Review was changed in July, 1909, the country was subdivided into twelve climatological districts (see section 357), and the observations are now summarized for these districts as a whole. These districts are also adhered to as far as practicable in matters of administration. In 1910 was commenced the publication of the summaries of the climatological data of the United States, by sections, and for this purpose the country was divided into

106 districts. Thus in discussing the climate of the United States, it would be best to consider the country divided into 21, 12, or 106 districts, following the subdivisions of the U. S. Weather Bureau.

414. Detailed treatment of certain subdivisions. — The scope of this book prevents the complete discussion of the climate of even one place or district. The reader who is interested in the climate of any particular locality must be referred to the literature of the subject for information.

The general question, however, could well be raised as to what would constitute a full and complete discussion of the climate of a place. Such a complete treatise might conveniently consist of the following nine parts: (I) A map of the locality and surrounding regions is usually presented first. This map should contain the usual geographical features and the elevations. A brief description of the cities, mountains, etc., usually accompanies the map. (II) Next a full description of the surface topography might be given. This would include the characteristics of the rivers, the nature of the soil, whether forested or not, and the like. (III) Next the climatological data might be presented as tables of statistics or as charts and graphs. The source of the data, the length of the records, and their probable accuracy might also be discussed. (IV) Next would come a discussion of the data, perhaps considering first the meteorological elements in order (temperature, pressure, wind, moisture, cloud, precipitation), and then the other meteorological and phenological occurrences, such as frost, fog, evaporation, time of harvest, migration of birds, etc. (V) Next might come a discussion of the types of storms, their prevalence and severity. (VI) Next the influence of the climate on certain diseases and the general healthfulness of the climate might be discussed. (VII) Next the influence of the climate on agriculture and vegetation might be discussed. Here would be considered such questions as the kind of crops which could be best grown, the possibility of fruit growing, the kind of forest trees which would be most common, etc. (VIII) Next the general influence of the climate on the industries and habits of life of the people might be discussed. (IX) A bibliography might be added. These nine subdivisions will serve to indicate what one might expect to find in a full, complete discussion of the climate of a locality.

[side note:] What a full, complete treatise on the climate of a place would contain.

The Climate of Other Countries and Places

415. It is entirely impossible to present, in a limited space, the complete discussion of the climate of even one country or place. This will not be attempted, but the chief characteristics of the three great zones, the torrid, temperate, and frigid, should perhaps be stated. The zones are, however, of such large extent that there are but few characteristics which are common to an entire zone.

The climate of the torrid zone is characterized by great uniformity, small irregular changes in the meteorological elements, high temperature, and a small yearly variation in temperature. Nowhere else in the world is the weather so nearly the same day after day. This means a very uniform climate, and the reason for it is the type of storms. Tropical cyclones and thundershowers are the only storms. Tropical cyclones are few in number, occur at certain times of year, and cover a very small area. Thundershowers are very prevalent, occurring at many places almost daily. Large, irregular changes in the meteorological elements following each other in quick succession are thus almost unknown. The noonday sun always stands high in the sky and the change in temperature during the year is small. Monsoons are well marked in many countries in the torrid zone, and the rainfall occurs either during the rainy season, caused by a monsoon, or almost daily, usually with a thundershower in the afternoon, when convection is most powerful. The seasons thus depend more on the general wind system and the rainfall than on changes in temperature.

The characteristics of the torrid zone.

The temperate zone is characterized by a very variable climate and large changes in temperature between summer and winter. The temperatures are, of course, lower than in the torrid zone. The temperate zone is constantly being traversed by passing highs and lows, and, as a result, the changes in the meteorological elements are abrupt and large. This makes the climate very variable. The rainfall results both from the general wind system and the passing storms.

The temperate zone.

The frigid zone is characterized by greater uniformity and much lower temperature than the temperate zone. The changes in the meteorological elements, particularly temperature, between summer and winter, are large. The daily changes at times are very small. Large, irregular changes are also present.

The frigid zone.

The Constancy of Climate

416. The question of the constancy of climate must be discussed for three different time intervals. First, has the climate remained constant during the recent past, say the last hundred years? Secondly, has the climate remained constant during historic times, say the last 7000 years? Thirdly, has the climate remained constant during recent geologic ages, say the last 10,000,000 years?

There are many stations where meteorological observations have been made for more than a hundred years. In fact, a few records cover more than three hundred years. Based upon these observations, the statement can confidently be made that the climate is essentially the same now as it was many years, or even a hundred years ago. This is largely contrary to popular belief. It means that, taking one year with another, the snowfall is just as large now as then. It means that sleighing lasts just as long now as then. It means that the winters are no milder now than then. It means that our summers are no hotter now than then. The constant statements by the older people, that the climate is different now than it used to be when they were much younger, are due to the tendency to magnify and remember the unusual while the ordinary is forgotten. Thus, in time, it is only the unusual snowfall or the extremely low temperatures that are well remembered, and unconsciously the abnormal has thus been substituted for the normal. These statements are also due to the fact that the attitude towards life, the amount of energy, the daily occupations, and perhaps the place of residence of the older people are very different now than when they were much younger.

The climate has remained unchanged during the last hundred years.

In discussing possible changes in climate during the last 7000 years, inference must be drawn from such recorded facts as the dates of harvest, the kind and amount of crops raised, the kind of clothing worn by the people, the habits of life of the people, the existence of certain wild animals and forest trees, the size of rivers, the height of lakes and inclosed seas, etc. From evidence of this kind, the conclusion has been drawn that there have been no marked changes in climate during historic times. It has been often thought that certain climatic cycles have been detected. The 11-year and 35-year cycles have been well investigated, while cycles of much longer duration have also been suspected. The 11-year cycle corresponds to the sun spot cycle and is very poorly

The climate has remained essentially the same during historic times.

marked. Brückner's 35-year cycle is on the contrary fairly well marked in Europe, both in temperature and precipitation. None of these cycles, with the possible exception of the 35-year cycle, are at all regular or well marked. (See section 397.)

There can be no doubt that the climate has changed greatly during recent geologic ages. Almost tropical vegetation has existed in Greenland, and glaciation has extended many times far towards the equator. Various explanations of these changes have been advanced. Among them are a change in the location of the earth's axis, a change in the eccentricity of the earth's orbit, the precession of the equinoxes which brings the long cold winter to the northern (land) hemisphere every 25,000 years, a change in the energy emitted by the sun, a change in the composition of the earth's atmosphere, a change in the elevation of the place, a change in the distribution of land and water, and thus the ocean currents.

Great changes during geological ages and the possible causes.

THE SNOW LINE

417. The temperature in general grows less with increasing altitude, and thus there are regions on high mountains even in the torrid zone near the equator where the snow which falls during one winter has not sufficient time to melt entirely during the following summer before the advent of the snows of the next winter. These are regions of perpetual snow, and the lower boundary of these regions is called the snow line.

Definition of the snow line.

On the equatorial Andes, the height of the snow line is about 5000 meters, roughly three miles. With increasing south latitude, the height of the snow line increases somewhat due to scantier precipitation, and reaches a maximum height of about 6000 meters in 25° south latitude. The height of the snow line then decreases rapidly with increasing latitude. Its height is about 3500 meters in latitude 32°; 1600 meters in latitude 42°; 800 meters in latitude 50°; 400 meters in latitude 55°; and according to the observations made by certain Antarctic expeditions, it reaches sea level in from 67° to 70° south latitude.

Its height in South America.

In the northern hemisphere, the height of the snow line is somewhat greater than in South America. In the Alps in Europe (latitude about 46°) it is about 2900 meters. In Norway (latitude about 61°) it is about 1500 meters. In the Himalaya Mountains the height is about 5000 meters. In North America, at

Its height in other parts of the world.

latitude 19° N., in Mexico, the height of the snow line is about 4600 meters. On Mt. Shasta the height is 2400 meters; in the Cascade Mountains, at the northern boundary of the United States, it is 2000 meters; on Vancouver Island it is 1700 meters; on Mt. St. Elias (lat. 60° N.) it is 800 meters. Even in polar regions it does not everywhere absolutely reach sea level.

The height of the snow line does not depend on latitude alone, and thus is often somewhat different on two mountains in the same latitude. The other factors which influence the height of the line are The factors the amount of snow during the winter, the change in tem- which deter-perature between winter and summer, the steepness of the height. mountains, the exposure of its slopes, and the general wind system.

TOPICS FOR INVESTIGATION

(1) What a full list of climatic data would consist of.
(2) The factors which determine climate.
(3) Some system of climatic subdivision and its basis.
(4) The contrast between marine and littoral climate in the same latitude.
(5) The climatic characteristics of the continents.
(6) Climatic changes during historic times.
(7) Geological changes in climate.
(8) The change in the height of the snow line between winter and summer.
(9) The temperature of the snow line.
(10) Effect of climate on the mental characteristics of the inhabitants.

PRACTICAL EXERCISES

(1) Prepare tables of climatic data.
(2) Express these tables graphically.
(3) Summarize the observations of some station with a long record to show the constancy of climate.
(4) Treat fully the climate of some place.

REFERENCES

The following books and pamphlets deal with climate or climatology in general:

BEBBER, W. J. VAN., *Hygienische Meteorologie*, Stuttgart, 1895.
BONACINA, L. C. W., *Climatic Control*, viii + 167 pp., London, 1911.
CULLIMORE, D. H., *The Book of Climates for all Lands*, London, 1890.
HANN, JULIUS, *Handbuch der Klimatologie*, 3 vols., 2d ed., Stuttgart, 1897; 3d ed. of Vol. I, 1908; of Vol. II, 1910; of Vol. III, 1911.
HERBERTSON, A. J. and F. D., *Man and his Work*, London, 1899.
KÖPPEN, *Klimakunde*, 2d ed., Leipsig, 1906.
MEYER, HUGO, *Anleitung zur Bearbeitung meteorologischer Beobachtungen für die Klimatologie*, 8°, viii + 187 pp., Berlin, 1891.

MOORE, WILLIS L., Climate. Bulletin No. 34 of U. S. Weather Bureau (W. B. publication No. 311).

RATZEL, *Anthropogeographie*, 2d ed., Stuttgart, 1899.

STOCKMAN, WILLIAM B., Invariability of our Winter Climate, (W. B. publication No. 312).

SUPAN, A., *Grundzüge der physischen Erkunde*, 4th ed., Leipzig, 1908.

WARD, R. DEC., *Climate*, G. P. Putnam's Sons, New York, 1908.

WARD, R. DEC., *Hann's Handbook of Climatology* (translation of Vol. I), The Macmillan Co., 1903.

WOEIKOF, A., *Die Klimate der Erde*, Jena, 1887.

(Of the books just mentioned, the two by Ward are by far the most complete and readable in English. Taken together they cover the subject of climate in a most complete and interesting way. The three-volume work by Hann in German is by far the most complete and useful treatise on the subject. It is a veritable mine of information.)

For a description of the climate of the United States as a whole and of various places in the country, see:

BLODGET, LOUIS, *Climatology of the United States*, xvi + 536 pp., Philadelphia 1857.

FASSIG, OLIVER L., *The Climate and Weather of Baltimore*, Baltimore, 1907.

GREELY, A. W., *Report on the Climate of Colorado and Utah*, Washington, 1891

HAZEN, HENRY A., The Climate of Chicago. Bulletin No. 10 of U. S. Weather Bureau.

HENRY, ALFRED J., Climatology of the United States, 1012 pp., Washington, 1906. Weather Bureau, Bulletin Q. (The standard descriptive and statistical work on this subject.)

M'ADIE, ALEX. G., and WILLSON, GEORGE H., The Climate of San Francisco Bulletin No. 28 of the U. S. Weather Bureau, (W. B. publication No. 211).

M'ADIE, ALEX. G., Climatology of California Bulletin L of U. S. Weather Bureau, (W. B. publication No. 292).

SMOCK, JOHN C., *Climate of New Jersey*, Trenton, 1888. (Part of the final report of the state geologist.)

Summary of the Climatological Data for the United States, by Sections (106 are to be issued, covering the whole country).

WALDO, FRANK, *Elementary Meteorology* (Chapter 13).

For a description of the climate of various countries and place, outside of the United States, see:

ABBE, CLEVELAND, JR., The Climate of Alaska; Washington, 1906. (Extract from Professional Paper No. 45, U. S. Geological Survey.)

ABBOT, HENRY L., Climatology of the Isthmus of Panama (W. B. publication No. 201).

ALEXANDER, WILLIAM H., Climatology of Porto Rico, Monthly Weather Review, July, 1906.

ALGUÉ, JOSÉ, *The Climate of the Philippines*, 103 pp., Manila, 1904.

BEHRE, OTTO, *Das Klima von Berlin*, 8°, 158 pp., Berlin, 1908.

BLANFORD, *Climates and Weather of India*, 8°, 382 pp., London, 1889.

DAVIS, WALTER G., *Climate of the Argentine Republic*, vi + 154 pp., Buenos Aires, 1902.

ELIOT, SIR JOHN, *Climatological Atlas of India*, Edinburgh, 1906.

KNOX, ALEXANDER, *The Climate of the Continent of Africa*, 8°, xii + 552 pp., Cambridge, 1911.

MOORE, JOHN W., *Meteorology*, 2d ed. London, 1894. (Climate of the British Islands, Chapters XXV and XXVI.)

NAKAMURA, K., *The Climate of Japan*, 109 pp., Tokio, 1893.

PHILLIPS, W. F. R., Climate of Cuba, Bulletin 22 of U. S. Weather Bureau, (W. B. publication No. 163).

QUETELET, A., *Sur le climat de la Belgique*, 4°, 2 vols., Bruxelles, 1849.

RIZZO, G. R., *Il Clima di Torino.*

ROSTER, GIORGIO, *Climatologia dell' Italia nelle sue attineze con l'igiene e con agricoltura*, 8°, xxix + 1040 pp., Torino, 1909.

For a treatment of the variations in climate, see:

BUCKNER, EDWARD, *Klimaschwankungen seit 1700*, viii + 324 pp., Wien, 1890.

ECKARDT, WILHELM R., *Das Klimaproblem der geologischen Vergangenheit und historischen Gegenwart*, 8°, vi + 183 pp., Braunschweig, 1909.

Die Veränderungen des Klimas seit dem Maximum der letzten Eiszeit (Pub. by 11th International Geological Congress), 4°, lviii + 459 pp., Stockholm, 1910.

For climatic charts see the references in connection with the charts of the meteorological elements at the end of the various chapters.

CHAPTER X

FLOODS AND RIVER STAGES

DEFINITION OF THE TERMS USED IN CONNECTION WITH RIVERS

418. There are several terms used in connection with rivers and floods which require at the outset exact definition and brief consideration.

Drainage area. — By the drainage area of a river is meant the tract of country from which the water drains into the river. This is some-
Drainage times called the catchment basin or the watershed. The
area. area drained by some of the rivers of the world is enormous, that by the Amazon probably being the largest. The drainage area of the Mississippi River and its tributaries probably stands next in size. The Missouri River drains about 527,150 square miles; the Ohio River, about 201,700; the Arkansas River, about 186,300; the Red River, about 90,000; so that the Mississippi River and its tributaries drain about 1,240,000 square miles. Even the smallest brook with a name usually has a drainage area of a good many square miles. A complete description of the drainage area of a river would include an account of its topography, meteorology, and climate.

Inland drainage area. — If a tract of country is landlocked, that is, surrounded on all sides by land of greater elevation, the rivers will flow

442

to the lowest point and form a lake or sea without an outlet to the ocean. Such an area is called an inland drainage area, and the evaporation must here equal the precipitation. The sea or lake is often below sea level and is usually salt. The reason is because the rivers always carry down a small amount of salt and other minerals, and these are left behind when evaporation takes place. The Caspian Sea, with an area of 180,000 square miles, is the largest sea of this kind in the world. Great Salt Lake, Utah, contains 17 per cent of salt and other minerals, and is the best known inland lake in the United States. In Australia, 52 per cent of the whole country consists of inland drainage areas; in Africa, 31 per cent; in Europe and Asia, 28 per cent; in South America, 7.2 per cent; in North America, 3.2 per cent.

Inland drainage area.

Run-off. — By run-off is meant the percentage of the precipitation which eventually drains into a river, and this varies all the way from a few per cent to nearly 90 per cent in some extreme cases. The run-off depends chiefly upon the characteristics of the drainage area, but also upon the amount and rapidity of the rainfall. Ground is ordinarily classified as permeable or impermeable. If rock or a stratum of material impervious to water is near the surface, the ground is impermeable, and the run-off, in this case, will be large and occur very shortly after the precipitation. If the ground is permeable, the run-off is much smaller and much more gradual and regular. Most watersheds have also a decided seasonal change in characteristics. At one time of year vegetation may be luxuriant. At another time of year the ground may be frozen hard or covered with a layer of snow and ice. All these seasonal changes make a tremendous difference in the run-off. The run-off also depends, to a large extent, on the amount and rapidity of the rainfall. It increases both with the amount of the rainfall and the rapidity with which it falls. One of the hardest problems in connection with rivers is to try to estimate the run-off when, over a drainage area in a certain condition, a certain amount of rain falls in a certain time interval. The average run-off for all watersheds in the world is between 20 and 30 per cent.

Run-off.

Thalweg. — The term thalweg is sometimes used to designate the valley bottom through which a river runs.

Thalweg.

Regimen. — The term regimen is used to designate the characteristics of a river. Its normal height, its greatest and least height, its normal discharge of water, its normal velocity of flow, its cross section — all these things go to make up its regimen.

Regimen.

River stage. — By river stage is meant the height of the surface of a river above some arbitrarily chosen zero point. The zero point may
River stage. be mean sea level, or the lowest point reached by the river or the normal height of the river, or any arbitrarily chosen point. Thus, if a river stage is stated as 36 feet, it simply means that the surface of the river is 36 feet above the definite zero point from which all heights are reckoned.

Flood line. — The flood line is some definite river stage so chosen
Flood line. because a greater height than this can be considered a flood. Thus, if the flood line is 40 feet, it means that a river stage above 40 feet would result in an overflow and a damage-causing flood.

River slope. — By river slope is meant the change in elevation of the river surface with distance. It is usually expressed as so many inches
River slope. per mile. In the case of great rivers, it is never more than a few inches per mile. In rapidly flowing streams it is much more and may amount to several feet.

Wetted perimeter. — The length of a line from one side of a river
Wetted perimeter to the other measured along the bottom is called the wetted perimeter.

Mean hydraulic depth. — The area of the cross section of a river
Mean hydraulic depth. divided by the wetted perimeter is called the mean hydraulic depth.

THE MEASUREMENTS MADE IN CONNECTION WITH RIVERS

419. Velocity of flow. — The velocity of flow of a river is determined ordinarily by means of a current meter, which consists essentially of a
A current meter. propeller wheel which revolves faster the greater the velocity of the current. The Price current meter as made by W. and L. E. Gurley of Troy, N.Y., is pictured in Fig. 146. This instrument is used by the U. S. Coast and Geodetic Survey and by many hydraulic engineers in different parts of the country. It consists essentially of five conical buckets so arranged that they turn easily with the slightest current. They are provided with a rudder consisting of four light metal wings or vanes in order to keep the wheel in line with the current. A heavy weight (about sixty pounds) with a wooden rudder is attached for deep-water work or where the current is particularly swift. The instrument is so constructed that electrical contact is made after every revolution of the wheel, and the number of revolutions is counted auto-

matically by an electric register. A reduction table is furnished with
the instrument for finding the velocity which corresponds to a given

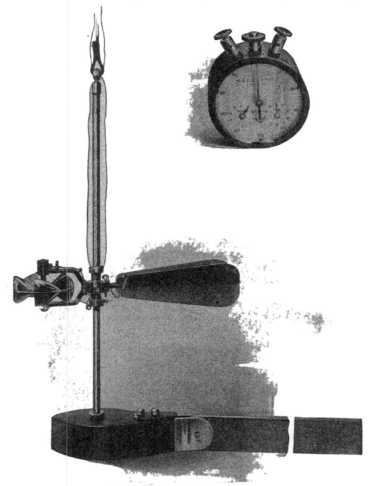

FIG. 146. — The Price Current Meter.

number of revolutions per second. It would be a little more accurate
to determine the reduction table for each instrument separately. This

is done by dragging it at a known rate of speed through still water. Sometimes, instead of an electric register for counting the number of revolutions, the wheel is so constructed that a hammer strikes against a diaphragm after every ten revolutions, and the sound is conveyed to the ear of the observer. These are called acoustic current meters.

In making a complete determination of the velocity of flow of a river, observations must be made at different depths and in different parts of a river. This practically amounts to determining the velocity of flow at all points in the cross section.

If only the surface velocity is desired, it can be determined roughly by watching some floating object and determining the time required for it to be carried a known distance.

The velocity of flow can also be determined by computation from the slope of the river and the mean hydraulic depth. In order to determine the slope, the difference in level of the river at Other methods of determining velocity of flow. points several miles apart must be accurately determined by surveying methods. Only approximate results can, however, be obtained, as the constants in the formula for computing velocity from slope and mean hydraulic depth are too uncertain and depend upon too many things.

420. **River stage.** — The river stage or the height of the river surface above some arbitrarily chosen point is determined by means of a river The river gauge. gauge. This consists ordinarily of a heavy plank, 8 or 10 inches wide and a couple of inches thick, and of sufficient length to cover the greatest possible fluctuations in the height of the surface of the river. The gauge is ordinarily divided into feet, possibly inches or tenths of a foot, and the foot marks are usually numbered. The gauge is placed vertically and securely fastened to a bridge pier, the end of a wharf, or the like. The height of the surface of the river can thus be readily read off. Sometimes the gauge is not placed vertically, but is inclined to follow the river bank. It should always be graduated to show vertical heights, however.

Every river gauge should be provided with a permanent bench mark near by on shore. A bench mark is simply a very stable, permanent The bench mark. point whose elevation above sea level is not supposed to change. A copper bolt in the stone foundation of a building, the water table of a firmly placed building, the surface of some large stone in a building, all serve well as bench marks. By surveying methods, the height of the bench mark above mean sea level should be determined, and also the difference in level between the bench mark and the zero of-the river gauge. If a river gauge is then repaired or carried away by a flood, the new one can be placed with its zero mark at exactly the same level as before. The zero of a river gauge may be placed anywhere on the gauge, but it is customary to put it so low that the lowest water will never reach it. Negative values of river stages are thus avoided.

The following descriptions of the river gauges at Albany, N.Y., and New Orleans, La., taken from W. B. publication No. 227, will serve as illustrations.

Albany, New York

" Albany, N.Y., is on the Hudson River, 150 miles from its mouth.

" The gauge is a self-registering tide gauge, patterned after those used by the United States Coast and Geodetic Survey in former years. It is the property of the United States Engineer Corps, and is located on the east side of the State Street Bridge.

" The bench mark was established in 1896 by the United States Engineer Corps, is on the southeast corner of the east basement window on the south or State Street front of the United States Government building near Dean Street. It is 18 feet above the zero of the gauge and 18.2 feet above mean sea level.

" The highest water was 21.4 feet on February 9, 1857. It was due to back water. On October 4 and 5, 1869, the water reached a stage of 18.5 feet, the highest stage due to rainfall alone. The lowest water was −1.2 feet on Sept. 30, 1867."

New Orleans, Louisiana

" New Orleans, La., is on the Mississippi River, 108 miles above the Gulf. The river is 2400 feet wide. The drainage area above the station is 1,235,500 square miles.

" The river gauge is the property of the city and is situated at the foot of Canal Street among a cluster of piles in rear of ferry wharf. It is made of cypress, and is painted white with markings in black.

" Bench mark at corner of Common and Delta Streets, on iron cornice, 6 inches above sidewalk at E. Conery's store, is 16.5 feet above zero of gauge, and 14 feet above mean sea level. Curbstone under third window of customhouse from Decatur Street, and on Customhouse Street, is 11 feet above zero of gauge, and 8.5 feet above mean sea level.

" Graduation is from zero to 17 feet above. Highest water was 19.5 feet on May 13, 1897; lowest, −0.2 feet on December 27, 1872. Danger line is at 16 feet."

421. Cross section of a river. — The determination of the cross section of a river belongs to hydrographic surveying. It is necessary to determine by sounding the depth of the water every few feet all the way across the river. These observations can then be plotted to scale and a cross section of the river determined.

The cross section of a river.

422. River discharge. — River discharge may be found in two ways: by means of a weir or dam, and by computation from the cross section and the velocity of flow.

The two ways of determining river discharge.

The weir method is by far the most accurate, but can be applied only to small streams. It consists in forcing the water to flow over a weir or dam. By measuring the width of the stream and the depth of the water flowing over the dam, fairly exact values of the discharge can be found by computation.

If the cross section of a river is known and the average velocity of flow has been determined, the discharge can be computed. The product of the area of the cross section in square feet by the velocity in feet per second gives the discharge in cubic feet per second.

RIVER DATA

423. A complete description of a river and its characteristics would include material and data concerning both the watershed and the river itself. In connection with the watershed or drainage area, a full treatment would include an account of its topography, meteorology, and climate. The most important items are a map of the watershed showing the elevations and the character of the surface, and normal values and data concerning the precipitation, snowfall, temperature, evaporation, and run-off. All other facts in connection with the topography, meteorology, or climate would be of interest and value, but of secondary importance.

What a complete description of a river would include.

A treatment of the river itself would include a detailed description of the course of the river and data in connection with the flow of water. The cross section of the river at various points, its length, the height of its banks, and the area which would be covered by a rise of a given amount should all be known. In connection with the flow of water, normal values and data for the river stages, the velocity of flow at different stages, the river discharge at various stages, the causes of floods, and the velocity of progression of a flood wave should all be known.

Complete data for a river and its drainage area are probably available for very few rivers, but the important items are known for most rivers.

THE DIFFERENT KINDS OF FLOODS

424. Floods in rivers may be caused in a variety of ways. (1) Floods may be caused by the breaking of a dam, by the breaking of levees, or by a sudden change of course by a river. (2) Floods may be caused by the temporary blocking of a river by an avalanche, landslide, or glacier. This choking of the river channel would cause a temporary lake back of the obstruction, and *The six kinds of floods in rivers.* the giving way of this barrier might cause disastrous floods along the lower course of the river. (3) Floods can be caused by the luxuriant growth of vegetation which may choke the river channel and thus cause a rise of water. (4) Floods may be caused by the formation of ice dams when the ice breaks up in the spring. These cause floods both above and later below the dam. (5) Floods are often caused by rather sudden melting of the snow and ice over a considerable portion of a watershed. (6) Floods due to excessive precipitation over the watershed are the most common of all floods. Of these various kinds of floods, those caused in the first two ways are of unusual occurrence, and are far from common. Those caused by vegetation and ice dams are fairly common in certain rivers. Those caused by the melting of snow and excessive precipitation are the commonest of all floods.

The best-known flood due to the breaking of a dam is the one which occurred at Johnstown, Pa., June 1, 1889. A reservoir about 3 miles long and 1 mile wide and perhaps 100 feet deep was held by a dam 1000 feet wide. The sudden breaking of this dam precipitated the water upon Johnstown, 18 miles below, and *Illustrations of the various kinds.* caused the loss of nearly 3000 lives.

The breaking of levees along the lower Mississippi River, particularly when the river is in flood, is of fairly common occurrence and often floods large areas.

. The sudden changing of the course of a river usually occurs in mountainous regions where the river is small and swift, or at the delta where a very large river flows into the ocean. In the first instance, very little damage is usually done. Disastrous results ordinarily follow when a large river changes its course. It is said that the Hwangho river in China is particularly prone to change its channel, and that the place where it empties into the ocean has varied as much as 300 miles during the last 4000 years. These great changes have caused the loss of many millions of lives.

The temporary blocking of a river by an avalanche or landslide or by

2 G

the forward movement of a glacier occurs only in mountainous regions near the source of a river. This blocking of the channel may cause, however, a fairly large lake; and if the barrier breaks suddenly, disastrous floods are sure to occur along the course of the river.

Floods due to vegetation are said to be common along the Upper Nile and in the Parana River in South America.

Ice dam floods are common in all rivers which freeze over to a considerable depth in winter. When the ice breaks up in the spring, it is apt to form a dam at a narrow part of the channel or where it is obstructed by a bridge. Such a dam causes floods above it, and, if it breaks suddenly, floods below it are likely to occur.

Floods due to the sudden melting of the snow and ice on the watershed or to excessive precipitation over the watershed are common in all parts of the world. These two causes are also likely to occur together during the late winter and early spring, in which case the rise of the rivers will be particularly large.

Of these six kinds of floods, the last two are the only ones which can be predicted.

The Characteristics of Individual Rivers

425. Limited space does not permit the detailed treatment of the characteristics of even one particular river. The reader must be referred to the literature on the subject for the characteristics of any river in which he may be especially interested. What would be included in such a complete treatise has already been stated in connection with river data.

A few facts about the more important rivers of the world may be of interest. In the Nile the lowest water occurs ordinarily in June and the highest in September or October. At Cairo, Egypt, the rise is from 15 to 30 feet, and the country is inundated annually to a considerable distance on each side of the river. The fertility of Lower Egypt is due to the water gained in this way, and to the alluvium brought down by the river. The rise in the Nile is due to large rainfall in Abyssinia of the monsoon type.

A few facts about individual rivers.

In the Yangtse-Kiang, the Amur, and the Hwangho rivers, the rise occurs during the late summer. This is due to monsoon rains over the interior of Asia and shows to what extent the monsoon penetrates the continent. Floods during the summer are particularly damage-causing as they occur during the time of growing crops. A summer rise due to monsoon rains is also true of the Congo, the Ganges, and the Brahmaputra.

The Amazon River changes but little during the year, as the rainfall is more uniform, and when the tributaries on one side are in flood, those on the other side are usually not.

In the Mississippi and Ohio rivers, and in the Rhine, Seine, and Elbe in Europe, the floods occur during the winter and spring. They are caused by large rainfall over the watershed while the ground is frozen, or by the sudden melting of large quantities of snow. Very often the two causes operate together. The rivers of New England and near-by states also have floods during the winter and spring for the same reasons. In these rivers ice dams often form when the ice breaks up in the spring.

The Prediction of River Stages and Floods

426. The prediction of river stages and the height and time of occurrence of a flood crest may be made in two different ways. One is on the basis of what has occurred on the watershed, and the other is on the basis of river stages which have been observed on gauges higher up the river and on its various tributaries. *The two methods of predicting river stages.*

If the first method is followed, the condition of the watershed must be known; that is, its extent, whether the ground is frozen or not, the amount of snow which may rest on it, etc. A measured amount of rain at a known temperature has fallen on the whole or a part of the watershed in a certain time. The problem is to determine the resulting rise in the river due to the run-off. It might seem that the problem could be solved *In the first method occurrences on the watershed are used.* theoretically; that is, that it could be determined what the run-off would be for the watershed in a known condition due to the rainfall in question. As a matter of fact, the problem is too complex to be treated with any accuracy in this way. If, however, records for a considerable time are available, it is possible to determine from past occurrences about what the resulting rise in a river will be for certain happenings on the watershed. But even when done in this way, the results are so uncertain that this method is never used by the U. S. Weather Bureau in forecasting.

The second method of predicting floods and river stages is by means of gauges placed higher up the river and on its various tributaries.

When a flood occurs the water rises, attains its greatest height, and then falls. A flood may thus be considered as a wave which progresses down a river at a certain speed. By means of the river gauges placed higher up the river and on its tributaries the location and height of the different flood waves may be determined. Based on past experience, the height of the flood wave at the station in question and its time of occurrence can be predicted. The whole prediction of floods rests upon a series of rules or tables derived by studying critically the records of previous floods. This method can, of course, be applied only to the lower part of the course of a river and not to its sources.

In the second method gauge readings are used.

427. The prediction of river stages and floods is part of the regular work of the U. S. Weather Bureau, and belongs to the river and flood section. It was formerly under one man, but the predictions are now made by many near the various rivers. Daily river gauge readings are made at 8 A.M. at many stations located on the various rivers. These observations are telegraphed to 32 centers, and the preparation of forecasts and warnings is, in most cases, intrusted to the officials of the Weather Bureau at these centers, under the supervision of the forecast official at Washington. The regulations concerning the issuing of forecasts have already been stated in section 391.

The work of the Weather Bureau.

SUDDEN RISES OF OCEANS AND LAKES

428. Sudden rises of the ocean are caused either by earthquakes or tropical cyclones. There are several notable examples where tidal waves more than 50 feet high have been caused by earthquakes and great loss of life has resulted. The rise of water caused by the Galveston cyclone (see section 271) is a good example of the second kind.

Sudden rise of the ocean.

A sudden increase in the height of the surface of a lake at a certain point is called a seiche, and these are common in the Great Lakes of the United States and in some lakes in Switzerland, and other parts of the world. They are probably caused by thundershowers or a sudden change in wind direction or velocity. They sometimes amount to several feet. It is said that the surface of the water on the south side of Lake Erie is always several feet higher when the north wind blows than when the south wind blows. Thus a sudden change in wind direction might readily cause a seiche. After the surface of a lake had once been changed from a level surface by the action of the wind, if

Seiches.

the wind should die down to a calm, the surface would then oscillate about a nodal line near the center of the lake until finally brought to rest by friction.   The continuance of seiches after the cessation of wind can thus be explained.

TOPICS FOR INVESTIGATION

(1) The run-off under various conditions.
(2) Inland drainage basins.
(3) The various kinds of current meters.
(4) The complete description of some river.
(5) The seiche.

PRACTICAL EXERCISES

(1) Investigate critically some small brook, determining the characteristics of the watershed and the discharge and behavior of the brook under all conditions.
(2) Determine the cross section of some river and the velocity at all points in it.

REFERENCES

FRANKENFIELD, H. C., The Floods of the Spring of 1903 in the Mississippi Watershed.   Bulletin M ; W. B. publication No. 303.
HENRY, ALFRED J., Wind Velocity and the Fluctuations of Water Level on Lake Erie.   Bulletin J ; W. B. publication No. 262.
MOORE, WILLIS L., *The Influence of Forests on Climate and on Floods.*
MORRILL, PARK, Floods of the Mississippi River.   Bulletin E ; W. B. publication No. 143.
RUSSELL, THOMAS, *Meteorology*, New York, 1895.   (Chapters IX and X cover rivers and floods and river stage predictions.)
Daily River Stages at River Gauge Stations on the principal Rivers of the United States (6 parts have been issued.   The Publication was commenced by the Signal Service and continued by the Weather Bureau.)
Monthly Weather Review. (The condition of the rivers and the occurrence of floods are here summarized monthly.   The hydrographs of the seven principal rivers are also given.)

CHAPTER XI

ATMOSPHERIC ELECTRICITY

INTRODUCTION — HISTORY

429. Lightning is an electric spark on a tremendous scale. Vague, indefinite opinions that this might be the case were expressed by various scientists from 1600 on, but the first definite assertion was made by J. H. Winkler in Leipzig, in 1746, and he attempted to prove it by analogy. Benjamin Franklin proposed experimental proofs in 1749, and in 1752 sent a communication to the Royal Society in London, recommending the use of rods with points as lightning conductors. D'Alibard, at Marly sur Ville, near Paris, through translating Franklin's communication, received the incentive to carry out a series of experiments. An iron rod some forty feet long and provided with a point was attached to an insulated support. When a thundershower approached, sparks an inch or more in length could be drawn from the rod. Franklin's famous kite experiment was performed a little later. In a letter dated October 19, 1752, he describes it as follows:

Early opinions and experiments.

454

" Make a small cross of light sticks of cedar, the arms so long as to reach to the four corners of a large, thin silk handkerchief when extended. Tie the corners of the handkerchief to the extremities of the cross, so you have the body of a kite which, being properly accommodated with a tail, loop, and string, will rise in the air like those made of paper, but being made of silk is better fitted to bear the wet and wind of a thunder gust without tearing. To the top of the upright stick of the cross is to be fixed a very sharp-pointed wire, rising a foot or more above the wood. To the end of the twine next the hand is to be tied a silk ribbon, and where the silk and twine join a key may be fastened. This kite is to be raised when a thunder gust appears to be coming on, and the person who holds the string must stand within a door or window, or under some cover, so that the silk ribbon may not be wet ; and care must be taken that the twine does not touch the frame of the door or window. As soon as the thunder clouds come over the kite, the pointed wire will draw the electric fire from them, and the kite, with all the twine, will be electrified, and stand out every way and be attracted by an approaching finger. And when the rain has wet the kite and twine, you will find the electric fire stream out plentifully from the key on the approach of your knuckle."

Franklin's kite experiment.

De Ramas, in France, a few years later (1757) was able to get sparks ten to twelve feet long by means of a kite. A few days after D'Alibard's experiment, Le Monnier was able, by means of better constructed and insulated apparatus, to prove that there was a difference of potential between a point at a given height in the air and the earth at all times, even when the sky was clear. From this time on, long series of observations were made by many investigators in different parts of the world. The digest of all this experimental and observational material led to the formation of what we may now call the old conceptions concerning atmospheric electricity. This old conception has been considerably modified during the last twenty years, but still deserves careful consideration.

The basis of the old conception.

The earth itself is a conductor, and is surrounded by a non-conducting medium, the atmosphere. The earth is highly charged with negative electricity and in the non-conducting atmosphere around it, there is thus a field of force. There would thus be a difference of potential between a point at a given height in the atmosphere and the earth, and this potential difference would be greater, the greater the elevation of the point in question. Since the earth has a charge of negative electricity, lines of

A negatively charged earth surrounded by a non-conducting medium.

force must start from the earth and extend outward. The great question was, where these lines of force ended, in other words, where the corresponding charges of positive electricity were located. Some said in the clouds, others in the outer regions of the atmosphere, or on the heavenly bodies, or in the remote depths of space. The origin of the earth's negative charge was never satisfactorily explained, it being generally supposed that it had had its charge from the beginning.

Floating in the atmosphere and carried from one point to another, are innumerable dust particles and minute water drops. These are con-
The charged ductors and charged with either negative or positive elec-
particles. tricity. Various ways in which these particles may have become charged have been suggested. (1) They may have become electrified by friction. Carried rapidly by the wind, these particles might strike against material objects or each other. Ice crystals and snowflakes in the upper air may strike against each other. (2) At the moment of evaporation, a particle may have become charged negatively by induction. (3) If a cloud forms, the lower side would have a positive charge induced on it, and the upper side would have a negative charge. If such a cloud should be suddenly broken through horizontally by the wind, the two portions would be charged and with opposite kinds of electricity. (4) Every particle floating in the atmosphere would have positive electricity on its lower side and negative electricity on its upper side due to induction by the charged earth. Now ultra-violet light readily discharges negative electricity. Thus the negative electricity on these particles might be discharged and scattered, which would leave the particle charged with positive electricity. There are thus at least four different ways in which these little conducting particles might become charged with negative or positive electricity. When a cloud forms, these particles become nuclei of condensation, and thus the cloud particles become highly electrified. When these cloud particles further unite to form raindrops, the quantity of electricity steadily increases, until finally there is a lightning flash between two clouds charged differently or between a cloud and the earth.

This whole conception, then, briefly summarized is as follows. The earth is a nearly spherical conductor charged with negative electricity
Summary. and surrounded by a non-conducting medium. It will be surrounded by equipotential surfaces and there will be lines of force going out from it. In the non-conducting surrounding medium there are conducting particles which become positively or negatively electrified. These serve as nuclei of condensation when a cloud forms,

and thus cloud particles and raindrops collect electricity until a lightning flash occurs.

The first difficulty with this explanation was encountered when it was found that the atmosphere was not a non-conductor, but a poorly conducting medium. It was found that a charged body in the atmosphere was slowly discharged. This was at first laid to poor insulation of the body, or the presence in the atmosphere of dust and water particles. It was soon found that this discharging of a charged body could not be accounted for in this way, as the rate of discharge was slower when the moisture was high or even when it was foggy. Since the atmosphere conducted the electricity more like an electrolyte than a metallic conductor it was soon assumed that there were present in the atmosphere, numerous small particles or portions of molecules, called gas ions, which were charged some positively, some negatively. The actual existence of these ions has since been experimentally demonstrated. The presence of the charged ions at once raised new questions. What was the origin of these ions, and how had they become charged, some positively, some negatively? How did the earth retain its negative charge; in other words, why was not the earth discharged? It was also soon found that these ions, particularly the negatively charged ones, also served as nuclei of condensation. Another source of the electrification of the cloud particles and raindrops had thus been found. A readjustment of the old conception was thus necessary, and this will be shortly given.

The difficulties with the old conception.

Readjustment necessary.

THE ELECTRICITY OF THE EARTH, AIR, CLOUDS, AND RAINDROPS

430. The measurement of potential differences due to the earth's electric field. — Since the earth is charged with negative electricity, it must be surrounded by an electric field of force and a potential difference must exist between any point in the atmosphere and the charged earth. In order to measure the potential differences between the earth and a given point in the atmosphere, an electrometer and a " collector " are necessary. In the early determinations, a simple crude electroscope was used as the electrometer. It consisted of a glass globe or case, containing a metal rod to which was attached two pith balls or two light straws, or two leaves of thin gold foil. By the divergence of these light objects the charge and thus the potential difference could be judged.

Two pieces of apparatus necessary.

The electrometer.

In Fig. 147, two simple electroscopes which may serve as electrometers
are shown.   One is a gold leaf electroscope which has been made more
sensitive and accurate and has been provided with a scale.   The other
is Braun's electrometer.   Here a light aluminum pointer moves over a
scale and indicates the potential difference.   For more precise measurements, some form of quadrant electrometer must be used.   For the
description, theory, and use of these well-known pieces of electrical
apparatus, the reader must be referred to text-books on physics.

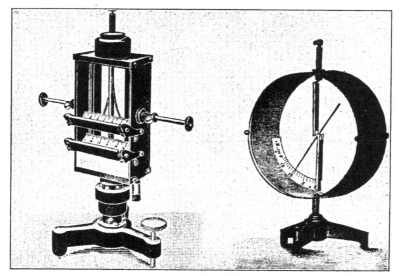

FIG. 147. — Two Simple Electroscopes.
(From GOCKEL'S *Die Luftelektrizität.*)

The so-called collector is placed at the point in the atmosphere for
which the potential difference as regards the earth is to be determined.
The   In the early experiments, it consisted of an insulated point
collector.   or points connected by means of a wire with the electrometer.
It would take up the potential of the point where it was located, and the
difference of potential between this point and the earth would thus be
indicated by the electrometer.   Later, the flame of a lamp placed on an
insulated support or some slowly burning substance was used.   Still
later, the water-dropping collector was devised.   This consists simply of
a vessel of water on an insulated support, from which the water is allowed.

to fall drop by drop. This last collector has probably been the most widely used of any. In still more recent experiments, a small plate or rod covered with a radioactive substance has been used.

The method of determining the potential difference is thus to connect one part of the electrometer with the earth, and the other part with the collector which is placed at the point, and the electrometer reading indicates the potential difference. *The method of determining potential differences.*

431. The electric field of the earth. — Since the earth is a conductor highly charged with negative electricity, it must be surrounded by an electric field of force, extending out indefinitely through the atmosphere, and an equipotential surface could be drawn through any point in this field of force. By an equipotential surface is meant a surface containing all points which have the same potential difference as regards the earth. If the earth were a perfectly smooth conductor, the equipotential surfaces would be parallel to the earth's surface, that is, *Equipotential surfaces and their form.*

concentric with the earth. As a matter of fact, the surface of the earth is far from smooth and level, and the irregularities greatly distort the equipotential surface. Numerous investigations have been made to determine the effect of a hill, mountain, tree, building, or the like, on the equipotential surfaces. It has been found that the general effect

FIG. 148. — The Equipotential Surfaces over an Irregular Surface.

(From HANN's *Lehrbuch der Meteorologie*.)

of projections is to warp the equipotential surfaces upward and cause them to be closer together. This is represented roughly in Fig. 148. It will be seen from this that the change in potential with elevation would be very small beside a building or hill. On the other hand, above a tree or bill it would be particularly large. This must be held in mind in choosing a point for which to determine the potential difference as regards the earth. The most typical and usual values would be found by choosing a point over a level plain. *The geographical, daily, annual, and irregular variations in the potential difference.*

The change in potential with elevation amounts ordinarily to about 100 volts per yard, and a point in the atmosphere is positive as compared with the earth. This change in potential with elevation is by no means a constant. It grows rapidly less with altitude. It. is very different in different parts of the world. It has a periodic daily and annual variation and very large irregular fluctuations, which are closely correlated with the meteorological elements and storms. Near the earth's surface, as just stated, the change in potential amounts to about 100 volts for an ascent of one yard. At a height of two or three miles, the change per yard has decreased to nearly one half its value at the earth's surface and there is some evidence that, at the height of five miles or more, a change with elevation practically ceases to exist. This proves that the earth is not the only charged conductor, but that there are charges of electricity in the atmosphere itself. If the earth alone were charged, the change in potential with elevation would not cease to exist. The values found at various places on the earth's surface are very different. This may be, in a large part, due to irregularities in the surface, but the larger values seem to be found in middle latitudes. Smaller values seem to be found for cold or dry places. The daily variation is very complicated, and seems to show two maxima and two minima very similar to the daily variation in barometric pressure. The maxima occur in the middle of the morning and in the early evening. The minima occur in the early afternoon and before sunrise. The graph which represents the daily variation is very different for different places and sometimes has only one maximum and minimum. The annual variation shows a maximum in winter and a minimum in summer. The values of potential difference grow less with higher temperatures. Under long-continued bright sunshine, the values are usually less. The values also grow less with increasing cloudiness. With increasing dampness and during foggy weather, the values are usually larger. The effect of wind and pressure is extremely small and has never been definitely determined. During a snowstorm or thundershower tremendous irregular fluctuations occur. The positive potential difference often becomes negative and may attain values as high as 10,000 volts per yard.

If under normal fair weather conditions, the change in potential per yard is taken as 100 volts, and if it is furthermore assumed that this state

The charge of the earth.

of things is the same over the whole earth, the negative potential to which the earth must be charged can be computed. The value would be about 600,000,000 volts.

432. **The conductivity of the atmosphere.** — Until recent times, the atmosphere was always considered a non-conductor. The

The rise of the conception that the atmosphere is a conductor.

old conception was that the atmosphere was a non-conducting gaseous medium surrounding a highly charged conductor, the earth. Coulomb, in 1785, was the first to note that an insulated charged body exposed to the free atmosphere lost

its charge, and he stated definitely that this was not due to poor insulation of the supports of the charged body, but that the charged body
gave up its electricity, in part at least, to the air itself. This was at
once ascribed to the conducting particles, dust particles, and minute
water drops which were present in the atmosphere. More than a hundred years passed without any noticeable progress or change in opinion.
In 1887 Linss commenced the quantitative investigation and measurement of the conductivity of the atmosphere. From that time on, an
immense amount of research work has been done on the conductivity
of the atmosphere; and among the numerous investigators, the names
of Elster and Geitel, Wilson, Ebert, and Gerdien are particularly worthy
of mention. It was soon found that the atmosphere was not a non-
conductor, but a poor conductor, and that it conducted like an electrolyte, rather than like a metallic conductor. The presence of positively
or negatively charged gas ions as the cause of the conductivity was thus
suspected in the atmosphere. The actual presence of the   The ions.
ions was later experimentally demonstrated, and it is now
possible, by means of rather elaborate experimentation, to determine
the number present, their velocity of motion, and the charges which
they carry. High conductivity thus indicates a large number of these
charged ions present in the atmosphere, and low conductivity a small
number. It has also been found that there are two sizes of ions,
and these may be designated as large and small ions. The small ions
are of about the same size as molecules, while the large
ions are very much larger. The large ions correspond very   The two kinds of ions
likely to the dust and moisture particles and are thus made   and their
up of perhaps millions of molecules. The number of large   relative im-portance.
ions was found by Langevin in 1905 to be about fifty times
as large as the number of small ions, but the small ions were about 3000
times as rapid in their movements. This means that in the conductivity
of the atmosphere the small ions play by far the larger part, while in
serving as nuclei for condensation, the large ions are by far the most
important.

The number of ions present in the atmosphere, that is, its conductivity,
is by no means constant. It is very different in different places, changes
markedly with elevation, has a well marked periodic daily   The varia-
and annual variation, and large irregular fluctuations which   tions in the
are closely correlated with the meteorological elements and   conductivity.
storms. The conductivity at various places on the earth's surface is
very different and seems to be due to local influences such as the presence

of mountains, the presence of radioactive material in the soil, etc. The conductivity near the earth's surface and in caves is large. It then decreases somewhat with altitude, but grows very large again at the height of a few miles. The daily variation is very different at different places, but ordinarily there are minima at sunrise and sunset, and maxima in the early afternoon and about midnight. The annual variation shows higher values in summer, and lower values in winter. The conductivity of the atmosphere is particularly small during hazy, foggy weather, when the moisture is large, when the sky is cloud-covered, or when precipitation is falling. This is directly contradictory to the old view that it was the presence of dust and moisture particles in the air which gave it its conductivity. The conductivity is particularly large when it is very clear and dry. The influence of temperature, pressure, and wind has never been definitely determined when the influence of these has been separated from the influence of the other things, which usually change in value at the same time.

The change in potential with elevation and the conductivity of the atmosphere usually vary in value at the same time but in opposite directions. That is, when the conductivity increases, the change in potential with elevation decreases, and *vice versa*. The reason for this will be seen later.

433. The source of the charged ions and the earth's negative charge. — There are several possible sources of the ions which are found in the atmosphere and which import to it its conductivity. The

The various sources of the charged ions.

most important cause of ionization is without doubt the presence of radioactive material or the decomposition products and emanations of such material. These substances are found in minute quantities in the soil, in the rocks of the earth's crust, in the water from springs, in the air in caves and below the earth's surface, and in the lower layers of the atmosphere itself. These substances possess the power of forming ions, and, in fact, one of the best methods of determining the amount of radioactive material present and its strength is by means of the number of ions which it produces in the air. Experimental determinations of the number of ions present in the air in caves, or closed cellars, or in the air taken from below the surface of the ground shows that it always contains a particularly large number of ions, and this is exactly what would be expected. Another cause of ions is ether waves of very short wave length, that is, ultra-violet light. This cause of ionization would probably be most effective in the upper layers of the atmosphere where most of these rays are absorbed. Another source of ions is the breaking up of air molecules into positive and negative ions either spontaneously or possibly due to impact or friction. Again, charged dust particles and electrons pushed away from the sun by the pressure of light waves may make their way into the earth's atmosphere and cause the formation of ions. It has also been found that when raindrops are broken up into smaller ones, both positive and negative ions are given off. The number of negative ions is usually much larger than the number of positive, so that the broken raindrop is left positively charged. If these are the causes of the ions in the atmosphere, it will be seen at once that the number of these ions and thus the conductivity of the atmosphere ought to be very different in different localities. Furthermore one would expect a change with altitude and both periodic and irregular fluctuations.

434· Since the atmosphere which surrounds the negatively charged earth is not a non-conductor, but a poor conductor, due to the ions present in it, the maintenance of the negative charge of the earth must be explained. As long as the atmosphere was considered a non-conductor, it could readily be supposed that the earth had received its negative charge in the beginning, and it would always retain it. Since the atmosphere is *The two explanations of the negative charge of the earth.* a poor conductor due to the ions in it, the earth would attract the positive ions and repel the negative ones and soon lose its charge. The continuance of the negative charge must thus be explained. One explanation is that it is due to precipitation. It has been determined by experiments that raindrops and snowflakes are frequently negatively charged. This would mean that precipitation would bring to the earth negative electricity, and thus the negative charge of the earth might be maintained. Recent experiments, however, seem to show that precipitation brings to the earth positive electricity more often and in as great quantity as negative electricity. The larger raindrops seem to be charged positively and the smaller ones negatively. Another explanation is that the earth's negative charge is due to the ionization of the air in the cracks and crevices of the earth's surface itself. The radioactive material ionizes the air in the cracks and crevices below the surface of the earth. This air eventually makes its way out, particularly when the barometric pressure is lessening. The negative ions have much greater rapidity of motion than the positive ions. As a result, as this air makes its way out, the negative ions strike the sides of the vents in greater number, and thus charge the earth negatively while the positive ions escape in greater numbers. Both of the causes may be operative even at the same time.

435. **The source of the electricity of clouds and raindrops.** — The nine methods of cloud formation have been fully discussed in a previous chapter. In eight of these methods, condensation is caused by the lowering of the temperature. The air grows colder, reaches the dew point, and becomes saturated with moisture.

The methods of cloud formation.

If the cooling continues, the air must become supersaturated or condensation must take place. In the one method moisture is added to the air which is already saturated, and thus either condensation or supersaturation must result. When condensation takes place, the dust and other particles serve as nuclei of condensation.

It has been shown experimentally that if particles are absent, the ions can serve as nuclei of condensation, and it has been furthermore found that the negatively charged ions will promote condensation when supersaturation has been carried to a much less extent than is necessary if the positive ions must be used as nuclei. The negative ions are also much more rapid in their movements than positive ions, so that they will join themselves to small cloud particles in greater numbers than the positive ions.

How raindrops may become negatively charged.

As a result, cloud particles are usually charged with negative electricity. These cloud particles join together to form raindrops and snowflakes, and then, under the action of gravity, commence their fall to the earth's surface. As a result, an excess of positive ions has been left behind, and it is a fact of observation that the cirrus, the highest cloud, is usually positively electrified. As the negatively charged raindrops or snowflakes grow in size, they become more and more highly charged with electricity, and as they approach the earth, the positive difference of potential between a point in the atmosphere and the earth must lessen. In fact, it often becomes zero, changes to a negative difference of potential, and may attain values so large that a lightning flash is the result. The raindrops and snowflakes thus bring down large quantities of negative electricity to the earth's surface.

On the other hand, the raindrops, particularly the large ones, must be frequently broken up by the wind. When this occurs, more negative than positive ions are released, so that the drop becomes positively electrified. This may be repeated many times. Thus there may be positively charged raindrops as well as negatively charged ones, and both kinds of electricity may be brought to the earth by precipitation.

How rain drops may become positively charged.

436. Air currents. — The atmosphere contains these positively and negatively charged ions in large numbers· and is a conductor for this reason. The atmosphere is also an electric field of force, since the earth always has a negative charge, and clouds are charged sometimes positively, sometimes negatively. As a result, the charged ions must move along the lines of force at right angles to the equipotential surfaces. The negatively charged bodies will attract the positive ions and repel the negatively charged ions, and a positively charged body will do the opposite. Furthermore, these charged ions will be carried about by the wind and carried up by rising air currents. As a result, there ought to be electric currents in the atmosphere, usually vertical ones, but sometimes in any direction, particularly in the upper atmosphere and among the clouds. The existence of such currents has been experimentally verified.

The reasons for the existence of air currents.

Since precipitation must bring down more electricity to some parts of the earth than others, there ought to be earth currents to neutralize these discrepancies. These, again, have been found experimentally; and, in fact, it is their presence in long telegraph and telephone lines which at times causes great inconvenience.

Earth currents.

437. The view concerning the electricity of the atmosphere, earth, clouds, and raindrops has thus in very recent times changed from what might be called a statical conception to a dynamical conception. The statical conception was that the earth had received its negative charge once for all, and since it was surrounded by a non-conducting atmosphere, it would always retain it. In the present conception also, the earth has a negative charge, but this must be constantly replenished, and two possible and probable sources have been discussed. The atmosphere is a poor conductor by reason of the charged ions which exist in it. The ions are constantly coming into existence in several different ways. Radioactive material and ultra-violet light are probably the two chief causes. The ions go out of existence by joining themselves to the earth and other masses, and by serving as nuclei of condensation for the moisture. As condensation takes place, the cloud particles gain charges of electricity, and as these fall later to the earth as raindrops or snowflakes, great changes in potential are caused which may result in lightning. These ions, by their movements, also give rise to air currents and earth currents. The picture is thus not one of static equilibrium, but of constant motion and interchange which has reached a steady stage.

The old and modern conception of atmospheric electricity contrasted.

2 H

The Nature and Kinds of Lightning

438. The cause and kinds of lightning. — The origin and nature of the electrification of the cloud particles and raindrops have already been

The cause of a lightning flash.

fully discussed. As the raindrops or snowflakes begin their fall to the earth's surface under the action of gravity, they steadily increase in size until the bottom of the cloud is reached. This is due to impact with other cloud particles or condensation on their cold surface. As they increase in size, they just as steadily become more and more highly charged. If the cloud is a thick one, or if condensation is particularly copious, or if the raindrops are especially large, the charge may become very large. Thus it is that thick clouds with copious rainfall and large raindrops are especially likely to be attended by lightning. As these highly charged raindrops come towards the earth under the action of gravity, great changes take place in the earth's electric field of force. Ordinarily, the positive potential difference between a point in the atmosphere and the earth grows less, becomes zero, and then negative, and may reach enormous values. Eventually the potential difference becomes sufficient to cause an electric spark to pass between the cloud and the earth, and a lightning flash is the result. It has been found experimentally that in about 60 per cent of the cases a lightning flash conveys negative electricity to the earth, and in the remaining 40 per cent, positive. Different parts of the same cloud or different clouds may be charged with different kinds of electricity. If these are brought nearer together, a lightning flash may result. Lightning occurs between two clouds probably much more frequently than between a cloud and the earth.

There are several different kinds of lightning. The commonest form by far is the lightning flash which occurs between two clouds, or between

The kinds of lightning.

a cloud and the earth. This is usually classified as zigzag lightning. Four other kinds are usually recognized, viz. sheet lightning, heat lightning, ball lightning, and beaded lightning.

439. Zigzag lightning. — The ordinary lightning flash which occurs between two clouds, or a cloud and the earth, is usually considered zig-

Description.

zag in appearance. Sinuous would probably better describe its appearance, as it resembles a river course more than a series of straight lines joining each other at an angle. The exact

Photographing a lightning flash.

appearance can be much better studied by photography than by observing directly, and many good lightning photographs are now in existence. They can best be secured by point-

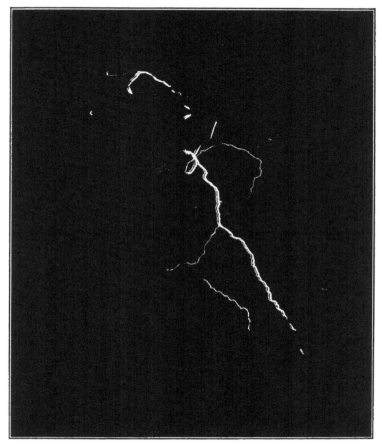

Fig. 149. — A Lightning Flash.

ing the camera at night towards an approaching thundershower and leaving the shutter open several minutes in the hope that a favorably placed lightning flash may occur during the interval. The same plate can be exposed as long as twenty minutes if the night is dark, and there are no artificial lights near to fog the plate. A lightning flash from an "untouched" negative is reproduced in Fig. 149, and the sinuous or zigzag appearance is easily seen. There are several causes which contribute to giving the lightning flash this appearance. It was formerly supposed that the lightning flash always followed the line of least resistance, that is, that the electric discharge sought out the pockets of air which, on account of the presence of dust or moisture, or some other cause, had the greatest conductivity. This is probably to some extent true, but the sinuous appearance is due largely to other things. If a photograph of a lightning flash is studied carefully, it will be seen that wherever there is a particularly sharp bend, a branch extends off to one side. When a lightning flash occurs, the eye is somewhat blinded by the glare, so that only the trunk flash is seen, and this, furthermore, so claims the attention that the side branches are overlooked. It is this branching structure of lightning, however, which is largely responsible for its sinuous appearance. Again, a long lightning flash is not seen through a uniform, homogeneous medium. There are various layers in the atmosphere of different temperature and different density, and these are being moved about and mixed by the wind. As a result, refraction is very different in some directions than in others, and the broken, sinuous appearance is the result. This is exactly the same as the appearance of telegraph wires seen through the windowpane of a moving train. Again the lightning flash might have a slightly spiral form. This seen in projection against a dark background would have a sinuous appearance. Probably all these causes work together to give the lightning flash the appearance which it has.

The reasons for the zigzag or sinuous appearance.

The length of a lightning flash between a cloud and the earth is not usually more than three quarters of a mile or a mile at most. Between two clouds, however, lightning flashes as long as twenty miles have been observed. The color of a lightning flash is usually white, although red, yellow, blue, and violet tinges have at times been observed. The spectrum of lightning is a bright line spectrum, showing generally the lines of nitrogen, although the lines of oxygen, hydrogen, and some of the other constituents of the atmosphere are sometimes glimpsed.

The length, color, and spectrum of a lightning flash.

Formerly it was thought that a lightning flash was simply an electric spark which passed from the cloud to the earth, or possibly from the earth to the cloud. It is now known that the discharge is oscillatory like the discharge of a Leyden jar. That is, instead of a single spark or discharge passing from the cloud to the earth or from the earth to the cloud, the discharge goes backward and forth many times, growing constantly weaker. This oscillation occurs many times, perhaps in some cases scores of times, in an extremely short time interval, a time interval certainly less than a thousandth of a second. This has been determined by photographing a lightning flash with a rapidly revolving camera. It has thus been possible to separate and photograph as separate the successive oscillations. Such a photograph shows the lightning flash, not as a single sinuous line, but as a series of sinuous lines side by side and nearly parallel. There is some evidence that the beginning of a flash does not consist of oscillations, but of a series of impulses going, say, from the cloud towards the earth, and each penetrating a greater distance, until finally the earth's surface is reached. That a lightning flash is practically instantaneous can be shown in a variety of ways. A fast moving object like a railway train is seen by the light of a lightning flash as stationary. Trees swayed by the wind are also as stationary, and a photograph of them shows no blurring due to moving. A simple experiment can also be tried. Cover a bicycle wheel with pieces of colored paper, and when a thundershower occurs at night, spin the wheel rapidly. Each lightning flash will reveal the wheel as stationary, and the colored pieces of paper will be seen distinctly and without blurring or running together. A variously colored spinning top will also serve as well as the bicycle wheel. It is a well-known fact of experience, however, that a lightning flash seems many times to last at least nearly a second. This is due partly to the observer. A lightning flash is a rather unexpected, startling, dazzling occurrence. The various details which have impressed themselves upon the eye are slow in coming to the conscious attention of the observer, and the duration is thus estimated as much longer than it really is. But photography has again revealed the cause of this long duration in many cases. After an oscillatory discharge lasting say a thousandth of a second has taken place over a certain path, there is a decided tendency for another discharge to take place over the same path a few tenths of a second later, and even a third or fourth discharge may follow, so that the duration of the whole occur-

Lightning is an oscillating discharge.

A lightning flash is practically instantaneous.

Multiple flashes.

rence may be nearly a second, but we have to do here with a multiple, not a single, flash. The reason why a second discharge follows over the same track is probably because the resistance has been lessened by the first discharge. This can often be photographed without a moving camera. The path will drift enough with the wind between the first and second discharge, to allow them to be photographed separately. This was the case in the photograph which is reproduced as Fig. 149.

A lightning flash is thus an oscillatory discharge lasting but an extremely short time and followed often by subsequent discharges a few tenths of a second later over nearly the same path. Attempts *The amount* have been made to estimate the number of amperes of elec- *of current.* tricity in a lightning flash. The basis for these estimations is the induction effects of the discharge. Twenty thousand amperes is the figure usually given for an ordinary lightning flash. Attempts have also been made to reproduce on a small scale, by means of electrical machines, the characteristics and effects of lightning and with fair success.

440. Other kinds of lightning. — Sheet lightning, heat lightning, ball lightning, and beaded lightning are the other kinds of lightning which deserve a passing consideration.

The term sheet lightning is applied to the sudden, brief lighting up of á whole cloud. Sometimes the edges of the cloud appear more brilliantly illuminated than the center. At times it appears as if a cur- *Sheet light-* tain were suddenly drawn away, disclosing the bright cloud. *ning and its* It occurs only during a thundershower, and thunder is often *nature and cause.* heard. The duration is fairly long, perhaps a second or two, and the illumination often occurs in pulses. The explanation which most naturally suggests itself is that it is due to a regular lightning flash which is hidden from view by the cloud. The spectrum, however, consists of bands due to nitrogen rather than lines. It may thus be of the nature of a brush discharge rather than a spark discharge. It will be remembered that when the poles of an electrical machine are pulled too far apart, the spark discharge changes to a brush discharge.

Heat lightning is the term applied to the sudden lighting up of the atmosphere, usually when it is hazy and misty but when no thundershower is visible. It is generally so indefinite that it is hard *Heat* to localize it and thunder is never heard. It is usually ex- *lightning.* plained as the reflection from the hazy air of the lightning which accompanies a thundershower below the horizon of the observer. It may also be a brush discharge and thus of the same nature as sheet lightning.

Ball lightning is the least known and the hardest to explain of all, and many even deny its existence. It sometimes appears as a ball of fire

Ball lightning. descending from the clouds but more usually it suddenly makes it appearance on some object near the ground. It appears often on objects inside of houses. The ball of fire is variously described as being as small as an egg, or in some extreme cases even as large as a man's head. It always moves slowly, sometimes in a zigzag line, and is often accompanied by a hissing, sputtering sound. Its path is sometimes indicated by charring, but it may leave no trace. It sometimes silently disappears, sometimes goes into the ground, sometimes explodes with a loud report. A lightning flash of the usual kind and loud thunder usually accompany it. As has just been stated, there are many who deny the existence of ball lightning and would explain it as imagination or an optical illusion. There are, however, rather too many carefully described authentic cases to deny its existence. Gockel in his book *Das Gewitter* gives twenty-four good examples, and if all literature were searched, hundreds of cases would be found. The things observed can probably be explained as the result of an ordinary lightning flash together with induction effects, and a peculiar mind state on the part of the observer caused perhaps by the lightning.

Beaded lightning is of very rare occurrence and is the term applied to a lightning flash which appears, not of the same intensity throughout, **Beaded lightning.** but like a series of luminous beads strung together. There are not enough cases on record to make an attempt at an explanation possible or worth while. It might be an ordinary lightning flash seen through a layer of rippled clouds.

DANGER FROM LIGHTNING AND PROTECTION FROM LIGHTNING

441. Loss of life and property due to lightning. — It is no easy matter to get accurate statistics for a whole country as to the loss of life and **Sources of information.** property due to lightning. Vital statistics and the accounts which appear in newspapers must be relied upon for information concerning the loss of life. The records of insurance companies and newspaper accounts must be used to determine the property loss. The work of determining the loss was begun by the Weather Bureau in the United States in 1890 and the work ended with the year **Loss of life in the United States.** 1900. During this last year a particular effort was made to get as complete and accurate statistics as possible. The average number of persons killed in the United States each year for the period 1891 to 1900 inclusive was 377. This means that on the average six persons are killed each year out of every million inhabitants. During the single year 1900, when the particular effort was made to secure complete statistics, 713 were killed. This would be an average of about 10 per million. Of the 713 persons killed during 1900, 291 were killed in the open, 158 in houses, 57 under trees, and 56 in barns. The circumstances attending the death of the remaining 151 were not known. Of the 973 persons who were more or less injured by lightning during the same year, 327 were injured in houses, 243 in the open, 57 in barns, and 29 under trees. The circumstances attending the injury of the remaining 317 were not known. The number killed and injured in the open is a rather remarkable showing. Most of those killed in the open were either raised above their surroundings because they were on horseback or on a load of grain or on a wagon or agricultural implement or they had some metallic tool in their hands. Statistics as to the number killed yearly per million of inhabitants have been prepared by various investigators for various countries Loss of life in Europe. For England and Wales the number killed in Europe. annually is about 1 in a million; for France about 3; for Belgium about 2; for Sweden about 3; for Prussia about 6; for Hungary about 16.

All of the figures just given are certainly too small rather than too large. There is but a slight chance that the same case would be recorded and counted twice, while a great many cases, particularly in sparsely populated districts, must escape notice entirely. The figures are without doubt sufficiently accurate, however, to give a fair idea as to the danger from lightning.

The property loss due to lightning is even harder to determine than the loss of life. During 1898 in the United States there were 1866 property losses, aggregating $2,000,000. Certain kinds of The kind of buildings are struck more often than others. A summary buildings for Schleswig-Holstein for the years 1874 to 1883 shows that struck. each year, out of every million buildings, 163 ordinary buildings with hard roofs, 386 ordinary buildings with soft roofs, 6277 churches, 8524 windmills, and 306 factories were struck by lightning. The character of the soil also makes a difference with the number of lightning strokes. The order of frequency of lightning strokes on the various The effect of soils in percentages, deduced from 380 reports in the United the soil. States, is as follows; loam, 26 per cent; sand, 24 per cent; clay, 19 per cent; prairie, 19 per cent; scattering, 12 per cent. According to Hellmann, for North Germany, if the liability for chalk formation is considered 1, then it is 2 for marl, 7 for clay, 9 for sand, and 22 for loam. If lightning strikes sandy soil, a glazed tube called a fulgurite, sometimes two or three feet in length, is formed. If lightning strikes a rock, the surface is sometimes glazed in spots. There is also a great difference in the kind of trees that are struck by lightning. Observations made

The kinds of trees that are struck. on an area of about 45,000 acres in a German forest showed that the various kinds of trees were struck as follows : oaks, 159 times ; beeches, 21 times ; pines, 20 times ; firs, 59 times ; birches, 4 times ; larches, 7 times ; ashes, 5 times. And this occurred in a forest which was composed approximately as follows : beech, 70 per cent ; oak, 11 per cent ; pine, 13 per cent ; fir, 16 per cent. Using these and similar observations as a basis, the statement can be made that the oak is particularly liable to be struck, that the elm, chestnut, and pine stand next, and that the beech, birch, and maple are almost never struck. When a tree is struck, it is usually the trunk which is injured, not the branches.

The question is often asked if the danger from lightning is increasing. It is a question which it is very hard to answer from the point of view of

Is the danger from lightning increasing? statistics. The actual number of deaths and injuries from lightning and the actual number of property losses is, of course, increasing. But the number of inhabitants and the number of buildings is also increasing very rapidly. Again, the means of gathering news and the wide publicity given to each death or each property loss have become much greater. It is not to be wondered at, then, that the number of deaths and the number of property losses seem very much larger than they used to be. Statistics on the subject give no very definite answer, but it would seem from them that there has been no marked increase in the danger from lightning.

442. Protection from lightning. — The first suggestion to protect buildings by means of lightning rods was made by Benjamin Franklin.

History of lightning rods. It is contained in a communication to the Royal Society in London, and dated July 29, 1752. The first lightning rod was probably erected by Franklin on his own house in September, 1752, about a month after his famous kite experiment, and consisted of a pointed iron rod which extended about nine feet above the chimney of the building and was connected by means of a thick iron wire with a well. Lightning rods soon became quite common in the States, but in Europe their introduction was slow. In 1760 a lightning rod was placed on the Eddystone lighthouse at Plymouth, and this was probably the first one in Europe. A lightning rod on the tower of the

Fig. 151. — A Black Walnut struck by Lightning.
(Henry, U. S. Weather Bureau.)

Fig. 150. — An Oak struck by Lightning.
(Henry, U. S. Weather Bureau.)

Jakobikirche in Hamburg, in 1796, was probably the first in Germany. Up to thirty years ago, the lightning rod was very common and most large buildings were protected by them. During the last thirty years, however, the use of lightning rods has fallen off greatly. This is not only a fact of popular observation, but it is born out by the statistics of fire insurance companies and by the reports of the companies engaged in the manufacture of lightning rods. The reasons for this are probably the changed ideas as to the nature of lightning and the kind of protection furnished by rods, and also the fact that nearly all buildings are now insured and that covers loss due to lightning as well as fires started in other ways. It is sometimes answered that it is cheaper to insure them than to put up lightning rods.

Lightning rods now less common.

The damage caused by lightning is mechanical as well as thermal. If a tree is struck, it may be completely shattered or the bark may be torn from the trunk in a long strip. Figures 150 and 151 illustrate an oak and a black walnut which were struck by lightning. If a building is struck, bricks may be torn from the wall or chimney, masonry may be cracked, doors may be splintered, and objects broken to pieces. A church spire struck by lightning is shown in Fig. 152. All these are the mechanical effects of lightning; and it is sometimes thought that the sudden expansion of inclosed pockets of air or the sudden evaporation of inclosed moisture is responsible for a large part of the damage. Woodwork may be charred, inflammable or easily ignited material may be set on fire, gas pipes may be punctured and the gas ignited. All these are the thermal effects of lightning; and, in fact, the greatest damage comes from the fires which are started. Not only is damage caused by the main discharge, but currents are induced in near-by metal objects and conductors, and these are often damage-causing. Many of the vagaries and unusual things done by lightning are due to these induction effects.

Mechanical damage.

Thermal effects.

It was formerly supposed that a lightning rod afforded protection in two ways. The pointed rod was supposed to carry the accumulation of electricity to the ground, so that a disruptive discharge would never occur; and if one did occur, it would lead it to the ground, as it would be the best conductor. It is now known that the potential difference between a point in the air and the ground changes so rapidly during a thundershower that a lightning rod could not possibly equalize the potential difference and prevent a discharge. Furthermore, the quantity of electricity in a lightning flash is so enormous

Are lightning rods worth while?

Fig. 151. — A Black Walnut struck by Lightning.

FIG. 152. — A Church Spire Struck by Lightning. Spire (198 feet, brick and brownstone) struck July 29, 1894.

(From *Scientific American*, F. J. MOULTON.)

that only a very large, well-constructed and grounded lightning rod could carry it all safely to the ground. It is now also known that lightning is an oscillatory discharge, and there are many induced currents which may be as damage-causing through their mechanical and thermal effects as the main flash. For example, lightning has perhaps struck the lightning rod of a house and has been conveyed to the ground. A little induced current in a gas pipe may have set on fire some easily ignited substance, and the building is destroyed by fire in spite of the fact that the lightning rod did its full duty. The question is thus a very pertinent one as to whether it is worth while putting lightning rods on a building. In a city where buildings are close together, where iron is much used in construction, where there are many metallic gutters, and where fire companies are efficient, it is very likely not worth while to instal lightning rods. In the country, however, where buildings stand alone and conditions are very different, a considerable protection would probably be gained from lightning rods.

The question is often asked as to what kind of lightning rods should be used, and how they should be attached to the building. The best answer to such questions is contained in the

RULES FOR THE ERECTION OF LIGHTNING CONDUCTORS, AS ISSUED BY THE LIGHTNING ROD CONFERENCE IN 1882, WITH OBSERVATIONS THEREON BY THE LIGHTNING RESEARCH COMMITTEE, 1905

(NOTE. — Paragraphs beginning with odd numbers refer to Lightning Rod Rules, 1882; those with even numbers to Lightning Research Committee's observations, 1905.)

(1) *Points.* — The point of the upper terminal should not be sharp, not sharper than a cone of which the height is equal to the radius of its base. But a foot lower down a copper ring should be screwed and soldered on to the upper terminal, in which ring should be fixed three or four sharp copper points, each about six inches long. It is desirable that these points be so platinized, gilded, or nickel plated as to resist oxidation.

(2) It is not necessary to incur the expense of platinizing, gilding, or electroplating. It is desirable to have three or more points beside the upper terminal, which can also be pointed ; these points must not be attached by screwing alone, and the rod should be solid and not tubular.

(3) *Upper terminals.* — The number of conductors or points to be specified will depend upon the size of the building, the material of which it is constructed, and the comparative height of the several parts. No general rule can be given for this, but the architect must be guided by the directions given. He must, however, bear in mind that even ordinary chimney stacks, when exposed, should be protected by short terminals connected to the nearest rod, inasmuch as accidents often occur owing to the good conducting power of the heated air and soot in the chimney.

(4) This is dealt with below in suggestion 3.

(5) *Insulations.* — The rod is not to be kept from the building by glass or other insulators, but attached to it by metal fastenings.

(6) This regulation stands.

(7) *Fixing.* — Rods should preferentially be taken down the side of the building which is not exposed to rain. They should be held firmly, but the holdfast should not be driven in so tightly as to pinch the rod or prevent the contraction and expansion produced by changes of temperature.

(8) In most cases it would be advantageous to support the rods by holdfasts (which should be of the same metal as the conductor) in such a manner as to avoid all sharp angles. The vertical rods should be carried a certain distance away from the wall to prevent dirt accumulating and also to do away

with the necessity of their being run around projecting masonry or brickwork.

(9) *Factory chimneys.* — These should have a copper band around the top, and stout, sharp, copper points, each about one foot long, at intervals of two or three feet throughout the circumference, and the rod should be connected with all bands and metallic masses in or near the chimney. Oxidation of the points must be carefully guarded against.

(10) As an alternative, the rods above the band might, with advantage, be curved into an arch provided with three or four points. It is preferable that there should be two lightning rods from the band carried down to the earth in the manner previously described. Oxidation of the points does not matter.

(11) *Ornamental ironwork.* — All vanes, finials, ridge ironwork, etc., should be connected with the conductor, and it is not absolutely necessary to use any other point than that afforded by such ornamental ironwork, provided the connection be perfect and the mass of ironwork considerable. As, however, there is a risk of derangement through repairs, it is safer to have an independent upper terminal.

(12) Such ironwork should be connected as indicated below in suggestion 3. In the case of a long line of metal ridging, a single main vertical rod is not sufficient, but each end of the ridging should be directly connected to earth by a rod. Where the ridge is non-metallic, a horizontal conductor (which need not be of large sectional area) should be run at a short distance above the ridge and be similarly connected to earth.

(13) *Material for rod.* — Copper, weighing not less than 6 ounces per foot run, and the conductivity of which is not less than 90 per cent of that of pure copper, either in the form of tape or rope of stout wires — no individual wire being less than No. 12 B. W. G. Iron may be used, but should not weigh less than 2¼ pounds per foot run.

(14) The dimensions given still hold good for main conductors. Subsidiary conductors for connecting metal ridging, etc., to earth may with advantage be of iron and of a smaller gauge, such as No. 4 S. W. G. galvanized iron. The conductivity of the copper used is absolutely unimportant, except that high conductivity increases the surges and side flashes, and therefore is positively objectionable. It is for that reason that iron is so much better.

(15) *Joints.* — Although electricity of high tension will jump across bad joints, they diminish the efficacy of the conductor, therefore every joint, besides being well cleaned, screwed, scarfed, or riveted, should be thoroughly soldered.

(16) Joints should be held together mechanically as well as connected electrically, and should be protected from the action of the air, especially in cities.

(17) *Protection.* — Copper rods to the height of 10 feet above ground should

be protected from injury and theft by being inclosed in an iron pipe reaching some distance into the ground.

(18) This regulation stands.

(19) *Painting.* — Iron rods, whether galvanized or not, should be painted; copper ones may be painted or not, according to the architectural requirements.

(20) This regulation stands.

(21) *Curvature.* — The rod should not be bent abruptly round sharp corners. In no case should the length of the rod between two points be more than half as long again as the straight line joining them. Where a string course or other projecting stonework will admit it, the rod may be carried straight through, instead of around the protection. In such a case, the hold should be large enough to allow the conductor to pass freely, and allow for expansion, etc.

(22) The straighter the run the better. Although in some cases it may be necessary to take the rod through the projection, it is better to run outside, keeping it away from the structure by means of holdfasts, as described above.

(23) *Extensive masses of metal.* — As far as practicable, it is desired that the conductor be connected to extensive masses of metal, such as hot-water pipes, etc., both internal and external; but it should be kept away from all soft metal·pipes, and from internal gas pipes of every kind.¹ Church bells inside well-protected spires need not be connected.

(24) It is advisable to connect church bells and turret clocks with the conductors.

(25) *Earth connections.* — It is essential that the lower extremity of the conductor be buried in permanently damp soil; hence proximity to rain water pipes and to drains is desirable. It is a very good plan to make the conductor bifurcate close below the surface of the ground, and adopt two of the following methods for securing the escape of the lightning into the earth. A strip of copper tape may be led from the bottom of the rod to the nearest gas or water main — not merely to a lead pipe — and be soldered to it; or a tape may be soldered to a sheet of copper, 3 feet by 3 feet and $\frac{1}{16}$ inch thick, buried in permanently wet earth, and surrounded by cinders or coke; or many yards of the tape may be laid in a trench filled with coke, taking care that the surfaces of copper are, as in the previous cases, not less than 18 square feet. Where iron is used for the rod, a galvanized-iron plate of similar dimensions should be employed.

(26) The use of cinders or coke appears to be questionable owing to the chemical or electrolytic effect on copper or iron. Charcoal or pulverized carbon (such as ends of arc light rods) is better. A tubular earth consisting of a perforated steel spike driven tightly into moist ground and lengthened up to the surface, the conductor reaching to the bottom and being packed with granulated charcoal, gives as much effective area as a plate of larger surface, and can easily be kept moist by connecting it to the nearest rain water pipe. The resistance of a tubular earth on this plan should be very low and practically constant.

(27) *Inspection.* — Before giving his final certificate, the architect should have the conductor satisfactorily examined and tested by a qualified person, as injury to it often occurs up to the latest period of the work from accidental causes, and often from carelessness of workmen.

(28) Inspection may be considered under two heads:

(*A*) The conductor itself.

(*B*) The earth connection.

(*A*) Joints in a series of conductors should be as few as possible. As a rule, they should only be necessary where the vertical and horizontal conductors are connected, and the main conductors themselves should always be continuous and without artificial joints. Connections between the vertical and horizontal conductors should always be in places readily accessible for inspection. Visible continuity suffices for the remainder of the circuit. The electrical testing of the whole circuit is difficult and needless.

(*B*) The electrical testing of the earth can, in simple cases, be readily effected. In complex cases, where conductors are very numerous, tests can be effected by the provision of test clamps of a suitable design.

(29) *Collieries.* — Undoubted evidence exists of the explosion of fire damp in collieries through sparks from atmospheric electricity being led into the mine by the wire ropes of the shaft and the iron rails of the galleries. Hence the head gear of all shafts should be protected by proper lightning conductors.

Suggestions of the Committee

The investigations of the committee warrant them in putting forward the following practical suggestions:

(1) Two main lightning rods, one on each side, should be provided, extending from the top of each tower or high chimney stack by the most direct course to earth.

(2) Horizontal conductors should connect all the vertical rods (*a*) along the ridge, or any other suitable position on the roof; (*b*) at or near the ground.

(3) The upper horizontal conductor should be fitted with aigrettes or points, at intervals of 20 or 30 feet.

(4) Short vertical rods should be erected along minor pinnacles and connected with the upper horizontal conductor.

(5) All roof metals, such as finials, ridging, rain water and ventilating pipes, metal cowls, lead flashing, gutters, etc., should be connected with the horizontal conductors.

(6) All large masses of metal in the building should be connected either directly or by means of the lower horizontal conductor.

(7) Where roofs are partially or wholly metal lined, they should be connected to earth by means of vertical rods at several points.

(8) Gas pipes should be kept as far away as possible from the position occupied by lightning conductors, and as an additional protection, the service mains,

to the gas meter should be metallically connected with house services leading from the meter.

(Signed)   JOHN SLATER, Chairman,
E. ROBERT FESTING,
OLIVER LODGE,
J. GAVEY,
W. N. SHAW,
A. R. STENNING,
ARTHUR VERNON,
KILLINGWORTH HEDGES, Honorary Secretary,
G. NORTHOVER, Secretary.

These rules were formulated in England, and the most active member of the Research Committee was without doubt, Sir Oliver Lodge, F.R.S.

There are a few illusions and prevalent misapprehensions with regard to lightning which should receive passing mention. It is not true that lightning never strikes twice in the same place. Nearly A few popular misapprehensions. every person knows of houses or barns which have been struck more than once. It is also foolish to believe that a few inches of glass or a few feet of air or a quarter of an inch of rubber will serve to bar the progress of a lightning flash which has forced its way through perhaps a mile of air. It is also not the highest point which is always struck.

OTHER MANIFESTATIONS OF ATMOSPHERIC ELECTRICITY

443. There are two other manifestations of atmospheric electricity which deserve brief consideration. These are the aurora borealis and St. Elmo's fire.

The aurora borealis, or northern lights, as seen in middle latitudes, usually consists of a whitish arc of light or quivering, rapidly moving beams. Sometimes a faint illumination without definite Description of the various appearances of the aurora. form is seen, and again it takes the form of clouds or patches of light. The rays or beams seem sometimes to form a curtain or to radiate from a crown. The arc may also take the form of a vertically suspended curtain and sometimes appears folded or convoluted. The simple arc and the quivering rays are, however, the most common forms in middle latitudes. The arc is usually seen in the north, but at times it may pass through the zenith or even be seen in the south. The quivering beams generally come out of the north, and usually follow the direction of the magnetic north and south line. The color of the light is usually pale white, although reddish, yellowish, and greenish tinges have often been noticed. It has sometimes been stated that there is a peculiar odor which has been noticed, but this has never been determined with certainty. An auroral display usually commences in the early evening and lasts a few hours, although it has been observed to last several days. Auroras do not occur in all parts of the world with the same frequency. The belt of maximum frequency in the northern hemisphere extends from about

Belt of maximum frequency. 65 to 80° north latitude. In middle latitudes they are much less frequent and also in the immediate vicinity of the north pole. This belt extends farther south over North America than over Europe and Asia, and thus seems to be related to the magnetic north pole, which it seems to surround rather than the geographical pole. Auroras are of very rare occurrence in the torrid zones. There is a poorly marked daily period. An auroral display generally commences in the early evening and lasts a few hours, although auroras have

Daily and annual variation. been observed at any time of the day or night. There is a well-marked annual periodicity, the maximum number occurring in March and again in October. The minima occur in midwinter and midsummer. There are also several longer cycles in connection with their frequency of occurrence which are well marked. The sun spot cycle of a little more than eleven years is extremely well marked and easily noticeable. In fact, the auroras are closely related with all solar disturbances. A particularly large sun spot or an outburst of solar activity which causes earth currents and magnetic disturbances is almost sure to be attended by brilliant auroral displays as well. The various determinations of the height of auroral arches or streamers are

Height. very discordant. Values of from 10 to 15 up to nearly 1000 miles have been found. The average height seems to be a hundred miles or more, and the height is much less in higher than in middle latitudes. This is, without doubt, due to the fact that the auroral streamers follow the lines of magnetic force which actually enter the earth in the region of the magnetic north pole. It was once thought that the light of the aurora was due to the reflection of sunlight from the ice crystals of the upper atmosphere or to the reflection of

Spectrum. sunlight from the snow and ice in the polar regions. It is now known that it is not reflected light at all, because it shows no trace of polarization due to reflection, and the spectrum is not that of sunlight. More than fifty bright lines or bands have now been mapped in the spectrum of the aurora, and many have thought to identify them with the lines in the spectra of the rarer constituents of the atmosphere, such as argon, neon, and krypton. The aurora then is caused by electrical discharges in the rare upper atmosphere, and is of **Nature of** very much the same character as cathode rays. All agree **the aurora.** that the rarefied upper atmosphere is in an ionized condition, and is thus a fairly good conductor, so that these discharges are easily possible. These ions may be negatively charged particles pushed away from the sun by the pressure of light waves, or they may have originated in the earth's atmosphere in ways which have already been discussed. Such long since disproved theories as to the cause of the aurora as that it is due to cosmic dust which strikes the atmosphere and becomes luminous, or that it is due to the phosphorescence of snow or ice or dust, need only be mentioned in passing. A complete theory of the aurora which will explain all its varied forms, its place and time of occurrence, its periodicity, and the many facts observed in connection with it has hardly yet been worked out.

St. Elmo's fire consists of brushlike tufts of light which sometimes appear on all pointed objects or objects with sharp angles, during a thundershower or snowstorm. It has been occasionally **St. Elmo's** observed at low levels, but it is of commonest occurrence at **fire.** high elevations on mountains. It has also appeared on the masts of ships at sea. A hissing sound is usually heard, and sometimes an odor is noticed. It is particularly visible and noticeable just before a lightning flash, when the potential difference is very large. It is simply a brush discharge of electricity due to the large change in potential with height.

TOPICS FOR INVESTIGATION

(1) Benjamin Franklin and atmospheric electricity.
(2) The history of atmospheric electricity to 1760.
(3) Electrometers.
(4) Form of the equipotential surfaces over projections.
(5) The variations in potential difference.
(6) The variations in conductivity.
(7) The earth's negative charge and its maintenance.
(8) Earth currents.
(9) Photography of lightning.
(10) Ball lightning.
(11) Heat lightning.
(12) Effects of lightning on trees.
(13) The kinds of trees struck by lightning.
(14) Loss of life due to lightning.
(15) Lightning rods.
(16) The aurora.
(17) St. Elmo's fire.

2 I

PRACTICAL EXERCISES

(1) Determine the potential difference between a point in the atmosphere and the earth under various conditions.

(2) Determine whether raindrops are positively or negatively electrified.

(3) If suitable apparatus is available, determine the conductivity of the atmosphere under different conditions, and repeat some of the experiments in connection with the ions.

(4) Photograph some lightning flashes.

(5) Determine the duration of lightning flashes.

REFERENCES

For an admirable presentation of the modern point of view in connection with atmospheric electricity, see:

GOCKEL, ALBERT, *Die Luftelectrizität*, 8°, vi + 206 pp., Leipzig, 1908.

MACHE, H., UND SCHWEIDLER, E. VON, *Die atmospherische Electrizität*, 8°, xi + 247 pp., Braunschweig, 1909.

For a treatment of the kinds, nature, and effects of lightning, see:

FLAMMARION, CAMILLE, *Thunder and Lightning*, translated by Walter Mostyn, 12°, 281 pp., Boston, 1906.

GOCKEL, ALBERT, *Das Gewitter*, 8°, 264 pp., Köln, 1905.

For a treatment of lightning rods and protection from lightning, see:

ANDERSON, RICHARD, *Lightning Conductors*, 8°, xv + 470 pp., London, 1885.

HEDGES, KILLINGWORTH, *Modern Lightning Conductors*, 4°, vi + 119 pp., London, 1905.

LODGE, SIR OLIVER J., *Lightning Conductors and Lightning Guards*, 12°, xii + 544 pp., London, 1892.

SPANG, HENRY W., *A Practical Treatise on Lightning Protection*, 12°, 63 pp., New York, 1883.

In connection with the aurora borealis, see:

ANGOT, ALFRED, *The Aurora Borealis*, 8°, xii + 264 pp., London, 1896.

ARRHENIUS, SVANTE A., *Lehrbuch der cosmischen Physik*, Chapter XVII, Leipzig, 1903.

CAPRON, J. RAND, *Auroræ; their Characters and Spectra*, London, 1879.

LEMSTRON, *L'aurore boreale*, Paris, 1886.

Chronological list of Auroras 1870 to 1879. Professional Papers of the Signal Service, No. 3, by A. W. GREELY.

Catalog der in Norwegen bis June 1878 beobachteten Nordlichter (SOPHUS TROMHOLT).

Six pamphlets have been published by the U. S. Weather Bureau on atmospheric electricity and lightning:

McADIE, ALEXANDER, Protection from Lightning (Circular of Information).

McADIE, ALEXANDER, Protection from Lightning (Bulletin No. 15, 1895).

McADIE AND HENRY, Lightning and Electricity of the Air (Bulletin No. 26; W. B. publication No. 197).

HENRY AND McADIE, Property Loss by Lightning, 1898 (W. B. publication No. 199).

HENRY, ALFRED J., Loss of Life in the United States by Lightning (Bulletin No. 30; W. B. publication No. 256).

HENRY, ALFRED J., Recent Practice in the Erection of Lightning Conductors (Bulletin No. 37; W. B. publication No. 349), 1906.

CHAPTER XII

ATMOSPHERIC OPTICS

INTRODUCTION

444. The light of the sun, moon, planets, and stars must pass through the earth's atmosphere before reaching the eye of an observer. Under atmospheric optics are considered the effects of the atmosphere on this light and the optical phenomena to which its passage through the atmosphere gives rise. The phenomena are varied and complex, and can be conveniently grouped under three heads : (1) those phenomena which are due to the gases of the atmosphere themselves ; (2) those due to the particles sometimes present in the atmosphere ; (3) those due to the small particles always present in the atmosphere. The first group includes refraction, twinkling, mirage, and looming. The second group includes halos, the rainbow, and cloud shadows. The third group includes the blue color of the sky, sunrise and sunset colors, and twilight. At the end of the chapter, under the head of miscellaneous

What is included under atmospheric optics.

The subdivisions of the chapter.

483

optical phenomena, cloud colors and the transparency of the atmosphere will be briefly treated.

The Optical Phenomena due to the Gases of the Atmosphere

445. **Refraction.**—It is a well-known law of optics that when a ray of light passes from a medium of one optical density into that of another, it is bent from its course, being bent toward the normal

The definition of refraction.

to the bounding surface when passing from a rarer to a denser medium, and *vice versa*. Thus a ray of light entering the earth's atmosphere from space and passing through layers of air of steadily increasing density must be continuously bent towards the

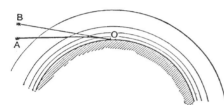

FIG. 153. — The Effect of Refraction.

The effect of the earth's atmosphere.

normal. As a result, the ray of light follows a curved path as indicated by *AO* in Fig. 153. An object is seen, however, in the direction from which the rays enter the eye, that is, in the direction *OB* in the figure. The effect of refraction is thus to raise an object or increase its altitude above the horizon. The amount of refraction is zero at the zenith or point directly overhead, and increases steadily toward the horizon, where it has a maximum value which, on the average, is about 35′ of arc, or a little more than half a degree. The amount of refraction is not constant, but depends upon the density of the air. This in turn depends upon the temperature,

Amount of refraction depends upon three things.

the pressure, and the amount of water refraction for any given altitude and a known condition of vapor present in it. In order to find the exact value of the atmosphere, elaborate tables or formulæ are used, and these may be found in books on Practical Astronomy. There are several approximate formulæ for computing rough values of refraction. The best one, due to Professor Comstock, is probably $r = \dfrac{983b}{460 + t} \cdot \tan \zeta$, where r is the value of refraction in seconds of arc, b is the barometric pressure in inches, t is the temperature Fahrenheit, and ζ is the zenith distance, that is, the angular distance from the zenith. This is not applicable to an object very near to the horizon.

The angular diameter of both the sun and the moon is just about half a degree, while the value of refraction at the horizon is a little more than half a degree. As a result, both sun and moon come into view before they have really come (geometrically risen) above the horizon and are still visible after they have really set. The day is lengthened in these latitudes from 4 to 8 minutes by this effect of refraction. The value of refraction is, on the average, 35′ at the horizon, while at an altitude of only one half a degree above it, the value has already lessened to 29′. Thus when the sun or moon is on the horizon, the lower edge or limb is raised 35′ while the upper limb is raised only 29′. As a result, the disk appears not circular, but decidedly flattened, the flattening amounting to about one fifth of the diameter.

The effect of refraction on the time of rising and setting.

The effect of refraction on the form of the disk.

Briefly put, the effects of refraction are to raise all objects, to lengthen the day, and to flatten the disk of the sun and moon when rising and setting.

446. Twinkling (steadiness of the atmosphere). — The twinkling or, as it is technically called, the scintillation of the stars is a well-known phenomenon which is particularly conspicuous on cold winter nights, and near the horizon. When critically considered, it consists of a change in position, a change in brightness, and a change in color. These three components will be considered in order.

Of what twinkling consists.

The atmosphere is always far from homogeneous, but consists of numerous layers and pockets of air of very different temperature and moisture content, and thus with different densities. These layers and pockets of air are moved about and mixed by the wind. As a result, the condition of the air through which a ray of light comes to the eye of an observer is different each succeeding moment and the amount of refraction is constantly changing. It is this constantly changing amount of refraction which causes the small change in position which is one of the components of twinkling. When a star is viewed through a telescope, this change in position sometimes becomes so marked that the star fairly " dances " in the field of view.

The cause of change in position.

The change in brightness is due to what may be called the lens effect of the atmosphere. If at night the light from an arc light shines through a window pane upon the opposite white wall of a dark room, it will be found that the wall is not uniformly illuminated, but that it is covered with dark and light mottlings. This is because the window pane is not perfectly

The cause of changes in brightness.

smooth and of the same thickness at all points. It acts like a jumble of convex and concave lenses concentrating the light at some points, and diverting it from others. The atmosphere acts in exactly the same way on the light which passes through it. As the various layers and pockets of air are wafted past the line of sight of the observer, at one moment the light is concentrated, while the next it may be diverted. This constant change in brightness is the result.

The change in color is due to optical interference. The rays of light which reach the eye of the observer at the same instant may have come

The cause of the change in color. by paths of slightly different length. As a result, the ether waves are out of phase and may interfere, thus causing the destruction of certain wave lengths or colors. Since the colors eliminated by interference will be different at successive moments, the star will appear to change color.

Twinkling is much more violent near the horizon, because the thickness of the air through which the rays of light come is there much

Why planets do not twinkle. greater. The planets, except when near the horizon, seldom appear to twinkle. The reason is because they have disks and are not mere points of light like the stars. Each point on the disk twinkles, but the twinklings do not synchronize, so that the average condition of the whole disk is much more nearly constant.

The steadiness of the atmosphere is, in a certain sense, the antithesis of twinkling. On a cold winter night when the sky is brilliant and the

Steadiness of the atmosphere. stars sparkle, but little astronomical work can be done. Steadiness on the part of the atmosphere is the great desideratum for astronomical work. This means an atmosphere which is as uniform and constant as possible. It is a condition just the opposite of this which causes the greatest twinkling.

447. **Mirage and looming.** — A description and explanation of mirage has already been given in section 52. It was treated there as

The cause of a mirage. illustrating the fact that conditions were then suitable for local convection. A layer of very warm air is next to the surface of the ground, and above it is a layer of colder and thus denser air. If an observer is situated a little distance above this warm layer, the rays of light may be so bent by refraction that total reflection finally takes place, and the observer sees an inverted image of the object as if reflected from a horizontal body of water, and all intervening objects are invisible. This appearance is called a mirage and occurs chiefly during the hot hours of the day when the air is quiet, and in level desert regions. It occurs also over water surfaces near the land. The height of the observer above the warm layer usually makes a great difference in the mirage.

Looming is, in a certain sense, the opposite of mirage. Here the cold, dense layer of air is next the surface of the ground and the warmer, less dense layer is above. The rays of light passing upward from an object may be so bent by refraction that total reflection again takes place, and the observer sees an inverted image above the object. Objects even below the horizon may be brought into view in this way, and nearer objects seem much raised and elongated. For this reason the term looming is applied to the phenomenon. It occurs chiefly over the ocean near the seashore, and in the Arctic regions.

The explanation of looming.

THE OPTICAL PHENOMENA DUE TO THE PARTICLES SOMETIMES PRESENT IN THE ATMOSPHERE

448. Halos and related phenomena. — The sun during the day and the moon during the night are often surrounded by rings or circles of light which are of different diameters, and are sometimes colored. These rings may be divided into two classes: the coronæ and the halos (Greek ἄλως = disk). These two classes of rings differ not only in size and coloring, but are formed in entirely different ways. The coronæ are due to water drops, and are caused by diffraction and interference, while the halos are due to ice crystals and are caused by refraction and reflection.

Halos and coronæ contrasted.

The coronæ are the smallest rings which may appear around the sun or moon, and several concentric rings may be visible at the same time. The radius of the circle varies from 1° to 10°, and they are usually colored with the red on the outside and shading off to a whitish blue on the inside. Portions only of a coronal ring are usually seen; a full colored circle is of very rare occurrence. A corona is formed when a thin cloud covers the sun or moon, or when one is very near to them. The light is diffracted by the water drops and by interference causes the colored rings to appear. The larger the drops, the smaller the ring. Thus, when several rings are seen at the same time, water drops of several sizes must be present. The characteristics and explanation of the corona are exactly the same as in the case of the colored rings which are seen when a strong source of light is viewed through a piece of glass which has been coated with moisture by breathing upon it.

Description of a corona.

Its explanation.

Two kinds of halos are recognized; one has a radius of about 21° 50′ and the other a radius of 45° 46′. These are usually spoken of as the **The two** 22° and 46° halo. The radius of a halo may be measured **halos.** by means of a sextant or surveyor's transit. If a star or planet happens to be on the edge of a halo and the time of observation is noted, the angular distance between the moon and the stars can be computed and thus the radius determined. The different determinations vary by perhaps 20′. A halo about the moon is usually so little colored that it appears essentially white, while one about the sun is generally colored with the red on the inside and shading off to a whitish blue on the outside. Halos are much more common than coronæ, and the 22° halo is far more common than the 46° halo. Halos are formed when the sky is covered with a thin veil of cirro-stratus or alto-stratus clouds. **The cause** These clouds consist of snowflakes and ice needles as has been **of the halo.** abundantly proved by observations made on mountains and in balloons. It is the refraction and reflection of light from these ice particles which cause the halo. This correct explanation was first given by Mariotte in 1686, but it was long neglected and another explanation even supplanted it in part for a time. During recent times, the theory has been much improved, and numerous investigations have been carried out to determine how the ice particles must be oriented to give rise to the various phenomena.

There are several other optical phenomena which are closely related with halos and are occasionally seen. Sometimes a ring of white light is observed parallel to the horizon and at the altitude of the sun. **Related phenomena.** Where this circle crosses the halos, patches of light, sometimes colored, appear which are known as sun dogs or mock suns. Arcs tangential to the halos and convex to the sun are also occasionally seen. Also columns of light and crosses extending vertically and horizontally through the sun are sometimes glimpsed. All these are represented in Fig. 154. They can all be explained as the refraction and reflection

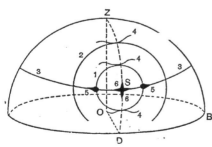

FIG. 154. — Halos and Related Phenomena.

ABD, horizon; *ZSD*, vertical circle through the sun; *Z*, zenith; *O*, position of the observer; *S*, the position of the sun; 1, 22° halo; 2, 46° halo; 3, white ring parallel to the horizon; 4, tangential arcs; 5, mock suns or sun dogs; 5, columns of light extending vertically and horizontally.

of light from the ice particles which are either haphazard in arrangement, or are oriented in a definite way because they are rising or falling.

449. Rainbow. — The rainbow, or arc of prismatic colors, is too well known to need description. It is formed if the sun is shining and at the same time it is raining in a direction opposite to the sun. The location The sun, the observer's eye, and the center of the circle of of a rainbow. which the bow is a part are always in a straight line. As a result, if the sun is exactly on the horizon, the length of the bow is 180°, while in most cases less than a semicircle is seen. Furthermore, each observer sees his own rainbow and in a slightly different place. The six spectrum colors (violet, blue, green, yellow, orange, and red) are so arranged that the red is on the outside and the violet on the inside. The radius of the red part is 42° 2′, while the radius of the violet The size of part is 40° 17′. Thus, if the sun is more than 42° above the a rainbow. horizon, no rainbow can be visible. Rainbows, therefore, always occur during the early morning hours or during the hours of the late afternoon. Rainbows are by far the most common during the later part of a summer afternoon. The reason is because they are nearly always When they associated with thundershowers. The steady cyclonic rains occur. of winter are followed by large cloud areas, so that the sun seldom shines shortly after it has ceased to rain. The summer thundershowers, on the contrary, usually occur during the afternoon, and the clouds often break through quickly, allowing the necessary sunshine.

There is often a fainter secondary bow concentric with the first and somewhat larger. The red is here on the inside, and the violet on the outside. The radius of the red part is 50° 59′, while the The secondary bow. radius of the violet part is 54° 9′.

The rainbow is caused by the refraction and reflection of sunlight in the falling drops of water. In the case of the primary bow, there are two refractions and one total reflection. In the case of the secondary bow, there are two refrac- The explanations and two total reflec- nation of the tions. The course of the rainbow. sun's rays is pictured in Fig. 155 in the case of both the primary and the secondary bow. It will be seen that the raindrops which send any particular wave length of light to the eye of the observer are all located in a conical surface extending from the

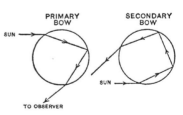

Fig. 155. — The Formation of the Rainbow.

observer to the particular arc of color in the rainbow. Light of this wave length sent back by other raindrops will not be perceived by the observer, as it will pass above or below his eye. The first theory of the rainbow was given by Descartes in 1637 and seems very simple. If all the factors are, however, taken into account, the theory is by no means so simple and, in its modern form, it dates from Airy in 1836.[1] The truth of the theory concerning rainbows can be tested experimentally by means of glass balls or glass globes, filled with water.

Three, four, or even five or more total reflections are possible as well as one or two. Thus more rainbows are a possibility and more have occasionally been glimpsed. The purity of color of a rainbow depends upon the size of the raindrops and their uniformity. When examined critically, it will be found that rainbows differ in the width and prominence of certain colors.

450. **Cloud shadows.** — In connection with cloud shadows only the shadows of clouds on the atmosphere need consideration. These occur when the atmosphere is hazy or misty, that is, when it contains an unusual number of dust or moisture particles. If the sky is covered with broken clouds, the sun shines through the openings in the clouds and illuminates the dust and moisture particles beneath. Light streaks radiating from the sun and extending down from the clouds are then seen. Between the light streaks are dark bands due to the unilluminated parts of the atmosphere. These dark bands are thus simply cloud shadows on the atmosphere. This whole phenomenon is called popularly the " sun drawing water." It is caused simply by the sun shining through rifts in the clouds and illuminating portions of the atmosphere while other portions are in the shadows of the clouds. This is illustrated in Fig. 156.

Cloud shadows and the appearances to which they give rise.

THE OPTICAL PHENOMENA DUE TO THE SMALL PARTICLES ALWAYS PRESENT IN THE ATMOSPHERE

451. **The blue color of the sky.** — The colors of the clear sky when the sun is not near the horizon are deeper or lighter shades of blue. Near the sun the blue fades out and becomes more and more whitish. Near the horizon the blue color becomes so faint that it practically disappears and is replaced by gray. The clearer the sky, the purer and more intense the blue. If the atmosphere

Description.

[1] *Monthly Weather Review*, November, 1904, p. 506.

Fig. 156. — Cloud Shadows; "the Sun drawing Water."

becomes gradually hazy, the blue becomes more and more whitish and is finally almost lost.

The blue color of the sky is due primarily to the selective scattering of sunlight by the myriads of particles which are always present in the atmosphere. Some of these particles are smaller than the wave length of light, and some are larger. If the atmosphere were entirely gaseous and contained no particles, there would be practically no scattering, and but little light would come to us from the sky. The sky would then appear nearly black, and the sun, moon, planets, and stars would be seen resplendent against a dark background even in the daytime. Since the atmosphere contains so many particles, it must be considered a turbid medium, and the **The cause of the blue color of the sky.**

problem is to determine the influence of this turbid medium on the light passing through it. This is illustrated in Fig. 157. The longer ether waves, that is, the red and yellow waves, get through the turbid medium with the greatest facility, while the shorter ether waves are more scattered. Diffuse reflection, diffraction, and perhaps a

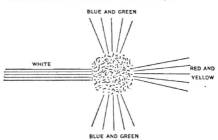

FIG. 157. — Selective Scattering by a Turbid Medium.

certain amount of interference play the chief part in causing this scattering. It will be seen that this scattering is selective. The reds and yellows are allowed to pass through, while the blues and greens predominate in the light which is scattered. This can be easily illustrated experimentally by filling a flask with a turbid medium (possibly a soap solution). On viewing a source of light through the flask reds and yellows predominate, while on viewing the flask from the side blues and greens are chiefly in evidence. Exactly this same thing occurs in connection with the atmosphere. If one looks in a direction away from the sun, the light that is received is the light which has been scattered by the myriads of particles which happen to be in the line of sight at the moment, and the predominating color is then blue. On looking in a direction nearer the sun much white light which has been reflected by the dust particles near the line of sight is added to the blue, so that it becomes whitish and pale. Near the horizon the thickness of the atmosphere through which one is looking is very great, and the dust particles again add much white light. The smaller the particles, the smaller the amount of light that is scattered, but the purer will be the blue. The larger the particles, the greater the amount of scattered light, but all wave lengths will be present and the selection will be much less definite.

The first explanation of the blue color of the sky was attempted by Leonardo da Vinci. The modern correct explanation begins with Lord Rayleigh in 1871.

452. Sunrise and sunset colors. — As the sun approaches the horizon in setting, its color often becomes yellow or orange. In fact, if the atmosphere is particularly dusty and hazy, its color may be decidedly red. The reason for this is evident from the explanation of the blue color of the clear sky which has just been given. The atmosphere contains myriads of particles, and these exercise a selective scattering on the sunlight which passes through it. The short waves are scattered, while the long waves (red, orange, and yellow) predominate in the light which is allowed to pass through. When the sun is near the horizon, the thickness of the atmosphere through which the light must come is large, so that the sorting of the wave lengths has been particularly efficient, and only the red, or the red, orange, and yellow light gets through.

The color of the sun at sunset.

From the time the sun gets near the horizon until it is some 16° below, a long series of changing sunset glows and colors are seen. The presence of a little more or a little less haze, or the presence of a few clouds, makes a tremendous difference in the sunset colors. In order to determine exactly the sequence of changes in the sunset colors and glows, it would be necessary to observe critically sunset colors during many absolutely clear sunsets. This has been done by a few observers, and the reader must be referred to their work for detailed descriptions. The colors are mostly reds, yellows, and purples, and selective scattering is the agency which is the cause of most of the coloring.

Sunset glows and colors.

Sunset colors are also visible in the east as the sun is setting. These are due to the colored light coming from the west and reflected to the observer by the dust particles in the east. The pink twilight arch is seen rising in the east as the sun goes farther and farther below the horizon. Beneath this arch is a blue black patch. This is the shadow of the earth on its atmosphere.

During sunrise the colors are essentially the same as during sunset, only they occur in the inverse order.

Sunrise colors.

453. Twilight. — Twilight has already been briefly treated in section 24 in connection with the height of the atmosphere. It is caused by the reflection and diffraction of sunlight by the many The cause particles in the upper atmosphere after the sun has gone of twilight. below the horizon of the observer and is no longer directly visible. This was illustrated in Fig. 2. By noting the duration of twilight at a certain place and on a given date, it is possible to compute the height of the particles which sent the light to the observer. It has been found there are enough particles above a height of 100 miles to send a perceptible amount of light to an observer.

The duration of twilight or the transition period from daylight to darkness is very different at different times of year and at different places. It is ordinarily stated that twilight lasts until the Duration of sun has gone 18° below the horizon. This means that at the twilight. equator, where the sun always sets perpendicularly to the horizon, the twilight would last an hour and ten minutes. In higher latitudes, where the sun makes a small angle with the horizon when rising and setting, the duration would be much longer. In the polar regions it lasts for months.

MISCELLANEOUS OPTICAL PHENOMENA

454. Cloud colors. — So much light is returned to the observer by reflection and diffraction in connection with the myriads of particles which constitute a cloud, that the color of a cloud is usually Cloud that of the light by which it is illuminated. Thus a cloud colors. upon which the light of the sun is falling usually appears white. If a thick cloud comes between the sun and an observer, the central portion which is in the shadow usually appears gray or even black. The brilliant white edge is due chiefly to diffraction. Clouds seen near the horizon or at the time of sunrise or sunset are usually reddish in color. This is because the light which falls upon them contains chiefly the long wave lengths.

455. Transparency of the atmosphere — haze. — Haze has already been fully treated in section 229, and, as was there stated, there are two kinds of haze. The haze which exists high up in the atmosphere The two and gives the sky a whitish appearance by day and dims the kinds of stars at night is due to the presence of ice particles in the haze. upper atmosphere. The low haze which renders distant objects indistinct in outline is optical as well as due to foreign particles in the atmosphere. The mixture of layers and pockets of air of different temperature and moisture content, and thus of different density, renders distant objects indistinct. This is purely an optical effect and may be called optical haze. The dust and moisture particles present also cut down the distance to which one can see.

It is a fact of everyday observation, that the air is much more transparent at certain times than others. Distant hills and mountains are

The cause of the indistinctness of distant objects. sometimes said to look near because they are so sharp and distinct. When distant objects are indistinct, it is usually said that it is hazy. Now the light coming from a distant object is weakened in two ways. Some of the light waves are absorbed by the gases of the atmosphere. This absorption is usually selective, as certain wave lengths are much more readily absorbed than others. The light is also scattered by the dust and moisture particles which are present. The distance to which one can see from a smoky city is usually very much greater in the direction from which the wind comes than that towards which it is blowing. But the indistinctness of distant objects is not due alone to the weakening of light by absorption and scattering, but to the optical haze as well. Transparency then depends, not only on the absence of dust and moisture particles, but also on the steadiness and uniformity of the atmosphere. Several ways of measuring the transparency of the atmosphere have been devised. The usual method makes use of two white disks with

The measurement of transparency. black crosses of unequal size, say, in the ratio of 12 to 1. If the air were perfectly transparent, the two disks would appear the same to the eye when the distance of the larger one was 12 times the distance of the smaller one. It will be found by observation that the ratio of distance, when the two appear the same, is always less than 12, and from the observed ratio values of the transparency can be computed. From such measurements it has been found that the transparency is greater in mountains than in the lowlands, in the morning than in the evening, and during the winter than during the summer.

In order to determine the exact cause of changes in transparency, the values of the meteorological elements at the earth's surface and above, and the number of dust particles per cubic centimeter, should be determined as well as the transparency.

TOPICS FOR INVESTIGATION

(1) Approximate formulæ for the values of refraction.
(2) The sequence of colors during sunset.

PRACTICAL EXERCISES

(1) If familiar with astronomical apparatus and methods, determine the altitude of a star, compute its value, and thus determine the amount of refraction.

(2) Measure the radius of a halo.

· (3) Study critically the width and pureness of the colors of a rainbow.

(4) Note the duration of twilight.

REFERENCES

Besson, Louis, *Sur la théorie des halos*, Paris, 1909.

Davis, William M., *Elementary Meteorology*, Ginn & Co., 1894.   (Chapter IV deals with atmospheric optics.)

Mascart, E., *Traité d'optique*, Paris, 1889–1894 (especially Vol. 3).

Pernter-Exner, *Meteorologische Optik*, Wien und Leipzig, 1910.   (This is an extensive treatise and an invaluable compendium of information.)

CHAPTER XIII

ATMOSPHERIC ACOUSTICS

SOUND — ITS ORIGIN, NATURE, AND PROPAGATION

456. Sound may be defined as the sensation produced when a disturbance or wave motion in the air reaches the ear and excites the auditory nerves. Sound is caused by a vibrating body and conveyed to the ear by some elastic medium, usually air.

Definition.

That sound is always produced by a vibrating body can be readily determined. If the finger is placed lightly on the edge of a bell which has been struck and which is emitting a sound, the tremulous motion of the edge can be felt. If greater pressure is applied, the vibration of the edge is stopped and the sound ceases. In the case of the piano, or harp, or any stringed instrument, the actual vibration of the string which is emitting the musical sound can be seen. That an elastic medium is necessary for the conveyance of sound can also be readily proved experimentally. If a sound-emitting body, like a watch or a bell, is placed under the receiver of an air pump and the air is exhausted, the sound grows much fainter and would cease entirely if it were not conveyed to some extent by the imperfect vacuum and the supports of the sound-emitting body. If the air is again admitted, the sound returns to its customary loudness. The action of the vibrating body and the elastic medium in producing and conveying sound is thus as follows. The vibrating body moves in and out and thus causes rarefactions and compressions in the elastic medium. This wave motion is then conveyed by the elastic medium with a speed which depends on the density and elasticity. The velocity of transmission increases with greater elasticity and smaller

The origin of sound.

The propagation of sound.

density. For air, the velocity is 1090 feet per second at a temperature of 32° F. For water, the velocity is a little more than four times as large. When these waves of compression and rarefaction reach the ear, the sensation of sound is produced.

Since sound is a wave motion, it may be reflected by a rigid barrier, just as a water wave is turned back from a wall, or a light wave is reflected from a mirror. The echo is the familiar example of The reflecthe reflection of sound waves, and many places have become tion of famous on account of their echoes. In mountainous regions, sound. where there are many reflecting surfaces, the reverberations which follow any sharp report are particularly noticeable. Sound waves may be reflected by a cloud, or by the bounding surface of two layers of air of different density, as well as by a barrier. In this last instance, the process is very similar to the cause of mirage and looming where the light waves are finally totally reflected from such a bounding surface between media of different temperature.

THUNDER

457. A near-by lightning flash is always followed by thunder, which is simply the familiar snap of the electric spark in a much modified form. If the observer is very near the lightning flash, only a single The cause tremendous crash is heard. The lightning flash suddenly of thunder. heats the filament of air which marks its path and causes it to expand quickly. This causes a wave of compression which travels outward in every direction from the path of the flash. Thus the observer near the flash receives but the single impulse and hears but a single crash. If an observer is at a considerable distance from the The characlightning flash, the well-known rolling and reverberation of the thunder and thunder is heard, and the sound may continue for some time. the explanation. There are two main causes which contribute to this. In the first place, the observer is at very different distances from different parts of the lightning flash. This is particularly the case if the lightning flash is several miles long. Since sound travels but 1090 feet per second, it will require a very different time for the sound to reach the observer from different parts of the flash. Thus the long continuance of the thunder is in part explained. In the second place, the sound is reflected by surrounding objects, particularly in mountainous regions, and also by the clouds and the bounding surface of layers of air of different density. All the characteristics of thunder can be explained as a combination of

2 K

these two things, varying distance from different parts of the flash, and the reflection of sound.

Since sound travels 1090 feet per second, it requires very nearly five seconds to cover a mile. By counting the number of seconds which elapse between a lightning flash and the resulting thunder, and then dividing by five, the distance in miles of the lightning is at once determined.

Time interval between lightning and thunder.

The Effect of the Meteorological Elements on the Character and Carrying Distance of Sounds

458. It is a well-known fact of observation that certain sounds, such as those coming from a distant whistle or foghorn, or a railway train passing over a bridge, are much more distinctly heard and are heard at much greater distances at certain times than at others. In fact, this is such a common matter of experience that these things are sometimes used popularly as weather prognostics. The wind direction, of course, plays the major part in this. A sound can be much better heard if the sound-emitting body is in the direction from which the wind comes. If the sound waves must make their way against the moving air to the observer, they are more mixed up and are reflected away from the observer. This reflection is due to the fact that the upper layers of air are moving at a higher velocity than those nearer the ground. They thus act in the same way in reflecting sound as if they were at a different temperature. The presence of a small amount of fog also serves to make sounds more distinctly heard. This may be due partly to the fact that the air is usually very quiet at the time of fog. A fog also usually occurs when there is an inversion of temperature, and this means layers of air of different temperature and thus density. Sound waves are readily reflected from the bounding surfaces of these layers. Critical observation of the distinctness and carrying distance of sounds and the values of the meteorological elements would lead to interesting generalizations.

The elements which affect the distinctness and carrying distance of sounds.

It has also been found that near a foghorn, for example, there are patches where no sound can be heard and these areas of silence or sound shadows may be differently located at different times. They are probably caused by a pocket of air of peculiar characteristics which reflects away the sound waves and thus causes a sound shadow.

Sound shadows.

There are certain sounds produced in the atmosphere by its passage over obstacles as, for example, the howling of the wind and the singing of telegraph wires. Here compressions and rarefactions are produced in the moving air and these constitute the sound. The phenomena are analogous to the formation of waves in a stream passing over a rough bed. In the case of telegraph wires, it is usually a wind of a particular velocity, not necessarily high, which is especially efficacious in producing the sound. The direction of the wind also plays a part. The pitch of the sound depends on the size of the wire, the length between the poles, and also the pole which through resonance emphasizes certain tones.[1]

Sounds produced by air in motion.

TOPICS FOR INVESTIGATION

(1) Sound shadows.
(2) Famous echoes.

PRACTICAL EXERCISES

(1) Observe critically the characteristics of thunder and attempt to explain them in each case.
(2) Observe critically the distinctness and carrying distance of a particular sound and attempt to correlate it with the values of the meteorological elements.
(3) Determine the distance at which thunder is audible.

REFERENCES

ARRHENIUS, SVANTE AUGUST, *Lehrbuch der kosmischen Physik*, Leipzig, 1903.
(Chapter XIV deals with meteorological acoustics.)

[1] *Monthly Weather Review*, March, 1904, p. 230.
Meteorologische Zeitschrift, 1902, p. 525.

APPENDICES

Appendix I
THE METRIC SYSTEM

Appendix II
A COMPARISON OF THE THREE THERMOMETRIC SCALES

Appendix III
A GRAPHICAL COMPARISON OF THE ENGLISH AND METRIC BAROMETRIC
SCALES AND THE FAHRENHEIT AND CENTIGRADE THERMOMETRIC SCALES

Appendix IV
THE REDUCTION OF BAROMETRIC READINGS TO SEA LEVEL

Appendix V
THE TABLES IN THE TEXT

Appendix VI
THE ALBANY DATA GIVEN IN THE TEXT

Appendix VII
THE REGULAR STATIONS OF THE U. S. WEATHER BUREAU AND THE
CANADIAN STATIONS

Appendix VIII
TEACHING METEOROLOGY

Appendix IX
THE LITERATURE OF METEOROLOGY

    (A) General directions.

    (B) A list of books.

    (C) The U. S. government publications on meteorology.

    (D) Bibliography of Bibliography.

    (E) Digest of literature.

    (F) Periodicals.

        (1) Where lists may be found.

        (2) A partial list.

APPENDIX I

THE METRIC SYSTEM

Length

10	millimeters = 1 centimeter.
10	centimeters = 1 decimeter.
10	decimeters = 1 meter.
1000	meters = 1 kilometer.

1 centimeter = 0.393700 inch.

1 meter = 39.3700 inches
or 3.28083 feet.

1 kilometer = 3280.83 feet
or 0 621370 mile.

1 centigram = 0.154323 grain.

1 gram = 15.4323 grains
or 0.0352739 ounces.

1 kilogram = 35.2739 ounces
or 2.20462 pounds (av.).

Mass

10	milligrams = 1 centigram.
10	centigrams = 1 decigram.
10	decigrams = 1 gram.
1000	grams = 1 kilogram.

1 inch = 2.54000 centimeters.

1 foot = 0.304801 meter.

1 mile = 1.60935 kilometers.

1 ounce = 28.3495 grams.

1 pound = 0.453592 kilogram.

For rough ratios it is sufficient to remember that:

1 centimeter = .4 inch.		1 inch = $2\frac{1}{2}$ centimeters.
1 meter = $3\frac{1}{4}$ feet.		1 foot = .3 meter.
1 kilometer = .6 mile.		1 mile = 1.6 kilometers.
1 gram = $\frac{1}{30}$ ounce.		1 ounce = 28 grams.
1 kilogram = $2\frac{1}{4}$ pounds.		1 pound = $\frac{4}{9}$ kilogram.

APPENDIX II

A Comparison of the Three Thermometric Scales

$$\frac{C}{5} = \frac{F - 32^\circ}{9} = \frac{R}{4}.$$

Fahrenheit	Centigrade	Réaumur	Fahrenheit	Centigrade	Réaumur	Fahrenheit	Centigrade	Réaumur
$-459\frac{2}{5}$	-273	$-218\frac{2}{5}$	-16	$-26\frac{2}{3}$	$-21\frac{1}{3}$	92	$33\frac{1}{3}$	$26\frac{2}{3}$
-148	-100	-80	-13	-25	-20	95	35	28
-139	-95	-76	-10	$-23\frac{1}{3}$	$-18\frac{2}{3}$	98	$36\frac{2}{3}$	$29\frac{1}{3}$
-130	-90	-72	-7	$-21\frac{2}{3}$	$-17\frac{1}{3}$	101	$38\frac{1}{3}$	$30\frac{2}{3}$
-121	-85	-68	-4	-20	-16	104	40	32
-112	-80	-64	-1	$-18\frac{1}{3}$	$-14\frac{2}{3}$	107	$41\frac{2}{3}$	$33\frac{1}{3}$
-103	-75	-60	2	$-16\frac{2}{3}$	$-13\frac{1}{3}$	110	$43\frac{1}{3}$	$34\frac{2}{3}$
			5	-15	-12	113	45	36
-100	$-73\frac{1}{3}$	$-58\frac{2}{3}$	8	$-13\frac{1}{3}$	$-10\frac{2}{3}$	116	$46\frac{2}{3}$	$37\frac{1}{3}$
-97	$-71\frac{2}{3}$	$-57\frac{1}{3}$	11	$-11\frac{2}{3}$	$-9\frac{1}{3}$	119	$48\frac{1}{3}$	$38\frac{2}{3}$
-94	-70	-56	14	-10	-8	122	50	40
-91	$-68\frac{1}{3}$	$-54\frac{2}{3}$	17	$-8\frac{1}{3}$	$-6\frac{2}{3}$	125	$51\frac{2}{3}$	$41\frac{1}{3}$
-88	$-66\frac{2}{3}$	$-53\frac{1}{3}$	20	$-6\frac{2}{3}$	$-5\frac{1}{3}$	128	$53\frac{1}{3}$	$42\frac{2}{3}$
-85	-65	-52	23	-5	-4	131	55	44
-82	$-63\frac{1}{3}$	$-50\frac{2}{3}$	26	$-3\frac{1}{3}$	$-2\frac{2}{3}$	134	$56\frac{2}{3}$	$45\frac{1}{3}$
-79	$-61\frac{2}{3}$	$-49\frac{1}{3}$	29	$-1\frac{2}{3}$	$-1\frac{1}{3}$	137	$58\frac{1}{3}$	$46\frac{2}{3}$
-76	-60	-48	32	0	0	140	60	48
-73	$-58\frac{1}{3}$	$-46\frac{2}{3}$						
-70	$-56\frac{2}{3}$	$-45\frac{1}{3}$	35	$1\frac{2}{3}$	$1\frac{1}{3}$	149	65	52
-67	-55	-44	38	$3\frac{1}{3}$	$2\frac{2}{3}$	158	70	56
-64	$-53\frac{1}{3}$	$-42\frac{2}{3}$	41	5	4	167	75	60
-61	$-51\frac{2}{3}$	$-41\frac{1}{3}$	44	$6\frac{2}{3}$	$5\frac{1}{3}$	176	80	64
-58	-50	-40	47	$8\frac{1}{3}$	$6\frac{2}{3}$	185	85	68
-55	$-48\frac{1}{3}$	$-38\frac{2}{3}$	50	10	8	194	90	72
-52	$-46\frac{2}{3}$	$-37\frac{1}{3}$	53	$11\frac{2}{3}$	$9\frac{1}{3}$	203	95	76
-49	-45	-36	56	$13\frac{1}{3}$	$10\frac{2}{3}$	212	100	80
-46	$-43\frac{1}{3}$	$-34\frac{2}{3}$	59	15	12			
-43	$-41\frac{2}{3}$	$-33\frac{1}{3}$	62	$16\frac{2}{3}$	$13\frac{1}{3}$	392	200	160
-40	-40	-32	65	$18\frac{1}{3}$	$14\frac{2}{3}$	572	300	240
			68	20	16	752	400	320
-37	$-38\frac{1}{3}$	$-30\frac{2}{3}$	71	$21\frac{2}{3}$	$17\frac{1}{3}$	1832	1000	800
-34	$-36\frac{2}{3}$	$-29\frac{1}{3}$	74	$23\frac{1}{3}$	$18\frac{2}{3}$	3632	2000	1600
-31	-35	-28	77	25	20	5432	3000	2400
-28	$-33\frac{1}{3}$	$-26\frac{2}{3}$	80	$26\frac{2}{3}$	$21\frac{1}{3}$	7232	4000	3200
-25	$-31\frac{2}{3}$	$-25\frac{1}{3}$	83	$28\frac{1}{3}$	$22\frac{2}{3}$	9032	5000	4000
-22	-30	-24	86	30	24	10832	6000	4800
-19	$-28\frac{1}{3}$	$-22\frac{2}{3}$	89	$31\frac{2}{3}$	$25\frac{1}{3}$	12632	7000	5600

CENTIGRADE SCALE TO FAHRENHEIT

CENTI-GRADE	.0	.1	.2	.3	.4	.5	.6	.7	.8	.9
	°F.	°F.	°F.	°F.	°F.	°F.	°F.	°F.	°F.	°F.
0	+32.00	+31.82	+31.64	+31.46	+31.28	+31.10	+30.92	+30.74	+30.57	+30.38
1	30.20	30.02	29.84	29.66	29.48	29.30	29.12	28.94	28.76	28.58
2	28.40	28.22	28.04	27.86	27.68	27.50	27.32	27.14	26.96	26.78
3	26.60	26.42	26.24	26.06	25.88	25.70	25.52	25.34	25.16	24.98
4	24.80	24.62	24.44	24.26	24.08	23.90	23.72	23.54	23.36	23.18
5	+23.00	+22.82	+22.64	+22.46	+22.28	+22.10	+21.92	+21.74	+21.56	+21.38
6	21.20	21.02	20.84	20.66	20.48	20.30	20.12	19.94	19.76	19.58
7	19.40	19.22	19.04	18.86	18.68	18.50	18.32	18.14	17.96	17.78
8	17.60	17.42	17.24	17.06	16.88	16.70	16.52	16.34	16.16	15.98
9	15.80	15.62	15.44	15.26	15.08	14.90	14.72	14.54	14.36	14.18
−10	+14.00	+13.82	+13.64	+13.46	+13.28	+13.10	+12.92	+12.74	+12.56	+12.38
11	12.20	12.02	11.84	11.66	11.48	11.30	11.12	10.94	10.76	10.58
12	10.40	10.22	10.04	9.86	9.68	9.50	9.32	9.14	8.96	8.78
13	8.60	8.42	8.24	8.06	7.88	7.70	7.52	7.34	7.16	6.98
14	6.80	6.62	6.44	6.26	6.08	5.90	5.72	5.54	5.36	5.18
−15	+ 5.00	+ 4.82	+ 4.64	+ 4.46	+ 4.28	+ 4.10	+ 3.92	+ 3.74	+ 3.56	+ 3.38
16	+ 3.20	+ 3.02	+ 2.84	+ 2.66	+ 2.48	+ 2.30	+ 2.12	+ 1.94	+ 1.76	+ 1·58
17	+ 1.40	+ 1.22	+ 1.04	+ 0.86	+ 0.68	+ 0.50	+ 0.32	+ 0.14	− 0.04	− 0.22
18	− 0.40	− 0.58	− 0.76	− 0.94	− 1.12	− 1.30	− 1.48	− 1.66	− 1.84	− 2.02
19	− 2.20	− 2.38	− 2.56	− 2.74	− 2.92	− 3.10	− 3.28	− 3.46	− 3.64	− 3.82
−20	− 4.00	− 4.18	− 4.36	− 4.54	− 4.72	− 4.90	− 5.08	− 5.26	− 5.44	− 5.62
21	5.80	5.98	6.16	6.34	6.52	6.70	6.88	7.06	7.24	7.42
22	7.60	7.78	7.96	8.14	8.32	8.50	8.68	8.86	9.04	9.22
23	9.40	9.58	9.76	9.94	10.12	10.30	10.48	10.66	10.84	11.02
24	11.20	11.38	11.56	11.74	11.92	12.10	12.28	12.46	12.64	12.82
−25	−13.00	−13.18	−13.36	−13.54	−13.72	−13.90	−14.08	−14.26	−14.44	−14.62
26	14.80	14.98	15.16	15.34	15.52	15.70	15.88	16.06	16.24	16.42
27	16.60	16.78	16.96	17.14	17.32	17.50	17.68	17.86	18.04	18.22
28	18.40	18.58	18.76	18.94	19.12	19.30	19.48	19.66	19.84	20.02
29	20.20	20.38	20.56	20.74	20.92	21.10	21.28	21.46	21.64	21.82
−30	−22.00	−22.18	−22.36	−22.54	−22.72	−22.90	−23.08	−23.26	−23.44	−23.62
31	23.80	23.98	24.16	24.34	24.52	24.70	24.88	25.06	25.24	25.42
32	25.60	25.78	25.96	26.14	26.32	26.50	26.68	26.86	27.04	27.22
33	27.40	27.58	27.76	27.94	28.12	28.30	28.48	28.66	28.84	29.02
34	29.20	29.38	29.56	29.74	29.92	30.10	30.28	30.46	30.64	30.82
−35	−31.00	−31.18	−31.36	−31.54	−31.72	−31.90	−32.08	−32.26	−32.44	−32.62
36	32.80	32.98	33.16	33.34	33.52	33.70	33.88	34.06	34.24	34.42
37	34.60	34.78	34.96	35.14	35.32	35.50	35.68	35.86	36.04	36.22
38	36.40	36.58	36.76	36.94	37.12	37.30	37.48	37.66	37.84	38.02
39	38.20	38.38	38.56	38.74	38.92	39.10	39.28	39.46	39.64	39.82
−40	−40.00	−40.18	−40.36	−40.54	−40.72	−40.90	−41.08	−41.26	−41.44	−41.62
41	41.80	41.98	42.16	42.34	42.52	42.70	42.88	43.06	43.24	43.42
42	43.60	43.78	43.96	44.14	44.32	44.50	44.68	44.86	45.04	45.22
43	45.40	45.58	45.76	45.94	46.12	46.30	46.48	46.66	46.84	47.02
44	47.20	47.38	47.56	47.74	47.92	48.10	48.28	48.46	48.64	48.82
−45	−49.00	−49.18	−49.36	−49.54	−49.72	−49.90	−50.08	−50.26	−50.44	−50.62
46	50.80	50.98	51.16	51.34	51.52	51.70	51.88	52.06	52.24	52.42
47	52.60	52.78	52.96	53.14	53.32	53.50	53.68	53.86	54.04	54.22
48	54.40	54.58	54.76	54.94	55.12	55.30	55.48	55.66	55.84	56.02
49	56.20	56.38	56.56	56.74	56.92	57.10	57.28	57.46	57.64	57.82
−50	−58.00	−58.18	−58.36	−58.54	−58.72	−58.90	−59.08	−59.26	−59.44	−59.62

APPENDIX II

CENTIGRADE SCALE TO FAHRENHEIT

CENTI-GRADE	.0	.1	.2	.3	.4	.5	.6	.7	.8	.9
°	° F.	° F.	° F.	° F.	° F.	° F.	° F.	° F.	° F.	° F.
+50	+122.00	+122.18	+122.36	+122.54	+122.72	+122.90	+123.08	+123.26	+123.44	+123.62
49	120.20	120.38	120.56	120.74	120.92	121.10	121.28	121.46	121.64	121.82
48	118.40	118.58	118.76	118.94	119.12	119.30	119.48	119.66	119.84	120.02
47	116.60	116.78	116.96	117.14	117.32	117.50	117.68	117.86	118.04	118.22
46	114.80	114.98	115.16	115.34	115.52	115.70	115.88	116.06	116.24	116.42
+45	+113.00	+113.18	+113.36	+113.54	+113.72	+113.90	+114.08	+114.26	+114.44	+114.62
44	111.20	111.38	111.56	111.74	111.92	112.10	112.28	112.46	112.64	112.82
43	109.40	109.58	109.76	109.94	110.12	110.30	110.48	110.66	110.84	111.02
42	107.60	107.78	107.96	108.14	108.32	108.50	108.68	108.86	109.04	109.22
41	105.80	105.98	106.16	106.34	106.52	106.70	106.88	107.06	107.24	107.42
+40	+104.00	+104.18	+104.36	+104.54	+104.72	+104.90	+105.08	+105.26	+105.44	+105.62
39	102.20	102.38	102.56	102.74	102.92	103.10	103.28	103.46	103.64	103.82
38	100.40	100.58	100.76	100.94	101.12	101.30	101.48	101.66	101.84	102.02
37	98.60	98.78	98.96	99.14	99.32	99.50	99.68	99.86	100.04	100.22
36	96.80	96.98	97.16	97.34	97.52	97.70	97.88	98.06	98.24	98.42
+35	+ 95.00	+ 95.18	+ 95.36	+ 95.54	+ 95.72	+ 95.90	+ 96.08	+ 96.26	+ 96.44	+ 96.62
34	93.20	93.38	93.56	93.74	93.92	94.10	94.28	94.46	94.64	94.82
33	91.40	91.58	91.76	91.94	92.12	92.30	92.48	92.66	92.84	93.02
32	89.60	89.78	89.96	90.14	90.32	90.50	90.68	90.86	91.04	91.22
31	87.80	87.98	88.16	88.34	88.52	88.70	88.88	89.06	89.24	89.42
+30	+ 86.00	+ 86.18	+ 86.36	+ 86.54	+ 86.72	+ 86.90	+ 87.08	+ 87.26	+ 87.44	+ 87.62
29	84.20	84.38	84.56	84.74	84.92	85.10	85.28	85.46	85.64	85.82
28	82.40	82.58	82.76	82.94	83.12	83.30	83.48	83.66	83.84	84.02
27	80.60	80.78	80.96	81.14	81.32	81.50	81.68	81.86	82.04	82.22
26	78.80	78.98	79.16	79.34	79.52	79.70	79.88	80.06	80.24	80.42
+25	+ 77.00	+ 77.18	+ 77.36	+ 77.54	+ 77.72	+ 77.90	+ 78.08	+ 78.26	+ 78.44	+ 78.62
24	75.20	75.38	75.56	75.74	75.92	76.10	76.28	76.46	76.64	76.82
23	73.40	73.58	73.76	73.94	74.12	74.30	74.48	74.66	74.84	75.02
22	71.60	71.78	71.96	72.14	72.32	72.50	72.68	72.86	73.04	73.22
21	69.80	69.98	70.16	70.34	70.52	70.70	70.88	71.06	71.24	71.42
+20	+ 68.00	+ 68.18	+ 68.36	+ 68.54	+ 68.72	+ 68.90	+ 69.08	+ 69.26	+ 69.44	+ 69.62
19	66.20	66.38	66.56	66.74	66.92	67.10	67.28	67.46	67.64	67.82
18	64.40	64.58	64.76	64.94	65.12	65.30	65.48	65.66	65.84	66.02
17	62.60	62.78	62.96	63.14	63.32	63.50	63.68	63.86	64.04	64.22
16	60.80	60.98	61.16	61.34	61.52	61.70	61.88	62.06	62.24	62.42
+15	+ 59.00	+ 59.18	+ 59.36	+ 59.54	+ 59.72	+ 59.90	+ 60.08	+ 60.26	+ 60.44	+ 60.62
14	57.20	57.38	57.56	57.74	57.92	58.10	58.28	58.46	58.64	58.82
13	55.40	55.58	55.76	55.94	56.12	56.30	56.48	56.66	56.84	57.02
12	53.60	53.78	53.96	54.14	54.32	54.50	54.68	54.86	55.04	55.22
11	51.80	51.98	52.16	52.34	52.52	52.70	52.88	53.06	53.24	53.42
+10	+ 50.00	+ 50.18	+ 50.36	+ 50.54	+ 50.72	+ 50.90	+ 51.08	+ 51.26	+ 51.44	+ 51.62
9	48.20	48.38	48.56	48.74	48.92	• 49.10	49.28	49.46	49.64	49.82
8	46.40	46.58	46.76	46.94	47.12	47.30	47.48	47.66	47.84	48.02
7	44.60	44.78	44.96	45.14	45.32	45.50	45.68	45.86	46.04	46.22
6	42.80	42.98	43.16	43.34	43.52	43.70	43.88	44.06	44.24	44.42
+ 5	+ 41.00	+ 41.18	+ 41.36	+ 41.54	+ 41.72	+ 41.90	+ 42.08	+ 42.26	+ 42.44	+ 42.62
4	39.20	39.38	39.56	39.74	39.92	40.10	40.28	40.46	40.64	40.82
3	37.40	37.58	37.76	37.94	38.12	38.30	38.48	38.66	38.84	39.02
2	35.60	35.78	35.96	36.14	36.32	36.50	36.68	36.86	37.04	37.22
1	33.80	33.98	34.16	34.34	34.52	34.70	34.88	35.06	35.24	35.42
+ 0	+ 32.00	+ 32.18	+ 32.36	+ 32.54	+ 32.72	+ 32.90	+ 33.08	+ 33.26	+ 33.44	+ 33.62

APPENDIX III

A Graphical Comparison of the English and Metric Barometric Scales and the Fahrenheit and Centigrade Thermometers

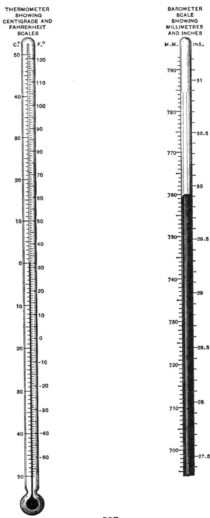

THERMOMETER
SHOWING
CENTIGRADE AND
FAHRENHEIT
SCALES

BAROMETER
SCALE
SHOWING
MILLIMETRES
AND INCHES

APPENDIX IV

The Reduction of Barometric Readings to Sea Level

As stated in section 111, the correction to be applied to reduce to sea level is not a constant but depends on the pressure, temperature, and moisture. In order to determine this correction precisely, elaborate tables must be used. In the following table, moisture has been neglected, or better, the values given hold for an average amount of moisture. The values of the correction are given for only a few pressures and temperatures. The table should not be used for the reduction of a long series of well-taken observations, but is intended simply to show roughly the amount of the correction in a few instances. Enough values have, however, been given so that interpolation is not troublesome. The table is based on the *Smithsonian Meteorological Tables*.

Reduction to Sea Level
(correction additive)

HEIGHT IN FEET	PRESSURE 30 INCHES							PRESSURE 29 INCHES						
	Temperature — Fahrenheit (external air)							Temperature — Fahrenheit (external air)						
	-20	0	20	40	60	80	100	-20	0	20	40	60	80	100
100	0.13	0.12	0.12	0.11	0.10	0.10	0.10	0.12	0.12	0.11	0.11	0.10	0.10	0.10
200	0.26	0.25	0.23	0.23	0.22	0.21	0.20	0.25	0.23	0.22	0.22	0.21	0.20	0.19
300	0.38	0.37	0.35	0.34	0.32	0.31	0.30	0.37	0.36	0.34	0.33	0.31	0.30	0.29
400	0.51	0.49	0.47	0.45	0.43	0.42	0.40	0.50	0.47	0.46	0.44	0.42	0.40	0.38
500	0.65	0.62	0.59	0.57	0.54	0.52	0.50	0.63	0.60	0.57	0.55	0.53	0.51	0.48
600	0.78	0.74	0.71	0.68	0.65	0.63	0.60	0.75	0.72	0.69	0.66	0.63	0.61	0.58
700	0.91	0.87	0.83	0.79	0.76	0.73	0.70	0.88	0.84	0.80	0.76	0.73	0.71	0.68
800	1.04	0.99	0.95	0.91	0.87	0.83	0.80	1.00	0.96	0.92	0.88	0.84	0.80	0.77
900	1.17	1.12	1.07	1.03	0.98	0.94	0.91	1.13	1.08	1.03	0.99	0.95	0.91	0.87
1000	1.31	1.24	1.19	1.14	1.10	1.05	1.01	1.26	1.20	1.15	1.10	1.05	1.01	0.97
1100	1.44	1.38	1.32	1.26	1.21	1.15	1.11	1.39	1.33	1.27	1.22	1.17	1.11	1.07
1200	1.57	1.50	1.43	1.38	1.32	1.26	1.21	1.52	1.45	1.39	1.33	1.27	1.22	1.17
1300	1.70	1.64	1.56	1.49	1.43	1.37	1.32	1.65	1.58	1.51	1.44	1.38	1.32	1.27
1400	—	—	—	1.62	1.54	1.48	1.41	1.78	1.70	1.63	1.56	1.49	1.43	1.37
1500	—	—	—	—	—	1.59	1.52	1.91	1.82	1.74	1.67	1.60	1.53	1.47
1600	—	—	—	—	—	—	1.63	2.04	1.95	1.86	1.79	1.71	1.64	1.58
1700								2.17	2.07	1.99	1.90	1.81	1.74	1.67
1800								2.31	2.20	2.11	2.02	1.93	1.84	1.77
1900								2.44	2.33	2.23	2.13	2.04	1.95	1.88
2000								2.57	2.45	2.35	2.24	2.15	2.06	1.98
2100								—	—	2.47	2.36	2.26	2.17	2.07
2200								—	—	—	2.48	2.37	2.27	2.18
2300								—	—	—	—	2.49	2.38	2.29
2400								—	—	—	—	—	2.49	2.38
2500								—	—	—	—	—	—	2.49
2600														
2700														
2800														
2900														
3000														

REDUCTION TO SEA LEVEL

(correction additive)

HEIGHT IN FEET	PRESSURE 28 INCHES							PRESSURE 27 INCHES						
	Temperature — Fahrenheit *(external air)*							Temperature — Fahrenheit *(external air)*						
	-20	0	20	40	60	80	100	-20	0	20	40	60	80	100
100	0.12	0.11	0.11	0.10	0.10	0.10	0.10							
200	0.24	0.23	0.22	0.21	0.20	0.19	0.18							
300	0.36	0.34	0.32	0.31	0.30	0.29	0.28							
400	0.48	0.46	0.44	0.42	0.40	0.39	0.37							
500	0.61	0.58	0.55	0.53	0.51	0.49	0.47							
600	0.73	0.70	0.66	0.63	0.61	0.59	0.56							
700	0.85	0.81	0.77	0.74	0.71	0.69	0.65							
800	0.97	0.93	0.88	0.85	0.81	0.77	0.75							
900	1.09	1.04	1.00	0.96	0.92	0.87	0.84							
1000	1.22	1.16	1.11	1.06	1.02	0.97	0.94	1.18	1.12	1.07	1.02	0.98	0.94	0.90
1100	1.34	1.29	1.23	1.17	1.13	1.07	1.03	1.29	1.24	1.18	1.13	1.08	1.03	0.99
1200	1.46	1.40	1.34	1.28	1.23	1.17	1.13	1.41	1.35	1.29	1.24	1.19	1.13	1.09
1300	1.59	1.52	1.45	1.39	1.33	1.28	1.23	1.53	1.47	1.40	1.34	1.28	1.23	1.19
1400	1.72	1.64	1.57	1.50	1.43	1.38	1.32	1.66	1.59	1.51	1.45	1.39	1.33	1.27
1500	1.84	1.76	1.68	1.61	1.54	1.48	1.42	1.78	1.70	1.62	1.55	1.49	1.43	1.37
1600	1.97	1.88	1.80	1.73	1.65	1.58	1.51	1.90	1.81	1.73	1.67	1.59	1.52	1.46
1700	2.10	2.00	1.91	1.83	1.75	1.68	1.61	2.02	1.93	1.84	1.76	1.69	1.62	1.55
1800	2.23	2.13	2.04	1.95	1.86	1.78	1.71	2.15	2.05	1.96	1.88	1.80	1.72	1.64
1900	2.36	2.25	2.15	2.06	1.97	1.88	1.81	2.28	2.17	2.07	1.98	1.90	1.82	1.74
2000	2.49	2.37	2.27	2.17	2.07	1.99	1.91	2.40	2.29	2.19	2.09	2.00	1.92	1.84
2100	2.62	2.50	2.38	2.28	2.19	2.09	2.00	2.52	2.41	2.30	2.20	2.11	2.02	1.93
2200	2.75	2.62	2.51	2.39	2.29	2.19	2.11	2.65	2.53	2.41	2.31	2.21	2.12	2.03
2300	2.88	2.75	2.62	2.51	2.40	2.30	2.21	2.78	2.65	2.53	2.42	2.31	2.22	2.13
2400	3.01	2.87	2.75	2.62	2.51	2.40	2.30	2.90	2.77	2.65	2.53	2.42	2.32	2.22
2500	3.14	3.00	2.86	2.74	2.62	2.51	2.40	3.03	2.89	2.76	2.64	2.52	2.42	2.32
2600	3.27	3.12	2.99	2.85	2.73	2.62	2.51	3.16	3.01	2.88	2.75	2.63	2.52	2.42
2700	3.41	3.25	3.10	2.97	2.83	2.72	2.61	3.28	3.14	2.99	2.86	2.73	2.62	2.51
2800	3.54	3.38	3.23	3.08	2.95	2.82	2.71	3.42	3.26	3.11	2.97	2.84	2.72	2.61
2900	3.68	3.51	3.34	3.20	3.06	2.93	2.81	3.55	3.38	3.22	3.09	2.95	2.82	2.71
3000	3.82	3.63	3.47	3.32	3.17	3.04	2.90	3.68	3.50	3.35	3.20	3.06	2.93	2.80

APPENDIX V

THE TABLES IN THE TEXT

The following tables have been given in the text and are indexed here for convenience of reference :

APPENDIX VI

THE ALBANY DATA GIVEN IN THE TEXT

Many tables of data for the Weather Bureau station at Albany, N.Y., have been given in the text to illustrate various points. These are indexed here for convenience of reference.

Section 28, page 22.   Number of thundershowers each year from 1884 to 1910.

Section 73, page 78.   The adjusted normal daily temperatures.

Section 74, page 79.   Average and normal monthly and yearly temperatures.

Section 79, page 87.   Average and normal values of daily range of temperature.

Section 79, page 90.   Average and normal values of variability of temperature.

Section 79, page 91.   The number of zero days, days above 90°, and days above 100°.

Section 251, page 247.   The amount of precipitation for the various months and for the year for several years, and the normal values.

Section 253, page 251.   Snowfall for the various months and for the year for several years, and the normal values.

APPENDIX VII

The Regular Stations of the U. S. Weather Bureau and the Canadian Stations

The following list contains the regular stations of the U. S. Weather Bureau during 1911. But little change occurs from year to year.

Abilene, Tex.
Albany, N.Y. M
Alpena, Mich. M
Amarillo, Tex.
Anniston, Ala.
Asheville, N.C.
Altanta, Ga.*†
Atlantic City, N.J.†
Augusta, Ga.

Baker City, Ore.
Baltimore, Md.†
Bentonville, Ark.
Binghamton, N.Y. M
Birmingham, Ala.
Bismarck, N. Dak.†
Block Island, R.I.
Boise, Idaho.†
Boston, Mass.†M
Buffalo, N.Y. M
Burlington, Vt. M

Cairo, Ill. M
Canton, N.Y.
Cape Henry, Va.

Cape May, N.J.
Charles City, Iowa.
Charleston, S.C.
Charlotte, N.C.
Chattanooga, Tenn.
Cheyenne, Wyo.†M
*Chicago, Ill.*M
Cincinnati, Ohio. M
Cleveland, Ohio. M
Columbia, Mo.†
Columbia, S.C.†
Columbus, Ohio.†M
Concord, N.H. M
Concordia, Kan.
Corpus Christi, Tex. M

Davenport, Iowa.
Del Rio, Tex.
Denver, Colo.†M
Des Moines, Iowa.*†M
Detroit, Mich. M
Devils Lake, N. Dak.
Dodge City, Kan.
Dubuque, Iowa.
Duluth, Minn.
Durango, Colo.

* Center of one of the twelve climatological districts.
† Center of one of the old forty-five sections.
M The station issues weather maps.
The six forecast centers are printed in italic type.

Eastport, Me.
Elkins, W. Va.
El Paso, Tex.
Erie, Pa. M
Escanaba, Mich.
Eureka, Cal.
Evansville, Ind.

Flagstaff, Ariz.
Fort Smith, Ark.
Fort Worth, Tex. M
Fresno, Cal. M

Galveston, Tex.
Grand Haven, Mich.
Grand Junction, Colo. M
Grand Rapids, Mich.†M
Green Bay, Wis.

Hannibal, Mo.
Harrisburg, Pa.
Hartford, Conn. M
Hatteras, N.C.
Havre, Mont.
Helena, Mont.†M
Honolulu, Hawaii.†
Houghton, Mich.
Houston, Tex.*†
Huron, S. Dak.†

Independence, Cal.
Indianapolis, Ind.†
Iola, Kan.
Ithaca, N.Y.*†M

Jacksonville, Fla.†
Jupiter, Fla.

Kalispell, Mont.
Kansas City, Mo. M
Keokuk, Iowa.
Key West, Fla.
Knoxville, Tenn.

La Crosse, Wis.
Lander, Wyo.
Lansing, Mich.
La Salle, Ill.
Lewiston, Idaho. M
Lexington, Ky.
Lincoln, Neb.†
Little Rock, Ark.†M
Los Angeles, Cal. M
Louisville, Ky.*†
Lynchburg, Va.

Macon, Ga. M
Madison, Wis.
Manteo, N.C.
Marquette, Mich.
Memphis, Tenn.
Meridian, Miss.
Miles City, Mont.
Milwaukee, Wis.†M
Minneapolis, Minn.†M
Mobile, Ala.
Modena, Utah.
Montgomery, Ala.†M
Moorhead, Minn.
Mount Tamalpais, Cal.
    (Through San Francisco Sta.)
Mount Weather, Va.
    (*Via* Bluemont, Va.)

Nantucket, Mass.
Narragansett Pier, R.I.
Nashville, Tenn.†
New Haven, Conn. M
New Orleans, La.*†M
New York, N.Y. M
Norfolk, Va. M
Northfield, Vt. M
North Head, Wash.
    (*Via* Ilwaco, Wash.)
North Platte, Nebr.

Oklahoma, Okla.†M
Omaha, Neb.
Oswego, N.Y.

Palestine, Tex.
Parkersburg, W. Va.† M
Pensacola, Fla.
Peoria, Ill. M
Philadelphia, Pa.†
Phœnix, Ariz. † M
Pierre, S. Dak.
Pittsburg, Pa. M
Pocatello, Idaho.
Point Reyes Light, Cal.
   (Through San Francisco Sta.)
Port Crescent, Wash.
Port Huron, Mich.
Portland, Me.
Portland, Ore.*†M
Providence, R.I.
Pueblo, Colo.

Raleigh, N.C.†
Rapid City, S. Dak.
Red Bluff, Cal.
Reno, Nev.†
Richmond, Va.†M
Rochester, N.Y.
Roseburg, Oreg.
Roswell, N. Mex. M

Sacramento, Cal.
St. Joseph, Mo.
St. Louis, Mo.*M
St. Paul, Minn.
Salt Lake City, Utah*†
San Antonio, Tex.
San Diego, Cal. M
Sand Key, Fla.
   (Through Key West Sta.)
Sandusky, Ohio.
San Francisco, Cal.*†M
San José, Cal.
San Juan, Porto Rico, W.I.†

San Luis Obispo, Cal.
Santa Fé, N. Mex. †
Sault Sainte Marie, Mich. M
Savannah, Ga. M
Scranton, Pa. M
Seattle, Wash.† M
Sheridan, Wyo.
Shreveport, La.
Sioux City, Iowa. M
Southeast Farallon, Cal.
   (Through San Francisco Sta.)
Spokane, Wash.
Springfield, Ill.†M
Springfield, Mo.
Syracuse, N.Y.

Tacoma, Wash.
Tampa, Fla.
Tatoosh Island, Wash.
Taylor, Tex. M
Thomasville, Ga.
Toledo, Ohio.
Tonopah, Nev.
Topeka, Kan.†

Valentine, Neb.
Vicksburg, Miss.†

Wagon Wheel Gap, Colo.
Walla Walla, Wash. M
Washington, D.C. M
Wichita, Kan.
Williston, N. Dak.
Wilmington, N.C.
Winnemucca, Nev.
Wytheville, Va.

Yankton, S. Dak.
Yellowstone Park, Wyo.
Yuma, Ariz.

The following list contains the Canadian stations from which observations are received by telegraph at Toronto for the construction of the weather map:

Dawson City	Battleford	Pt. Stanley	Yarmouth
Atlin	Prince Albert	Toronto	Halifax
Prince Rupert	Qu'Appelle	Kingston	Sydney
Victoria	Minnedosa	Stonecliffe	Charlottetown
New Westminster	Winnipeg	Ottawa	Sable Island
Kamloops	The Pas	Montreal	St. Johns
Barkerville	Port Arthur	Quebec	Burin
Calgary	White River	Father Point	Port Aux Basques
Edmonton	Cochrane	Anticosti	Fogo
Medicine Hat	Parry Sound	Chatham	Belle Isle
Swift Current	Southampton	St. John	

APPENDIX VIII

TEACHING METEOROLOGY

The method of teaching meteorology and the character of the course to be given depend upon the age and advancement of the students, the time allotted to the subject, and the standpoint from which it is taught. Instruction in meteorology is given in grammar and high schools as well as in colleges and universities. The time allotted to the subject is only a few minutes each day or·week in some schools, while in many colleges and universities systematic semester or year courses are given. Meteorology is sometimes taught with the ability to forecast as the chief aim; occasionally it is given with the emphasis on the mathematical side; sometimes the application of the laws and principles of physics is chiefly emphasized; and often the laboratory method of presentation is used. Each teacher must thus work out for himself the most suitable course and the best way of teaching it.

Appendix A, pages 171 to 185, in WARD'S *Practical Exercises in Elementary Meteorology*, contains suggestions to the teacher who is giving a course in a grammar or high school from the laboratory standpoint.

The New York State Education Department at Albany, N.Y., issues a pamphlet which contains an outline in syllabus form of the topics which the department considers should be included in a course on Physical Geography. Meteorology is included under this head, and the pamphlet is entitled *Syllabus for Secondary Schools; Physical Geography*. Attention should also be called to Bulletin No. 3 (1906) of the Geographic Society of Chicago, edited by Cox and Goode, and entitled *Lantern Slide Illustrations for the Teaching of Meteorology*. The National Educational Association has also considered from time to time the matter of teaching Physical Geography (usually including meteorology), and the Report of the Committee of Ten, 1894, should be consulted.

APPENDIX IX

THE LITERATURE OF METEOROLOGY

(A) General Directions

The lists of books and periodicals, and the bibliographical material here added are to enable two entirely different classes of students to gain more information than can be acquired by reading this book. The first class of students consists of those who wish to gain somewhat more detailed information on any given subject. The second class consists of the research students who wish to find every word which has been written on a certain very small and definite topic.

In the case of those who simply desire more information about a given subject, the best method of procedure is probably to read first whatever may be given on the subject in the twenty-five best books in the following list and also to look up all of the references to the literature at the end of each chapter of this book as far as they apply to the subject. If still more information is desired, it would probably be best to go over more books in the following lists and also to look up any references to the literature which may be given in what has been read. If still more information is desired, the student is essentially in the same position as the research student.

In the case of the research student who desires to find every word which has been written on a certain definite topic, the best method of procedure would probably be first to see what is given in each of the books in the following lists and also to look up any references at the end of the various chapters of this book which bear on the topic. The student should note every reference to the literature in the books, pamphlets, and articles read, and follow these up. Next the student would naturally look in the bibliography of bibliography to see if there is any bibliography on the topic in question and, if so, he would look up all the references contained in it. The student should then take the two digests of meteorological literature and, starting with the current year, work backwards to the beginning of the publications, or as far as seems advisable. If all the references obtained in all these ways are followed up, the student will probably have found nearly every word on the topic, and will have a clear idea how much has been done and to what extent the periodical literature has been incorporated in the latest books on meteorology.

(B) A List of Books

The following list contains nearly three hundred books which cover the whole subject of meteorology or some phase or part of it. Publications by the U. S. government and publications by the weather bureaus of other countries have in nearly all cases been excluded. All pamphlets and reprints of articles in the periodical literature have also been excluded. These will be found in the references at the end of each chapter. The books have been grouped. The first group contains general works on the whole field. The second group contains books which cover the whole field of meteorology, but from a special standpoint, or for a special purpose. The remaining groups follow the chapters of this book. If a library of exactly one hundred books on meteorology were to be chosen from the list, those marked with one star would be suggested as most suitable. If a library of exactly twenty-five were to be chosen from the list, those marked with two stars would be suggested. In this list of twenty-five, if a treatise is divided into several parts or volumes, it is considered as a single treatise. Books in foreign languages have not been excluded, as this list contains six German and one French book. This does not mean that the beginner could start with any one of these twenty-five books and find it understandable. The list contains, however, several elementary books which start at the beginning. It rather means that the possessor of these books would have a fairly complete treatment of meteorology from the foundation up.

(1) General Books

**ABBE, CLEVELAND, *The Aims and Methods of Meteorological Work*, 4°, pp. 219 to 330, Baltimore, 1899. (Part IIIa of Vol. I of Maryland Weather Service.)

*ABBE, CLEVELAND, *Treatise on Meteorological Apparatus and Methods*, 8°, 392 pp., Washington, 1888. (Annual report of the Chief Signal Officer for 1887 ; Appendix 46.)

ABBE, CLEVELAND, "Meteorology," in the *Encyclopædia Britannica*, Vol. XXX, 1902. (Revised by him for 1911 or 11th edition.)

*ABERCROMBY, RALPH, *Weather*, 12°, xix + 472 pp., London, 1887 (5th impression 1902). (International Scientific Series.),

*ALLINGHAM, WILLIAM, *A Manual of Marine Meteorology*, 12°, xvi + 182 pp., London, 1900.

ANDRE, CH., *Relations des phénomènes météorologiques*, 4°, 168 pp., Lyon, 1892.

**ANGOT, ALFRED, *Traité élémentaire de météorologie*, 8°, vi + 417 pp., Paris, 1899. (2d ed., 1907.)

ARAGO, FRANÇOIS, *Meteorological Essays*, 8°, 540 pp., London, 1855.

**ARCHIBALD, DOUGLAS, *The Story of the Atmosphere*, 16°, 210 pp., London, 1901.

*ARRHENIUS, SVANTE AUGUST, *Lehrbuch der kosmischen Physik*, 8°, viii + 1026 pp., Leipzig, 1903. (pp. 473 to 925 deal with meteorology.)

*BARNES, HOWARD T., *Ice Formation*, 8°, x + 260 pp., New York, 1906.

BEBBER, W. J. VAN, *Katechismus der Meteorologie*, 3d ed., 12°, xii + 259 pp., Leipzig, 1891.

*BEBBER, W. J. VAN, *Lehrbuch der Meteorologie*, 8°, xii + 391 pp., Stuttgart, 1890.

*BERGET, ALPHONSE, *Physique du globe et météorologie*, 8°, v + 353 pp., Paris, 1904. (pp. 162 to 343 deal with meteorology.)

*BEZOLD, WILHELM VON, *Gesammelte Abhandlungen aus den Gebieten der Meteorologie und des Erdmagnetismus*, 4°, viii + 448 pp., Braunschweig, 1906.

*BLANFORD, H. F., *Indian Meteorologists Vade Mecum*, in three parts, 8°, 85, 185, 81 pp., Calcutta, 1877.

*BÖRNSTEIN, R., *Leitfaden der Wetterkunde*, 2d ed., 8°, xi + 230 pp., Braunschweig, 1906.

BROCKLESBY, JOHN, *Elements of Meteorology*, 12°, xii + 240 pp., New York, 1848.

*BUCHAN, ALEXANDER, *A Handy Book of Meteorology*, 12°, 204 pp., London, 1867.

*BUCHAN, ALEXANDER, *Report on Atmospheric Circulation*, f°, iv + 263 pp., 52 maps, London, 1889. (Report on the scientific results of the voyage of H.M.S. *Challenger*.)

BUTLER, THOMAS BELDEN, *The Philosophy of the Weather*, 12°, xviii + 414 pp., New York, 1856.

*CHAMBERS, G. F., *The Story of the Weather*, 24°, 234 pp., London, 1897.

CHASE, PLINY EARLE, *Elements of Meteorology*, 2 vols., 128 and 256 pp., Philadelphia, 1884.

CORDEIRO, FREDERICK, J. B., *The Atmosphere; its Characteristics and Dynamics*, 4°, viii + 129 pp., New York, 1910.

CORNELIUS, C. S., *Meteorologie*, 8°, x + 614 pp., Halle, 1863.

Cyclopedia of American Agriculture, New York, 1907. Chap. XVII, "Weather Terms and Weather Knowledge," by WILFORD M. WILLSON; Chap. XVIII, "The Atmosphere and its Phenomena," by CLEVELAND ABBE, JR.

DANIELL, J. F., *Elements of Meteorology*, 3d ed., 2 vols., 8°, xxxv + 341, and viii + 389 pp., London, 1845.

**DAVIS, WILLIAM MORRIS, *Elementary Meteorology*, 8°, xi + 355, New York, 1894.

DICKSON, H. N., *Meteorology*, 12°, viii + 192 pp., London, 1893.

DREW, JOHN, *Practical Meteorology*, 12°, xi + 291 pp., London, 1855.

DUCLAUX, E., *Cours de physique et de météorologie professé à l'institut agronomique*, 8°, iv + 504 pp., Paris, 1891.

*DUNN, E. B., *The Weather*, 8°, viii + 356 pp., New York, 1902.

*FERREL, WILLIAM, *Recent Advances in Meteorology, systematically arranged in the Form of a Text-book*, 8°, 440 pp., Washington, 1886. (Annual Report of the Chief Signal Officer, 1885; Appendix 71.)

FINDLAY, ALEXANDER GEORGE, *Text-book of Ocean Meteorology*, 8°, 259 pp., London, 1887.

FLAMMARION, CAMILLE, *L'Atmosphère, Météorologie populaire*, 8°, **453 pp.**, London, 1873. (Translated by J. Glaisher.)

FLAMMARION, CAMILLE, *L'Atmosphère et les grands phénomènes de la nature,* 4°, 370 pp., Paris, 1905.

GEROSA, GIUSEPPE, *Elementi di Meteorologia,* 8°, x + 316 pp., Livorno, 1909.

*GIBERNE, AGNES, *The Ocean of Air,* 8°, xiv + 340 pp., London, 1903.

GILBERT, OTTO, *Die meteorologischen Theorien des Griechischen Altertums,* 8°, iv + 746 pp., Leipzig, 1907.

**GREELY, A. W., *American Weather,* 8°, xii + 286 pp., New York, 1888.

**HANN, JULIUS, *Lehrbuch der Meteorologie,* 4°, xi + 642 pp., 1st ed. Leipzig, 1901; 2d ed., Leipzig, 1906.

*HARRINGTON, MARK W., *About the Weather,* 12°, xx + 246 pp., New York, 1899.

HELLMANN, G., *Meteorologische Volksbücher,* 2d ed., 4°, 68 pp., Berlin, 1895.

HELLMANN, G., *Neudrucke von Schriften und Karten über Meteorologie und Erdmagnetismus,* 15 vols., Berlin, 1893-1904.

*HENKEL, F. W., *Weather Science,* 8°, 336 pp., London, 1911.

HERSCHEL, SIR J., *Meteorology,* 16°, vii + 288 pp., Edinburgh, 1862.

HILDEBRANDSSON AND HELLMANN, Codex of Resolutions adopted at International Meteorological Meetings, 1872-1907, British Meteorological Office, London, 1909.

*HILDEBRANDSSON, H., ET TEISSERENC DE BORT, LÉON, *Les bases de la météorologie dynamique historique,* 4°, 3 vols. (2 have appeared), Paris, 1898.

HORNBERGER, J., *Grundriss der Meteorologie und Klimatologie,* 8°, ix + 233 pp., Berlin, 1891.

*HOUSTON, EDWIN J., *The Wonder Book of the Atmosphere,* 8°, x + 326 pp., New York, 1907.

HOUZEAU, J. C., ET LANCASTER, A., *Traité élémentaire de météorologie,* 324 pp., Mons, 1883.

KÄMTZ, LUDWIG FRIEDRICH, *Lehrbuch der Meteorologie,* 8°, 3 vols., 526, 615, 563 pp., Leipzig, 1831.

KÄMTZ, LUDWIG FRIEDRICH, *Complete Course of Meteorology,* 12°, xxii + 598 pp., London, 1845. (Translated by C. V. Walker.)

*KASSNER, CARL, *Das Wetter und seine Bedeutung für das praktische Leben,* 12°, vi + 148 pp., Leipzig, 1908.

KASTNER, K. W. G., *Handbuch der Meteorologie,* 8°, 3 vols., 502, 655, 638 pp., Erlangen, 1823-1830.

KINDLER, P. FINTAN, *Das Wetter,* 16°, viii + 142 pp., Köln, 1909.

KLEIN, HERMANN F., *Allgemeine Witterungskunde,* 12°, 259 pp., Leipzig, 1884. (2d ed., Wien, 1905.)

*KOPPEN, W., *Grundlinien der maritimen Meteorologie,* 12°, vi + 83 pp., Hamburg, 1899.

LOMMEL, G., *Wind und Wetter,* 16°, vii + 344 pp., München, 1880.

LOOMIS, E., *Contributions to Meteorology,* rev. ed., 4°, 232 pp., New Haven, 1885-1887.

*LOOMIS, ELIAS, *A Treatise on Meteorology,* 8°, viii + 305 pp., New York, 1868.

MEYERS, *Konversationslexikon.* (See the various articles on meteorological subjects.)

M'PHERSON, J. G., *Meteorology; or Weather Explained*, 12°, 126 pp., London, 1905.

*MOHN, H., *Grundzüge der Meteorologie*, 5th ed., 8°, xii + 419 pp., Berlin, 1898.

**MOORE, JOHN WILLIAM, *Meteorology*, 8°, xvi + 445 pp., London, 1894. (New edition, revised, enlarged, and somewhat changed, 1910.)

**MOORE, WILLIS L., *Descriptive Meteorology*, 8°, xviii + 344 pp., New York, 1910.

*MULLER, JOH., *Lehrbuch der kosmischen Physik.* (5th ed. by C. F. W. Peters), 8°, xxiii + 907 pp., Braunschweig, 1894. (Atlas as separate vol.)

OLIVER, MIGUEL CORREA, *Tratado elemental de Meteorología*, 8°, 303 pp. (a vol. of plates), Madrid, 1909.

*PHIPSON, THOMAS LAMB, *Researches on the Past and Present History of the Earth's Atmosphere*, 12°, xii + 194 pp., London, 1901.

POWERS, EDWARD, *War and Weather*, 8°, 202 pp., Delavan, Wis., 1890.

RENK, FRIEDRICH, *Die Luft*, 8°, vi + 242 pp., Leipzig, 1886.

**RUSSELL, THOMAS, *Meteorology*, 8°, xxiii + 277 pp., New York, 1895.

*SALISBURY, ROLLIN D., *Physiography*, 8°, New York, 1907. (pp. 506–705 treat of meteorology.)

SCHMID, ERNST ERHARD, *Lehrbuch der Meteorologie*, 8°, xvi + 1009 pp., Leipzig, 1860.

*SCOTT, ROBERT H., *Elementary Meteorology*, 4th ed., 12°, xiv + 410 pp., London, 1887 (reprinted 1903). (International Science Series.)

*SPRUNG, A., *Lehrbuch der Meteorologie*, 8°, xii + 407 pp., Hamburg, 1885.

STEINMETZ, ANDREW, *Sunshine and Shadows*, 8°, xvi + 432 pp., London, 1867.

SYMONS, G. J., *The Eruption of Krakatoa and Subsequent Phenomena*, 4°, xvi + 494, London, 1888.

TAYLOR INSTRUMENT CO., *Weather and Weather Instruments*, 12°, 175 pp., Rochester, 1908.

TISSANDIER, GASTON, *L'Ocean aérien*, 8°, viii + 312 pp., Paris, 1883.

*TRABERT, WILHELM, *Meteorologie* (Sammlung Göschen), 24°, 150 pp., Leipzig, 1896 (3d ed., Leipzig, 1909).

TRABERT, WILHELM, *Lehrbuch der kosmischen Physik*, 8°, x + 662 pp., Leipzig, 1911.

UMLAUFT, FRIEDRICH, *Das Luftmeer*, 8°, viii + 488 pp., Wien, 1891.

**WALDO, FRANK, *Elementary Meteorology*, 12°, 373 pp., New York, 1896.

**WALDO, FRANK, *Modern Meteorology*, 12°, xxiii + 460 pp., London, 1893.

**WARD, ROBERT DE COURCY, *Practical Exercises in Meteorology*, 8°, xiii + 199 pp., Boston, 1899.

WEBER, LEONHARD, *Wind und Wetter*, 12°, v + 130 pp., Leipzig.

(2) **Miscellaneous Books covering the Whole Field of Meteorology, but with a Special Point in View**

(A) Composition of the Atmosphere

*RAMSAY, SIR WILLIAM, *The Gases of the Atmosphere; the History of their Discovery,* 3d ed., 8°, xii + 296 pp., London, 1905.

(B) Instructions to Observers

*ANGOT, ALFRED, *Instructions météorologiques,* 5th ed., 8°, vi + 161 pp , Paris, 1911.

JELINEK, CARL, *Jelinek's Anleitung zur Ausführung meteorologischer Beobachtungen nebst einer Sammlung von Hilfstafeln,* 5th ed., 4°, 124 and 94 pp., Wien, 1905 and 1910.

*LOISEL, JULIEN, *Guide de l'amateur météorologiste,* 8°, vi + 101 pp., Paris, 1906.

*MARRIOTT, WILLIAM, *Hints to Meteorological Observers,* 8°, 69 pp., London, 1906.

Prussia (K. preussisches met. Institut), *Anleitung zur Anstellung und Berechnung meteorologischer Beobachtungen,* 2 pts., 2d ed., 4°, Berlin, 1904–1905.

**The Observer's Handbook,* pub. by the Meteorological Office, London. (Revised almost annually.)

(C) Tables

Aspirations — Psychrometer — Tafeln (vom Kön. Preussischen meteorologischen Institut), 4°, xiv + 90 pp., Braunschweig, 1908.

HAZEN, *Handbook of Meteorological Tables,* 8°, vi + 127 pp., Washington, 1888.

JELINEK, CARL, *Jelinek's Psychrometer Tafeln,* 5th ed., f°, xiii + 107 pp., Leipzig, 1903.

***Smithsonian Meteorological Tables.* (Smithsonian misc. collections — 1032.) Rev. ed., lix + 274 pp., Washington, 1896. (3d revised edition, 1907.)

Tables Météorologiques internationales, 4°, Paris, 1890.

(D) Upper Air Investigation

*ASSMANN, R., BERSON, A., (and others), *Wissenschaftliche Luftfahrten,* 3 vols., f°, 150, 706, and 313 pp., Braunschweig, 1899–1900.

*HILDEBRANDT, A., *Airships Past and Present,* 8°, xvi + 364 pp., London, 1908. (Translated by W. H. Story.)

LINKE, FRANZ, *Aeronautische Meteorologie,* 8°, viii + 133 pp., Frankfurt, 1911.

*MOEDEBECK, HERRMANN W. L., *Pocket-book of Aeronautics,* 16°, xiii + 496 pp., London, 1907. (Translated by W. Mansergh Varley.)

NIMFÜHR, RAIMUND, *Leitfaden der Luftschiffart und Flugtechnik,* 2d ed., 8°, xvi + 528 pp., Wien, 1910.

**ROTCH, A. LAWRENCE, *Sounding the Ocean of Air,* 16°, viii + 184 pp., London, 1900.

ROTCH, A. L., and PALMER, ANDREW H., *Charts of the Atmosphere for Aeronauts and Aviators,* 4°, 96 pp. + 24 chts., New York, 1911.

ZAHM, ALBERT F., *Aerial Navigation,* 8°, xvii + 497 pp., New York, 1911.

(E) Charts

**BARTHOLOMEW, J. G., AND HERBERTSON, A. J., *Physical Atlas; Meteorology*, f°, 40 + xiv pp., 34 plates, London, 1899.　(3d vol. of Physical Atlas.)

*BERGHAUS, *Physical Atlas*, f°, Gotha, 1887.　(The met'l part by Hann can be bought separately.)

DENISON, CHARLES, *Climates of the United States*, 4°, 47 pp., Chicago, 1893.

ELIOT, SIR JOHN, *Climatological Atlas of India*, f°, xxxii pp. and 120 maps, Edinburgh, 1906.

Russia, *Atlas Climatologique de l'empire de Russie*, f°, xiv + 61 pp., (text), St. Petersbourg, 1900.

(F) Relation of meteorology to plants and animals — phenology

GÜNTHER, SIEGMUND, *Die Phänologie, ein Grenzgebiet zwischen Biologie und Klimakunde*, 8°, 51 pp., Münster, 1895.

(G) Relation of Meteorology to Medicine

*BEBBER, W. J. VAN, *Hygienische Meteorologie*, 8°, x + 330 pp., Stuttgart, 1895.

BELL, AGRIPPA NELSON, *Climatology and Mineral Waters of the United States*, 8°, vii + 386 pp., New York, 1885.

*DEXTER, EDWIN GRANT, *Weather Influences*, 8°, xxxi + 286 pp., New York, 1904.

GILES, G. M., *Climate and Health in Hot Countries*, 8°, xviii + 188, 109 pp., New York, 1905.

*HUGGARD, WILLIAM R., *A Handbook of Climatic Treatment*, 8°,. xiii + 536 pp., London, 1906.

*SOLLY, S. EDWIN, *A Handbook of Medical Climatology*, 8°, xii + 470 pp., Philadelphia, 1897.

WEBER, F. PARKER, AND HINSDALE, GUY, *Climatology; Health Resorts; Mineral Springs*, 2 vols., 8°, ix + (10) + 336 and x + (11) + 420 pp., Philadelphia, 1902.

*WEBER, SIR HERRMANN, AND WEBER, F. PARKER, *Climatotherapy and Balneotherapy*, 4°, 833 pp., London, 1907.

ZUNTZ, N. (and others), *Höhenklima und Bergwanderungen in ihrer Wirkung auf den Menschen*, 4°, xvi + 494 pp., Berlin, 1906.

(3) The Observation and Distribution of Temperature

*BOLTON, HENRY CARRINGTON, *Evolution of the Thermometer, 1592–1743*, 16°, 98 pp., Easton, Pa., 1900.

GUILLAUME, CH. ED., *Traité practique de la thermométrie de Precision*, 8°, xv + 336, Paris, 1889.

(4) The Pressure and Circulation of the Atmosphere — including the Winds and the General Theory of Atmospheric Circulation

*ABBE, CLEVELAND, *The Mechanics of the Earth's Atmosphere*, 8°, 324 pp., Washington, 1891. (Smithsonian Miscellaneous Collections — 843.) (Second Collection of Translations.)

*ABBE, CLEVELAND, *The Mechanics of the Earth's Atmosphere*, 8°, iv + 616 pp., Washington, 1910. (Third Collection of Translations.)

ANSART-DEUSY, A., *Théorie des mouvements de l'atmosphère et de l'océan*, 8°, 272 pp., Paris, 1877.

ASSMANN, RICHARD, *Die Winde im Deutschland.*

*BRILLOUIN, MARCEL, *Mémoires originaux sur la circulation générale de l'atmosphère*, 8°, xx + 165 pp., Paris, 1900.

BUYS-BALLOT, *Les courants de l'air et de l'atmosphère*, 8°, 39 pp., Bruges, 1891.

CHATLEY, HERBERT, *The Force of the Wind*, 8°, viii + 83 pp., London, 1909.

*COFFIN, J. H., *The Winds of the Globe*, 4°, 768 pp., Washington, 1876.

Deutsche Seewarte, *Segelhandbuch für den Atlantschen Ozean; Hierzu ein Atlas.* (A similar treatise is also issued for the Pacific and Indian Oceans.)

**FERREL, WILLIAM, *Popular Treatise on the Winds*, 2d ed., 8°, vii + 505 pp., New York, 1889.

GILBERT, G. K., *A New Method of Measuring Heights by Means of the Barometer.* (U. S. Geological Survey — annual report, 1881 — pp. 403 to 566.)

LIZNAR, J., *Die Barometrische Höhenmessung*, 8°, 48 pp., Leipzig, 1904.

LOISEL, JULIEN, *Le Baromètre anéroide*, 12°, 24 pp., Paris, 1905.

MOHN, H., AND GULDBERG, C. M., *Études sur les mouvements de l'atmosphère.* 4°, 39 and 53 pp., 2 parts, Christiania, 1876–1880.

PLYMPTON, GEORGE W., *The Aneroid Barometer*, 8°, 126 pp., New York, 1907.

SCHREIBER, PAUL. *Handbuch der barometrischen Höhenmessungen*, 2d ed., 8°, xiv + 480 pp., London, 1863.

SHAW, WILLIAM H., *The Life History of Surface Air Currents*, 4°, 107 pp., London, 1906.

SUPAN, A., *Statistik der unteren Luftströmungen*, 8°, vii + 296 pp., Leipzig, 1881.

WEGENER, ALFR., *Thermodynamik der Atmosphäre*, 8°, viii + 331 pp., Leipzig, 1911.

WHYMPER, EDWARD, *How to use the Aneroid Barometer*, 8°, 61 pp., London, 1891.

(5) The Moisture of the Atmosphere — Dew, Frost, Fog, Clouds, Precipitation

ABERCROMBY, R., *Seas and Skies in Many Latitudes*, xvi + 447 pp., London, 1888.

*BARBER, SAMUEL, *The Cloud World*, 8°, xii + 139 pp., London, 1903.

BARUS, CARL, *A Continuous Record of Atmospheric Nucleation*, f°, xvi + 226 pp., Washington, 1905.

BIGELOW, FRANK HAGAR, *Report on the International Cloud Observations*, 4°, 787 pp., Washington, 1900. (Vol. III of the Report of the Chief of the Weather Bureau, 1898–1899.)

**CLAYDEN, ARTHUR W., *Cloud Studies*, 8°, xiii + 184 pp., New York, 1905.
Cloud Crystals, wide 8°, 158 pp., New York, 1864.
COLLINSON, JOHN, *Rainmaking and Sunshine*, 12°, xvi + 280 pp., London, 1894.
FRITZSCHE, RICHARD, *Niederschlag, Abfluss, und Verdunstrung auf den Land-flächen der Erde*, 8°, 54 pp., Halle, 1906.
HELLMANN, G., *Die Niederschläge in den norddeutschen Stromgebieten*, 4°, 3 vols., Berlin, 1906.
HELLMANN, G., *Schneekrystalle*, 4°, 66 pp., Berlin, 1893.
HERBERTSON, *The Distribution of Rainfall over the Land*, 8°, 70 pp., London, 1901.
HOUDAILLE, F., *Les orages à grêle et le tir des canans*, 244 pp., Paris, 1901.
*International Cloud Atlas, (*Atlas International des Nuages*), by HILDERBRANDS-SON, and others, f°, 30 pp., xiv plates, Paris, 1896. (A new slightly changed edition; 4°, viii + 24 pp., appeared in 1911.)
**KASSNER, CARL, *Das Reich der Wolken und Niederschläge*, 12°, 160 pp., Leipzig, 1909.
LANCASTER, A., *La Pluie en Belgique*, 8°, 224 pp., Bruxelles, 1894.
**LEY, CLEMENT, *Cloudland*, 8°, ix + 208 pp., London, 1894.
MCADIE, ALEX., *The Clouds and Fogs of San Francisco*, 8°, 106 pp., San Francisco, 1912.
RUSSELL, ROLLO, *On Hail*, 8°, xv + 224 pp., London, 1893.
SCHARF, EDMUND, *Der Hagel*, 12°, vi + 195 pp., Halle a. S., 1906.
**SUPAN, ALEXANDER, *Die Verteilung des Niederschlags auf der festen Erdober-fläche*, 4°, iv + 103 pp., Gotha, 1898.
THOMPSON, MRS. JEANETTE MAY, *Water Wonders Every Child Should Know*, xvii + 233 pp., New York, 1907.
VINCENT, J., *Atlas des nuages*, f°, 29 pp., Bruxelles, 1907.
VOSS, ERNST LUDWIG, *Die Niederschlagsverhältnisse von Südamerika*, 4°, iv + 59 pp., Gotha, 1907.
WETHERELL, HENRY EMERSON, *Hygromedry*, 82 pp. (pub. by author).
WILD, H., *Die Regenverhältnisse des Russischen Reiches*, f°, 120 + 95 + 286 pp., St. Petersburg, 1887.
WILSON-BARKER, D., *Clouds and Weather Signs*, 8°, 31 pp., London.

(6) The Secondary Circulation of the Atmosphere — Storms

**ALGUÉ, JOSÉ, *The Cyclones of the Far East*, 2d ed., 4°, 283 pp., Manila, 1904.
**BERGHOLZ, PAUL, *Die Orkane des fernen Ostens*, xii + 260 pp., Bremen, 1900.
BLASIUS, WILLIAM, *Storms*, 8°, 342 pp., Philadelphia, 1875.
DAVIS, WILLIAM MORRIS, *Whirlwinds, Cyclones, and Tornadoes*, 24°, 90 pp., Boston, 1884.
DE PENNING, GEORGE A., *Meteorology and the Laws of Storms*, 276 pp., Calcutta, 1897.
DOBERCK, W., *The Law of Storms in the Eastern Seas*, 4th ed., 8°, 44 pp., Hongkong, 1904.

DOVE, HEINRICH WILHELM, *Law of Storms*, 8°, 331 pp., London, 1862.

*ELIOT, JOHN, *Hand-book of Cyclonic Storms in the Bay of Bengal*, iv + 212 pp., Calcutta, 1890. (2d ed., 1900–1901, 2 vols.)

ESPY, JAMES, *The Philosophy of Storms*, 8°, 592 pp., Boston, 1841.

FAYE, HERVÉ A. E. A., *Nouvelle étude sur les tempêtes, cyclones, trombes ou tornados*, 8°, 142 pp., Paris, 1897.

FINLEY, JOHN P., *Tornadoes*, 12°, 196 pp., New York, 1887.

FISCHER, ALFRED, *Die Hurricanes oder Drehstürme Westindiens*, 4°, 70 pp., Gotha, 1908. (Petermann's Mitteilungen; Erganzungscheft Nr. 159.)

**GOCKEL, ALBERT, *Das Gewitter*, 8°, 204 pp., Köln, 1905.

HANZLIK, STANISLAV, *Die räumliche Verteilung der meteorologischen elemente in den Antizyklonen*, 4°, 94 pp., Wien, 1898.

*HAZEN, H. A., *The Tornado*, 12°, ii + 143 pp., New York, 1890.

HOUDAILLE, F., *Les orages à grêle et le tir des canons*, 8°, 244 pp., Paris, 1901.

LOCKYER, WILLIAM J. S., *Southern Hemisphere Surface Air Circulation*, 4°, iii + 109 pp., 1910.

PERLEY, SIDNEY, *Historic Storms in New England*, 8°, x + 341 pp., Salem, 1891.

PIDDINGTON, HENRY, *The Sailor's Horn-book for the Law of Storms*, 6th ed., 8°, xviii + 408 pp., London, 1876.

PLUMADON, J. R., *Les orages et la grêle*, 8°, 192 pp., Paris, 1901.

REID, WILLIAM, *Attempt to develop the Law of Storms*, 8°, 538 pp., London, 1850.

*STREIT, A., *Das Wesen der Cyklonen*, vi + 125 pp., Wien, 1906.

TOMLINSON, CHARLES, *The Thunder-storm*, 3d ed., 16°, xii + 381 pp., London, 1877.

(7) Weather Bureaus and their Work

POLIS, P., *Der Wetterdienst und die Meteorologie in den Vereinigten staaten von America und in Canada*, 4°, 43 pp., Berlin, 1908.

(8) Weather Prediction — including Weather Proverbs and Prognostics

ABBE, CLEVELAND, *Preparatory Studies for Deductive Methods in Storm and Weather Predictions*, 8°, 165 pp., Washington, 1890. (Annual Report of Chief Signal Officer for 1889; Appendix 15.)

ABERCROMBY, RALPH, *Principles of Forecasting by Means of Weather Charts*, 8°, 122 pp., London, 1885.

BEBBER, W. J. VAN, *Beurteilung des Wetters auf mehrere Tage varaus*, 8°, 32 pp., Stuttgart, 1896.

**BEBBER, W. J. VAN, *Die Wettervorhersage*, 2d ed., xvi + 219 pp., Stuttgart, 1898.

*BEBBER, W. J. VAN, *Handbuch der ausübenden Witterungskunde*, 8°, 2 parts, x + 392 and x + 503 pp., Stuttgart 1885–1886.

BENDEL, T., *Wetterpropheten*, 8°, 166 pp., Regensburg, Manz, 1904.

DALLET, G., *La Prévision du temps*, 16°, 336 pp., Paris.

*DUNWOODY, H. H. C., *Weather Proverbs*, 8°, 148 pp., Washington, 1883. (Signal Service Notes, No. IX.)

FITZROY, ROBERT, *The Weather Book*, 2d ed., 8°, xiv + 480 pp., London, 1863.

*FREYBE, OTTO, *Praktische Wetterkunde*, 8°, vi + 173 pp., Berlin, 1906.

FRIESENHOF, GREGOR, *Wetterlehre oder praktische Meteorologie*, 8°, viii + 680 pp., Nedanocz, 1883.

GRANGER, FRANCIS S., *Weather Forecasting*, 8°, xii + 121 pp., Nottingham, 1909.

*GUILBERT, GABRIEL, *Nouvelle méthode de prévision du temps*, 8°, xxxiii + 343 pp., Paris, 1909.

**INWARDS, RICHARD, *Weather Lore*, 8°, xii + 233 pp., London, 1898.

KINDLER, FINTAN, *Das Wetter*, 24°, 142 pp., Köln, 1909.

KLEIN, H., *Wettervorhersage für Jedermann*, 12°, vi + 164 pp., Stuttgart, 1907.

KUHLENBÄUMER, TH., *Unser Wetter und seine Vorherbestimmung*, 12°, x + 164 pp., Münster, 1909.

PERNTER, J. M., *Wetterprognose in Osterreich*, 16°, 61 pp., Wien, 1907.

SCOTT, A. C., *Notes on Meteorology and Weather Forecasting*, 8°, 40 pp., London, 1909.

*SCOTT, ROBERT H., *Weather Charts and Storm Warnings*, 3d ed., 8°, vi + 229 pp., London, 1887.

*SHAW, W. N., *Forecasting Weather*, 8°, xxvii + 380 pp., London, 1911.

STEINMETZ, ANDREW, *Weather Casts and Storm Prognostics*, 8°, 208 pp., London, 1866.

*SWAINSON, REV. C., *A Handbook of Weather Folk-lore*, 12°, x + 275 pp., London, 1873.

TIMM., H., *Wie gestaltet sich das Wetter?* 8°, viii + 175 pp., Leipzig, 1892.

(9) Climate

ALGUÉ, JOSÉ S. J., *The Climate of the Philippines*, 8°, 103 pp., Manila, 1904.

ARMAND, *Traité de climatologie générale du globe*, 8°, xx + 868 pp., Paris, 1873.

BEHRE, OTTO, *Das Klima von Berlin*, 8°, 158 pp., Berlin, 1908.

BLANFORD, S. M., *Climates and Weather of India*, 8°, 382 pp., London, 1889.

*BLODGET, LORIN, *Climatology of the United States*, 4°, xvi + 536 pp., Philadelphia, 1857.

BONACINA, L. C. W., *Climatic Control*, viii + 167 pp., London, 1911.

BRÜCKNER, EDWARD, *Klimaschwankungen seit 1700*, 8°, viii + 324 pp., Wien, 1890.

CROLL, JAMES, *Climate and Time*, 8°, xvi + 577 pp., New York, 1887.

CROLL, JAMES, *Discussions on Climate and Cosmology*, 8°, xii + 327 pp., New York, 1886.

CULLIMORE, D. H., *The Book of Climate*, 12°, x + 279 pp., London, 1891.

DAVIS, WALTER G., *Climate of the Argentine Republic*, 8°, 154 pp., Buenos Aires, 1910. (Argentina Dept. of Agriculture.)

ECKARDT, WILHELM R., *Das Klimaproblem der geologischen Vergangenheit und historischen Gegenwart*, 8°, xi + 183 pp., Braunschweig, 1909.

*FASSIG, OLIVER L., *The Climate and Weather of Baltimore*, 8°, 515 pp., Baltimore, 1907.

FRITSCHE, H., *The Climate of Eastern Asia.* (Journal of the North-China Branch of the Royal Asiatic Society, Vol. XII, 1877, pp. 127–335.)

**HANN, JULIUS, *Hand-book of Climatology*, (translated by Robert De Courcy Ward), 8°, xiv + 437 pp., New York, 1903.

**HANN, JULIUS, *Handbuch der Klimatologie*, 2d ed., 3 vols., 404, 384, and 576 pp., Stuttgart, 1897; 3d ed., vol. 1, xiv + 394 pp., Stuttgart, 1908; 3d ed., vol. 2, xii + 426 pp., Stuttgart, 1910.

HERBERTSON, A. J. and F. D., *Man and his Work*, London, 1899.

HERZ, NORBERT, *Die Eiszeiten und ihre Uhrsachen*, 4°, iv + 306 pp., Leipzig, 1909.

KNOX, ALEXANDER, *The Climate of the Continent of Africa*, 8°, xii + 552 pp., Cambridge, 1911.

KÖPPEN, W., *Klimakunde*, 2d ed., 16°, 132 pp., Leipzig, 1906.

MARCHI, L. DE, *Climatologia*, 8°, x + 204 pp., Milano, 1890.

*MEYER, HUGO, *Anleitung zur Bearbeitung meteorologischer Beobachtungen für die Klimatologie*, 8°, viii + 187 pp., Berlin, 1891.

MÜHY, A., *Klimatographische Übersicht der Erde*, 8°, xvi + 744 pp. (supplement xii + 320 pp.), Leipzig, 1862.

QUETELET, A., *Sur la climat de la Belgique*, 4°, 2 vols., about 500 pp., Bruxelles, 1849.

RATZEL, *Anthropogeographie*, 2d ed., Stuttgart, 1899.

ROSTER, GIORGIO, *Climatologia dell Italia nelle sue attinenze con l' ingiene e con agricoltura*, 8°, xxix + 1040 pp., Torino, 1909.

SUPAN, A., *Grundzüge der physischen Erdkunde*, 4th ed., 8°, ix + 934 pp., Leipzig, 1908.

Die Veranderungen des Klimas seit dem Maximum des letzten Eisgeit, 4°, lviii + 459 pp., Stockholm, 1910. (Pub. by 11th International Geological Congress.)

**WARD, ROBERT DE COURCY, *Climate*, 8°, xiv + 372 pp., London, 1909.

WOEIKOF, A., *Die Klimate der Erde*, 2 parts, 396, 445 pp., Jena, 1887.

(10) Atmospheric Electricity

ANDERSON, RICHARD, *Lightning Conductors*, 8°, xv + 470 pp., London, 1885.

ANGOT, ALFRED, *The Aurora Borealis*, 8°, xii + 264 pp., New York, 1897.

CAPRON, J. RAND, *Auroræ: their Characters and Spectra*, 4°, xv + 207 pp., London, 1879.

CHAUVEAU, A. B., *L'Electricité Atmosphérique*, f°, 70 pp., Paris, 1902.

FLAMMARION, CAMILLE, *Thunder and Lightning*, 12°, 281 pp., Boston, 1906. (Translated by Walter Mostyn.)

*GOCKEL, ALBERT, *Die Luftelektrizität*, vi + 206 pp., Leipzig, 1908.

GREELY, A. W., *Chronological List of Auroras 1870 to 1879*, 4°, 76 pp., Washington, 1881. (Prof. Papers of the Signal Service, No. 3.)

HARRIS, WILLIAM SNOW, *On the Nature of Thunderstorms*, xvi + 226 pp., London, 1843.

*HEDGES, KILLINGWORTH, *Modern Lightning Conductors*, 4°, vi + 119 pp., London, 1905. (New ed, 1910.).

2 M

LEMSTROM, *L'aurore boreale*, xii + 179 pp., Paris, 1886.

LODGE, SIR OLIVER J., *Lightning Conductors and Lightning Guards*, 12°, xii + 544 pp., London, 1892.

**MACHE, H., and SCHWEIDLER, E. v., *Die Atmospherische Elektrizität*, 8°, xi + 247 pp., Braunschweig, 1909.

SCHROETER, J. FR., *Catalog der in Norwegen biz Juni 1878 beobachteten Nordlichter*, f°, 422 pp., Kristiania, 1902.

SPANG, HENRY W., *A Practical Treatise on Lightning Protection*, 12°, 63 pp., New York, 1883.

(11) Atmospheric Optics

BESSON, LOUIS, *Sur la théorie des halos*, 8°, 89 pp., Paris, 1909.

MASCART, E., *Traité d'optique* (especially vol. 3), 8°, Paris, 1889–1894.

**PERNTER, J. M., UND EXNER, FELIX M., *Meteorologische Optik*, 8°, xvii + 799 pp., Leipzig, 1910.

(C) The U. S. Government Publications on Meteorology

The publications issued by the U. S. Weather Bureau may be divided into two groups, the periodical publications, and those which appear at irregular intervals.

The periodical publications are: —

Daily Weather Map, from Jan., 1871, to date. These are now published once daily, based on the 8 A.M. observations at Washington and many regular stations. Many newspapers also publish daily weather maps based on both the 8 A.M. and the 8 P.M. observations. The maps were formerly issued twice and three times daily. Maps for Sunday and the holidays are issued at Washington only.

Daily Forecast Cards. These cards contain the forecast only, and are issued at Washington, by many regular stations, and from some post offices on receipt of telegraphic information from some Weather Bureau station.

Monthly Weather Review, 4°, 1872 to date. Prior to July, 1891, it was published by the U. S. Signal Service. Originally it was only a bulletin of current meteorological conditions in the United States and Canada, but later it included all the features of a general meteorological journal. January, 1908, the brief summaries of the observations at the coöperative stations were discontinued. Since July, 1909, its form has again been changed. Few research articles are now published and full monthly climatological reports from all stations have been included.

Bulletin of the Mount Weather Observatory, 8°, 1908 to date. This contains the results of the observations at the Mount Weather Observatory, and also research articles. Research articles are now being published in this Bulletin rather than in the Monthly Weather Review.

National Weather Bulletin. This is a large single sheet (19 by 24 inches), printed on one side only and issued weekly during the summer and monthly during the winter. In addition to the text, it contains these four charts: the average temperature, the amount of precipitation, and the departure from normal in the case of both temperature and precipitation.

Snow and Ice Bulletin, 1893 to date. This is a single sheet (12 by 19 inches), printed on one side only, and issued weekly during the winter. It contains, in addition to the tables and text, a chart showing the depth of snow on the ground.

Meteorological Chart of the Great Lakes, October, 1897, to date.

Annual Report of the Chief, 1870 to date. This is a large volume and contains, in addition to the administrative report, a summary of the observations of the year. Special reports and research articles have been added frequently as appendices.

All regular stations of the Weather Bureau publish a monthly and yearly summary of their observations.

The publications which appear from time to time are: —

Numbered Bulletins. These are special articles and nearly fifty have now appeared. The list is given below.

Lettered Bulletins. These are usually large volumes and contain very valuable material. Bulletin V has recently appeared. The list is given below.

W. B. Publications. Nearly all of the publications of the U. S. Weather Bureau now receive a serial number. The Monthly Weather Review, the Bulletin of the Mount Weather Observatory, the numbered bulletins, and the lettered bulletins are included in these. There are many other publications, however, which are not in any one of the four series which have a W. B. number. These numbers have now reached nearly 500.

Miscellaneous Publications. There are many publications of the U. S. Weather Bureau which unfortunately are not numbered or designated in any way.

Summary of the Climatological Data for the United States by Sections. One hundred and six summaries are to be published, and, when complete, this will be a veritable mine of information about the climate of the various parts of the United States. It will be complete in 1911.

Many of the publications have changed their form, name, and characteristics, but a full account of these changes is impossible here. There are also publications which have been discontinued. *Climate and Health* was published from July, 1895, to March, 1896. Forty-four of the forty-five sections published a monthly climatological report until July, 1909, and a weekly weather bulletin during the summer until 1909. Only three sections, Iowa, Hawaiian Islands, and Porto Rico, continue these publications at present. They all, however, continue to publish an annual summary.

Previous to 1891, while the weather service was part of the U. S. Signal Service, the meteorological articles were published as: —

Signal Service Notes; Professional Papers of the Signal Service; Reports of the Chief Signal Officer; Publication without any special designation.

A complete list of the meteorological publications of the U. S. Signal Service will be found in the Report of the Chief Signal Officer for 1891. This bibliography covers pages 389 to 409, and lists 119 publications by the Signal Service. This includes the 18 Professional Papers and the 23 Signal Service Notes. This bibliography also contains all books, pamphlets, or articles published anywhere by any person while he was connected with the Signal Service.

The four following lists contain the Professional Papers of the Signal Service, the Signal Service Notes, the lettered bulletins of the U. S. Weather Bureau, and the numbered bulletins of the U. S. Weather Bureau:

U. S. Signal Service Professional Papers

No. 1 ABBE, CLEVELAND, *Report on the Solar Eclipse of July, 1878.* 4°, 186 pp., 34 pls., Wash., 1881.

No. 2 GREELY, A. W., *Isothermal Lines of the United States, 1871–1880,* 4°, 1 p., 12 pls., Wash., 1881.

No. 3 GREELY, A. W., *Chronological List of Auroras Observed from 1870 to 1879,* 4°, 76 pp., Wash., 1881.

No. 4 FINLEY, J. P., *Report of the Tornadoes of May 29 and 30, 1879, in Kansas, Nebraska, Missouri, and Iowa.* 4°, 116 pp., 29 chs., Wash., 1881.

No. 5 *Information Relative to the Construction and Maintenance of Timeballs,* 4°, 31 pp., 5 pls., Wash., 1881.

No. 6 HAZEN, H. A., *The Reduction of Air-pressure to Sea Level at Elevated Stations West of the Mississippi River,* 4°, 42 pp., 20 maps, Wash., 1882.

No. 7 FINLEY, J. P., *Report on the Character of Six Hundred Tornadoes,* 4°, 29 pp., 3 chs., Wash., 1884.

No. 8 FERREL, WILLIAM, *Recent Mathematical Papers Concerning the Motions of the Atmosphere,* Part I, "The Motions of Fluids and Solids on the Earth's Surface," reprinted with notes by Frank Waldo, 4°, 51 pp., Wash., 1882.

No. 9 DUNWOODY, H. H. C., *Charts and Tables showing Geographical Distribution of Rainfall in the United States,* 4°, 29 pp., 3 chs., Wash., 1883.

No. 10 *Tables of Rainfall and Temperature Compared with Crop Production,* 4°, 15 pp., Wash., 1882.

No. 11 SHERMAN, O. T., *Meteorological and Physical Observations on the East Coast of British America,* 4°, 202 pp., 1 ch., Wash., 1883.

No. 12 FERREL, WILLIAM, *Popular Essays on the Movements of the Atmosphere,* 4°, 59 pp., Wash., 1882.

No. 13 FERREL, WILLIAM, *Temperature of the Atmosphere and Earth's Surface,* 4°, 69 pp., Wash., 1884.

No. 14 FINLEY, J. P., *Charts of Relative Storm Frequency for a Portion of the Northern Hemisphere,* 4°, 9 pp., 13 chs., Wash., 1884.

No. 15 LANGLEY, S. P., *Researches on Solar Heat and its Absorption by the Earth's Atmosphere.* (A Report of the Mount Whitney Expedition.) 4°, 139 pp., 22 pls., Wash., 1884.

No. 16 FINLEY, J. P., *Tornado Studies for 1884*, 4°, 15 pp., 72 chs., 72 tables, Wash., 1885.

No. 17 FERREL, WILLIAM, *Recent Advances in Meteorology*. Published as Part 2, Appendix No. 71, of the annual report of the Chief Signal Officer for 1885, 8vo, 440 pp., Wash., 1886.

No. 18 HAZEN, H. A., *Thermometer Exposure*, 4°, 32 pp., Wash., 1885.

U. S. Signal Service Notes

No. 1 BAILEY, W. O., *Report on the Michigan Forest Fires of 1881*, 8vo, 16 pp., 6 chs., Wash., 1882.

No. 2 BIRKHIMER, W. E., *Memoir on the Use of Homing Pigeons for Military Purposes*, 8vo, 27 pp., Wash., 1882.

No. 3. ALLEN, JAMES, *To Foretell Frost*, 8vo. 11 pp., Wash., 1882.

No. 4 UPTON, WINSLOW, *The Use of the Spectroscope in Meteorological Observations*, 8vo, 7 pp., 3 chs., Wash., 1883.

No. 5 *Work of the Signal Service in the Arctic Regions*, 8vo, 40 pp., 1 ch., Wash., 1883.

No. 6 HAZEN, H. A., *Report on the Wind Velocities at the Lake Crib and at Chicago*, 8vo, 20 pp., 1 ch., Wash., 1883.

No. 7 HAZEN, H. A., *Variation of the Rainfall West of the Mississippi River*, 8vo, 8 pp., Wash., 1883.

No. 8 WALDO, FRANK, *The Study of Meteorology in the Higher Schools of Germany, Switzerland, and Austria*, 8vo, 148 pp., 9 pp., Wash., 1883.

No. 9 DUNWOODY, H. H. C., *Weather Proverbs*, 8vo, 148 pp., 1 map, Wash., 1883.

No. 10 GARLINGTON, E. A., *Report on Lady Franklin Bay Expedition of 1883*, 8vo, 52 pp., 1 map, Wash., 1883.

No. 11 WARD, F. K., *The Elements of the Heliograph*, 8vo, 12 pp., Wash., 1883.

No. 12 FINLEY, J. P., *The Special Characteristics of Tornadoes, with Practical Direction for the Protection of Life and Property*, 8vo, 19 pp., Wash., 1884.

No. 13 CURTIS, G. E., *The Relation between Northers and Magnetic Disturbances at Havana, Cuba*, 8vo, 16 pp., Wash., 1885.

No. 14 LAMAR, W. H., Jr., AND ELLIS, F. W., *Physical Observations During the Lady Franklin Bay Expedition of 1883*, 8vo, 62 pp., 14 pls., 1 map, Wash., 1884.

No. 15 HAZEN, H. A., *Danger Lines and River Floods of 1882*, 8vo, 30 pp., Wash., 1884.

No. 16. CURTIS, G. E., *The Effect of Wind Currents on Rainfall*, 8vo, 11 pp., 2 pls., Wash., 1884.

No. 17 MORRILL, PARK, *A First Report upon Observations of Atmospheric Electricity at Baltimore, Md.*, 8vo, 8 pp., 6 chs., Wash., 1884.

No. 18 McADIE, ALEXANDER, *The Aurora in its Relations to Meteorology*, 8vo, 21 pp., 14 chs., Wash., 1885.

No. 19 GLENN, S. W., *Report on the Tornado of August 28, 1884, near Huron, Dak.*, 8vo, 10 pp., 11 chs., Wash., 1885.

No. 20 HAZEN, H. A., *Thunderstorms of May, 1884*, 8vo, 8 pp., 2 chs., Wash., 1885.

No. 21 *How to Use Weather Maps.* Not published as Signal Service Notes.

No. 22 RUSSELL, THOMAS, *Corrections of Thermometers*, 8vo, 11 pp., Wash., 1885.

No. 23 WOODRUFF, T. M. *Cold Waves and their Progress, A preliminary study*, 8vo, 21 pp., Wash., 1885.

The Lettered Bulletins of the U. S. Weather Bureau

A DUNWOODY, H. C. *Summary of International Meteorological Observations* (1878–1887), x pp. and 59 charts, 1893.

B HARRINGTON, MARK WALROD, *Surface Currents of the Great Lakes*, xiv pp., 1895.

C HARRINGTON, MARK W., *Rainfall and Snow of the United States compiled to the end of 1891*, 80 pp., 1894.

D HENRY, ALFRED J., *Rainfall of the United States*, 58 pp., 1897.

E MORRILL, PARK, *Floods of the Mississippi River*, 79 pp., 1897.

F FRANKENFIELD, H. C., *Report on the Kite Observations of 1898*, 71 pp., 1899.

G VERY, FRANK W., *Atmospheric Radiation*, 132 pp., 1900.

H GARRIOTT, E. B., *West Indian Hurricanes*, 69 pp., 1900.

I BIGELOW, FRANK H., *Eclipse Meteorology and Allied Problems*, 166 pp., 1902.

J HENRY, ALFRED J., *Wind Velocity and Fluctuations of Water Level on Lake Erie*, 22 pp., 1902.

K GARRIOTT, E. B., *Storms of the Great Lakes*, 9 pp. and 968 charts, 1903.

L MCADIE, ALEXANDER G., *Climatology of California*, 270 pp., 1903.

M FRANKENFIELD, H. C., *The Floods of the Spring of 1903 in the Mississippi Watershed*, 63 pp., 1904.

N STOCKMAN, WILLIAM B., *Periodic Variation of Rainfall in the Arid Region*, 15 pp., 1905.

O STOCKMAN, WILLIAM B., *Temperature and Relative Humidity Data*, 29 pp., 1905.

P GARRIOTT, EDWARD B., *Cold Waves and Frosts in the United States*, 22 pp. and 328 charts, 1906.

Q HENRY, ALFRED JUDSON, *Climatology of the United States*, 1012 pp., 1906.

R BIGELOW, FRANK H., *The Daily Normal Temperature and the Daily Precipitation in the United States*, 186 pp., 1908.

S BIGELOW, FRANK H., *Report on the Temperatures and Vapor Tensions in the United States*, 302 pp., 1909.

T COX, HENRY C., *Frost and Temperature Conditions in the Cranberry Marshes of Wisconsin*, 121 pp., 1910.

U BIGELOW, FRANK H., *Temperature Departures, Monthly and Annual, in the United States, January, 1873, to June, 1909*, inclusive, 5 pp., 474 charts, 1911.

V DAY, P. C., *Frost Data of the United States and Length of the Crop Growing Season*, 5 pp., 5 charts, 1911.

Numbered Bulletins of the U. S. Weather Bureau

No. 1 HARRINGTON, MARK W., *Notes on the Climate and Meteorology of Death Valley, California*, 50 pp., 1892.

No. 2 BIGELOW, FRANK H., *Notes on a New Method for the Discussion of Magnetic Observations*, 40 pp., 1892.

No. 3 HILGARD, E. W., *A Report on the Relations of Soil to Climate*, 59 pp., 1892.

No. 4 WHITNEY, MILTON, *Some Physical Properties of Soils in their Relation to Moisture and Crop Distribution*, 90 pp., 1892.

No. 5 KING, FRANKLIN H., *Fluctuations in the Level and Rate of Movement of Ground-water*, 75 pp., 1892.

No. 6 COLE, FRANK N., *The Daily Variation of Barometric Pressure*, 32 pp., 1892.

No. 7 *Report of the First Annual Meeting of the American Association of State Weather Services*, 49 pp., 1893.

No. 8 MELL, P. H., *Report on the Climatology of the Cotton Plant*, 68 pp., 1893.

No. 9 CONGER, N. B., *Report on the Forecasting of Thunderstorms during the Summer of 1892*, 54 pp., 1893.

No. 10 HAZEN, HENRY A., *The Climate of Chicago*, 137 pp., 1893.

No. 11 FASSIG, OLIVER L., *Report of the International Meteorological Congress Held at Chicago, Ill., Aug. 21–24, 1893*, 1896.

No. 12 BARUS, CARL, *Report on the Condensation of Atmospheric Moisture*, 104 pp., 1895.

No. 13 WILLIAMS, H. E., *Temperatures injurious to Food Products in Storage and during Transportation*, 20 pp., 1894.

No. 14 *Report of the Third Annual Meeting of the American Association of State Weather Services*, 31 pp., 1894.

No. 15 McADIE, ALEXANDER, *Protection from Lightning*, 26 pp., 1895.

No. 16 JEWELL, L. E., *The Determination of the Relative Quantities of Aqueous Vapor in the Atmosphere by Means of Absorption Lines of the Spectrum*, 12 pp., 1896.

No. 17 MOORE, WILLIS L., *The Work of the Weather Bureau in Connection with the Rivers of the United States*, 106 pp., 1896.

No. 18 *Report of the Fourth Annual Meeting of the American Association of State Weather Services*, 55 pp., 1896.

No. 19 HENRY, ALFRED J., *Report on the Relative Humidity of Southern New England and Other Localities*, 23 pp., 1896.

No. 20 BIGELOW, FRANK H., *Storms, Storm Tracks, and Weather Forecasting*, 87 pp., 1897.

No. 21 BIGELOW, FRANK H., *Abstract of a Report on Solar and Terrestrial Magnetism in their Relations to Meteorology*, 176 pp., 1898.

No. 22 PHILLIPS, W. F. R., *Climate of Cuba*, 23 pp., 1898.

No. 23 HAMMON, W. H., *Frost: When to Expect it, and How to Lessen the Injury therefrom*, 37 pp., 1899.

No. 24 *Proceedings of the Convention of Weather Bureau Officials held at Omaha, Neb., October 13–14, 1898*, 184 pp., 1899.

No. 25 MOORE, WILLIS L., *Weather Forecasting: Some Facts Historical, Practical and Theoretical*, 16 pp., 1899.

No. 26 MCADIE, ALEXANDER G., AND HENRY, ALFRED J., *Lightning and the Electricity of the Air*, 74 pp., 1899.

No. 27 BIGELOW, FRANK H., *The Probable State of the Sky along the Path of Total Eclipse of the Sun, May, 28, 1900, Observations of 1899*, 23 pp., 1899.

No. 28 MCADIE, ALEXANDER G., AND WILLSON, GEORGE H., *The Climate of San Francisco, California*, 30 pp., 1899.

No. 29 MCADIE, ALEXANDER G., *Frost Fighting*, 15 pp., 1900.

No. 30 HENRY, ALFRED J., *Loss of Life in the United States by Lightning*, 21 pp., 1901.

No. 31 BERRY, JAMES, AND PHILLIPS, W. F. R., *Proceedings of the Second Convention of Weather Bureau Officials*, 246 pp., 1902.

No. 32 ALEXANDER, WILLIAM H., *Hurricanes*, 79 pp., 1902.

No. 33 GARRIOTT, EDWARD B., *Weather Folk-lore and Local Weather Signs*, 153 pp., 1903.

No. 34 MOORE, WILLIS L., *Climate: its Physical Basis and Controlling Factors,* 19 pp., 1904.

No. 35 GARRIOTT, E. B., *Long-range Weather Forecasts*, 68 pp., 1904.

No. 36 ABBE, CLEVELAND, *A First Report on the Relations between Climates and Crops*, 386 pp., 1905.

No. 37 HENRY, ALFRED J., *Recent Practice in the Erection of Lightning Conductors*, 20 pp., 1906.

No. 38 EMERY, SAMUEL C., *Mississippi River Levees and their Effect on River Stages during Flood Periods*, 21 pp., 1910.

(D) Bibliography of Bibliography[1]

Group I. General

ABBE, CLEVELAND, *A First Report on the Relation Between Climates and Crops*, Bulletin 36, U. S. Weather Bureau. (Pages 365–375 contain a bibliography.)

BARTHOLOMEW, J. G., AND HERBERTSON, A. J., *Physical Atlas; Meteorology*. (It contains a general bibliography.)

BÖRNSTEIN, R., *Leitfaden der Wetterkunde*. (Pages 209–222 contain references to the literature.)

Brussels (Observatoire Royal de Belgique), *Catalogue des ouvrages d'astronomie et de météorologie qui se trouvent dans les principales bibliothèques de la Belgique*, 8°, xxiii + 645 pp., Bruxelles, 1878.

[1] See Appendix IX, *B*, for further details about the books mentioned.

FASSIG, OLIVER, L., *Bibliography of Meteorology: A Classified Catalogue of the Printed Literature of Meteorology to 1887*, Washington, 1889–1891. (Mimeographed; 4 parts only have been published. These parts cover temperature, moisture, wind, and storms.)

HARDING, J. S., JR., *Catalogue of the Library of the Royal Meteorological Society to 1890*, 8°, viii + 214 pp., London, 1891.

HELLMANN, G., *Repertorium der deutschen Meteorologie bis 1881*, 8°, xxii + 995 pp., Leipzig, 1883.

HELLMANN, G., " Contribution to the Bibliography of Meteorology and Terrestrial Magnetism in the 15th, 16th, and 17th Centuries," Rep. of Chicago Meteor. Congress, 1893, Part II, Washington, 1894.

HERSCHEL, SIR J., *Meteorology*, Edinburgh, 1862. (Critical bibl. to date.)

Index to the Publ. of the English Meteorological Societies, 1839–1881, 32 pp., London, 1881.

Katalog der Bibliothek der Deutschen Seewarte zu Hamburg, 8°, x + 619 pp., also Nachtrag, 1–8, Hamburg, 1890.

Library of Congress. (The Card Catalogue of this Library can be obtained.)

LOOMIS, ELIAS, *A Treatise on Meteorology*. (Pages 296–300 contain a list of the works on meteorology before 1868.)

Meteorological Annuaire of the Royal Observatory of Belgium for 1905. (A bibliography by J. Vincent of treatises on meteorology, containing about 200 books.)

Meteorologische Zeitschrift Namen- und Sachregister, Vols. 1–25, 1884–1908, 4°, 231 pp., Braunschweig, 1910.

MOORE, JOHN WILLIAM, *Meteorology*. (Pages 410–419 contain publications of the U.S. Weather Bureau and Signal Service. Omitted in the new edition.)

MOORE, WILLIS L., *Descriptive Meteorology*. (A bibliography at the end of each chapter.)

Quarterly Journal of the Royal Meteorological Society, Index to vols. VIII–XXVI, 1882–1900, 8°, 37 pp., London, 1901.

SYMONS, G. J., "English Meteorological Literature, 1337–1699," Report of the Chicago International Meteorological Congress, 1893, Part II, Washington, 1894.

SYMONS's *Monthly Meteorological Magazine*, Index to vols. I–XXX, 1886–1895, 8°, iv + 84 pp., London, 1897.

SZALAY, LADISLAUS V., *Namen- und Sachregister der Bibliothek der Kön. Ungarischen Reichsanstalt für Meteorologie und Erdmagnetismus*, 8°, viii + 423 pp., Budapest, 1902.

TALMAN, C. FITZHUGH, *Brief List of Meteorological Text-books and Reference Books*, 8°, 18 pp., U. S. Weather Bureau.

TRABERT, WILHELM, *Meteorologie*. (Pages 7 and 8 are on the literature.)

WALDO, FRANK, *Modern Meteorology*, London, 1893. (Pages 6–19 have to do with meteorological publications.)

WEBER, SIR HERMANN, AND WEBER, F. PARKER, *Climatherapy and Balneotherapy*. (Pages 745–774 extensive bibliography.)

Group II. Temperature

BOLTON, HENRY C., *Evolution of the Thermometer.* (Pages 92–96 contain a bibliography on the history of the thermometer.)

Group III. The Moisture of the Atmosphere — Dew, Frost, Fog, Clouds, Precipitation

HELLMANN, G., *Die Niederschläge in den norddeutschen Stromgebieten.* (Pages 31–36 bibliography.)

HELLMANN, G., *Schnee Krystalle*, 1893. (A bibliography of the subject.)

JELINEK, CARL, *Jelinek's Psychromter = Tofeln.* (Pages xi–xiii bibliograph of psychrometer and hair hygrometer.)

HERBERTSON, ANDREW J., *The Distribution of Rainfall over the Land.* (Bibliography of the subject.)

Monthly Weather Review. (An annotated bibliography of Evaporation, by Mrs. Grace J. Livingston, June, 1908, to June, 1909. Also reprinted in 1910 as a whole, 121 pp.)

VOSS, ERNST LUDWIG, *Die Niederschlagsverhältnisse von Sudamerika.* (Bibliography of the subject.)

Group IV. The Secondary Circulation of the Atmosphere — Storms

ALGUÉ, JOSÉ, *The Cyclones of the Far East.* (A bibliography at the end of each chapter.)

FERREL, WILLIAM, *Popular Treatise on the Winds.* (Pages 480–483 contain a list of books and articles on this general subject.)

POEY Y AGUIRRE, ANDRÉS, *Bibliographie Cyclonique*, 8°, 96 pp., Paris, 1866.

Group V. Weather Prediction — Including Weather Proverbs and Prognostics

BEBBER, W. J. VAN, *Handbuch der ausübenden Witterungskunde.* (Part I, pp. 367–392, Part II, pp. 480–494, references to articles on various meteorological subjects mentioned in the text.)

INWARDS, RICHARD, *Weather Lore.* (Pages 206–212 contain a bibliography of weather lore).

SWAINSON, REV. C., *A Handbook of Weather Folk-lore.* (A short list of works consulted.)

Group VI. Climate

ECKHARDT, WILHELM, R., *Das Klimaproblem der geologischen Vergangenheit und historischen Gegenwart*, Braunschweig, 1909. (Literaturangaben, pp. 176–183.)

RAMSAY, A., *A Bibliography, Guide, and Index to Climate*, 8°, 449 pp., London, 1884.

SUPAN, ALEXANDER, *Grundzüge der physichen Erdkunde*, 4th ed., Leipzig, 1908. (Bibliography on climatic changes, pp. 229–353.)

Group VII. Rivers and Floods

Pittsburg (Carnegie Library), *Floods and Flood Protection; References to Books and Magazine Articles*, 48 pp., Pittsburg, 1908.

Group VIII. Atmospheric Electricity

CHAUVEAU, A. B., *L'Electricité Atmospherique.* (Pages 61–70 bibliography of the subject.)

MACHE, H., UND SCHWIDLER, E. V., *Die atmosphärische Elektrizität.* (Pages 237–247 articles and books on atmospheric electricity.)

Group IX. Atmospheric Optics

Monthly Weather Review, Sept., 1900, pp., 386–389. (A bibliography of works on the intensity, color, and polarization of sky light.)

(E) Digest of Literature

The current literature on meteorology in the form of books, pamphlets, serial publications, or articles in scientific magazines, is now completely listed or abstracted in two very valuable publications. These are *Die Fortschritte der Physik* (3 volumes each year), and *The International Catalogue of Scientific Literature, Section F, Meteorology*. In each of these, the material is so classified and subdivided that it is comparatively easy to find the material on any given topic.

Die Fortschritte der Physik was first published in 1845 and the 67th volume covers the literature of 1911. Part III of the annual issue of this publication covers cosmical physics, and, of course, includes the whole of meteorology. *The International Catalogue of Scientific Literature* was begun in 1901–1902, with the literature for 1901, and the 8th annual issue covers the literature of 1908 (published in 1910).

The literature of meteorology is also partially abstracted in *Science Abstracts* (v. 14 for 1911), and *Beiblätter zu den Annalen der Physik* (v. 35 in 1911), but these cannot be relied upon for completeness. The *Monthly Weather Review*, the *Meteorologische Zeitschrift*, and the *Quarterly Journal of the Royal Meteorological Society* publish in each issue valuable comments on the current literature, but they do not aim at completeness. For popular articles on meteorological subjects the indices of Poole, Fletcher, and *Reader's Guide* should be consulted.

(F) Periodicals

In an appendix to BARTHOLOMEW AND HERBERTSON's *Physical Atlas—Meteorology*, (v. 3) will be found a list of the serial publications of the weather bureaus and weather services of the various countries. In the *International Catalogue of Scientific Literature*, (F, *Meteorology*) is given a list of scientific magazines and publications containing articles on meteorological subjects.

In the first of the following lists are given the more important periodicals which are devoted almost entirely to meteorology. In the second list are given those periodicals which regularly or occasionally contain meteorological articles. Periodicals in English, French, or German are the only ones which have been considered.

(i) Periodicals devoted entirely to Meteorology

Monthly Weather Review, 4°, July, 1872–date, Washington, D.C. 1 volume each year; v. 39 during 1911.

Bulletin of the Mount Weather Observatory, 8°, 1908–date, Washington, D.C. 1 volume each year; v. 4 during 1911.

American Meteorological Journal, 8°, May, 1884–April, 1896. v. 1, Detroit; v. 2–8, Ann Arbor; v. 9–12, Boston.

Quarterly Journal of the Royal Meteorological Society, 8°, November, 1871–date, London. 1 volume each year; v. 37 during 1911.

Symons's Meteorological Magazine, 8°, 1866–date, London. 1 volume each year; v. 46 during 1911.

Journal of the Scottish Meteorological Society, 8°, 3d series, 1864–date, Edinburgh and London. 3 years in one volume; v. 16 during 1911–1913.

Die Meteorologische Zeitschrift, 4°, January, 1884–date, Braunschweig (formerly Berlin and Wien). 1 volume each year; v. 28 during 1911.

Zeitschrift der Oesterreichischen Gesellschaft für Meteorologie, 4°, 1866–1884, Wien. 1 volume each year; 19 volumes in all.

Das Wetter, 8°, 1885–date. Berlin. 1 volume each year; v. 28 during 1911.

Himmel und Erde, 8°, 1888–date, Berlin. 1 volume each year; v. 22 during 1911.

Beiträge zur Physik der freien Atmosphäre, f°, later 4°, 1904–date, Strassburg. Several years in each volume.

Beobachtungen mit bemannten und unbemannten Ballons und Drachen sowie auf Berg- und Wolkenstationen (published by Internationale Komission für Wissenschaftliche Luftschiffahrten), 4°, 1900–date, Strassburg. 1 volume each year.

Annalen der Hydrographie und maritimen Meteorologie, 8°, 1873–date, Berlin. 1 volume each year; v. 39 during 1911.

Ciel et terre, 8°, 1880–date, Bruxelles. 1 volume each year; v. 31 in 1911–1912.

Annuaire de la Société météorologique de France, large 8°, 1849–date, Paris. 1 volume each year; v. 59 during 1911.

La Revue Nephologique, 8°, 1906–date.

(The publications of the weather services of the various countries should, perhaps, be added here.)

(ii) Periodicals which regularly or occasionally contain Meteorological Articles

The Astrophysical Journal, large 8°, 1895–date, Chicago. 2 volumes each year; v. 33 during the first half of 1911.

Science, 8°. New series, 1895–date, New York. 2 volumes each year; v. 33 during first half of 1911.

Terrestrial Magnetism and Atmospheric Electricity, 8°, 1896–date, Baltimore. 1 volume each year; v. 16 during 1911.

Scientific American, f°, 1845–date, New York. 2 volumes each year; v. 104 during first half of 1911.

Scientific American Supplement, f°, 1876–date, New York. 2 volumes each year; v. 71 during first half of 1911.

London, Edinburgh, and Dublin Philosophical Magazine, 8°, 1798–date, London. 2 volumes each year; v. 21, of 6th series, during first half of 1911.

Nature, large 8°, 1869–date, London. 3 volumes each year; v. 86 during first third of 1911.

Annalen der Physik, 8°, 1790–date, Leipzig. 3 volumes each year; v. 34 (4th series) during first third of 1911.

Physikalische Zeitschrift, large 8°, 1899–date, Leipzig. 1 volume each year; v. 13 during 1911.

Petermanns Mitteilungen, 4°, 1855–date, Gotha. 1 volume each year; v. 57 during 1911. (Many Ergänzungshefte, numbering 162 in 1908.)

Das Weltall, 4°, 1900–date, Treptow-Berlin. 1 volume each year; v. 11 during 1910–1911.

Comtes rendus, 4°, 1835–date, Paris. Two volumes each year; v. 152 during first half of 1911.

Bulletin de la société astronomique de France, 8°, 1887–date, Paris. 1 volume each year; v. 25 during 1911.

INDEX

[The numbers refer to pages.]

A

Abbe, C., 48, 354.
Abbot, C. G., solar constant, 40.
Absorption, effects of, 37.
    of insolation, 37.
Actinometry, 38.
Adiabatic cooling, 46.
Air (see atmosphere).
Aitken, J., 11.
Albany, N.Y., 86, 88, 125, 148.
    daily range of temperature at, 87.
    daily variation in temperature at, 84.
    graphical representation of station normals of temperature at, 80.
    normal absolute humidity at, 205.
    normal daily temperature at, 78.
    normal monthly temperature at, 79.
    normal relative humidity at, 209.
    normal sunshine at, 230.
    number of rainy days at, 253.
    precipitation at, 247, 248.
    river gauge at, 447.
    snowfall at, 251, 252.
    tables of data for, 501, 512.
    temperature data at, 91.
    thunder showers at, 21, 327.
    variability of temperature at, 90.
    weather prediction at, 389.
Altitude, barometric determination of, 130.
    pressure variation with, 132.
Alto-cumulus, 223.
Alto-stratus, 223.
Anemometer, 141.
    deflection, 141.
    Lind's, 142.
    pocket, 144.
    Robinson's, 142, 147.
Anemoscope, 138.
Aneroid barometer, 119.
Angot, 41.
Anomalies, thermal, 98.
Anticyclones (see Chapter VI, B).
    definition of, 294.
    description of, 294.
    distribution of elements about, 294.
    energy of, 314.
    origin of, 312.
    structure above earth's surface of, 279.
    tracks of, 305.
    velocity of motion of, 306.

Arago, 69.
Arctic winds, 177.
Argon, 7, 8, 15.
Aristotle, 3.
Assmann, R., 48, 53, 70.
Astronomy, 2, 6.
Atmosphere, composition of, 7, 9.
    convection in, 45.
    definition of, 6.
    dust in, 11.
    evolution of, 15, 16.
    future of, 16.
    heating and cooling of, 56.
    height of, 18.
    nocturnal stability of, 55.
    offices of, 13.
    of other heavenly bodies, 15.
    pressure of, 18 (see Chapter IV).
    temperature change between day and night, 42.
Atmospheric acoustics, Chapter XIII.
Atmospheric electricity, Chapter XI.
Atmospheric optics, Chapter XII.
Atom, 29.
Aurora borealis, 19, 479.
Australia, 176.
Avalanche winds, 181.

B

Backing of wind, 137.
Bacteria, 10.
Baguois, 266.
Balloons, 48.
Barograph, 121.
Barometer, accuracy of aneroid, 121.
    accuracy of mercurial, 119.
    aneroid, 115, 119.
    corrections to reading of mercurial, 117.
    history of, 115.
    kinds of, 115.
    mercurial, 115.
    mouth, 122.
Barometric gradient, 159.
Battles and rain, 243.
Beaufort wind scale, 139.
Bench mark, 446.
Bentley, 241.
Bibliography (see end of each chapter and Appendix IX).
Bigelow, 302, 311, 312.

2 N

K

Khamsin, 348.
Kites, 48.
Krakatoa, 12.
Krypton, 8.

L

Lake temperatures, 105.
Land breeze, 177.
Latent heat, 38, 191.
Leste, 347.
Leveche, 347.
Lightning, beaded, 470.
　　cause of, 466.
　　color of, 467.
　　damage by, 473.
　　danger from, 470.
　　heat, 469.
　　kinds of, 466.
　　protection from, 442.
　　rods, 472, 475.
　　sheet, 469.
　　spectrum of, 467.
　　zigzag, 466.
Lind's pressure anemometer, 142.
Linnæus, 62.
Literature of meteorology, 518.
London fogs, 217.
Looming, 487.
Lows (see extratropical cyclones).

M

Mars, 15.
Marvin,. 49.
Meniscus, 118.
Meteorology, definition, 2.
　　history, 3.
　　utility, 3–5.
Meteors, 12, 17, 19.
Metric system, 503.
Migration annual of the winds, 170.
Migration of isotherms, 97.
Mirage, 46, 486.
Mistral, 348.
Moisture (see Chapter V).
Molecule, 29.
Monsoon, 174.
Moon and weather, 414.
Moore, Willis L., 355, 440.
Mountain and valley breeze, 179.
Mountain observations, 146.
Mouth-barometer, 122.

N

Natural sciences, definition, 1.
　　enumeration of, 2.

Neon, 8.
Nephoscope, 226.
Nimbus, 222.
Nitrogen, 7, 14.
Normal values, 21, 76.
　　in connection with precipitation, 246–253.
　　of cloudiness, 229.
　　of fog, 218.
　　of frost, 216.
　　of moisture, 203–209.
　　of number of thunder showers, 327.
　　of pressure, 124–129.
　　of temperature, 76–91.
　　of wind, 147–155.
Northern lights, 19, 479.
Nuclei of condensation, 231.

N

Ocean currents, 95.
Ocean temperature, 105.
　　temperature change between day and night, 41.
Optics, atmospheric (see Chapter XII).
Oxygen, 7.
Ozone, 12, 55.

P

Pampero, 347.
Parallax in reading thermometer, 65.
Particles, 231, 491.
　　inorganic, 11.
　　organic, 10.
Pericyclonic ring, 268.
Periodicals, 540.
Physics, 2, 6.
Piche evaporimeter, 193.
Planetary winds, 165.
Planets, wind system of, 183.
Polar temperatures, 101.
Precipitation (see Chapter V, D; also rainfall and snowfall).
Prediction, accuracy of, 407.
　　by similarity, 398.
　　cold wave, 400.
　　flood, 404.
　　frost, 399.
　　general method of, 379.
　　long range, 407.
　　storm, 403.
　　system of verifying, 405.
　　terms used in, 404.
　　tornado, 339, 403.
　　when high dominates, 398.
　　when low dominates, 383.
Pressure anemometer, 141.
Pressure change with altitude, 129.

[The numbers refer to pages.]

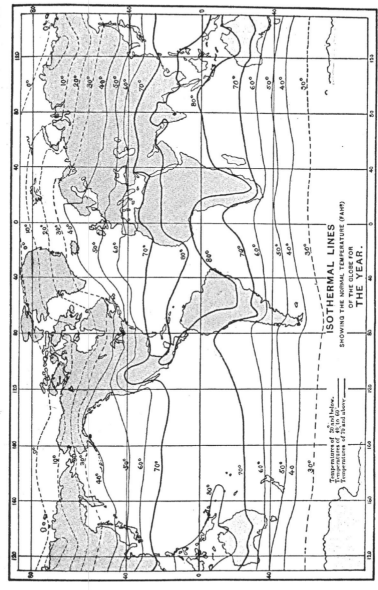

ISOTHERMAL LINES

SHOWING THE NORMAL TEMPERATURE (FAH°)
OF THE GLOBE FOR
THE YEAR.

Temperatures of 30° and below.
Temperatures of 40° to 60°
Temperatures of 70° and above

CHART I.—Isothermal Lines for the Year.

(Based on BUCHAN's *Challenger* Report; similar to the Chart in DAVIS's *Elementary Meteorology*.)

CHART II. — Ocean Currents.

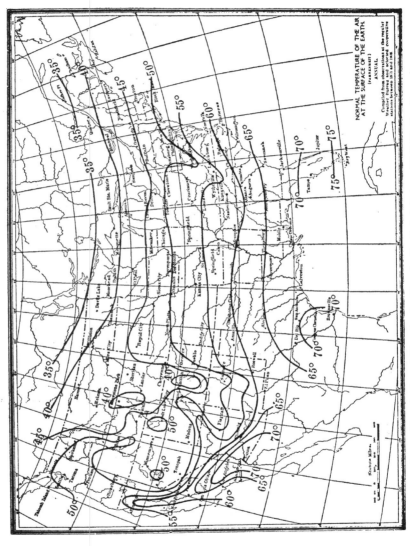

CHART III. — Isothermal Lines for the Year for the United States. (U. S. Weather Bureau.)

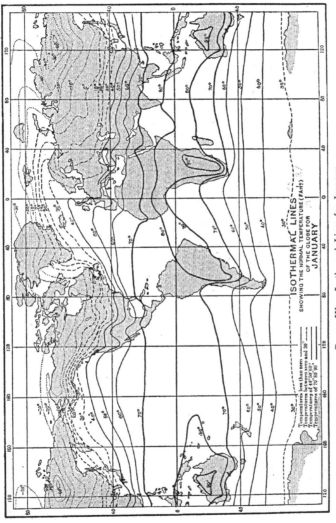

CHART IV. — Isothermal Lines for January.

(Based on BUCHAN's *Challenger* Report; similar to the Chart in DAVIS's *Elementary Meteorology*.)

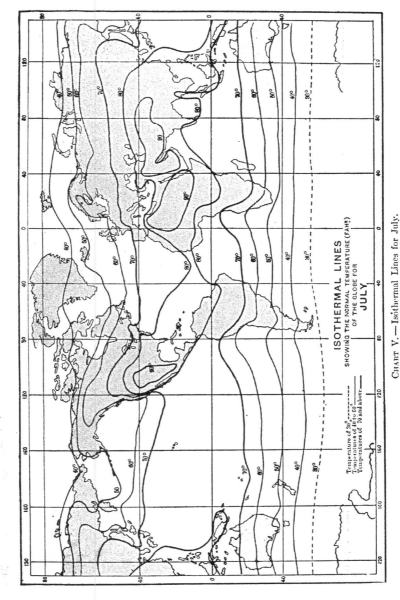

CHART V. — Isothermal Lines for July.

(Based on BUCHAN's *Challenger Report*; similar to the Chart in DAVIS's *Elementary Meteorology*.)

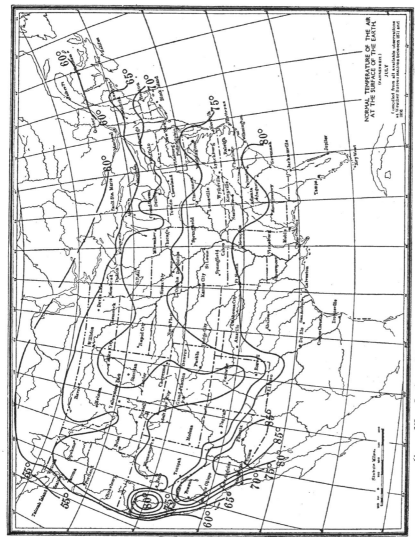

CHART VI. — Isothermal Lines for July for the United States. (U. S. Weather Bureau.)

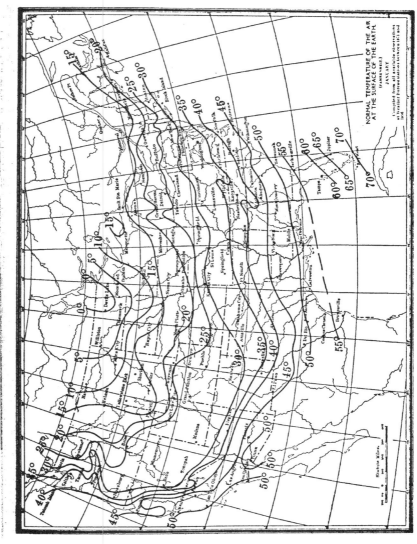

CHART VII. — Isothermal Lines for January for the United States. (U. S. Weather Bureau.)

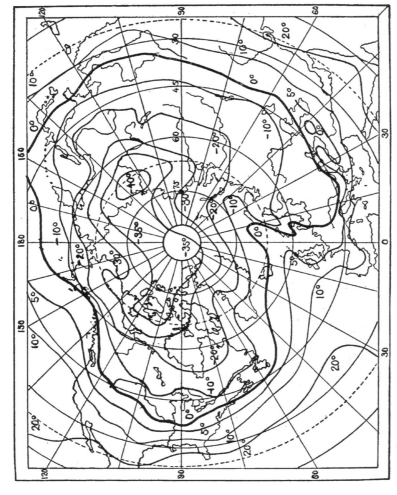

Chart VIII. — Isotherms for the North Polar Regions for January (Temp. C.). (After Angot.)

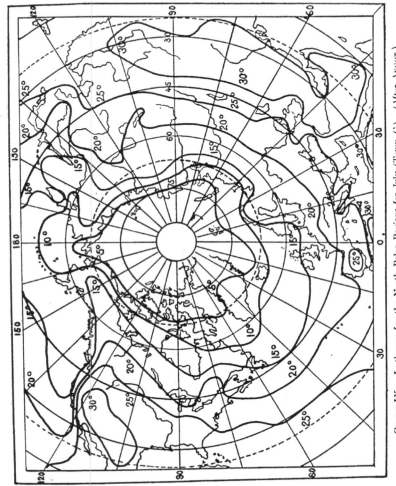

CHART IX. — Isotherms for the North Polar Regions for July (Temp. C). (After Angot.)

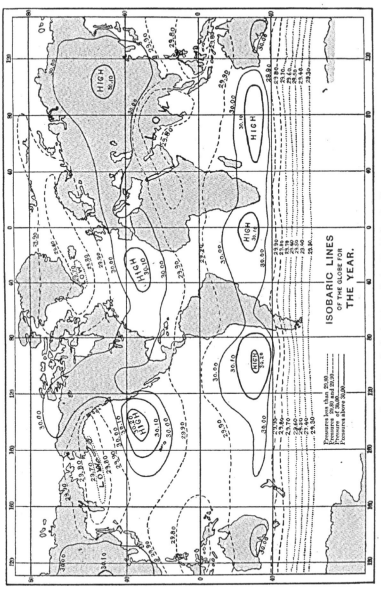

CHART X. — Isobaric Lines of the World for the Year.

(Based on BUCHAN's *Challenger* Report; similar to the Chart in DAVIS's *Elementary Meteorology*.)

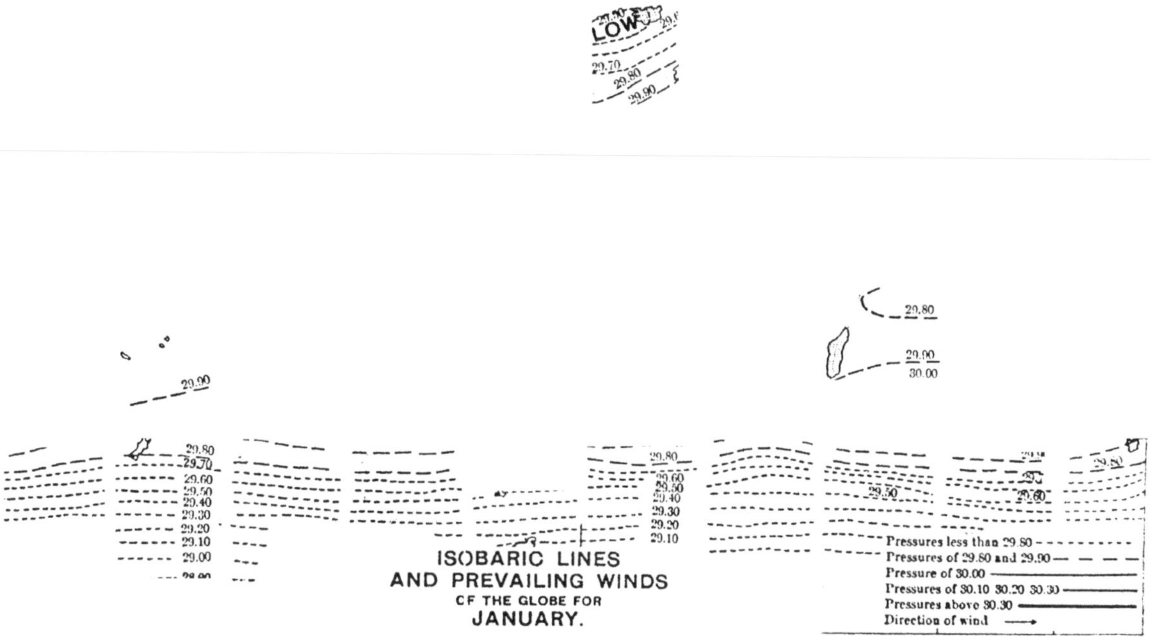
LOW
29.70
29.80
29.90
29.80
30.00
29.60
29.50
29.40
29.30
29.20
29.10
29.00
ISOBARIC LINES
AND PREVAILING WINDS
OF THE GLOBE FOR
JANUARY.
Pressures less than 29.80
Pressures of 29.80 and 29.90
Pressure of 30.00
Pressures of 30.10 30.20 30.30
Pressures above 30.30
Direction of wind

Chart XI. — Isobaric Lines of the World for January.

(Based on Buchan's *Challenger* Report; similar to the Chart in Davis's *Elementary Meteorology*.)

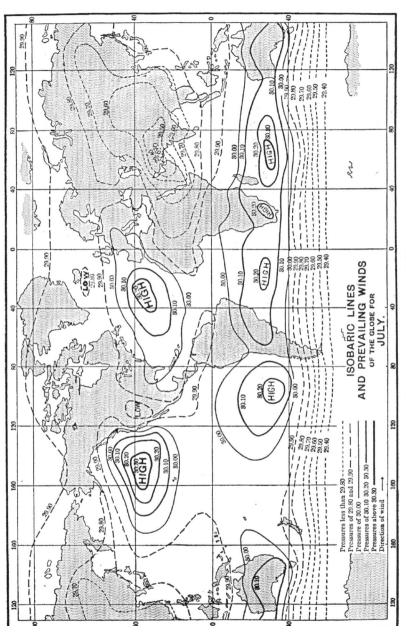

CHART XII. — Isobaric Lines of the World for July.

(Based on BUCHAN's *Challenger* Report; similar to the Chart in DAVIS's *Elementary Meteorology*.)

ISOBARIC LINES
AND PREVAILING WINDS
OF THE GLOBE FOR
JULY.

Pressures less than 29.90 ----------
Pressures of 29.80 and 29.90 ----------
Pressures of 30.00 ——————
Pressures of 30.10 30.20 30.30 ——————
Pressures above 30.30 ——————
Direction of wind ——————→

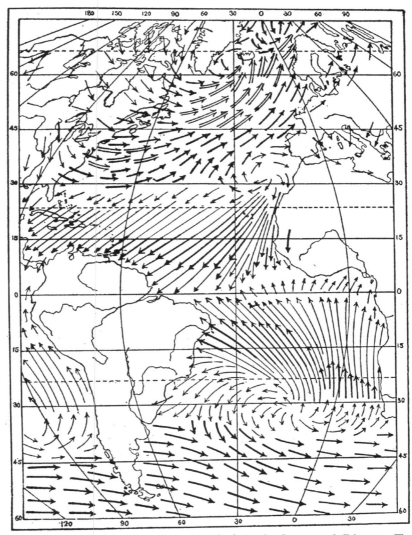

CHART XIII. — Air Circulation of the Atlantic Ocean for January and February. The length of the arrows represents the constancy of the winds and the thickness of the arrows the strength.

(Based on the *Segelhandbuch der Deutschen Seewarte;* similar to the Charts in ANGOT'S *Traité élémentaire de météorologie.*)

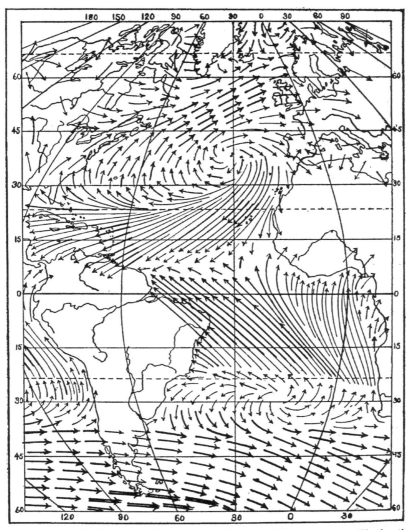

CHART XIV. — Air Circulation of the Atlantic Ocean for July and August. The length of the arrows represents the constancy of the winds and the thickness of the arrows the strength.

(Based on the *Segelhandbuch der Deutschen Seewarte;* similar to the Chart in ANGOT's *Traité élémentaire de météorologie.*)

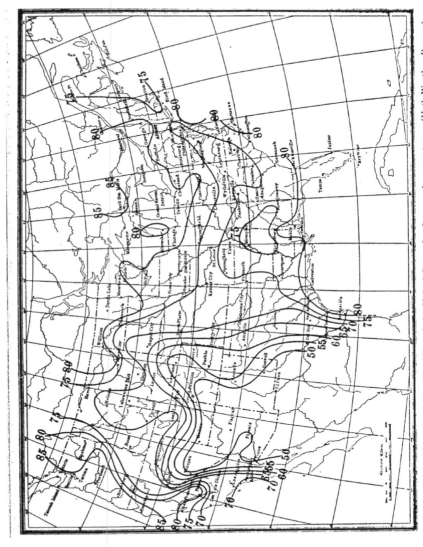

CHART XV. — Normal Relative Humidity of the United States for January, in percentage.   (U. S. Weather Bureau.)

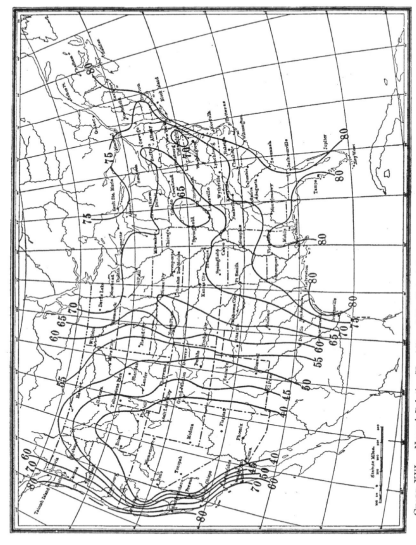

CHART XVI.—Normal Relative Humidity of the United States for July, in percentage. (U. S. Weather Bureau.)

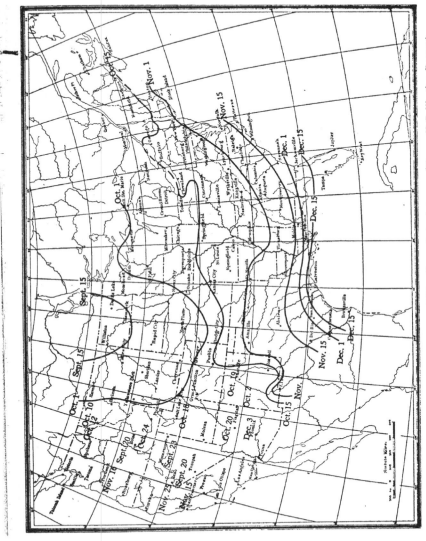

CHART XVII. — Normal Date of the First Killing Frost of the Autumn. (HENRY, U. S. Weather Bureau.)

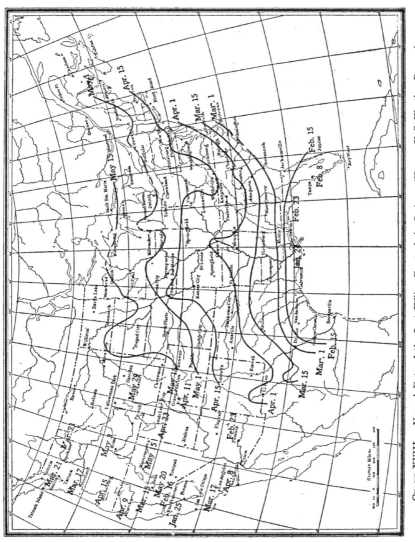

CHART XVIII.— Normal Date of the Last Killing Frost of the Spring.   (HENRY, U. S. Weather Bureau.)

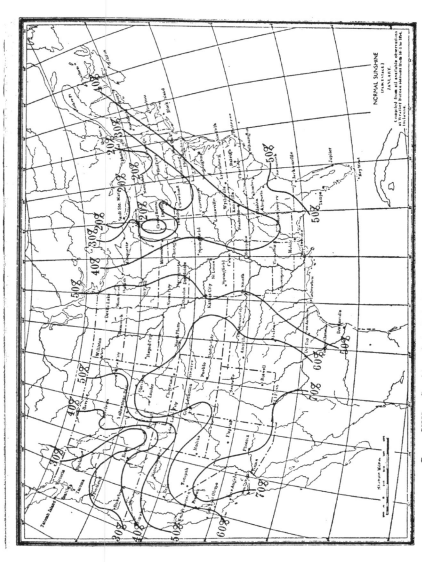

CHART XIX. — Sunshine of the United States for January. (U. S. Weather Bureau.)

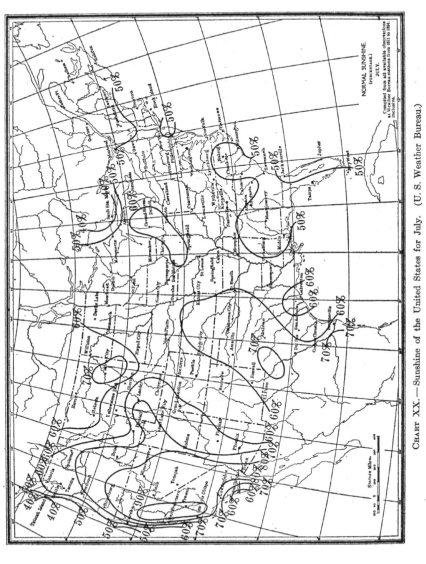

CHART XX. — Sunshine of the United States for July. (U. S. Weather Bureau.)

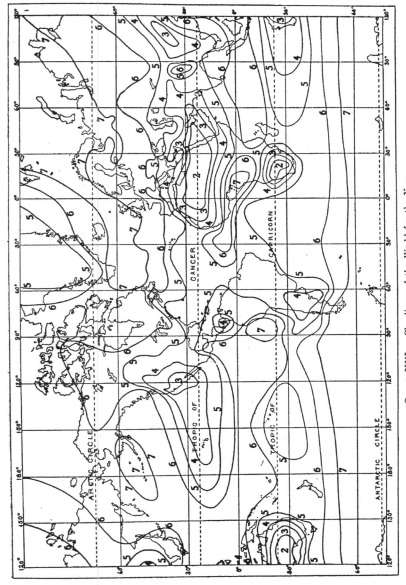

CHART XXI. — Cloudiness of the World for the Year.

(Similar to the Chart in BERGET's *Physique du globe et météorologie*.)

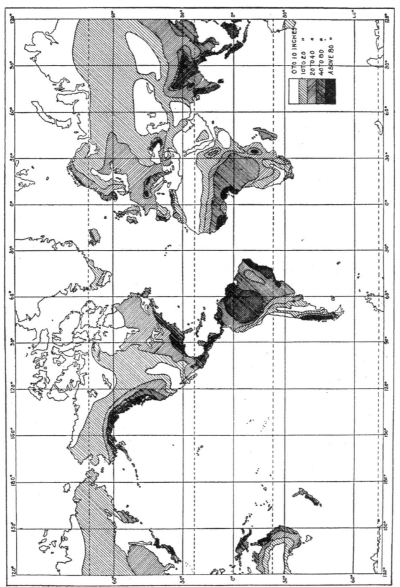

CHART XXII. — Normal Annual Precipitation for the World.

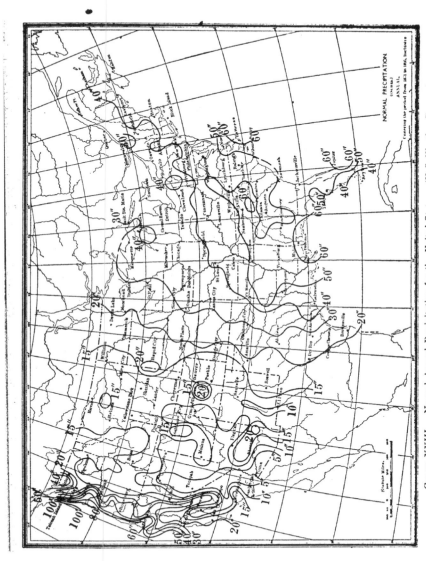

CHART XXIII. — Normal Annual Precipitation for the United States. (U. S. Weather Bureau.)

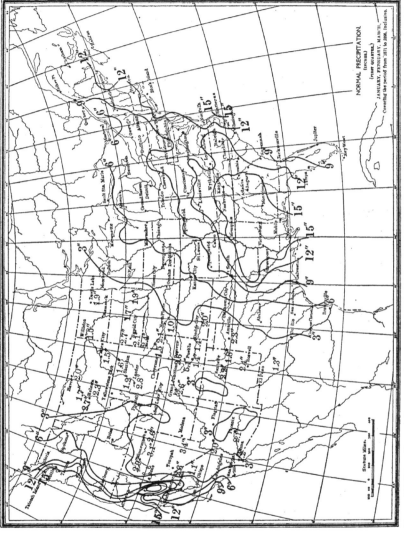

CHART XXIV. — Normal Precipitation for the United States for January, February, and March. (U. S. Weather Bureau.)

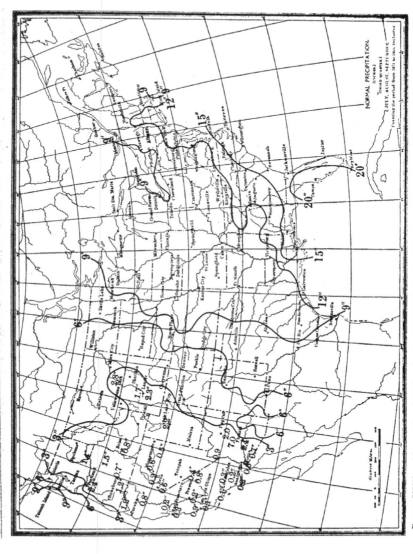

CHART XXV. — Normal Precipitation for the United States for July, August, and September. (U. S. Weather Bureau.)

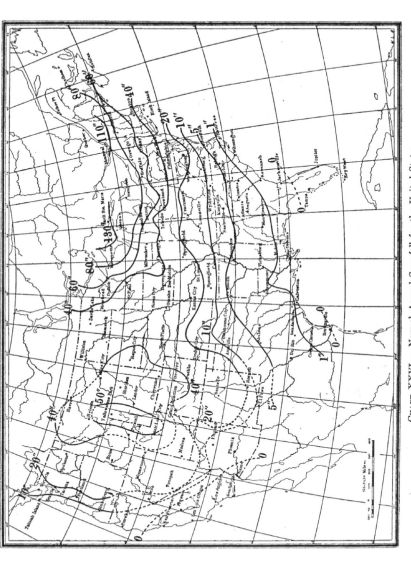

CHART XXVI.—Normal Annual Snowfall for the United States.

(FROM GREELEY'S *American Weather*.)

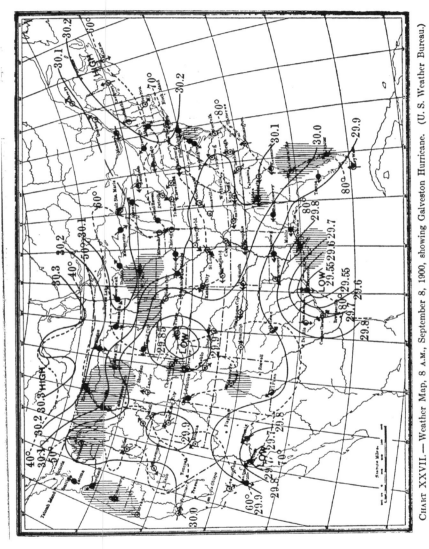

CHART XXVII.— Weather Map, 8 A.M., September 8, 1900, showing Galveston Hurricane. (U. S. Weather Bureau.)

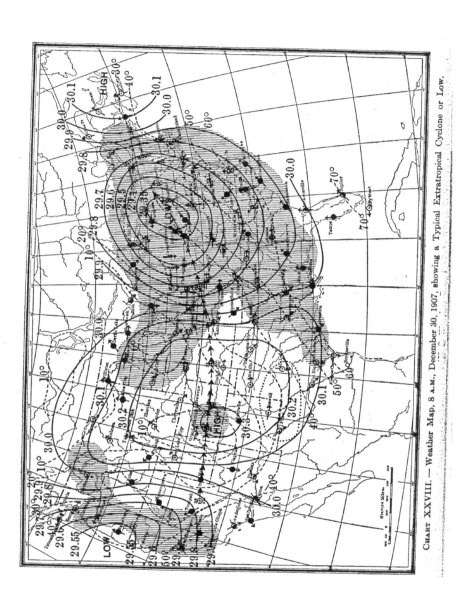

CHART XXVIII. — Weather Map, 8 A.M., December 30, 1907, showing a Typical Extratropical Cyclone or Low.

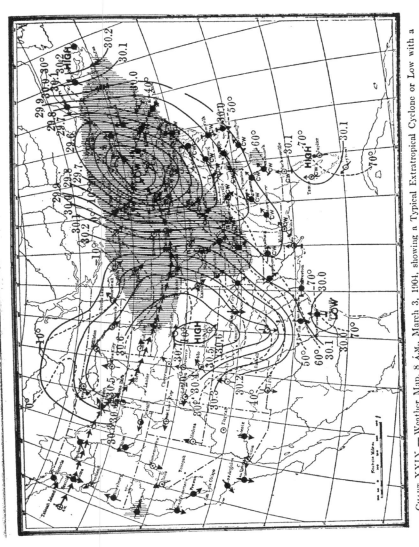

Снакт XXIX. — Weather Map, 8 a.m., March 3, 1904, showing a Typical Extratropical Cyclone or Low with a Pronounced Wind Shift Line. (U. S. Weather Bureau.)

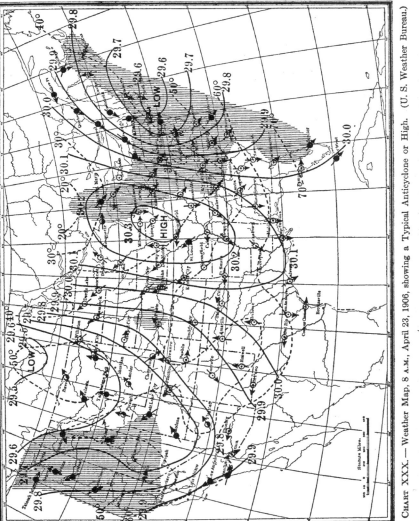

CHART XXX. — Weather Map, 8 A.M., April 23, 1906, showing a Typical Anticyclone or High. (U. S. Weather Bureau.)

ND - #0007 - 110123 - C0 - 229/152/35 [37] - CB - 9781527949232 - Gloss Lamination